Transport Phenomena in Partially Ionized Plasma

Transport Phenomena in Partially Ionized Plasma

V.A. Rozhansky and L.D. Tsendin
St Petersburg Technical University, Russia

CRC Press
Taylor & Francis Group
Boca Raton London New York

CRC Press is an imprint of the
Taylor & Francis Group, an **informa** business
A TAYLOR & FRANCIS BOOK

CRC Press
Taylor & Francis Group
6000 Broken Sound Parkway NW, Suite 300
Boca Raton, FL 33487-2742

First issued in paperback 2017

ISBN-13: 978-0-415-27187-5 (hbk)
ISBN-13: 978-0-367-39664-0 (pbk)

Visit the Taylor & Francis Web site at
http://www.taylorandfrancis.com

and the CRC Press Web site at
http://www.crcpress.com

British Library Cataloguing in Publication Data
A catalogue record for this book is available from the British Library

Library of Congress Cataloging in Publication Data
A catalogue record has been requested

Contents

Nomenclature

a	cylinder radius (m)
A	relative value of the density perturbation
b	mobility ($m^2/s\cdot V$)
$\vec{B}$	magnetic induction (T)
C	capacity; heat capacity
D	diffusion coefficient (m^2/s); thermodiffusion coefficient ($m^{-1}\cdot s^{-1}\cdot J^{-1}$)
e	unsigned charge of an electron ($1.602\cdot 10^{-19}$C)
e	the natural base (2.718)
$E, \vec{E}$	electric field (V/m)
f	distribution function ($m^{-6}\cdot s^{-3}$)
G	Green's function
I	electrical current (A); generation rate ($m^{-3}\cdot s^{-1}$)
i	$\sqrt{-1}$
j	electrical current density (A/m^2)
J	Bessel function of the zeroth kind
k	Boltzmann's constant ($1.381.10^{-23}$J/K); wave number or wave vector (m^{-1})
K	dimensionless functions which characterize deviations of diffusion and mobility from elementary theory; electron to ion temperature ratio in nonisothermal plasma; dimensionless wave vector
L, l	discharge or vessel length; sheath thickness; characteristic plasma scale (m)
m	particle mass (kg)
M	velocity moment of distribution function
n	particle density; densty of negative ions (m^{-3})
N	neutral gas density; density of electrons in electronegative plasma; donor or trap density (m^{-3})
p	pressure (N/m^2); density of positive ions (holes)
$\vec{q}$	heat flow vector (W/m^2); external ionization rate ($cm^{-3}\cdot s^{-1}$)
Q	collisional heat production rate (W/m^3)
r	radial position; r_d, Debye radius (m)
R	collisional friction force (N/m^3)

S	closed surface; intensity of particle sinks ($m^{-3}\cdot s^{-1}$)
St	collision integal ($m^{-6}.s^{-3}$); particle sink ($m^{-3}\cdot s^{-1}$)
t	time (s)
T	partial temperature (J)
u, $\bar{u}$	average velocity (m/s)
U	voltage (V)
V	particle velocity (m/s)
W	shock velocity (m/s)
x, y	rectangular coordinates (m); ratio of gyrofrequency to collision frequency
Y	spherical harmonics in the velocity space
z	rectangular coordinate; axial coordinate; coordinate along magnetic field (m); phase which corresponds to transition between plsma and space charge phases
Z	relative particle charge (in units of e)
α	the first Townsend coefficient (m^{-1}); dimensionless attachment rate; polar angle; dimensionless electron heat conductivity
β	dimensionless detachment rate; ion to electron mobility ratio
γ	secondary electron emission coefficient; dimensionless recombination rate
Γ	particle flux ($m^{-2}\cdot s^{-1}$); gamma-function
δ	Dirac delta function; Kroenecker symbol; thickess of the transition layer (m); fractional energy loss for quasielastic collisions
Δ	denotes the change of a quantity; Laplace operator
ε	energy; full energy; ε_0 - vacuum permittivity ($8.854\cdot 10^{-12}$F/m)
η	orthogonal coordinate (m)
κ	heat conductivity ($m^{-1}\cdot s^{-1}$)
λ	mean free path; relaxation length (m)
Λ	Coulomb logarithm
μ	reduced mass (kg); cosine of polar angle
ν	collision frequency (s^{-1})
π	3.1416; viscous stress tensor (N/m^2)
ρ	gyroradius (m)
σ	cross-section (m^2)
Σ	hight integrated ionosphere conductivity
θ	polar angle
τ	collision time; relaxation time (s)
φ	electrostatic potential (V)
Φ	dimensionless electrostatic potential
χ	heat conductivity
Ψ	electrostatic potential without Boltzmann's part

ξ	elliptic coordinate (m^2); coordinate along or normal to a surface (m)
ζ	orthogonal coordinate (m); root of Bessel function of the zeroth kind
ϑ	azimuth angle
ω	radian frequency (rad/s); plasma frequency; ω_c, gyration frequency
Ω	solid angle (sr)
$\vec{h}$	unit vector
$\hat{A}$	tensor
$\tilde{A}$	oscillating or RF part; dimensionless quantity
$\bar{A}, \langle A \rangle$	average or *dc* part
$A_{\parallel}$	tensor component along the magnetic field
$A_{\perp}$	tensor component orthogonal to the magnetic field along electric field or gradient (Pedersen component)
$A_{\wedge}$	tensor component orthogonal to the magnetic and to the electric field or gradient (Hall component)

Subscript and Superscript Abbreviations

a	ambipolar
A	anode
att	attachment
c	critical value; cold electrons; positive column
C	cathode
det	detachment
e	denotes electron
ef	electron distribution function relaxation
eff	effective
el	elastic; elementary; electrode
ex	excitation; charge exchange
fl	floating
h	hot electrons
i	denotes positive ion
ion	ionization
i, j, k	denote vector components
l, m	sub/superscripts of spherical harmonics in the expansion of electron distribution function
M	Maxwellian
n	denotes negative ion
N	denotes neutral gas
p	denotes positive ion; probe

r	radial
rec	recombination
s	denotes sheath edge; sound
S	area
sh	sheath
T	thermal
w	wall
α,β	particles species
ε	energy relaxation
0	denotes initial value, uniform value or central value
$\pm$	values at right and left shock sides
$\parallel$	parallel (to magnetic field) component
$\perp$	component along electric field or gradient orthogonal to magnetic field (Pedersen component)
$\wedge$	component in the Hall direction
$\vec{\nabla}$	operator of gradient

Chapter 1

Introduction

Transport phenomena in plasmas are the relatively slow processes of particle, momentum and energy transport in the situation when the system is in a state of the mechanical equilibrium, so that the sum of the forces applied to any element of its volume is zero. Even if the system is not in the state of thermodynamic equilibrium, there remain the transport processes caused by collisions. The most important of them are diffusion and thermal diffusion (particle transport), viscosity (momentum transport), heat conductivity (energy transport), conductivity (transport of charged particles). In contrast to the neutral gases, transport phenomena in plasmas are greatly influenced by the self-consistent fields, primarily by the electric fields. The latter are inherent to the inhomogeneous plasma. In addition to the particle and energy fluxes generated by the inhomogeneity of the plasma composition and temperature, fluxes produced by the electric fields inevitably arise. These fluxes result in the radical changes of the transport phenomena with respect to the well-known and thoroughly investigated phenomena in gaseous mixtures of neutral particles. As a result, the physical effects which accompany the transport phenomena in plasmas, are far more numerous and complicated, and the solution of the corresponding problems causes considerable difficulties. But the effects are usually far more interesting and sometimes even surprising, than in the case of neutral gases.

Generally speaking, the transport phenomena in partially ionized plasma are described by the following equation system.

a) The set of continuity equations for the plasma components:

$$\frac{\partial n_\alpha}{\partial t} + \vec{\nabla} \cdot \vec{\Gamma}_\alpha = I_\alpha - S_\alpha , \tag{1.1}$$

where n_α is the density, Γ_α is the flux of the plasma particles of the species α, terms I_α and S_α correspond to the sources and sinks of these particles.
b) The expressions for the fluxes of the charged particles:

$$\vec{\Gamma}_\alpha = -\hat{D}_\alpha \vec{\nabla} n_\alpha - \hat{D}_\alpha^T \vec{\nabla} T_\alpha + Z_\alpha \hat{b}_\alpha \vec{E} . \qquad (1.2)$$

These fluxes are proportional to the gradients of the macroscopic parameters, partial density n_α and temperature T_α, and to electric field strength. Here b_α is the mobility tensor, Z_α is the charge number, D_α is the diffusion tensor, D_α^t is the thermodiffusion tensor. The tensor coefficients $b_\alpha, D_\alpha, D_\alpha^t$ (and also the coefficients of viscosity, thermal conductivity, etc.) are often called the transport coefficients.
c) The Maxwell equations for the fields. In simple cases the latter can be reduced to the quasineutrality condition

$$\sum_\alpha Z_\alpha n_\alpha = 0 . \qquad (1.3)$$

The equation system (1.1)-(1.3) with appropriate boundary conditions describes the temporal evolution of arbitrary profiles of plasma density and composition. It represents the mathematical formulation of one of the most important transport problems.

The investigation of the transport processes in partially ionized plasma nowadays attracts considerable attention. It is greatly stimulated by the continuously growing number of important applications. In all these cases the problems of formation, evolution and control of the spatially inhomogeneous profiles of the plasma parameters are extremely significant.

The first historically and up to now the most important example of a partially ionized plasma represents the traditional gas discharge. In recent decades this branch of plasma physics has undergone specific renaissance due to the intensive advance of laser science, the development of novel light sources, significant progress in plasma chemistry, etc. The intensification of technology is accompanied by continuous growth in energy intensity. It implies that at least part of the material becomes ionized, and a corresponding plasma problem arises. This is the tendency now, growing the importance of a plasma dynamics in the modern and especially in the future technology. The wide use of the gas discharge plasmas in various methods of the surface treatment, such as sputtering, coating, and especially etching, also greatly stimulated interest in the processes in gas discharges. The spatial inhomogeneity of the plasma, which is caused by the walls of the discharge vessel and by the electrodes, determines in typical situations such important characteristics of the plasma as the current distribution, intensity and spatial distribution of the ionization, excitation and other plasmachemical processes, and homogeneity of fluxes of charged and neutral active particles over the target surface, etc..

Great and invariably increasing interest exists in numerous problems of structure and evolution of plasma inhomogeneities in the ionosphere and magnetosphere, such as problems of the global structure of these objects, the formation and dynamics of the current systems here, properties and development of the natural and artificial inhomogeneities of the space plasma, etc. Typically, the interstellar media consists of strongly inhomogeneous partially ionized plasma, where the transport processes are extremely important.

Of particular interest are the processes of plasma transport in connection with the problem of the controlled thermonuclear fusion. To begin with, even the feasibility of the effective fusion depends crucially on the efficiency of the plasma particles and energy isolation in the magnetic traps. In other words, the problem of stationary or quasistationary fusion is primarily the problem of plasma particles and energy transport across the magnetic field. The success of the nuclear fusion to a large extent depends on the plasma-surface interaction in the special divertor chambers, where plasma is partially ionized. Transport processes in these chambers and in the so-called scrape-off layer, where plasma is moving towards the plates along the magnetic field lines, determine the particle and energy fluxes to the surface. The latter are to be reduced up to the level which the surface can sustain for a sufficiently long time. Very important also are the plasma transport problems that arise in connection with research in alternative approaches to the controlled fusion.

A great number of important situations in the semiconductor physics and in physics of electrolytes also is described by equations which are similar to (1.1) - (1.3).

Different experimental and theoretical aspects of the transport problem in plasma have been treated in a great number of publications. In spite of the fact, that a vast amount of data has been collected, a great number of interesting problems have been solved, and many physically non-trivial mechanisms have been discovered, the general physical understanding up to now is far from comprehensive.

The problems of the derivation of the transport equation system, similar to (1.1)-(1.2), and of the calculation of transport coefficients - diffusion, thermodiffusion, mobility, viscosity and thermal conductivity - in terms of the characteristics of the binary collisions, were discussed in detail in numerous books (see, for example, [1-14]). By applying the widely known Chapman-Enskog method the calculations can be performed with a level of accuracy which is satisfactory for the most part of applications. It seems that the most serious limitations in the calculation of the collisional transport coefficients (sometimes they are called also classical ones) are connected mainly with the shortage and low accuracy of the atomic cross-sectional data.

But the situation with the influence of the self-consistent fields in plasma on the transport processes is far more complicated. The different scenaria are extremely diversified, and the creation of a rigorous general classification and theory in this field of plasma physics, in contrary to the case of the transport of neutral particles, is now only under construction. There are several difficulties which hamper this process.

First of all, the plasma parameters in different applications vary over an extremely wide range. The physical processes which determine the temporal and spatial evolution of the inhomogeneous plasma, can be totally different in different ranges of density, temperature, composition, magnitude of the electric field and depend crucially on the properties of the vessel walls, etc. Such factors as a direct external current, which flows through the plasma inhomogeneity, or application of an external magnetic field, totally change the character of the transport.

The second difficulty results from the self-consistent nature of the electromagnetic fields, which greatly influence the particles and energy fluxes. But in plasma physics these fields are determined not only by the external space charges and currents, but also (and even mainly) by the space charges and currents of the charged particles of the plasma. It means that, with a few exceptions, it is impossible to treat these fields as prescribed ones, but it is necessary to find them in the course of solution. The fields in the inhomogeneous plasma are determined by the inhomogeneous plasma profile itself, and are changing during its evolution. In other words, some sort of feedback control exists - the plasma inhomogeneity results in the development of self-consistent fields, these fields in turn create the field-driven fluxes of charged particles and of energy, which control the process of evolution of the initial plasma profile. A great variety of non-linear phenomena arise, most of which have no analogies in the traditional gas dynamics.

The third group of problems is connected with the fact that the inhomogeneous plasma, as a rule, represents a system which is rather far from thermodynamic equilibrium. It is widely known, that quite often such objects are macroscopically unstable. This situation is quite frequent in plasma research. As a result of these instabilities, turbulence often develops - an extremely complicated chaotic state, in which motions with different scales are excited to level, which by orders of magnitude exceeds that of the thermal fluctuations. The chaotic electromagnetic fields generated by these pulsations, cause so-called anomalous transport. The anomalous fluxes of the particles and energy in these stochastic fields sometimes dominate over the classical (collisional) ones. Such a situation is standard, for example, for the fusion plasmas. The quantitative analysis of the anomalous transport presents up to now an open problem. The main difficulty follows from the fact that the macroscopic fluxes are determined now not only by the average fields and by average profiles of the plasma parameters, but also by the properties and level of the non-thermal fluctuations. Up to now as a rule it is impossible even to write down a closed system of equations describing the anomalous plasma transport. In spite of numerous efforts, the results mainly consist in semi-quantitative non-linear estimates, and development of the quantitative theory of the anomalous transport phenomena remains one of the central problems of the plasma theory.

But numerous situations exist, when the inhomogeneous plasma remains stable, and the transport phenomena in it are determined by the binary collisions. It seems natural to start the investigation of transport from these cases. We can hope that qualitatively the obtained results will remain valid for anomalous transport too, at least if it is possible to treat the displacements of the charged particles in the turbulent plasma as small random steps.

The first ideas about the specifics of the transport processes of charged particles, when the self-consistent electric field dominates, were formulated as early as in the end of the last century, for the case of electrolytes. Nernst and Planck [15,16] were the first to formulate the equations of the type of (1.1), (1.2) for the description of the ion transport with account of the diffusion and of the drift in the electric field. They already widely used the idea of quasineutrality and the proportionality between the diffusion coefficient and the mobility, now widely known as the Einstein relation. A concept of the Debye radius, fundamental for plasma physics, was also introduced originally in the physics of electrolytes [17].

The investigation of the transport of the charged particles in gases was started by Rutherford [18], Townsend [19], Langevin [20], and Frank [21]. Townsend was the first to notice that the imposed external magnetic field B results in considerable reduction of transverse transport coefficients, which are inversely proportional to B^2.

One of the first striking examples of the fundamental role of the self-consistent electric field in the plasma transport processes represents the phenomenon of ambipolar diffusion, which was discovered by Schottky in 1924 [22]. The self-consistent electric field, which arises in the inhomogeneous plasma, hinders the electronic diffusion, and intensifies the ionic one. As a result, the rate of evolution is determined by the less mobile particles - by the ions. Across the relatively strong magnetic field the ions are more mobile than the electrons. Accordingly, the ambipolar electric field hinders the ionic motion, and increases the electronic one - the resulting evolution is determined mainly by the electrons. In the pure plasma (i.e., consisting of electrons and of one species of positive ions) with constant mobilities, the problem can be reduced to the standard diffusion equation. The evolution process is controlled by the less mobile particles - the ions along the magnetic field, and by the electrons - across it. The striking feature of the ambipolar diffusion process consists in the fact that at given source terms it is completely independent from the value of the electric current, which can flow through the plasma inhomogeneity.

But as early as the end of the last century, Kolrausch [23] and Weber [24] have demonstrated, that a such situation is degenerate, and this result is valid only in a pure plasma. In more complicated systems - in the multispecies plasma, or if the mobilities are density- or field-dependent - the pattern of evolution is far more complicated. Even in the absence of external magnetic field it depends crucially on the current, which flows through the plasma inhomogeneity. If the plasma density profile is smooth enough, it is possible to neglect the diffusion and to reduce the problem to the widely known Riemann's solution for the overturning wave of finite amplitude [25].

Haynes and Shockley [26] predicted and demonstrated experimentally, that in three-component semiconductor plasmas (electrons, holes and immobile charged centers) with d.c. current the small perturbations of the plasma density are propagating with the current-dependent velocity, which was called the velocity of ambipolar drift. In other words, in a current-carrying plasma the propagation of signals is possible. In [27], [28] it was shown that, in the presence of the external magnetic field, this velocity exists even in the simple two-component plasma. Its value and direction is determined by the electric current and by the plasma density. Because of this, the overturn of plasma density perturbations of finite amplitude occurs, and regions of steep variation of the plasma density arise, which are analogous to the shocks in conventional gas dynamics [29]. Contrary to the standard ambipolar diffusion scenario, when the initial density profile in the course of evolution is gradually smoothed in the whole space, now this process is accompanied by formation of the steep sections of the plasma profile.

For a considerable time, numerous attempts were undertaken to reduce multidimensional problem of the plasma diffusion in the magnetic field to some effective ambipolar diffusion. Simon [30] was the first to demonstrate, that even in the pure plasma it is in principle impossible. Trying to interpret the experiments of Bohm [31], he noticed, that in a vessel with conducting walls the separate diffusion of the charged plasma particles is possible. The electrons, which are more mobile along the magnetic field, are able in this case to propagate in this direction, the ions can move across $\vec{B}$, and the arising vortex current flows partially along the plasma, and partially along the conducting walls of the vessel. Due to this "short-circuiting" effect, the plasma diffusion occurs considerably faster than the ambipolar one. But the difficulties with formulation of the boundary conditions for the quasineutral plasma equations did not allow formulation rigorous solution of this problem in that time, and for a long time, despite the numerous attempts to prove it mathematically, this striking result remained on the qualitative level.

The rigorous solution of the fundamental problem of the evolution of small (with respect to the background plasma density) point plasma density perturbation on the uniform stationary unbounded background was obtained by Gurevich and Tsedilina [28]. The solution turned out to be surprisingly complicated, and corresponded to fast spreading of the plasma inhomogeneity. The decay rate was considerably higher than it was expected according to the ambipolar scenario. The underlying physical mechanism resembles Simon's effect. The main distinction consists in the fact, that the role of the conducting walls of the plasma-containing vessel plays the background plasma itself [32]. In other words, the vortex currents through the plasma significantly stimulate the evolution process, and essentially change its character. Similar ideas were formulated by Haerendel et al. [33] in relation to experiments with the barium clouds in the ionosphere. The concept of the "short-circuiting" of the vortex currents turned out to be very effective in the interpretation of numerous experiments in the ionospheric and laboratory plasmas [34].

All the above-mentioned effects, and numerous other important and interesting ones, have one important peculiarity - the self-consistent electric field, which arises in the inhomogeneous plasma, results in fundamental distinctions in the process of its evolution, with respect to the situation in the neutral gases. Analysis of interesting and instructive examples of such phenomena is scattered in plentiful publications in diverse fields of physics - in gas discharge physics, in physics of the ionosphere and space, in the controlled fusion, in physics of semiconductors and electrolytes, etc. Such a disconnection between physically related problems from different fields of physics resulted in considerable doubling of efforts and significantly retarded the progress in investigation. Accordingly, it seemed reasonable to systematize the existing material in the form of a monograph, in order to classify the main physical phenomena, and to stimulate further work in this important and interesting field. Therefore several years ago we have published the Russian book under the same title. It was favorably met and was, as we know, rather useful for Russian-speaking readers, but remained unknown to the international scientific community. On the other hand, after 1987, when the Russian version was prepared, many new results have been obtained, and our approaches to some topics have considerably changed. So we decided to bring to the attention of the English-speaking reader a new version of the above-mentioned book, which differs considerably from its Russian prototype.

In this book an attempt is undertaken to summarize and to systematize the existing knowledge and the state of art in the classical (collisional) transport processes in partially ionized plasma. Of course, in the restricted volume at the current state of research it is impossible to cover this vast field exhaustively. Therefore, problems have arisen regarding selection of the material and of the style of presentation.

First of all, we have decided not to describe in detail the derivation of the transport equations and the methods of calculating the transport coefficients, and restricted ourselves mainly to the detailed summary of the quantitative results and qualitative estimates of transport coefficients. The second extremely important topic, which we decided to ignore here totally, deals with the problems of the anomalous transport, which develops in the turbulent state. It seems that the properties of this state depend crucially on the rather subtle characteristics of the specific physical system, and the general approaches to these problems are not clear yet. In any case, an adequate treatment of this extensive issue deserves separate discussion.

In order to outline broadly the subject of the book, it is to be noted that in typical situations it is possible to subdivide the evolution process of an arbitrary plasma disturbance (as well, as the motions of an ordinary neutral gas) into the fast and slow stages. The fast motions correspond to the ideal magnetohydrodynamic processes, which occur with the ion sound velocity in the absence of the magnetic field, and with the Alfven and magnetic sound velocities - in the presence of the magnetic field. In these motions the mechanical equilibrium is absent, and they are, as a rule, non-dissipative. The small collisional fluxes of the particles, momentum and energy, which are accompanied by the entropy production, exist simultaneously with these fast motions, and result in their dissipation. After the extinction of the fast motions and establishment of the mechanical equilibrium, the profiles of the plasma macroscopic parameters remain non-uniform. During the slow stage, the entropy growth tends eventually to level off the values of the plasma macroscopic parameters - the partial densities, mean velocities and partial temperatures of the plasma components. These slow dissipative phenomena of levelling, which are accompanied by an increase of entropy, are called the transport processes. This division is, of course, to some extent conventional. For example, it is impossible to describe such fast processes as the reconnection of the magnetic field lines, or the shock formation, in the framework of the ideal magnetohydrodynamics. But these phenomena are not considered as transport ones. On the other hand, we shall classify the slow (with respect to the skin time) non-dissipative drift motions in the potential electric field, as the transport processes. Since the transport processes occur on a slow time scale, it is possible typically (with the exception of the thin boundary layers at the walls, electrodes etc.) to treat them in the quasineutral approximation and to consider the local distribution functions as close to the Maxwellian ones.

The principal distinction between the transport processes in the fully and in the partially ionized plasmas consists in the fact that in the latter cases, as a rule, the preferred reference frame exists. It is associated with the neutral gas in the gas discharges or in the ionosphere (or with the lattice in the semiconductors). In the majority of practically important cases the degree of ionization is low, and the influence of the plasma motion on the neutral gas is negligible. The motion of the neutral gas can be treated as prescribed. On the other hand, the fast exchange of momentum (and in most cases of the energy also) between the charged particles and the neutral gas (lattice) occurs. Consequently, it is possible to neglect the inertial terms in the momentum balance equations for every species of charged particles and to reduce them to the form Eq. (1.2). On the contrary, in the fully ionized plasma the collisions between the charged particles do not result in momentum losses, but only in redistribution of momentum between them. Hence, it is impossible to reduce the momentum equations to the form Eq. (1.2). This fact, and also the complicated geometry of the magnetic field, results in considerable peculiarities of the transport processes in the fusion plasma. These processes have been analyzed thoroughly, in the context of the problem of controlled fusion (see, for example, [35-41]). We decided to exclude here the discussion of this important problem. The only exception is made for the divertor and scrape-off layer partially ionized fusion plasma, when due to the considerable amount of neutral molecules the preferred reference frame exists, which is determined by them (and by the chamber walls). It is to be noted, that in spite of considerable differences, the transport phenomena in the fully and partially ionized plasmas have many characteristic features in common, such as the decisive role of the self-consistent electric fields, the current short-circuiting, the ambipolarity of the one-dimensional diffusion, etc..

Sometimes the transport processes in plasmas are greatly influenced not only by the self-consistent electric fields, but also by the magnetic ones. We shall not treat such complicated problems here, and consider the magnetic field as given, and in most cases stationary and spatially uniform.

Highly important are the transport processes in the physics of gas discharges [42], [43]. But at least equally important here are the processes of ionization and recombination. Numerous interesting problems arise, involving the mutual influence of the self-consistent electric field, transport and ionization. In the majority of these cases it is impossible to single out the role of the transport processes in such complicated situations. So in this topic we shall restrict ourselves to analysis of the situations, which are mainly determined by the transport phenomena, while the ionization and other plasmachemical processes are treated in a simplified manner.

In numerous problems, especially in astrophysics and in solid state physics the situations arise, where quantum effects are important [44], [45]. Description of these interesting phenomena goes far beyond the scope of the given book.

As for the style of presentation, it is impossible to analyze systematically in one book the great variety of involved phenomena and objects. First of all, a lot of important and interesting problems are not yet investigated in the necessary detail. On the other hand, in many cases manifestations of the general plasma transport processes depend crucially on the characteristics of the given concrete system. We have decided to compromise, and present a sort of mosaic. In the first part of every chapter the characteristic features of the transport processes in a given situation are outlined. Afterwards, the manifestations of these main physical mechanisms are illustrated, using instructive examples from different branches of physics. Of course, the choice is to a large extent determined by the individual taste and experience of the authors. We tried to select these examples in such a way, that they have a simple and, if possible, analytic solution, and also allow convincing and unambigous experiment. In order to present clear and impressive evidence of the existence and importance of the described physical phenomena, the main features are to be more or less insensitive to the detailed properties of a concrete system. From the other side, we preferred the situations of independent practical interest. The presented experimental material is essentially of illustrative character; detailed description of the experimental methods, as well as thorough analysis of the results, goes far beyond the scope of the book.

The extreme variety of the involved processes and the self-consistent character of the transport processes in plasma, result in the fact that the solution of most of the practically interesting problems can be obtained only numerically. It seems that an optimal strategy consists in combining analytical and numerical approaches. As an example of such analysis we have chosen the problem of the evolution of ionospheric inhomogeneities. In all the other cases we tried to avoid detailed description of the numerical calculations for specific experimental or technological systems, and stressed the qualitative interpretation of the numerical results.

We hope that such a book, where the different approaches, developed in the remote fields of plasma physics are systematized, will be useful to reasonable number of engineers and scientists. It will stimulate their research work and help in the mutual development of different domains of science. Every Chapter begins with introductory Sections, where the specifics of a given topic are outlined, the general approach is formulated and qualitative description of the main physical mechanisms is given. For these parts the standard engineer's knowledge in mathematics and physics is sufficient. For some special examples, which are treated in the subsequent sections, more detailed acquaintance with electrodynamics, fluid dynamics, kinetics and partial differential equations is necessary. Many topics and problems considered here were formulated and solved in the process of preparation and presentation of the lecture courses in plasma theory, plasma physics, gas discharge physics, and waves and oscillations in plasma, which the authors have taught for the last decade in St. Petersburg State Technical University and in St. Petersburg State University. We hope, that the book will be useful for similar courses, as well as for courses and seminars in semiconductor physics, physics of the ionosphere, the radiophysics, etc..

The second Chapter is mainly of reference character. Here the expressions for the transport coefficients are given, and qualitative estimates are presented.

In the third Chapter the processes in the space charge sheaths are discussed and boundary conditions for the plasma quasineutral equations are formulated.

In Chapter 4 the simplest diffusive problems in weakly ionized, currentless, unmagnetized plasmas are analyzed. The ambipolar diffusion equation for the pure plasma is derived, and the diffusion problem in multispecies plasmas is discussed. The expression for the specific effective ambipolar diffusion coefficient which arises in pure plasma in the presence of an external radiofrequency current is also presented.

Chapter 5 deals with simple 1D diffusion problem in pure magnetized plasma. It is shown that even in this simple case the vortex current in the plasma volume arises, which accompanies the evolution process and accelerates the diffusion. The diffusion of multispecies plasma across a magnetic field is analyzed.

In Chapter 6 the problem of diffusion of spatially restricted inhomgeneity on the stationary uniform currentless background is studied. It is shown that such a process is accompanied by flow of the vortex current through the inhomogeneity. This fact results in great enhancement of the evolution process, and in the formation of depletion regions in the ambient plasma. Some typical nonlinear problems are considered. In the regime of intermediate nonlinearity which corresponds to insignificant depletion of the ambient plasma, the evolution process is controlled by the same process of the short-circuiting of the vortex current by the ambient plasma. Only if the nonlinearity is very strong, and the perturbed plasma density greatly exceeds the ambient one, transition to the ambipolar regime occurs.

In the seventh Chapter the simplest boundary problems are analyzed. The solution of the problem of diffusive plasma decay in vessels with insulating and conducting walls is presented. The current-voltage characteristic of the probe in a magnetic field is also discussed.

The problems of thermal diffusion in magnetized plasma are considered in Chapter 8. Such a situation arises in the vicinity of a localized plasma heating source. This process is also dominated by the short-circuiting of the vortex currents through the ambient plasma or the conducting vessel walls.

The ninth Chapter deals with the influence of the net current through a plasma inhomogeniety on its evolution. The typical evolution scenario of relatively smooth inhomogenieties in the current-carrying plasmas corresponds to splitting of the initial plasma profile into several inhomogeneities which move with the ambipolar drift velocity. Their slow damping is caused by the diffusion processes. The main nonlinear effect corresponds to the overturn of these smooth profiles, and to the formation of shock-like narrow transition layers. In other words, the plasma density and composition in the course of evolution increases in these regions. The structure of these diffusive shocks is determined by the diffusion processes, but their position and propagation pattern can be found in the so-called drift approximation, ignoring the plasma diffusion. The only exception from such a scenario is the case of pure plasma with constant mobilities, when the ambipolar drift velocities are equal to zero.

In Chapter 10 it is shown that strong difference in the partial temperatures (and the strong inequality $T_e >> T_i$ is typical for weakly ionized plasmas) acts in some respect in the same way, as the flow of the net current through a plasma inhomogeneity. The formation of the shock-like structures in currentless, multispecies plasmas with $T_e >> T_i$ is illustrated for the case of a diffusion-dominated positive column in electronegative gas.

In the next Chapter some problems of inhomogeneity evolution in current-carrying magnetized plasma are considered. The situations are far more diverse in this case, than in the absence of the magnetic field, but the main result that the evolution is accompanied by the steepening of some density profile parts, and by the formation of shock-like regions, remains valid. For multidimensional evolution the generation of short-circuiting currents in the ambient plasma is also typical. These effects are considered with emphasis on the problem of evolution of the ionospheric inhomogeneities.

In the last Chapter several problems of energy transport are analyzed. This topic is, generally speaking, far more complicated than the problems of particles transport which were considered in all the preceding Chapters. This is because the energy transport is practically always accompanied with spatial redistribution of the particle densities too. We have restricted ourselves to the plasma in the absence of the magnetic field. As an example of thermal effects in currentless plasmas, the phenomenon of diffusive electron cooling during the plasma decay was chosen. The propagation of the ionization waves (striations) in a *dc* positive column in the fluid approximation presents an example of more complicated phenomena, in which the particle densities and temperatures are controlled by the energy transport processes in current carrying plasmas.

The authors are deeply grateful to Professors V. E. Golant and A. P. Zhilinsky, who have initiated the investigation of the plasma transport processes in the St.Petersburg State Technical University and have attracted our attention to this interesting field. We are grateful also to Dr. S. P.

Voskoboinikov and Dr. I. Yu. Veselova, who performed the main part of numerical calculations of evolution of the plasma inhomogeneities in a magnetic field, and to Dr. I. D. Kaganovich for assistance in preparing Chapter 10. This book would never have been written without the numerous stimulating discussions with all our colleagues - theoreticians, experimentalists and engineers. We would like to thank all who have participated in these discussions and have presented their results.

REFERENCES

1. S. Chapman, T. G. Cowling, *The Mathemathical Theory of Non-Uniform Gases*, 2nd ed. (Cambr.: University Press, 1952): 350.
2. B. M. Smirnov, *Physics of Weakiy Ionized Gases* (Moscow: Mir, 1981): 430.
3. I. Shkarofsky, T. Johnston, M. P. Bachinski, *The Particle Kinetics of Plasmas* (Reading: Addison-Wesley., 1966): 518.
4. E. A. Mason, E. W. McDaniel, *Transport Properties of Ions in Gases* (New York: Wiley, 1988): 560.
5. L. G. H. Huxley, R. W. Crompton, *The Diffusion and Drift of Electrons in Gases* (New York: Wiley, 1974): 669.
6. M. Mitchner, C. H. Kruger, *Partially Ionized Gases* (New York:Wiley-Interscience Publ., 1973): 518.
7. A. V. Gurevich, *Nonlinear Phenomena in the Magnetosphere* (New York: Springer, 1978): 370.
8. V. M. Zhdanov, (*Transport Phenomena in Multispecies Plasma*) (Moscow: Energoizdat, 1982): 176 (in Russian).
9. V. P. Silin, *Vvedenije v Kineticheskuju Teoriju Gasov* (*Introduction into the Kinetic Theory of Gases*) (Moscow: Nauka: 1971): 331 (in Russian).
10. H. E. Holt, R. E. Haskell, *Plasma Dynamics* (New York: Macmillan, 1965): 510.
11. H. Grad, *Comm. Pure Appl. Math.* **2**: 331-369 (1949).
12. L. Herdan, B. S. Liley, *Rev. Mod. Phys.* **32**: 731-741 (1960).
13. S. I. Braginskii, in *Rewievs of Plasma Physics*, **vol.1**, Editor: M.A. Leontovich (New York: Consultants Bureau, 1963): 183-271.
14. R. Balesku, *Transport Processes in Plasma* (Amsterdam: North Holland, 1988): 804.
15. W. Nernst, *Z. Phys. Chem.* **2**: 613-652 (1888).
16. M. Planck, *Ann. Phys. Chem.* **39**:161-186 (1890).
17. *P. J. W. Debye, E. Hueckel, Phys.* Zs. **24**: 305-345 (1923).

18. E. Rutherford, *Phil. Mag.* **543**: 241-282 (1897).
19. J. S. Townsend, *Electricity in Gases* (Oxford: Clarendon Press, 1915): 496; (London: Hutchinson, 1947): 166.
20. P. Langevin, *Ann. Chim. Phys.* **5**: 245-251 (1905).
21. J. Frank, W. H. Westfahl, *Vehr. Deutch. Phys. Ges.* **11**: 146-171; 276-302 (1909).
22. W. Schottky, *Phys. Zs.* **23**: 635-640; 25: 342-348 (1924).
23. F. Kolrausch, *Ann. Phys. Chem.* **62**: 209-231 (1897).
24. H. Weber, *Sitz. Akad. Wiss. Berlin* **44**: 936-945 (1897).
25. G. B. Whitham, *Linear and Nonlinear Waves* (New York: Wiley, 1974): 636.
26. J. Haynes, W. Schockley, *Phys. Rev.* **81**: 835-843 (1951).
27. P. C. Clemmow, M. A. Johnson, *J. Atm. Terr. Phys.* **16**: 21-35 (1959).
28. A. V. Gurevich, E. E. Tsedilina, *Sov.Phys. Uspekhi* **10**: 214-236 (1967).
29. A. P. Dmitriev, V. A. Rozhansky, L. D. Tsendin, *Sov. Phys.Uspekhi* **28**: 467-483 (1985).
30. A. Simon, *Phys. Rev.* **98**: 317-318; 100: 1557-1559 (1955).
31. D. Bohm, *The Characteristics of Electrical Discharges in Magnetic Field*, Editors: A.Guthrie, R. Wakerling, (New York:McGraw-Hill., 1949): 346. Chapt. 1, 2, 9.
32. V. A. Rozhansky, L. D. Tsendin, *Sov. J. Plasma Phys.* **1**: 516-521 (1975).
33. G. Haerendel, R. Lust, E. Rieger, *Planet. Space Sci.* **15**: 1-18 (1967).
34. A. P. Zhilinsky, L. D. Tsendin, *Sov.Phys. Uspekhi* **23**: 331-355 (1980).
35. A. A. Galeev, R. Z. Sagdeev, in *Rewievs of Plasma Physics*, vol.7, Editor: M.A. Leontovich (New York: Consultants Bureau, 1979): 257-344.
36. F. L. Hinton, R. D. Hazeltine, *Rev. Mod. Phys.* **48**: 239-308 (1976).
37. D. D. Rutov, G. V. Stupakov, in *Reviews of Plasma Physics*, vol. **13**, Editor: M.A. Leontovich (New York: Consultants Bureau, 1987): 93-202.
38. S. P. Hirshman, D. J. Sigmar, *Nuclear Fusion* 21: 1079-1201 (1981).
39. B. B. Kadomtsev, *Tokamak Plasma: a Complex Physical System* (Bristol: Institute of Phys. Publishing, 1992): 208.
40. R. D. Hazeltine, J. D. Meiss, *Plasma Confinement* (Reading: Addison-Wesley, 1992): 412.
41. R. G. Goldstone, P. H. Rutherford, *Introduction to Plasma Physics* (Bristol: Institute of Phys. Publishing, 1995): 491.
42. Yu. P. Raiser, *Gas Discharge Physics*, (New York: Springer, 1991): 449.
43. M. A. Lieberman, A. J. Lichtenberg, *Principles of Plasma Discharges and Materials Processing* (New York: Wiley, 1994): 572.
44. P. M. Platzman, P. A. Wolf, *Waves and Interactions in Solid State Plasmas* (New York: Academic, 1973): 382.

45. N. M. March, M. Parinello, *Collective Effects in Solids and Liquids*(Bristol: Hilger, 1982): 270.

Chapter 2

Transport Equations

2.1 FLUID APPROXIMATION

For the transport processes comparatively long time scales and smooth spatial profiles of plasma parameters and fields are typical. For such conditions the collisions of plasma particles result in the formation of some sort of equilibrium state between the particles and the local and instantaneous values of the external fields. This partial equilibrium can significantly differ from thermodynamic equilibrium. In other words, an average particle is representative for description of the behavior of all the plasma particles in a given place and moment. The particles of each species can be described in terms of their density, mean velocity (or momentum), and mean energy. Such averaged description is usually called a fluid approximation, and the closed set of equations for these values known as fluid, or transport equations.

In this chapter we shall briefly discuss the main results and modern state of art in this extensive field. The main approaches to the derivation of the transport equations and to the calculation of the transport coefficients are presented here, and the typical situations are shortly outlined. The material in this chapter is mainly of reference character. Some new results are presented for electron transport coefficients, in the case when electron-ion and electron-neutral collision frequencies are comparable, and in strong electric fields, when the electron energy balance is dominated by inelastic collisions.

Microscopic detailed description of a plasma is characterized by a set of distribution functions for each species $f_\alpha(\vec{r},\vec{V},t)$. The six coordinates in the six-dimensional phase space which correspond to the particle positions

$\vec{r}$, the velocities $\vec{V}$ and the time t are considered as independent variables. As particles move and are accelerated by the forces, they flow from one volume in the phase space to another. Hence each distribution function should obey a continuity equation in the six-dimensional phase space. It is known as a Boltzmann kinetic equation [1-7]

$$\frac{\partial f_\alpha}{\partial t} + \vec{V}\frac{\partial f_\alpha}{\partial \vec{r}} + \frac{\vec{F}_\alpha}{m_\alpha}\frac{\partial f_\alpha}{\partial \vec{V}} + \frac{eZ_\alpha}{m_\alpha}(\vec{E} + [\vec{V} \times \vec{B}])\frac{\partial f_\alpha}{\partial \vec{V}} = St_\alpha = \sum_\beta St_{\alpha\beta}(f_\alpha, f_\beta), \tag{2.1}$$

where m_α and Z_α are the mass and the charge number of the particle, E, B electric and magnetic fields, and $\vec{F}$ represents other non-electromagnetic external forces such as gravitational force. Subscript α refers both to the different species and to the different quantum states of the ions, atoms or molecules (rotational, vibration, electronic). In the absence of collisions this equation is known as Vlasov's equation. Due to the collisions, particles practically instantaneously change their velocities and can appear and disappear in remote regions of the phase space, forming sources and sinks. This effect is taken into account by the right-hand side of the Eq. (2.1).

The collision term or collision integral St_α for a given sort α is a sum, where each summand corresponds to the collisions between particles of a sort α with all other species (including α). The collisions are divided into elastic ones, when the internal state of the particles remains the same in the process of collision, and inelastic interactions. For the elastic processes the Boltzmann collision integral in a general form is discussed in many textbooks, such as [1-4]. It is given by

$$St_{\alpha\beta} = \iint \left|\vec{V}_\alpha - \vec{V}_\beta\right| \sigma_{\alpha\beta}(\Omega)[f_\alpha(\vec{V}_\alpha')f_\beta(\vec{V}_\beta') - f_\alpha(\vec{V}_\alpha)f_\beta(\vec{V}_\beta)]dV_\beta d\Omega\ , \tag{2.2}$$

where $\sigma_{\alpha\beta}(\Omega)$ is the differential cross-section for scattering into the solid angle $d\Omega$. The second term in brackets corresponds to the particles from the distribution functions f_α and f_β with velocities $\vec{V}_\alpha$ and $\vec{V}_\beta$ that are scattered to velocities $\vec{V}_\alpha'$ and $\vec{V}_\beta'$ and disappear from the differential volume $d\vec{V}_\alpha d\vec{V}_\beta$. The first term stands for the particles that appear in this differential volume due to the scattering from the volume $d\vec{V}_\alpha' d\vec{V}_\beta'$.

At equilibrium each collision integral $St_{\alpha\beta}$ is equal to zero. The Maxwellian distributions with a common temperature for all the species: $f_\alpha(\vec{V}_\alpha) = C_\alpha \exp(-m_\alpha V_\alpha^2 / 2T)$ satisfy this condition. Due to kinetic energy conservation the integrand of $St_{\alpha\beta}$ is

$$f_\alpha\left(\vec{V}_\alpha\right) f_\beta\left(\vec{V}_\beta\right) - f_\alpha\left(\vec{V}_\alpha'\right) f_\beta\left(\vec{V}_\beta'\right) = 0 .$$

Moreover, it can be shown by use of Boltzmann H-theorem [1-3] that the only distribution function that satisfies the condition $St_{\alpha\beta}=0$ is the Maxwellian distribution.

The Boltzmann form of the collision integral is very inconvenient for the Coulomb collisions, when the main contribution to the integral is caused by the small angle scattering of the particles with large impact parameters of the order of the Debye radius. Expanding Eq. (2.2) for small angle scattering, the Landau integro-differential form of the collision integral [8] can be obtained (for Coulomb logarithm Λ see Eq. (2.24)):

$$\begin{aligned} St_{\alpha\beta}(f_\alpha, f_\beta) &= -\frac{2\pi \Lambda e^4 Z_\alpha^2 Z_\beta^2}{m_\alpha (4\pi\varepsilon_0)^2} \times \\ &\times \frac{\partial}{\partial V_i} \int \left[\frac{f_\alpha(\vec{V})}{m_\beta} \frac{\partial f_\beta(\vec{V}')}{\partial V_k'} - \frac{f_\beta(\vec{V}')}{m_\beta} \frac{\partial f_\alpha(\vec{V})}{\partial V_k'} \right] U_{ik} d\vec{V}' , \end{aligned} \tag{2.3}$$

where

$$U_{ik} = \frac{\delta_{ik}}{\left|\vec{V} - \vec{V}'\right|} - \frac{(V_i - V_i')(V_k - V_k')}{\left|\vec{V} - \vec{V}'\right|^3} .$$

Here and below, the sum is implied over the repeating subscripts.

The inelastic collisions, which are connected with the variation of the internal state of the colliding particles, are very diverse. They include excitation, ionization, recombination and various plasmachemical reactions, etc.. The general form of the collision integral for inelastic processes is very complicated and inconvenient. Fortunately, in many applications these processes can be described in a simplified phenomenological manner. That's why we shall not discuss the general inelastic collision term, for further details see [9,10].

For the analysis of relatively slow large-scale transport processes the description of the plasmas using the set of distribution functions is unnecessarily detailed, and all the necessary information is contained in the averaged characteristics - in the velocity moments of the distribution function:

$$M_{j,k,\dots n}(\vec{r},t)=\int V_j V_k \dots V_n f_\alpha(\vec{r},\vec{V},t)d\vec{V}\ .$$

The most important averaged or macroscopic quantities are the particle density

$$n_\alpha(\vec{r},t)=\int f_\alpha d\vec{V}\ , \tag{2.4}$$

the particle flux

$$\vec{\Gamma}_\alpha(\vec{r},t)=n_\alpha\vec{u}_\alpha=\int f_\alpha \vec{V}d\vec{V}\ , \tag{2.5}$$

where $\vec{u}_\alpha(\vec{r},t)$ is the mean (fluid) velocity of the α species, and the mean chaotic energy density per unit volume

$$\frac{3}{2}n_\alpha T_\alpha(\vec{r},t)=\int\frac{m_\alpha(\vec{V}-\vec{u}_\alpha)^2}{2}f_\alpha d\vec{V}\ . \tag{2.6}$$

For the Maxwellian distribution the functions $T_\alpha(\vec{r},t)$ represent the partial temperatures measured in energy units. The mean thermal velocity is often defined as $V_{T\alpha}=\sqrt{8T_\alpha/(\pi m_\alpha)}$.

Formally, the equation system for the moments of the distribution function can be obtained by multiplying the kinetic equation by $V_j V_k \dots V_n$ and integrating it over the velocity space. As a result, one obtains the coupled infinite set of momentum equations. The first three of them have a simple interpretation as particle, momentum and energy conservation equations.

The lowest (the zero) moment of the kinetic equation is obtained by pure integration of the Eq. (2.1). The integration yields the continuity equation

$$\frac{\partial n_\alpha}{\partial t}+\vec{\nabla}\cdot\vec{\Gamma}_\alpha(\vec{r},t)=I_\alpha-S_\alpha\ . \tag{2.7}$$

The right-hand side of this equation is equal to zero when the elastic collision integral in the form (2.2), (2.3) is integrated over velocities. Only inelastic collisions that create or destroy charged particles result in the sources (I_α) and sinks (S_α) in the r.h.s. of Eq. (2.7).

The first moment equation is obtained by multiplying Eq. (2.1) by $m_\alpha V$ and integrating over velocities:

$$m_\alpha\left(\frac{\partial \Gamma_{\alpha j}}{\partial t}+\frac{\partial M_{\alpha jk}}{\partial x_k}\right)-Z_\alpha e n_\alpha\left(E_j+[\vec{u}_\alpha\times\vec{B}]_j\right)-F_{\alpha j}n_\alpha = \\ = R_{\alpha j}+\left(I^m_{\alpha j}-S^m_{\alpha j}\right), \tag{2.8}$$

where

$$R_{\alpha j}=\int m_\alpha V_j St_\alpha\, d\vec{V}\,, \tag{2.9}$$

and $I_{\alpha j}{}^m$, $S_{\alpha j}{}^m$ are the momentum sources and sinks. Combining this Eq. (2.8) with the continuity equation (2.7.), one obtains the alternative form of the momentum conservation equation (see e.g.[11]):

$$m_\alpha n_\alpha\left(\frac{\partial u_{\alpha j}}{\partial t}+u_{\alpha k}\frac{\partial u_{\alpha j}}{\partial x_k}\right)=-\frac{\partial p_\alpha}{\partial x_j}-\frac{\partial \pi_{\alpha jk}}{\partial x_k} \\ +Z_\alpha e n_\alpha\left(E_j+[\vec{u}_\alpha\times\vec{B}]_j\right)+F_{\alpha j}n_\alpha+R_{\alpha j} \\ +I^m_{\alpha j}-S^m_{\alpha j}-m_\alpha u_{\alpha j}(I_\alpha-S_\alpha). \tag{2.10}$$

The left-hand side is the species mass density multiplied by the convective derivative of the mean velocity, representing the acceleration of the fluid element. The right-hand side is the sum of the forces applied to the particles of the species α per unit volume. The first term is the gradient of the partial pressure

$$p_\alpha = m_\alpha\, n_\alpha \langle(\vec{V}-\vec{u}_\alpha)^2\rangle/3 = n_\alpha\, T_\alpha\,,$$

and the second term represents the divergence of the partial viscous stress tensor

$$\pi_{\alpha kj}=\pi_{\alpha jk}=m_\alpha n_\alpha\langle(V_j-u_{\alpha j})(V_k-u_{\alpha k})-\delta_{jk}(\vec{V}-\vec{u}_\alpha)^2/3\rangle\,.$$

The brackets < > denote the velocity averaging of the bracket quantity over f with the normalization $\langle f \rangle = 1$. The total stress tensor of the species α is defined as

$$P_{\alpha jk} = p_\alpha \delta_{jk} + \pi_{\alpha jk} \ ,$$

where δ_{jk} is the unit tensor.

In weakly ionized plasma the general form of the stress tensor is seldom used, and the simplified isotropic version $P_{\alpha jk} = p_\alpha \delta_{kj}$ is usually employed. The third term in the r.h.s. of Eq. (2.10) is the Lorentz force, the fourth one corresponds to other external forces (e.g. gravitational). The force $R_{\alpha j}$ represents the momentum transfer due to collisions with all other species. The last term gives the momentum transfer caused by the creation or destruction of the plasma particles. Particles are created at different velocities and disappear with the mean velocity $u_{\alpha j}$, hence a drag force on the moving fluid of the species α appears.

The energy conservation equation is obtained by multiplying the kinetic equation by $m_\alpha V^2/2$ and integrating over velocities:

$$\frac{m_\alpha}{2}\left(\frac{\partial M_{\alpha jj}}{\partial t} + \frac{\partial M_{\alpha jkk}}{\partial x_j}\right) - (Z_\alpha e E_j + F_j)\Gamma_{\alpha j} = \\ = \int \frac{m_\alpha V^2}{2} St_\alpha d\vec{V}_\alpha \ , \tag{2.11}$$

Introducing the temperature according to Eq. (2.6), we have (in the absence of inelastic collisions)

$$\frac{\partial}{\partial t}\left(\frac{m_\alpha n_\alpha}{2} u_\alpha^2 + \frac{3}{2} n_\alpha T_\alpha\right) \\ + \frac{\partial}{\partial x_j}\left[\left(\frac{m_\alpha n_\alpha}{2} u_\alpha^2 + \frac{5}{2} n_\alpha T_\alpha\right) u_{\alpha j} + \pi_{\alpha jk} u_{\alpha k} + q_{\alpha j}\right] \\ = (Z_\alpha e E_j + F_{\alpha j}) u_{\alpha j} n_\alpha + R_{\alpha j} u_{\alpha j} + Q_\alpha . \tag{2.12}$$

Here

$$\vec{q}_\alpha = \frac{m_\alpha n_\alpha}{2} \langle (\vec{V} - \vec{u}_\alpha)^2 (\vec{V} - \vec{u}_\alpha) \rangle \tag{2.13}$$

is the heat (chaotic energy) flux, i.e. the energy flux in the reference frame, where the mean velocity of the particles α is zero,

$$Q_\alpha = \int \frac{m_\alpha (\vec{V} - \vec{u}_\alpha)^2}{2} St_\alpha d\vec{V} = \sum_\beta Q_{\alpha\beta} \tag{2.14}$$

is the heat production due to collisions. In the case of elastic collisions $Q_{\alpha\beta} \neq 0$ only for α≠β. The first term in the left-hand side of Eq. (2.12) represents the change of the partial energy per unit volume, which consists of the energy of the directed motion and the chaotic energy. The second term is the sum of the divergence of the energy flux $[(m_\alpha n_\alpha u_\alpha^2 / 2 + 3n_\alpha T_\alpha / 2)\vec{u}_\alpha + \vec{q}_\alpha]$ and the work of the total pressure gradient $\partial(P_{\alpha jk} u_{\alpha k}) / \partial x_j$. The right-hand side of Eq. (2.12) represents the work of the electric field, external forces and frictional force, and the energy production by the sources.

Combining Eq. (2.12) with the continuity Eq. (2.7), one gets the equation for the internal (thermal) energy:

$$\begin{aligned} &\frac{3}{2}\frac{\partial\, n_\alpha T_\alpha}{\partial\, t} + \vec{\nabla} \cdot \left(\frac{3}{2} n_\alpha T_\alpha \vec{u}_\alpha \right) + n_\alpha T_\alpha\, \vec{\nabla} \cdot \vec{u}_\alpha \\ &+ \pi_{\alpha\, jk} \frac{\partial\, u_{\alpha\, j}}{\partial\, x_k} + \vec{\nabla} \cdot \vec{q}_\alpha = Q_\alpha . \end{aligned} \tag{2.15}$$

Here $n_\alpha T_\alpha \vec{\nabla} \cdot \vec{u}_\alpha$ is the part of the power performed by the partial pressure, which results in heating or cooling of the particles of a given species α due to compression or expansion of plasma volume, and $\pi_{\alpha\, jk} \partial\, u_{\alpha\, j} / \partial\, x_k$ represents the viscous heat production. Employing Eq. (1.6), Eq. (2.15) can be reduced to the equation for the species temperatures:

$$\begin{aligned} &\frac{3}{2} n_\alpha \left(\frac{\partial}{\partial\, t} + u_{\alpha\, j} \frac{\partial}{\partial\, x_j} \right) T_\alpha + n_\alpha T_\alpha \vec{\nabla} \cdot \vec{u}_\alpha + \vec{\nabla} \cdot \vec{q}_\alpha \\ &+ \pi_{\alpha\, jk} \frac{\partial\, u_{\alpha\, j}}{\partial\, x_k} = Q_\alpha . \end{aligned} \tag{2.16}$$

The equation system (2.7), (2.10), (2.15) is the rigorous sequence of the kinetic equation and, therefore, is an exact one. However, in such a form this system cannot be resolved because the equations are coupled. Indeed, to determine the density profile from the continuity equation it is necessary to know the first moment of the distribution function $u_{\alpha j}$, to obtain the mean velocity one should know the second moments T_α, $\pi_{\alpha jk}$ and the momentum transfer $R_{\alpha j}$, the moments $q_{\alpha j}$, $R_{\alpha j}$, Q_α enter the Eq. (2.16) for the temperature, etc. To close the system it is necessary to truncate the system and to express $\pi_{\alpha jk}$, $q_{\alpha j}$, $R_{\alpha j}$ and Q_α as functions of n_α, $u_{\alpha j}$, T_α and their derivatives. Such a procedure is possible for large spatial and time scales, when due to the frequent collisions the distribution functions are close to the local Maxwellians:

$$f_\alpha^{(0)} = f_\alpha^{(M)} = \frac{n_\alpha}{(2\pi T_\alpha / m_\alpha)^{3/2}} \exp\left[\frac{m_\alpha(\vec{V} - \vec{u}_\alpha)^2}{2T_\alpha}\right]. \qquad (2.17)$$

Such approximation is known as a fluid or transport approach. The corresponding criterion requires that the characteristic temporal and spatial scales of the processes, electric and magnetic fields τ, L must exceed temporal and spatial scales of the relaxation of the distribution functions to the Maxwellian ones. For the elastic collisions between particles of comparable masses, for example, this process occurs comparatively quickly - practically during one collision time. The electric field must be also small, so that the energy gain between the collisions is to be small with respect to the temperature; see below, Section 2.3. In this case, of course, the Maxwellian distributions of all the species with one common temperature are established.

In the absence of a magnetic field (or parallel to it) the Maxwellization length for the particles with comparable masses Δr is equal to the mean free path λ_α , while across the strong uniform magnetic field it is given by the particle gyroradius $\rho_{c\alpha}$. Therefore, the conditions $\lambda_\alpha \ll L$, $\rho_{c\alpha} \ll L$ are to be satisfied. In the nonuniform magnetic field or in $\vec{E} \times \vec{B}$ fields the displacement of the particles Δr across $\vec{B}$ can be much larger than $\rho_{c\alpha}$ owing to the drifts, and the corresponding criterion is more severe [2], [5-6].

In plasma the situation is far more complicated, since the meaning of collision frequencies often requires more precise definition. For example, in the fully ionized plasma the values of momentum and energy relaxation times differ by orders of magnitude. The fastest is the electron collision time, which corresponds to electron momentum relaxation in electron-ion collisions. The Maxwellization of the electron distribution by electron-electron collisions occurs on the same time scale. Considerably longer is the characteristic time for the ion Maxwellization and momentum exchange. Due to large mass difference, the longest time corresponds to the energy exchange between the electrons and ions. Accordingly, the Maxwellian distributions with different partial temperatures of the plasma components are established quickly, and the thermalization process to common plasma temperature is relatively slow. As a result, especially in the presence of external energy sources, in many problems it is reasonable to distinguish between the partial electron and ion temperatures. A similar situation is typical for partially ionized plasma. On the contrary, for ions it is usually unreasonable to introduce the partial temperatures, since all the characteristic times of Maxwellization of the heavy particles (ions and neutrals) are of the same order - of the order of the inverse ion-neutral collision frequencies. Therefore, the common temperature of the heavy particles is established. Only in the case of strongly different masses of heavy components, when the energy exchange between them is hindered, it is sometimes necessary to introduce different partial temperatures of the heavy species in mixture [12].

In order to obtain the closed set of fluid equations it is necessary to linearize the system of kinetic equations for the distribution functions, which are close to the Maxwellian distributions Eq. (2.17). As a result the deviation from the Maxwellian distributions $f_\alpha^{(1)}$ and the macroscopic quantities $\pi_{\alpha jk}$, $q_{\alpha j}$, $R_{\alpha j}$, Q_α are to be expressed as functions of the factors, which characterize the violation of the local thermodynamic equilibrium. The latter are: the temperature gradients $\vec{\nabla} T_\alpha$, the difference in the species temperatures T_α-T_β, the relative velocities $\vec{u}_\alpha - \vec{u}_\beta$, and velocity shear tensor

$$W_{\alpha jk} = \frac{\partial u_{\alpha j}}{\partial x_k} + \frac{\partial u_{\alpha k}}{\partial x_j} - \frac{2}{3}\delta_{jk}\vec{\nabla}\cdot\vec{u}_\alpha .$$

In thermodynamics of irreversible processes the macroscopic quantities $\pi_{\alpha jk}$, $q_{\alpha j}$, $R_{\alpha j}$, Q_α are called thermodynamic 'fluxes' and are denoted as I_n, while the factors $\vec{\nabla} T_\alpha$, T_α-T_β, $\vec{u}_\alpha - \vec{u}_\beta$, $W_{\alpha jk}$ are known as thermodynamic 'forces' and are denoted as X_n [8,11,13]. For the smooth spatial and temporal variation of the macroscopic parameters the 'fluxes' are linear functions of the 'forces'

$$I_m = \sum_n L_{mn} X_n ,$$

and the matrix L_{mn} is called the matrix of transport coefficients. Since the thermodynamic 'forces' and 'fluxes' are vectors or tensors, in anisotropic media the transport coefficients L_{mn} are tensors. The main reason for the plasma anisotropy is connected with the influence of the magnetic field. The lattice anisotropy in semiconductors, or strong external electric fields (see lower, Section 2.9) result in similar effects. The fluid equations (2.7), (2.10), (2.15) with the transport coefficients calculated from the kinetic equation are known as transport equations. In the next Section we shall show that for partially ionized plasma the momentum balance transport equation (2.10) can be reduced to form similar to Eq. (1.2). For partially ionized plasma the coefficients in the Eq. (1.2) - tensors of diffusion, thermodiffusion, mobility - are also often called the transport coefficients. They will be expressed in terms of the matrix L_{mn} in the next Section.

One must keep in mind that near the plasma boundaries the characteristic plasma scale L usually decreases. For example, for the diffusion of a weakly ionized plasma in the vicinity of an absorbing wall, this scale $L \sim |\nabla \ln n|^{-1}$ becomes of the order of the mean free path λ_α. Hence, near the boundaries the fluid approximation fails and the kinetic equation is to be solved. Fortunately, the kinetic analysis often results in the formulation of the effective boundary conditions imposed on the transport equations for the bulk plasma. Such effective boundary conditions are discussed in Chapter 3.

There are several ways of calculating the deviations from the Maxwellian distributions $f_\alpha^{(1)}$ and the corresponding transport coefficients. In fully ionized plasma the resulting two-temperature transport equations were obtained by Braginski [11]. The functions $f_\alpha^{(1)}$ were calculated by the Chapman-Enskog method [1] from the Boltzmann equation as an expansion in the Sonine-Laguerre polynomials. The coefficients of the expansion satisfy the infinite chain of the algebraic equations. After the chain is cut off, usually two or three first terms are used to obtain the transport coefficients with satisfactory accuracy. The transport coefficients obtained by Braginski are presented in Section 2.6.

In the Grad method of moments [14-16] the distribution function is expanded into the full system of tensor velocity polynomials. The usual 13 and 21 moment approximations are equivalent to leaving the first one or two Sonine-Laguerre polynomials in the Chapman-Enskog method. The Grad method is more convenient in the complicated situations, when the standard transport equations are not applicable, e.g. for calculation of the barodiffusion in the viscous flows [14], the thermostress convection [17], the 2D plasma motions with the strongly different scales along and across a magnetic field [18], etc.. The Grad method can sometimes be used even in plasmas with rare collisions if the macroscopic parameters are smooth. The corresponding 16 moment approximation with the biMaxwellian distributions is described in [19].

For the partially ionized plasmas the Chapman-Enskog or Grad methods are seldom used, because for the calculation of the plasma transport in the presence of the preferred neutral gas reference frame, these methods turn out to be unnecessarily complicated. The reason is that the density, velocity and temperature of the neutral gas (or of the lattice in case of semiconductors) can usually be treated as prescribed. In this case the neutral gas presents a natural background for the evolution of plasma inhomogeneity. For electrons the solution of the kinetic equation by expanding the distribution function in the spherical harmonics in this reference frame [20-22] is naturally more simple and convenient. This method is described in Section 2.4, where the corresponding transport coefficients are calculated.

The specifics of partially ionized plasma consist in the fact, that it is necessary to distinguish between the characteristic temporal and spatial scales of the momentum and energy relaxation, and of the relaxation of the distribution functions. Therefore, the possible scenaria of the fluid description are far more diverse.

This distinction manifests itself most vividly in the practically important case of plasma in a stationary electric field. For ions the situation is relatively simple, since all these characteristic scales are as a rule comparable. The above mentioned standard fluid approach is valid in the case of a weak electric field, when the energy acquired from it on the relaxation length is small with respect to the neutral temperature, and the variation of the fields on the relaxation scale is also small.

If the strong electric fields can be considered as uniform, the ion distribution functions are also local, i.e., they are close to the functions which correspond to spatially and temporally uniform plasma and the field. Starting from these distribution functions as from a zero approximation (in the strong electric fields they are extremely non-equilibrium and anisotropic), the equations for particle and momentum balance can be derived. In this case the separate equation for the ion energy transport is unnecessary. Due to the strong anisotropy of the distribution function, the expansion in spherical harmonics is inconvenient in this case, and other methods are used [23].

The behavior of electrons in an electric field is more complicated, since the characteristic scales of momentum and energy relaxation are, as a rule, extremely different. For the collisions between electrons and heavy particles the energy relaxation time significantly exceeds the momentum relaxation time, which equals the inverse transport collision frequency $\nu_{\alpha\beta}$. From now on, when mentioning the relaxation length or relaxation time scale we shall bear in mind the transport values. For elastic collisions the ratio of the energy relaxation frequencies to the transport frequencies is given by

$$\delta_{eN} = \nu_{eN}^{(\varepsilon)} / \nu_{eN} = 2m_e / m_N ;\ \delta_{ei} = \nu_{ei}^{(\varepsilon)} / \nu_{ei} = 2m_e / m_i , \qquad (2.18)$$

where subscripts e, i and N correspond to the electrons, ions and neutrals. The mean value is denoted as

$$\delta = \sum_j \delta_{ej} \nu_{ej} / \sum_j \nu_{ej} \ .$$

This value is extremely small, and the energy relaxation process can be considered as the continuous. It is worthwhile to introduce the electron energy relaxation time

$$\tau_e^{(\varepsilon)} = [\sum_j \delta_{ej} \nu_{ej}]^{-1} , \tag{2.19}$$

which characterizes the process of energy exchange between electrons and heavy particles. The corresponding length of energy relaxation $\lambda_e^{(\varepsilon)}$ for the elastic collisions is equal to the displacement of the electron after δ^{-1} collisions. Since the motion of an electron is a random walk process with mean step λ_e, we have for $\vec{B} = 0$:

$$\lambda_e^{(\varepsilon)} = \lambda_e / \sqrt{\delta} = V_{Te}[(\delta_{ei}\nu_{ei} + \delta_{eN}\nu_{eN})(\nu_{ei} + \nu_{eN})]^{-1/2} \ . \tag{2.20}$$

In the weak electric field, when the energy gain at the energy relaxation length is small with respect to the temperature of heavy particles, the electron distribution function is close to the Maxwellian with the same temperature as the temperature of the heavy particles. In the strong fields the situation depends on the ionization degree.

At relatively high ionization degree, when the electron-electron collisions are more frequent than $\nu_e^{(\varepsilon)}$, the distribution function is close to the Maxwellian with the electron partial temperature, and all three balance equations (for the electron density, momentum and temperature) are necessary. They can be derived, if the scales of the field variation exceed the time and length of Maxwellization, which is caused by the electron-electron collisions. This approximation corresponds to the standard fluid approach.

In the opposite case of the lower ionization degree, the situation is far more complicated and depends on such factors, as the degree of ionization, the characteristics of collisions and the electric field intensity. If the ionization degree is very low, so that $\nu_{ee} << \nu_e^{(\varepsilon)}$, the distribution significantly deviates from the Maxwellian in the strong electric fields. This situation is typical, for example, for discharges at relatively high pressure, and for the lower ionosphere. But still the expansion of the distribution over the spherical harmonics is valid. In this case even the special 'hydrodynamics' can be constructed for the time scales larger, than the inverse $\nu_{eN}^{(\varepsilon)}$, and for spatial scales, which exceed $\lambda_e^{(\varepsilon)}$; see Section 2.9. If the energy, which is acquired from the electric field on the energy relaxation length $\lambda_e^{(\varepsilon)}$, exceeds the temperature of the neutral gas, the field can be considered as strong, and the electron distribution is far from the Maxwellian. But since in such an approach the field intensity variation on the energy relaxation length and time is considered small, the electron distribution function corresponds to the local instantaneous field, and the energy balance equation of the type of Eqs. (2.15), (2.16) becomes extraneous.

The following classification of the plasmas according to the ionization degree is accepted below. Plasma with $\nu_{ee} \geq \nu_{eN}$, is usually called partially ionized. Since the Coulomb transport cross section for the energies typical for the partially ionized plasma is much larger than the cross section for the electron-neutral collisions, the ionization degree in the partially ionized plasma can be very small. For example, in the ionosphere the ratio of the cross sections is several orders of magnitude. Plasma with $\nu_{ee} << \nu_{eN}$ is called slightly or weakly ionized. Due to the fact that at electron energies $w \leq 1\, eV$ the Coulomb cross-sections greatly exceed the gas kinetic ones, ν_{ee} in the Earth ionosphere, for example, becomes of the order of ν_{eN} at a very low ionization degree $n_e/(n_N+n_i)$. The weakly ionized plasma can also be subdivided: the case $\nu_{eN}^{(\varepsilon)} << \nu_{ee} << \nu_{eN}$, when the distribution of electrons is close to Maxwellian, and the case $\nu_{ee} << \nu_e^{(\varepsilon)}$, when in the strong electric fields the distribution function can be far from Maxwellian. In this book the ionization degree is supposed to be small and the inertia and viscosity of the charged particles are ignored.

The situation with inelastic collisions is more complicated. If it is possible to characterize the electron distribution function in the electric field by a single characteristic energy scale (in plasmas of self-sustained gas discharges this energy is of the order of 1 eV), the inelastic collisions can be subdivided into two groups. The collisions with relatively small electron energy loss (this group usually corresponds to electron exchange with rotational and vibrational molecular levels with quanta of the order of 10^{-3} eV and 10^{-1} eV) can be treated as continuous [3, 24]. They can be taken into account by introducing, instead of Eqs. (2.18), (2.19), the effective value δ_{eff} which now becomes energy-dependent. As a rule, the inequality $\delta_{eff} << 1$ still holds, the transport equations can be derived, and the transport coefficients calculated on the same lines as described above.

At lower pressures in the gas discharges the field intensity increases, and the electron energy balance is determined by the inelastic electron collisions with excitation of electronic energy levels. These collisions are accompanied by considerable losses of electron energy of the order of several eV, and discreteness of these energy losses is sometimes crucial. In such a situation it is necessary to distinguish between the spatial and temporal scales of the energy relaxation, and the scales of the relaxation of the electron distribution function. Some aspects of this problem are discussed below. Frequently used is an approach when the effective value of δ, averaged over all the collisions, is introduced. The relaxation scales, calculated according to Eqs. (2.18) - (2.20), are identified with the scales in the criterion of applicability of the fluid approach. In reality, in the presence of inelastic collisions with large energy losses, the real criterion can be considerably more stringent.

At the higher field intensities the typical electron energies exceed the characteristic atomic values. In these conditions the well-known phenomenon of electron runaway takes place [25-26]. This situation in principle cannot be described in the framework of the fluid approach. It follows from the fact, that the properties of the runaway electrons depend on the potential difference between the source point and the point of observation but not on the local field intensity. However, any modification of the transport equations of the type of Eqs. (2.12) - (2.16) contains only the fields, but not potential.

2.2 TRANSPORT COEFFICIENTS. QUALITATIVE CONSIDERATION

The matrix of transport coefficients L_{mn} connects linearly the thermodynamic 'fluxes' in the macroscopic momentum and energy balance equations and the thermodynamic 'forces'. The rigorous kinetic calculations are, of course, necessary to obtain proper numerical coefficients in the expressions for the transport coefficients, and the corresponding quantitative results will be presented in the following sections. However, the dependence of the components of matrix L_{mn} on plasma parameters follows from a simple physical analysis. Here we shall focus on this qualitative aspect of the problem. In anisotropic plasma, when the transport coefficients are represented by tensors, we shall restrict ourselves to the case of a strong magnetic field, which is assumed to be uniform with the z axis parallel to $\vec{B}$.

Momentum transfer. This term represents the rate of partial momentum transfer due to collisions with the other species. The momentum transfer in plasma consists of two components - friction and thermal forces:

$$\vec{R}_{\alpha\beta} = \vec{R}^{(\vec{u})}_{\alpha\beta} + \vec{R}^{(T)}_{\alpha\beta} \,. \qquad (2.21)$$

Friction force. Since the particles lose their relative velocity in each collision, the relaxation of the directed motion is given by

$$\vec{R}^{(\vec{u})}_{\alpha\beta} = \hat{c}^{(\vec{u})}_{\alpha\beta} \nu_{\alpha\beta} \mu_{\alpha\beta} n_\alpha (\vec{u}_\beta - \vec{u}_\alpha) \ , \tag{2.22}$$

where $\mu_{\alpha\beta} = m_\alpha m_\beta / (m_\alpha + m_\beta)$ is the reduced mass, $\nu_{\alpha\beta}$ is the transport collision frequency averaged over the Maxwellian distribution function, and $\hat{c}^{(\vec{u})}_{\alpha\beta}$ is the dimensionless tensor . For the collisions between the charged particles and neutrals the averaged (over the Maxwellian distribution) transport collision frequency is defined as

$$\nu_{\alpha\beta}(T_\alpha, T_\beta) = \frac{16}{3}\left(\frac{2\pi}{\gamma_{\alpha\beta}}\right)^{1/2} N \int_0^\infty \xi^5 \exp(-\xi^2) \int_0^\pi \sigma_{\alpha\beta}(V,\chi)(1-\cos\chi)\sin\chi d\chi d\xi, \tag{2.23}$$

where N is the neutral density, $\gamma_{\alpha\beta} = \gamma_\alpha \gamma_\beta / (\gamma_\alpha + \gamma_\beta)$, $\gamma_\alpha = m_\alpha / T_\alpha$, and $\sigma_{\alpha\beta}(V,\chi)$ is the differential scattering cross section in the center of mass system, which depends on the relative velocity of the colliding particles $V = (2/\gamma_{\alpha\beta})^{1/2}\xi$ and corresponds to scattering angle χ. For electron-ion collisions [11]

$$\nu_{ei}(T_e) = \frac{4\sqrt{2\pi} e^4 n_e \Lambda Z_i^2}{3(4\pi\varepsilon_0)^2 m_e^{1/2} T_e^{3/2}} \ , \tag{2.24}$$

where $\Lambda = \ln(T_e r_d 4\pi\varepsilon_0 / e^2)$ is the Coulomb logarithm, $r_d = (\varepsilon_0 T_e / ne^2)^{1/2}$ is the Debye radius. In practical units

for $T_e < 50eV$ $\Lambda = 23.4 - 1.15 \lg n_e + 3.45 \lg T_e$;
for $T_e > 50eV$ $\Lambda = 25.3 - 1.15 \lg n_e + 2.3 \lg T_e$,

where temperature is measured in eV and density is in CGS units. The Coulomb logarithm depends also on the frequency of the electric field and on the magnetic field intensity [27]-[29].

The components of the dimensionless tensor $\hat{c}^{(\vec{u})}_{\alpha\beta}$ are determined by the dependence of the transport collision frequency (unaveraged one) on the relative velocity. If it is velocity independent, then $\hat{c}^{(\vec{u})}_{\alpha\beta}$ coincides with the unit tensor. For collisions between electrons and heavy particles, for example, the friction force in this case has a very simple form:

$\vec{R}_{e\beta}^{(\vec{u})} = -m_e n_e \nu_{\alpha\beta}(\vec{u}_e - \vec{u}_\beta)$, and the distribution function of electrons is given by a shifted Maxwellian, Eq. (2.17). In general, if under the influence of some external force electrons acquire the mean velocity, their distribution function deviates from a shifted Maxwellian distribution. The reason is that if the collision frequency decreases with electron velocity, as e.g. for Coulomb collisions $\nu_{ei} \sim V^{-3}$, the faster electrons are more intensively shifted by the applied force than the slower electrons. As a result, the distribution function is not simply shifted but also deformed, and the faster electrons contribute more to the mean velocity than the slower ones. In the absence of the magnetic field, or parallel to it, the friction force for the same drift velocity is smaller than for the pure shifted Maxwellian distribution, and the zz component of the tensor $\hat{c}_{\alpha\beta}^{(\vec{u})}$ is less than unity. On the contrary, for the collision frequency, which increases with velocity, the friction force is larger than in the case of the shifted Maxwellian distribution.

Across a strong magnetic field $\omega_{c\alpha} >> \nu_\alpha$, where $\omega_{c\alpha} = eB/m_\alpha$ is the species hyrofrequency, the deviation from the shifted Maxwellian distribution is of the order of $(\omega_{c\alpha}/\nu_\alpha)^{-2}$, and, hence, the friction force is simply $\vec{R}_{e\beta}^{(\vec{u})} = -m_e n_e \nu_{e\beta}(\vec{u}_e - \vec{u}_\beta)$. The nondiagonal elements of the tensor have different signs and are of the order of unity only if $\omega_{c\alpha} \sim \nu_\alpha$, while both in the strong and low magnetic fields these coefficients tend to zero.

Thermal force. The second component of the momentum transfer in Eq. (2.21) is known as thermal force. This force emerges due to the velocity dependence of the transport collision frequency, and therefore disappears for the shifted Maxwellian distribution function. Consider electrons with the zero mean velocity colliding with neutrals in the absence of a magnetic field. The first term in Eq. (2.21) in this case is equal to zero, and the fluxes of electrons Γ_+ and Γ_- in the opposite directions through the arbitrary unit cross-section at $z = z_0$ compensate each other. By order of magnitude they are given by an estimate $\Gamma_+ = \Gamma_- \sim n_e V_{Te}$. The collisions with neutrals results in the friction forces R_+, R_-, Fig. 2.1, proportional to the corresponding fluxes: $R_+ \sim R_- \sim m_e n_e V_{Te} \nu_e$. In the plasma with uniform temperature and density these forces are balanced. If the mean energies of electrons coming from right and left sides are different, then $R_+ \neq R_-$. In the presence of a temperature gradient, this difference in the energies of electrons, which contribute to the forces, is given by $\delta T_e = (dT_e/dz)\lambda_e$, where λ_e is the mean free path. If, for example, the temperature of electrons coming from the right side is larger, and transport collision frequency increases with velocity, then we have $R_+ > R_-$.

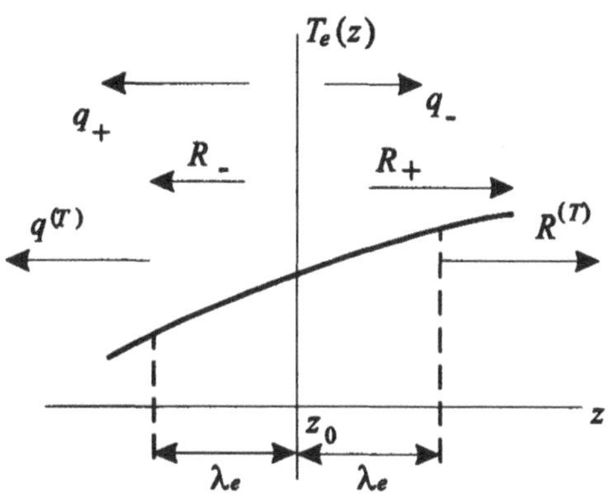

Figure 2.1.

The resulting unbalanced force is given by an estimate

$$R^{(T)} \sim m_e \frac{d\nu_e}{dT_e} \delta T_e n_e V_{Te} \sim n_e \frac{dT_e}{dz} \ . \tag{2.25}$$

Consequently, the value of the thermal force is determined by velocity dependence of the transport frequency, though the transport frequency itself does not enter Eq. (2.25) explicitly.

In general, the thermal force is a linear function of the temperature gradient:

$$\vec{R}^{(T)}_{\alpha\beta} = -\hat{C}^{(T)}_{\alpha\beta} n_\alpha \vec{\nabla} T_\alpha \ . \tag{2.26}$$

For collisions between particles with strongly different masses only the temperature of the light particles (electrons) results in the thermal force. The sign of the zz element of the tensor is positive if the transport frequency increases with velocity, and is negative in the opposite case. Therefore, for electron-ion collisions this coefficient is positive, and for electron-neutral collisions it is usually negative. In the strong magnetic field, $\omega_{c\alpha} >> \nu_\alpha$, the (xx) and (yy) components of the tensor are of the order of $(\nu_\alpha/\omega_{c\alpha})^2$, and only the nondiagonal elements of the tensor are important:

$$\hat{C}^{(T)}_{\alpha\beta xy} = -\hat{C}^{(T)}_{\alpha\beta yx} \sim \hat{C}^{(T)}_{\alpha\beta zz} \nu_{\alpha\beta} / \omega_{c\alpha} \ .$$

Let us estimate the thermal force across a magnetic field for electron-neutral collisions. Magnetic field is parallel to the z axis, and temperature gradient to the x axis, Fig. 2.2. Since electrons rotate over the Larmor orbits, they bring with them the temperature difference $\delta T_e \sim \rho_{ce} dT_e/dx$, and, as can be seen in Fig. 2.2, the friction forces in the y direction are unbalanced. On the contrary, the fluxes in the y direction come on the average from the same coordinate x, and, therefore, are balanced. The resulting force is perpendicular to the temperature gradient, and, similar to the Eq. (2.25),

$$\vec{R}^{(T)}_{\alpha\beta\perp} \sim n_\alpha \nu_{\alpha\beta} / \omega_{c\alpha} [\vec{h} \times \vec{\nabla} T_\alpha] , \qquad (2.27)$$

where $\vec{h} = \vec{B}/B$. The coefficient in this expression is positive for the transport frequency, which increases with velocity, and negative in the opposite case.

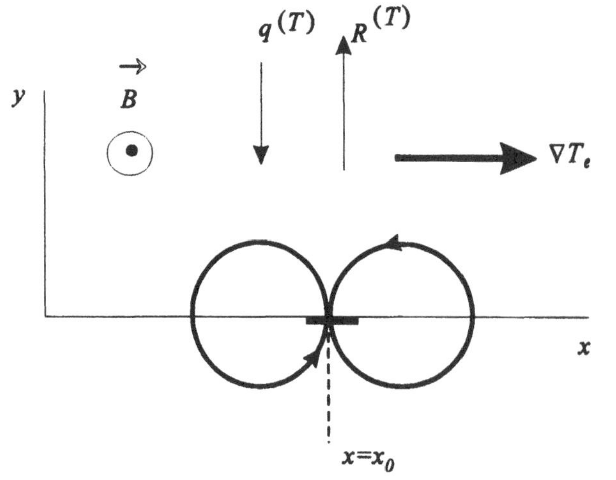

Figure 2.2.

In the multispecies plasmas the distribution function of the species α is determined by the collisions with all other particles. Therefore, the tensor $\hat{C}^{(T)}_{\alpha\beta}$ depends on all types of the collisions, and is a function of the densities of all other species. This is, of course, also true for the other transport coefficients. In such complicated situation it is convenient to introduce the new tensor

$$\hat{c}^{(T)}_{\alpha\beta} = \frac{\sum_j \nu_{\alpha j} \mu_{\alpha j}}{\nu_{\beta j} \mu_{\beta j}} \hat{C}^{(T)}_{\alpha\beta} , \qquad (2.28)$$

where $\mu_{\alpha j}$ is the reduced mass. When all the collisions excluding collisions between α and β species are negligible, these two tensors coincide. In the multispecies plasma the tensor $\hat{C}^{(T)}_{\alpha\beta}$ can be approximately calculated from Eq. (2.28) if one takes for $\hat{c}^{(T)}_{\alpha\beta}$ their values in the absence of all other collisions, see the next Section.

Heat flux. Heat conductivity. The expression for the heat flux, as for the momentum transfer, consists of two parts:

$$\vec{q}_\alpha = \vec{q}^{(T)}_\alpha + \vec{q}^{(\vec{u})}_\alpha \ . \tag{2.29}$$

The first term is a linear function of the temperature gradient:

$$\vec{q}^{(T)}_\alpha = -\hat{\kappa}_\alpha \vec{\nabla} T_\alpha = -C_{V\alpha}\hat{\chi}_\alpha \vec{\nabla} T_\alpha \ , \tag{2.30}$$

where $\hat{\kappa}_\alpha$ is the tensor of the thermal or heat conductivity, and $C_{V\alpha}$ is the heat capacity (for monatomic species $C_{V\alpha}$=3/2).

The estimates for the components of the heat conductivity tensor can be obtained from a simple gaseous kinetic consideration. Let us consider the temperature gradient parallel to the magnetic field in the absence of the particle fluxes. The particles bring the temperatures on the average from their mean free path, therefore the heat flux from one side is of the order of $q_+ \sim \lambda_\alpha v_\alpha n_\alpha T_\alpha$, see Fig. 2.1. The difference between right and left heat fluxes is of the order of $\lambda_\alpha (dT_\alpha/dz)/T_\alpha$ of each flux. Hence the resulting heat flux is of the order of $q_{\alpha z} \sim -\kappa_{zz} dT_\alpha/dz$, where

$$\kappa_{zz} \sim n_\alpha \lambda^2_\alpha v_\alpha \ . \tag{2.31}$$

Across a strong magnetic field the mean free path must be replaced by the gyroradius, and

$$\kappa_{xx} = \kappa_{yy} \sim n_\alpha \rho^2_{c\alpha} v_\alpha \ . \tag{2.32}$$

Along $\vec{B}$ the heat conductivity for electrons is much larger than the ion heat conductivity, but across a strong magnetic field the situation is reversed. For $T_e \sim T_i$, if the cross-sections are of the same order, one has: $\kappa_{ezz}/\kappa_{izz} \sim (m_i/m_e)^{1/2}$; $\kappa_{exx}/\kappa_{ixx} \sim (m_e/m_i)^{1/2}$.

In a magnetic field the heat fluxes in the $[\vec{B} \times \vec{\nabla} T]$ direction also arise. As can be seen from Fig. 2.2, there is unbalanced heat flux of the particles passing through the area element at $x=x_0$. The one-sided heat flux is of the order of nTV_T, and its unbalanced part is $\sim nTV_T(\rho_c/T)dT/dx$. The resulting heat flux for electrons can be estimated as

$$\vec{q}_{e\wedge}^{(T)} \sim -\frac{n_e T_e}{eB}[\vec{h} \times \vec{\nabla} T_e]. \tag{2.33}$$

For ions the sign is positive. However, in partially ionized plasma the heat conductivity of ions is usually unimportant due to the fast energy exchange with neutrals. The heat flux given by Eq. (2.33) is perpendicular to the temperature gradient, and, therefore, in the uniform magnetic field and homogeneous plasma this component of the heat flux is divergence-free. In the nonuniform magnetic field due to the drift of the leading centers of the Larmor circles [5]-[6] there arises the convective energy flux associated with this drift. This flux is contained implicitly in the expression for the heat flux (2.33), because the flux (2.33) is not divergence-free in the nonuniform magnetic field, and its divergence corresponds to this drift.

Heat flux connected with convective motion. The second term in the r.h.s. of the Eq. (2.29) represents the heat flux generated by the relative velocities of the particles:

$$\vec{q}_{\alpha}^{(\vec{u})} = \sum_{\beta} \hat{C}_{\alpha\beta}^{(T)} n_\alpha T_\alpha (\vec{u}_\alpha - \vec{u}_\beta). \tag{2.34}$$

Similar to the thermal force, origin of this flux is connected with the velocity dependence of the transport collision frequency. For example, consider the case of electrons along a magnetic field. If electron-neutral collision frequency rises with velocity, the distribution function deviates from a shifted Maxwellian, and the electron current is mainly transported by slower electrons. As a result, the unbalanced heat flux has the form

$$q_{\alpha z} \sim -n_\alpha T_\alpha (u_{\alpha z} - u_{\beta z}) \tag{2.35}$$

with a negative coefficient. It is positive for the transport frequency, which decreases with velocity. Across a magnetic field similar arguments lead to the existence of the flux

$$\vec{q}_{\alpha\wedge}^{(\vec{u})} \sim -\frac{n_\alpha T_\alpha m_\alpha \nu_{\alpha\beta}}{eB}[\vec{u}_\alpha \times \vec{h}] \tag{2.36}$$

with the negative coefficient if the transport frequency increases with velocity. The dimensionless tensor in the Eq. (2.34) coincides with the tensor in the Eq. (2.26), where thermal force is defined. This fact reflects the symmetry of the transport coefficients L_{mn} introduced in the previous Section. This statement is proved in the thermodynamics of irreversible processes, for further details see, for example, [13].

Collisional heat production. Several mechanisms are responsible for the heat production Q_{α}, Eq. (2.14). The contribution from the elastic collisions for electrons consists of two terms:

$$Q_e = Q_e^{(\Delta)} + Q_e^{(\vec{u})}. \tag{2.37}$$

In the first place the elastic collisions result in the equalization of the temperatures of electrons and heavy particles. In one collision the fraction of the order of $m_e/m_{i,N}$ of the total energy is transferred. For that reason significant energy exchange occurs on the time scale of $m_{i,N}/m_e$ collisions, and the first term in Eq. (2.37) is given by

$$Q_e^{(\Delta)} = \sum_{\beta=i,N} Q_{e\beta} = \sum_{\beta=i,N} \frac{3m_e}{m_\beta} n_e \nu_{e\beta}(T_\beta - T_e). \tag{2.38}$$

This result can be easily obtained by integrating Eq. (2.14) over the Maxwellian distribution function. For the heavy particles these terms enter the energy balance equations with the opposite sign. The second term in Eq. (2.37) corresponds to the conversion of the energy of the direct motion into the heat by the friction force. This source has the form:

$$Q_e^{(\vec{u})} = -\sum_{\beta=i,N} \vec{R}_{e\beta}(\vec{u}_e - \vec{u}_\beta). \tag{2.39}$$

In the absence of the temperature gradient and the thermal force it is the well-known Joule heating. The third mechanism of heat production is connected with the inelastic collisions and the corresponding additional heat sources and sinks must be added to the r.h.s. of the energy balance equation. Sometimes the temperature exchange owing to inelastic collisions with small energy loss can be taken into account by introducing δ_{eff} instead of $2m_e/m_i$ into Eq. (2.38) [3,24], see Section 2.9.

Mobility, diffusion and thermodiffusion. The transport equations (2.7) - (2.16) with the transport coefficients L_{mn} present the complete fluid description of plasmas. Still the shortcoming of such an approach lies in the fact that the relative velocities $(\vec{u}_\alpha - \vec{u}_\beta)$ appear as independent variables, while the friction forces are calculated as their functions. However, it is more convenient to have velocities expressed as functions of density and temperature gradients and electric field. In partially ionized plasma such a procedure is possible because the inertia and viscosity of the charged particles can be ignored for the motions with mean velocities much smaller than the sound, or Alfven velocity.

Neglecting these terms, assuming quasineutrality $n_e=n_i=n$ and $T_i=T_N$, in the case of one sort of neutrals and single charged ions, the momentum balance equations (2.11) for electrons and ions read (we neglect here also the term with the particle sources, because they are usually small compared to friction forces)

$$-\vec{\nabla}(nT_e) - en(\vec{E} - \vec{F}_e / e + [\vec{u}_e \times \vec{B}]) - m_e n \nu_{eN} \hat{c}_{eN}^{(\vec{u})}(\vec{u}_e - \vec{u}_N) -$$

$$-m_e n \nu_{ei} \hat{c}_{ei}^{(\vec{u})}(\vec{u}_e - \vec{u}_i) - \left(\frac{\nu_{eN}}{\nu_{ei} + \nu_{eN}} \hat{c}_{eN}^{(T)} + \frac{\nu_{ei}}{\nu_{ei} + \nu_{eN}} \hat{c}_{ei}^{(T)} \right) n \vec{\nabla} T_e = 0;$$

$$-\vec{\nabla}(nT_i) + en(\vec{E} + \vec{F}_i / e + [\vec{u}_i \times \vec{B}]) + m_e n \nu_{ei} \hat{c}_{ei}^{(\vec{u})}(\vec{u}_e - \vec{u}_i) -$$

$$-\mu_{iN} n \nu_{iN} \hat{c}_{iN}^{(\vec{u})}(\vec{u}_i - \vec{u}_N) - \frac{\mu_{iN} \nu_{iN}}{m_e \nu_{ei} + \mu_{iN} \nu_{iN}} \hat{c}_{iN}^{(T)} n \vec{\nabla} T_i +$$

$$+ \frac{\nu_{ei}}{\nu_{ei} + \nu_{eN}} \hat{c}_{ei}^{(T)} n \vec{\nabla} T_e = 0. \tag{2.40}$$

Here for the friction and thermal forces we use the simple mixture rule interpolation. This rule means that for the cases

$$\nu_{eN} << \nu_{ei} \text{ and } \nu_{eN} >> \nu_{ei}$$

the expressions for the forces approach the limiting expressions for fully or weakly ionized plasmas correspondingly. Hence, the tensors $\hat{c}_{\alpha\beta}^{(T)}$ and $\hat{c}_{\alpha\beta}^{(\vec{u})}$ are to be calculated for the pure fully or weakly ionized plasma. Due to the different dependencies of the ν_{eN} and ν_{ei} on velocities, the accuracy of this approach (especially in the weak magnetic fields) for $\nu_{eN} \sim \nu_{ei}$ can be of the order of unity. More complicated mixture rules are briefly discussed in Section 2.7

Resolving Eq. (2.40) with respect to velocities, we obtain

$$\begin{aligned}
&\vec{u}_e - \vec{u}_N = -\hat{b}_e(\vec{E} + [\vec{u}_N \times \vec{B}] - \vec{F}_e / e) - \hat{D}_e \vec{\nabla} n / n - \hat{D}_e^{(T)} \vec{\nabla} T_e / T_e - \\
&- \hat{D}_{ei}^{(T)} \vec{\nabla} T_i / T_i; \\
&\vec{u}_i - \vec{u}_N = \hat{b}_i(\vec{E} + [\vec{u}_N \times \vec{B}] + \vec{F}_i / e) - \hat{D}_i \vec{\nabla} n / n - \hat{D}_i^{(T)} \vec{\nabla} T_i / T_i - \\
&- \hat{D}_{ie}^{(T)} \vec{\nabla} T_e / T_e .
\end{aligned} \tag{2.41}$$

The tensors $\hat{b}_\alpha$ are known as mobility tensors, they are related to the well-known conductivity tensor

$$\hat{\sigma} = en(\hat{b}_e + \hat{b}_i). \tag{2.42}$$

The tensors $\hat{D}_\alpha, \hat{D}_\alpha^{(T)}$ are called the diffusion and thermodiffusion tensors, and $\hat{D}_{ei,ie}^{(T)}$ are the tensors of relative thermodiffusion. General expressions for these tensors are presented in Section 2.7. Here we shall discuss briefly the simplest case of weakly ionized plasma with the velocity independent transport frequencies. In this case the thermal force vanishes, and the tensor $\hat{c}_{\alpha N}^{(\vec{u})}$ coincides with the unit one. Such an approach is sometimes called quasifluid or elementary approximation.

The mobility tensor, as follows from Eqs. (2.41) - (2.42), has the form

$$\hat{b}_e = \begin{pmatrix} b_{e\perp} & -b_{e\wedge} & 0 \\ b_{e\wedge} & b_{e\perp} & 0 \\ 0 & 0 & b_{e\parallel} \end{pmatrix}; \quad \hat{b}_i = \begin{pmatrix} b_{i\perp} & b_{i\wedge} & 0 \\ -b_{i\wedge} & b_{i\perp} & 0 \\ 0 & 0 & b_{i\parallel} \end{pmatrix}, \tag{2.43}$$

where in the elementary approximation

$$b_{\alpha\parallel} = \frac{e}{\mu_{\alpha N} \nu_{\alpha N}}; \quad b_{\alpha\perp} = \frac{e\nu_{\alpha N}}{\mu_{\alpha N} \omega_{c\alpha}^2}; \quad b_{\alpha\wedge} = \frac{e}{\mu_{\alpha N} \omega_{c\alpha}}. \tag{2.44}$$

Here $\mu_{\alpha N}$ is the reduced mass (m_e for electrons). Here $b_{\alpha\perp}$ are called the Pedersen components, and $b_{\alpha\wedge}$ - the Hall components. The parallel component has a very simple physical meaning - electric force is balanced by the friction force resulting in the motion of the charged particles in the same direction with constant velocity. The Hall components in the strong magnetic field $\omega_{c\alpha} >> \nu_{\alpha N}$ correspond to the guiding center collisionless drift in the crossed fields with the velocity $[\vec{E} \times \vec{B}] / B^2$, which is independent of the species charge. The origin of the components $b_{\alpha\perp}$ is also connected with this drift. The friction force, which is proportional to the drift velocity in the crossed fields (the neutrals are assumed to be at rest) $\vec{R}^{(\vec{u})}_{\alpha N\perp} = -\mu_{\alpha N} n_\alpha \nu_{\alpha N} [\vec{E} \times \vec{B}] / B^2$ causes in its turn the drift in the direction of the electric field with the mean velocity $\vec{u}_\alpha = \pm[\vec{R}^{(\vec{u})}_{\alpha N\perp} \times \vec{B}] / n_\alpha |e| B^2$. As a result, the particles move also in the direction of the electric field with a velocity, which is proportional to the collision frequency. Its sign depends on the species charge.

In the elementary approximation the diffusion and thermodiffusion tensors coincide $\hat{D}^{(T)}_\alpha = \hat{D}_\alpha$, and

$$\hat{D}_\alpha = \frac{T_\alpha}{e} \hat{b}_\alpha . \tag{2.45}$$

This expression is known as Einstein's relation. The diagonal components of the diffusion tensor are determined by the same random walk process as the heat conductivity, therefore one has an estimate: $D_{zz} \sim \kappa_{zz}/n \sim \lambda^2_\alpha \nu_\alpha$, $D_{xx} = D_{yy} \sim \kappa_{xx}/n \sim \rho^2_{c\alpha} \nu_\alpha$.

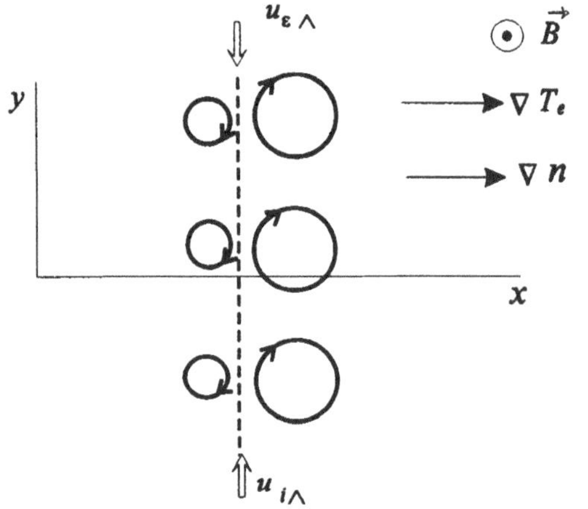

Figure 2.3.

The Hall components represent the diamagnetic drift of the charged particles. In the strong magnetic field $\omega_{c\alpha} >> \nu_{\alpha N}$ Eqs. (2.43) - (2.45) yield

$$\vec{u}_{i\wedge} = \frac{[\vec{B} \times \vec{\nabla} p_i]}{enB^2}; \ \vec{u}_{e\wedge} = -\frac{[\vec{B} \times \vec{\nabla} p_e]}{enB^2}; \ \vec{j}_\wedge = \frac{[\vec{B} \times \vec{\nabla} p]}{B^2} , \tag{2.46}$$

where $p=p_e+p_i=n(T_e+T_i)$ is the total pressure, $\vec{j}_\wedge = en(\vec{u}_{i\wedge} - \vec{u}_{e\wedge})$ is the diamagnetic current. The origin of the diamagnetic drifts is illustrated in Fig. 2.3.

Owing to the density and partial temperature gradients, the particles rotating over their gyroradii create an unbalanced flux. This flux arises as a difference in thermal fluxes 'brought' to a given point by the upward and downward moving particles from the vicinity of the order of gyroradius. Therefore, for example, for density gradient, $u_\wedge \sim V_T \rho_c dn/dx$, in accordance with Eq. (2.46). In the uniform magnetic field the diamagnetic fluxes are not connected with the real motion of the particle guiding centers. Hence these fluxes must be divergence-free, as it follows from Eq. (2.46). Therefore, they do not contribute to the particle's continuity equations. In the nonuniform magnetic field the divergences of the diamagnetic fluxes differ from zero. It corresponds to the real drift of the guiding centers caused by the magnetic field nonuniformity [5,6]. The friction force associated with the diamagnetic fluxes causes drift along the density or temperature gradient $\vec{u}_\alpha = \pm[\vec{R}^{(\vec{u})}_{\alpha N\wedge} \times \vec{B}]/n_\alpha|e|B^2$. Inserting $\vec{R}^{(\vec{u})}_{\alpha N\wedge} = -\mu_{\alpha N} n_\alpha \nu_{\alpha N} \vec{u}_{\alpha\wedge}$ we obtain again the expressions for the diagonal elements $D_{\alpha\perp}$ and $D^{(T)}{}_{\alpha\perp}$ in Eq. (2.45).

In the subsequent Sections we shall present the results of calculation of the transport coefficients $\hat{c}^{(\vec{u})}_{\alpha\beta}$, $\hat{c}^{(T)}_{\alpha\beta}$ and related tensors $\hat{D}_\alpha$, $\hat{b}_\alpha$, $\hat{D}^{(T)}_\alpha$, $\hat{D}^{(T)}_{\alpha\beta}$ and $\hat{\kappa}_\alpha$, $\hat{\chi}_\alpha$ from the kinetic theory. The detailed description of the corresponding methods can be found in the references cited below.

2.3 TWO-TERM EXPANSION FOR ELECTRON DISTRIBUTION FUNCTION IN WEAKLY IONIZED PLASMA

Consider the frame of reference moving with the mean velocity of homogeneous neutral gas. The ionization degree is assumed to be small so that the electron-ion transport frequency is negligible with respect to the electron-neutral one: $\nu_{ei} \ll \nu_{eN}$. The characteristic spatial and temporal scales are assumed to exceed the electron mean free path and the inverse transport collision frequency ν_{eN}. It is convenient to seek the solution of the Boltzmann kinetic equation in this case as expansion into the spherical harmonics $Y_l^{(m)}=P_l^{(m)}(\cos\theta)\exp(im\varphi)$ in the velocity space [20]-[22]:

$$f_e(\vec{r},\vec{V},t)=\sum_{l=0}^{\infty}\sum_{m=-l}^{l}f_l^{(m)}(\vec{r},V,t)Y_l^m(\theta,\varphi)\ , \qquad (2.47)$$

where the angles θ and φ determine the direction of electron velocity. Inserting this expansion into the Boltzmann equation (2.1), one obtains the coupled set of equations for the tensor coefficients $f_l^{(m)}$ [3]. If the characteristic spatial scale is large compared to the mean free path or gyroradius, the characteristic time scale is large compared to the inverse collision frequency, and the electron energy is much larger, than the energy gain between the collisions, the terms of the expansion Eq. (2.47) decrease fast, and account of two terms is sufficient:

$$f_e(\vec{r},\vec{V},t)=f_0(\vec{r},V,t)+(\vec{f}_1\vec{V})/V\ . \qquad (2.48)$$

Inserting Eq. (2.18) into Eq. (2.1) and averaging over the angles in the velocity space [3], one obtains equations for the isotropic (f_0) and vector ($\vec{f}_1$) components of the distribution function. Pure integration over solid angle yields

$$\frac{\partial f_0}{\partial t}+\frac{V}{3}\vec{\nabla}\cdot\vec{f}_1-\frac{e}{3m_eV^2}\frac{\partial}{\partial V}(V^2\vec{E}\,\vec{f}_1)=(St_{eN})_0+(St_{ee})_0\ . \qquad (2.49)$$

Multiplying Eq. (2.1) by $3V_i/(4\pi V)$, after integration over the solid angle one obtains the equation for the anisotropic part of the distribution function:

$$\frac{\partial\vec{f}_1}{\partial t}+V\vec{\nabla}f_0-\frac{e\vec{E}}{m_e}\frac{\partial f_0}{\partial V}+\frac{e}{m_ec}[\vec{B}\times\vec{f}_1]=\left\langle St_{eN}^{(1)}\right\rangle=-\nu_{eN}(V)\vec{f}_1\ . \qquad (2.50)$$

Here $\nu_{eN}(V)$ is the velocity-dependent transport frequency of electron-neutral collisions

$$\nu_{eN}(V)=2\pi N\int_0^{\pi}V\sigma_{eN}(V,\chi)(1-\cos\chi)\sin\chi d\chi\ . \qquad (2.51)$$

The averaged transport frequency ν_{eN} Eq. (2.23) is the result of the integration of Eq. (2.51) with the Maxwellian distribution function. For elastic collisions electron-neutral collision integral for isotropic distribution function is

$$(St_{eN})_0 = \frac{1}{V^2}\frac{\partial}{\partial V}[V^3 \delta_{eN} \nu_{eN}(V)(f_0 + \frac{T_N}{m_e V}\frac{\partial f_0}{\partial V})], \qquad (2.52)$$

where $\delta_{eN}=2m_e/m_N$. This electron-neutral collision integral corresponds to the energy relaxation, and, as was mentioned above, is of the order of $\delta_{eN}\nu_{eN}f_0$.

Since in the partially ionized plasma collisions between the electrons can also be important, the corresponding collision integral must be retained in the Eq. (2.49). However, this integral can be omitted in Eq. (2.50) since electron-electron collision frequency ν_{ee} is small compared to electron-neutral collision frequency ν_{eN}. The expression for electron-electron collision integral St_{ee} can be derived from the general Landau collision integral, Eq. (2.3), see, for example, [30]:

$$(St_{ee})_0 = \frac{1}{2V^2}\frac{\partial}{\partial V}\left\{V^2\left[A_1(f_0)Vf_0 + A_2(f_0)\frac{\partial f_0}{\partial V}\right]\right\},$$

$$A_1 = \frac{8\pi\nu_{ee}(V)}{n}\int_0^V W f_0(W)dW; \qquad (2.53)$$

$$A_2 = \frac{8\pi\nu_{ee}(V)}{n}\left(\int_0^V W^4 f_0(W)dW + V^3\int_V^\infty W f_0(W)dW\right);$$

The electron-electron collision frequency (unaveraged one) is defined as

$$\nu_{ee}(V) = \frac{4\pi n e^4 \Lambda}{(4\pi\varepsilon_0)^2 m_e^2 V^3}. \qquad (2.54)$$

The anisotropic part of the distribution function can be expressed through f_0 from the Eq. (2.50). Neglecting the time derivative for the slow transport processes, we have

$$\vec{f}_1 = -\hat{M}(V\vec{\nabla}f_0 - \frac{e\vec{E}}{m_e}\frac{\partial f_0}{\partial V}), \qquad (2.55)$$

where

$$\hat{M} = \begin{Bmatrix} \frac{\nu_{eN}(V)}{\omega_{ce}^2 + \nu_{eN}^2} & \frac{-\omega_{ce}}{\omega_{ce}^2 + \nu_{eN}^2} & 0 \\ \frac{\omega_{ce}}{\omega_{ce}^2 + \nu_{eN}^2} & \frac{\nu_{eN}(V)}{\omega_{ce}^2 + \nu_{eN}^2} & 0 \\ 0 & 0 & 1/\nu_{eN}(V) \end{Bmatrix}. \tag{2.56}$$

Inserting the distribution function Eq. (2.48) into the expression for the particle flux Eq. (2.5), we see that the isotropic part of f_e does not contribute to the flux:

$$\vec{\Gamma}_e = n\vec{u}_e = \frac{4\pi}{3}\int_0^\infty V^3 \vec{f}_1 dV. \tag{2.57}$$

The heat flux (2.13) is also calculated through the anisotropic part of f_e:

$$\begin{aligned} \vec{q}_e &= \frac{m_e}{2}\int_0^\infty (\vec{V}-\vec{u})^2(\vec{V}-\vec{u})[f_0 + (\vec{f}_1\vec{V}/V)dV = \\ &= \frac{2\pi m_e}{3}\int_0^\infty V^5 \vec{f}_1 dV - \frac{5}{2}nT_e\vec{u}_e. \end{aligned} \tag{2.58}$$

For the fluid description of plasma it is necessary to express these fluxes in terms of the density and temperature gradients and electric field.

2.4 ELECTRON TRANSPORT COEFFICIENTS IN WEAKLY IONIZED PLASMA FOR MAXWELLIAN DISTRIBUTION FUNCTION

Comparing terms in the kinetic equation (2.49), one can easily obtain an estimate for the spatial relaxation length of the distribution function λ_{ef}. This estimate can also be derived from the following simple arguments. The characteristic relaxation time of the distribution function is, obviously, $(\delta_{eN}\nu_{eN}+\nu_{ee})^{-1}$. The spatial relaxation is a random walk process with the step λ_{eN} along the magnetic field, and ρ_{ce} across the field with a characteristic time ν_{eN}^{-1} between the collisions. Hence, λ_{ef} corresponds to electron diffusion displacement with the diffusion coefficient $\lambda_{eN}^2\nu_{eN}$ or $\rho_{ce}^2\nu_{eN}$ during time $(\delta_{eN}\nu_{eN}+\nu_{ee})^{-1}$, which is given by

$$\lambda_{ef\parallel} = \lambda_{eN}\left(\frac{\nu_{eN}}{\delta_{eN}\nu_{eN} + \nu_{ee}}\right)^{1/2};$$
$$\lambda_{ef\perp} = \rho_{ce}\left(\frac{\nu_{eN}}{\delta_{eN}\nu_{eN} + \nu_{ee}}\right)^{1/2} \quad (2.59)$$

Here we consider the case $\nu_{ee} >> \delta_{eN}\nu_{eN}$. Under this condition the relaxation of the distribution function is determined by the electron-electron collisions, and the term $(St_{eN})_0$ in the r.h.s. of Eq. (2.49) is small with respect to $(St_{ee})_0$. Inserting Eq. (2.55) into Eq. (2.49), one obtains the equation for the isotropic part of the distribution function f_0. Since for the Maxwellian distribution with an arbitrary temperature the collision integral $(St_{ee})_0$ equals zero, the resulting equation describes the relaxation (in time and space) of f_0 to the Maxwellian distribution function. The electron temperature is to be found from the energy balance equation (2.15), (2.16).

If the characteristic scale L exceeds λ_{ef}, and $\nu_{ee} >> \delta_{eN}\nu_{eN}$, the distribution function is close to the Maxwellian distribution. Therefore, inserting $f_0 = f^{(M)}$ into Eq. (2.55), we have a final expression for the anisotropic part of the distribution function.

Now it is possible to calculate the macroscopic fluxes and the transport coefficients. Comparing Eqs. (2.57), (2.58), (2.55) with Eq. (2.41), we have the mobility tensor

$$\hat{b}_e = -\frac{4\pi e}{3m_e n}\int_0^\infty V^3 \hat{M}(V)\frac{\partial f^{(M)}}{\partial V}dV, \quad (2.60)$$

the diffusion tensor

$$\hat{D}_e = \frac{4\pi}{3n}\int_0^\infty V^4 \hat{M}(V) f^{(M)} dV, \quad (2.61)$$

and tensor of thermodiffusion

$$\hat{D}_e^{(T)} = \frac{4\pi}{3n}\int_0^\infty V^4 \hat{M}(V)\left(\frac{m_e V^2}{2T_e} - \frac{3}{2}\right) f^{(M)} dV = T_e \frac{d\hat{D}_e}{dT_e}. \quad (2.62)$$

As follows from equations (2.60) and (2.61), for the Maxwellian distribution the mobility and diffusion tensors are linked by the Einstein relation

$$\hat{D}_e = \hat{b}_e T_e / e. \tag{2.63}$$

Now we have the expressions for the particle and heat fluxes as functions of the density and temperature gradients and electric field, which are equivalent to the momentum balance equations. To close the system, one has to add continuity equation (2.7) and the energy balance equation (2.15) or (2.16) for T_e.

In the considered case it turned out to be more convenient to calculate the mobility, diffusion and thermodiffusion tensors directly from the kinetic equation. On the other hand, the dimensionless tensor coefficients in the friction and thermal forces $\hat{c}_{eN}^{(\vec{u})}, \hat{c}_{eN}^{(T)}$ introduced in Section 2.2 can now be expressed as the functions of mobilities. From Eqs. (2.40), (2.41) we obtain

$$\begin{aligned} c_{eN\parallel}^{(\vec{u})} &= c_{eN\,zz} = \frac{e}{b_{e\parallel} m_e \nu_{eN}}; \\ c_{eN\,xy}^{(\vec{u})} &= -c_{eN\,yx}^{(\vec{u})} = \frac{e b_{e\wedge}}{m_e \nu_{eN}}(b_{e\perp}^2 + b_{e\parallel}^2)^{-1} - \frac{\omega_{ce}}{\nu_{eN}}; \\ c_{eN\perp}^{(\vec{u})} &= c_{eN\,xx}^{(\vec{u})} = c_{eN\,yy}^{(\vec{u})} = \frac{e b_{e\perp}}{m_e \nu_{eN}(b_{e\perp}^2 + b_{e\parallel}^2)}, \end{aligned} \tag{2.64}$$

where components

$$b_{e\perp}=b_{exx}=b_{eyy},\ b_{e\parallel}=b_{ezz},\ b_{e\wedge}=b_{ezz}.$$

are defined according to Eq. (2.60). The averaged transport frequency ν_{eN} which is defined by Eq. (2.23) can be calculated for the power dependence of the transport frequency on the electron velocity. For $\nu_{eN}(V)=\nu_0 V^p$

$$\nu_{eN} = \frac{4}{3\sqrt{\pi}}\Gamma\left(\frac{p+5}{2}\right)\nu_0 \left(\frac{2T_e}{m_e}\right)^{p/2}. \tag{2.65}$$

The power law $\nu_{eN}(V)=\nu_0 V^p$ corresponds to the power dependence of the interaction potential on distance: $U_{eN}\sim r^{-q}$, where $q=4/(1-p)$. In the tensor form

$$\hat{c}_{eN}^{(\vec{u})} = e/(m_e \nu_{eN})\hat{b}_e^{-1} + (\omega_{ce}/\nu_{eN})\hat{K}, \tag{2.66}$$

where $K_{xy}=-K_{yx}=-1$, while the other components are equal to zero. The elements of the reversed mobility tensor are

$$(\hat{b}_e^{-1})_{zz} = \hat{b}_{e\,zz}^{-1}\,;\,(\hat{b}_e^{-1})_{xx} = (\hat{b}_e^{-1})_{yy} = \hat{b}_{e\perp}^2 / (\hat{b}_{e\perp}^2 + \hat{b}_{e\|}^2)\,;$$
$$(\hat{b}_e^{-1})_{xy} = -(\hat{b}_e^{-1})_{yx} = -b_{e\,xy} / (\hat{b}_{e\perp}^2 + \hat{b}_{e\wedge}^2)\,.$$

The components of the $\hat{c}_{eN}^{(T)}$ are given by

$$c_{eN\|}^{(T)} = c_{eN\,zz}^{(T)} = \frac{d\ln(D_{e\|}/T_e)}{d\ln T_e}\,;$$
$$c_{eN\perp}^{(T)} = c_{eN\,xx}^{(T)} = c_{eN\,yy}^{(T)} = \frac{D_{e\perp}D_{e\perp}^{(T)} + D_{e\wedge}D_{e\wedge}^{(T)}}{D_{e\perp}^2 + D_{e\wedge}^2} - 1\,; \tag{2.67}$$
$$c_{eN\,xy}^{(T)} = -c_{eN\,yx}^{(T)} = \frac{D_{e\perp}D_{e\,xy}^{(T)} - D_{e\,xy}D_{e\perp}^{(T)}}{D_{e\perp}^2 + D_{e\wedge}^2}\,,$$

where $D_{e\wedge}=eb_{e\wedge}/T_e=|D_{xy}|$; $D^{(T)}{}_{e\wedge}=|D^{(T)}{}_{xy}|$. In the tensor form

$$\hat{c}_{eN}^{(T)} = \hat{D}_e^{(T)}\hat{D}_e^{-1} - \hat{I}\,, \tag{2.68}$$

where $\hat{I}$ is the unit tensor.

The heat flux Eq. (2.58) is given by

$$\vec{q}_e = -\hat{\kappa}_e\vec{\nabla}T_e + \hat{c}_{eN}^{(T)}nT_e\vec{u}_e\,, \tag{2.69}$$

where the heat conductivity can be expressed through diffusion and thermodiffusion tensors according to

$$\hat{\kappa}_e = \frac{3}{2}n\hat{\chi}_e = n(\frac{3}{2}\hat{D}_e + \hat{D}_e^{(T)} - T_e\frac{d\hat{D}_e^{(T)}}{dT_e} + \hat{D}_e^{(T)}\hat{D}_e^{-1}\hat{D}_e^{(T)})\,. \tag{2.70}$$

In the elementary approximation when the collision frequency is velocity independent from Eqs. (2.61), (2.62), (2.63), (2.68), (2.70) we obtain

$$\hat{D}_e = \hat{D}_e^{(T)} = \frac{T_e}{e}\hat{b}_e = \frac{3}{5}\hat{\chi}_e\;\;;\;\;\hat{c}_{eN}^{(\vec{u})} = \hat{I}\;\;;\;\;\hat{c}_{eN}^{(T)} = 0\,, \tag{2.71}$$

where the components of the mobility tensor in the elementary approximation are defined by Eqs. (2.43), (2.44). Such an approach is also used for the weak dependence $\nu_{eN}(V)$ or when the high accuracy of the calculations is not important. In this case the mean transport collision frequency Eq. (2.23) can be inserted into Eqs. (2.43), (2.44) and (2.71) to obtain simplified expressions. Fortunately, in the strong magnetic field $\omega_{ce}>>\nu_{eN}$ the perpendicular components of the diffusion, thermodiffusion, mobility and heat conductivity tensors always coincide with the elementary approximation Eq. (2.44) for arbitrary dependence $\nu_{eN}(V)$. This fact follows from the expressions (2.60) - (2.62), where $\nu_{eN}(V)$ in the nominator is averaged over the Maxwellian distribution, and from definition of the averaged value of ν_{eN} given by Eq. (2.23) (for the parallel components the situation is more complicated since the inverse collision frequency is integrated). In the elementary approximation the transport collision frequency in the gaseous mixtures is given by the sum $\sum\nu_{e\alpha}$ over all neutral species. For $\omega_{ce}>>\nu_{eN}$ the perpendicular components are given by the sum over species, while for the parallel components the inverse elements are to be summed. For example, for mobility we have the well-known Blanc relation [4], [14]

$$\frac{1}{b_{e\parallel}}=\sum_{\alpha}\frac{1}{b_{e\alpha\parallel}}. \tag{2.72}$$

In the general case it is convenient to introduce functions K_σ and K_ε [3] representing the deviation from the elementary theory approximation

$$D_{e\perp}=D_{e\perp el}K_\sigma \ ; \ D_{e\wedge}=D_{e\wedge el}K_\varepsilon. \tag{2.73}$$

Here the subscript *el* corresponds to the elementary approximation Eqs. (2.71), (2.44). The coefficients K_σ and K_ε depend on the temperature and magnetic field, they have been calculated in [31], [4] for the power dependence of the collision frequency on electron velocity (2.65). These coefficients are presented in Table 1, as functions of the dimensionless parameter

$$x_e=\omega_{ce}/\nu_{eN} \ . \tag{2.74}$$

More detailed tables can be found in [31] for the functions $g_\sigma(x_e)$ and $h_\sigma(x_e)$ which are related to K_σ, K_ε according to

$$K_\sigma(x_e)=\frac{g_\sigma(x_e)(1+x_e^2)}{g_\sigma^2(x_e)+x_e^2h_\sigma^2(x_e)} \ ; \ K_\varepsilon(x_e)=-\frac{h_\sigma(x_e)(1+x_e^2)}{g_\sigma^2(x_e)+x_e^2h_\sigma^2(x_e)} \ . \tag{2.75}$$

The parallel diffusion coefficient can be calculated as $D_{e\parallel}=D_{e\perp}(x_{eN}=0)$. The velocity dependence of the collision frequency $p=1$ corresponds to hard sphere scattering. In gas discharge physics for the energies less than ~20*eV* the following approximations are often used: p=0 (hydrogen, helium), p=1 (neon), p=3 (argon, xenon). The scattering in the atmosphere for energies less than ~1eV corresponds to p=5/3 [30]. The thermodiffusion coefficients, in accordance with Eq. (2.62), are given by

$$D_{e\perp}^{(T)} = D_{e\perp el}^{(T)} K_\sigma (1 + \frac{p}{2}\frac{x_e^2 - 1}{x_e^2 + 1} - p x_e \frac{d \ln K_\sigma}{d x_e}) ;$$
$$D_{e\wedge}^{(T)} = D_{e\wedge el}^{(T)} K_\varepsilon (1 - \frac{p}{x_e^2 + 1} - p x_e \frac{d \ln K_\varepsilon}{d x_e}) . \tag{2.76}$$

The heat conductivity can be also expressed through K_σ, K_ε using Eq. (2.70).

Table 2.1 The values of coefficients K_σ and K_ε for the power dependence $\nu_{eN} \sim V^p$.

P	x= 0	x= 0.05	x= 0.1	x= 0.2	x= 0.5	x= 1.0	x= 2.0	x= 5.0	x= 10.0
-3	3.4 22.29	3.0 17.2	2.5 12.4	1.72 7.0	0.95 2.7	0.60 1.49	0.52 1.06	0.696 0.965	0.878 0.982
-2	1.67 3.9	1.64 3.8	1.58 3.56	1.41 2.98	1.02 2.87	0.78 1.23	0.77 1.0	0.92 1.0	0.98 1.0
-1	1.14 1.43	1.13 1.42	1.13 1.41	1.11 1.38	1.02 1.22	0.94 1.07	0.94 1.00	0.98 1.00	0.99 1.00
1	1.13 1.51	1.13 1.50	1.12 1.48	1.09 1.40	1.02 1.19	0.94 1.07	0.95 0.98	0.99 1.00	1.00 1.0
5/3	1.42 3.71	1.38 3.17	1.33 2.77	1.21 2.22	0.984 1.46	0.877 1.19	0.895 0.993	0.976 0.994	0.991 0.997
2	1.67 8.3	1.58 5.0	1.47 4.0	1.29 2.8	1.00 1.58	0.85 1.13	0.87 0.99	0.95 0.99	0.99 1.00
3	3.4 ∞	2.27 16.5	1.84 9.4	1.45 6.3	0.90 2.1	0.73 1.22	0.76 1.00	0.89 0.98	0.96 0.99

2.5 ION TRANSPORT COEFFICIENTS IN WEAKLY IONIZED PLASMA

A vast number of publications have been devoted to the calculation of mobility, diffusion and thermodiffusion coefficients for ions. We can mention here, for example, rather the complete book [23] and the references therein. For some other aspects see [32] and [33], where various interaction laws between different types of ions and molecules are considered and corresponding transport coefficients are calculated.

In the weak electric fields $eE\lambda_i << T_i$ the ion transport coefficients can be obtained by the Chapman-Enskog method. Since, as a rule, the velocity dependence of ion-neutral collision frequency $\nu_{iN}(V)$ is relatively weak, the deviation from the elementary theory for small electric fields is not too significant. Therefore, ion transport coefficients for many applications are often taken in the elementary theory approximation. According to the elementary theory tensors of diffusion and thermodiffusion coincide, and we have

$$\hat{D}_i = \hat{D}_i^{(T)} = \frac{T_i}{e}\hat{b}_i \ ; \ \hat{c}_{iN}^{(\vec{u})} = \hat{I} \ ; \ \hat{c}_{iN}^{(T)} = 0 \ ;$$
$$\hat{b}_i = \frac{e}{\mu_{iN}\nu_{iN}}\begin{pmatrix} \frac{1}{1+x_i^2} & \frac{x_i^2}{1+x_i^2} & 0 \\ \frac{-x_i^2}{1+x_i^2} & \frac{1}{1+x_i^2} & 0 \\ 0 & 0 & 1 \end{pmatrix}, \tag{2.77}$$

where

$$x_i = eB/(\mu_{iN}\nu_{iN}), \tag{2.78}$$

μ_{iN} is the reduced mass of the colliding ion and neutral molecule. For the interaction potentials $\sim r^{-q}$ for $q \geq 2$ the difference between the exact values of transport coefficients and Eq. (2.77) does not exceed 13%. It is maximal in the case $m_i << m_N$ when the two-term expansion Eq. (2.48) for the distribution function is valid, and the expressions similar to Eqs. (2.60), (2.61) are applicable. For the arbitrary mass ratio the more precise expression is [23]

$$\frac{D_{i\parallel}}{D_{i\parallel el}} = 1 + \frac{m_N^2(2\Omega^{(12)}/\Omega^{(11)} - 5)^2}{30m_i^2 + 10m_N^2 + (16/3)m_i m_N \Omega^{(22)}/\Omega^{(11)}}, \tag{2.79}$$

where

$$\Omega_{iN}^{(ls)} = \sqrt{\frac{2\pi(\gamma_i+\gamma_N)}{\gamma_i\gamma_N}} \times$$
$$\times \int_0^\infty d\xi \int_0^\pi \xi^{2s+3} \exp(-\xi^2)(1-\cos^l\chi)\sigma_{iN}(\xi,\chi)\sin\chi d\chi \tag{2.80}$$

are the well-known Chapman-Enskog integrals [1], [9], $\gamma_\alpha=m_\alpha/T_\alpha$. According to Eq. (2.79) the deviation from the elementary theory vanishes for $m_i/m_N \to \infty$. As in the case of electrons, in the strong magnetic fields $\omega_{ci}>>\nu_{iN}$ the ion transport coefficients are given by the elementary theory approximation.

An important case corresponds to motion of ions in their own gas, when the dominating collision process is the resonance charge exchange. Its cross section σ_{ex} depends weakly on the energy, and the elementary theory approximation Eq. (2.77) is applicable in this case with the transport collision frequency defined as

$$\nu_{iN} = \frac{32\sigma_{ex}}{3\pi^{1/2}}(T_i/m_i)^{1/2}n_N. \tag{2.81}$$

The results for strong electric fields $eE\lambda_i>>T_i$ can be found in [23], see also the paper [34]. Here we shall present only the transport coefficients in a strong electric field in the absence of the magnetic field for the case of charge exchange collisions [23]. The mobility is proportional to $E^{-1/2}$:

$$b_i(E) = (\frac{2e}{\pi m_i n_N \sigma_{ex} E})^{1/2}. \tag{2.82}$$

The diffusion coefficient is given by

$$D_i(E) = \frac{0.272}{2\sqrt{2}n_N\sigma_{ex}}(\frac{eE}{m_i n_N \sigma_{ex}})^{1/2}. \tag{2.83}$$

The approximating expression for mobility which is valid in arbitrary electric fields has the form:

$$b_i(E) = b_i(E=0)[1+0.13(\frac{eE}{2Tn_N\sigma_{ex}})^2]^{-1/4}. \tag{2.84}$$

2.6 TRANSPORT COEFFICIENTS IN FULLY IONIZED PLASMA

Transport coefficients in the fully ionized plasma were calculated by Braginski in [11] by the Chapman-Enskog method. We shall present here the results for the pure plasma consisting of electrons and ions with the charge number *Z=1*. Tensor coefficients in the friction and thermal forces defined according to Eqs. (2.22), (2.26), (2.28) to be inserted into the momentum balance equation (2.40), are (the collision frequency ν_{ei} is given by Eq. (2.24))

$$c_{ei\perp}^{(\vec{u})} = 1 - \frac{6.416x_e^2 + 1.837}{\Delta_e}; \quad c_{ei\parallel}^{(\vec{u})} = c_{ei\perp}^{(\vec{u})}(x_e = 0) = 0.5129;$$

$$c_{eixy}^{(\vec{u})} = -c_{eiyx}^{(\vec{u})} = \frac{x_e}{\Delta_e}(1.704x_e^2 + 0.7796); \quad c_{ei\perp}^{(T)} = \frac{5.101x_e^2 + 2.681}{\Delta_e}; \tag{2.85}$$

$$c_{ei\parallel}^{(T)} = c_{ei\perp}^{(T)}(x_e = 0) = 0.7110; \quad c_{eixy}^{(T)} = -c_{eiyx}^{(T)} = -\frac{x_e}{\Delta_e}(\frac{3}{2}x_e^2 + 3.053);$$

$$\Delta_e = x_e^4 + 14.79x_e^2 + 3.770; \quad x_e = \omega_{ce} / \nu_{ei}.$$

Electron heat flux has the form (2.29). The components of the electron heat conductivity tensor are

$$\kappa_{e\perp} = \frac{nT_e}{m_e \nu_{ei} \Delta_e}(4.664x_e^2 + 11.92);$$

$$\kappa_{e\parallel} = \kappa_{e\perp}(x_e = 0) = 3.162nT_e / (m_e \nu_{ei}); \tag{2.86}$$

$$\kappa_{exy} = -\kappa_{eyx} = -\frac{nT_e x_e}{m_e \nu_{ei} \Delta_e}\left(\frac{5}{2}x_e^2 + 21.67\right).$$

The ion heat flux is determined only by the heat conductivity ($\vec{q}_i^{(\vec{u})} = 0$)

$$\kappa_{i\perp} = \frac{nT_i(2x_i^2 + 2.645)}{m_i \nu_{ii} \Delta_i}; \quad \kappa_{i\parallel} = \kappa_{i\perp}(x_i = 0) = \frac{3.906nT_i}{m_i \nu_{ii}};$$

$$\kappa_{ixy} = \frac{nT_i(\frac{5}{2}x_i^2 + 4.65)x_i}{m_i \nu_{ii} \Delta_i}; \quad x_i = \omega_{ci} / \nu_{ii}; \tag{2.87}$$

$$\Delta_i = x_i^4 + 2.70x_i^2 + 0.677;$$

where the ion-ion collision frequency is defined as

$$\nu_{ii} = \frac{4\sqrt{\pi}\ \Lambda e^4 n}{3(4\pi\varepsilon_0)^2 \sqrt{m_i}\ T_i^{3/2}}. \tag{2.88}$$

2.7 TRANSPORT COEFFICIENTS IN PARTIALLY IONIZED PLASMA

The situation in partially ionized plasma when ν_{eN}~ν_{ei} is far more complicated. Up to now general expressions for transport coefficients which are valid for all range of ionization degrees and magnetic fields have not been formulated.

The rough approach corresponds to the approximation of the elementary theory, when

$$\hat{c}_{ei}^{(\vec{u})} = \hat{c}_{eN}^{(\vec{u})} = \hat{c}_{iN}^{(\vec{u})} = \hat{I}\ ;\ \hat{c}_{ei}^{(T)} = \hat{c}_{eN}^{(T)} = \hat{c}_{iN}^{(T)} = 0\ .$$

Unfortunately, this approach gives satisfactory results only across the strong magnetic field, while in the weak magnetic fields $\omega_{ce} \leq \max(\nu_{eN}, \nu_{ei})$ (or parallel to magnetic field) the error can reach several hundred percent. Therefore, different interpolation expressions have been proposed. The simple and reasonably accurate approach which we shall use here consists in the following. The momentum balance equation is taken in the form Eq. (2.40) with the tensors $\hat{c}_{e,iN}$, $\hat{c}_{ei}$ calculated for the weakly and fully ionized plasma respectively. In other words, these tensors are assumed to be independent from the ionization degree. In the limiting cases of weak or fully ionized plasma, such an approach, in contrast to the elementary theory approximation, leads to the exact values of transport coefficients. The structure of the factors in Eq. (2.40), such as

$$\nu_{ei}/(\nu_{eN}+\nu_{ei}),\ \nu_{eN}/(\nu_{eN}+\nu_{ei}),\ \mu_{iN}\nu_{iN}/(m_e\nu_{ei}+\mu_{iN}\nu_{iN}),$$

reflects the fact that the influence of the collisions with the additional species is important when the corresponding friction forces become comparable. Therefore, one can expect satisfactory accuracy of interpolation expressions, even in the worst case when ω_{ce}~ν_{eN}~ν_{ei}. The tensors of mobility, diffusion and thermodiffusion are obtained by resolving Eq. (2.40).

For the parallel components we have

$$
\begin{aligned}
b_{e\|} &= \frac{e}{m_e(c_{eN\|}^{(\bar{u})}\nu_{eN} + c_{ei\|}^{(\bar{u})}\nu_{ei})};\\
b_{i\|} &= \frac{e\,c_{eN\|}^{(\bar{u})}\nu_{eN}}{\mu_{iN}\nu_{iN}(c_{eN\|}^{(\bar{u})}\nu_{eN} + c_{ei\|}^{(\bar{u})}\nu_{ei})};\\
D_{e\|} &= \frac{T_e + (T_e + T_i)m_e c_{ei\|}^{(\bar{u})}\nu_{ei}/(\mu_{iN}\nu_{iN})}{m_e(c_{eN\|}^{(\bar{u})}\nu_{eN} + c_{ei\|}^{(\bar{u})}\nu_{ei})};\\
D_{i\|} &= \left(T_i + T_e\frac{c_{ei\|}^{(\bar{u})}\nu_{ei}}{c_{ei\|}^{(\bar{u})}\nu_{ei} + c_{eN\|}^{(\bar{u})}\nu_{eN}}\right)\Big/(\mu_{iN}\nu_{iN}).
\end{aligned}
\qquad (2.89)
$$

It should be noted that in contrast to the case of weakly ionized plasma, the ratio of the diffusion and mobility coefficients, both for electrons and ions, depends on the collision frequencies. Hence in the general case the Einstein relation fails even for Maxwellian distribution function. The parallel transport coefficients in the limiting case $\nu_{ei}\to 0$ coincide with those calculated for the weakly ionized plasma, Eqs. (2.60) - (2.62) and (2.77).

The situation is quite different in the opposite limiting case - there is no smooth transition to the case of fully ionized plasma. Indeed, for $\nu_{iN}\to 0$ the parallel diffusion coefficients $D_{e\|}$, $D_{i\|}\to\infty$. This means that there is no longitudinal diffusion in fully ionized plasma. This result can be obtained in a different way by summing up the momentum balance equations for electrons and ions. In the resulting equation the friction, thermal and electric forces are canceled, since these terms have opposite signs, and the total pressure gradient is balanced by the ion inertia and viscosity, see, for example, [5], [11]. Therefore, the characteristic velocity in fully ionized plasma in the absence of magnetic field is given by the ion sound speed $c_s=[(T_e+T_i)/m_i]^{1/2}$.

In contrast, in partially ionized plasma with a low degree of ionization the contribution of the charged particles to the total pressure is also small. Hence, the slow transport processes of mobility, diffusion and thermodiffusion occur while the total pressure (including the neutral gas pressure) remains constant: $p=\sum p_\alpha=const$. In other words, the partial pressure gradients of the charged species are balanced by the small perturbation of the neutral gas pressure. The latter can be neglected in calculating the transport coefficients.

The mobilities in Eq. (2.89) remain finite when electron-ion collision frequency when electron-ion collision frequency tends to zero: $\nu_{ei}\to 0$. This fact is connected with the assumption that the velocity of the neutral gas is zero. However, this assumption is violated for large ionization degree, since the neutral gas is dragged by the ionized particles. Therefore, the partial mobilities in fully ionized plasma do not make any sense. Indeed, in this case there is no preferable reference system, and only total conductivity can be introduced [11].

The thermodiffusion fluxes for electrons and ions depend both on ∇T_e and ∇T_i. From Eq. (2.40) we have

$$D_{e\|}^{(T)} = \frac{T_e[1+\beta_{e\|}+(1+\beta_{e\|}-C_{ei\|}^{(T)})m_e c_{ei\|}^{(\ddot{u})}\nu_{ei}/(\mu_{iN}\nu_{iN})]}{m_e(c_{eN\|}^{(\ddot{u})}\nu_{eN}+c_{ei\|}^{(\ddot{u})}\nu_{ei})};$$

$$D_{ei\|}^{(T)} = \frac{T_i c_{ei\|}^{(\ddot{u})}\nu_{ei}}{\mu_{iN}\nu_{iN}(c_{eN\|}^{(\ddot{u})}\nu_{eN}+c_{ei\|}^{(\ddot{u})}\nu_{ei})};$$

$$D_{i\|}^{(T)} = \frac{T_i}{\mu_{iN}\nu_{iN}}; \tag{2.90}$$

$$D_{ie\|}^{(T)} = \frac{T_e}{\mu_{iN}\nu_{iN}}\left[\frac{c_{ei\|}^{(\ddot{u})}(1+\beta_{e\|})\nu_{ei}}{c_{eN\|}^{(\ddot{u})}\nu_{eN}+c_{ei\|}^{(\ddot{u})}\nu_{ei}}-C_{ei\|}^{(T)}\right],$$

where

$$\hat{\beta}_e = \frac{\hat{c}_{eN}^{(T)}\nu_{eN}}{\nu_{ei}+\nu_{eN}}+\frac{\hat{c}_{ei}^{(T)}\nu_{ei}}{\nu_{ei}+\nu_{eN}} \equiv \hat{C}_{eN}^{(T)}+\hat{C}_{ei}^{(T)}.$$

The coefficients

$$C_{eN\|}^{(T)} = \frac{c_{eN\|}^{(T)}\nu_{eN}}{\nu_{ei}+\nu_{eN}},$$

$$C_{ei\|}^{(T)} = \frac{c_{ei\|}^{(T)}\nu_{ei}}{\nu_{ei}+\nu_{eN}}.$$

The general expressions for the transverse coefficients in an arbitrary magnetic field are very cumbersome. We shall restrict ourselves to the case of magnetized electrons

$$\omega_{ce} >> \max(\nu_{eN}, \nu_{ei}).$$

For ions the elementary theory approximation is used.

Employing

$$c_{ei\perp}^{(\ddot{u})} = c_{eN\perp}^{(\ddot{u})} = 1,\ c_{ei\wedge}^{(\ddot{u})} = c_{eN\wedge}^{(\ddot{u})} = 0$$

from Eq. (2.40) we find the transverse mobility coefficients:

$$b_{e\perp} = \frac{e}{m_e A}[\nu_{ei} + \nu_{eN}(1 + x_i^2)];$$

$$b_{e\wedge} = -\frac{e}{m_e A}\omega_{ce}\left(1 + x_i^2 + \frac{m_e \nu_{ei}}{\mu_{iN}\nu_{iN}}\right);$$

$$b_{i\perp} = \frac{e}{\mu_{iN}\nu_{iN}A}(\nu_{eN}^2 + \omega_{ce}^2 + \nu_{eN}\nu_{ei}); \tag{2.91}$$

$$b_{i\wedge} = \frac{e^2 B}{c\mu_{iN}^2\nu_{iN}^2 A}(\nu_{eN}^2 + \omega_{ce}^2 + \mu_{iN}\nu_{iN}\nu_{ei} / m_e);$$

$$A = (\nu_{eN} + \nu_{ei})^2 + \omega_{ce}^2\left(1 + \frac{2m_e\nu_{ei}}{\mu_{iN}\nu_{iN}} + x_i^2\right),$$

where x_i is defined according to Eq. (2.78):

$$x_i = eB/(\mu_{iN}\nu_{iN}).$$

The electron and ion diffusion coefficients perpendicular to magnetic field have the form

$$D_{e\perp} = \frac{T_e}{m_e A}(\nu_{eN} + \nu_{ei})\left(1 + \frac{m_e\nu_{ei}}{\mu_{iN}\nu_{iN}} + x_i^2\right) + \frac{T_i}{m_e A}\nu_{ei}\left[\frac{m_e(\nu_{eN} + \nu_{ei})}{\mu_{iN}\nu_{iN}} + x_i^2\right];$$

$$D_{exy} = -\frac{\omega_{ce}}{m_e A}\left[T_e\left(1 + \frac{2m_e\nu_{ei}}{\mu_{iN}\nu_{iN}} + x_i^2\right) + T_i\frac{m_e\nu_{ei}}{\mu_{iN}\nu_{iN}}\right];$$

$$D_{i\perp} = \left\{T_i\left[\omega_{ce}^2\left(1 + \frac{m_e\nu_{ei}}{\mu_{iN}\nu_{iN}}\right) + (\nu_{eN} + \nu_{ei})^2\right] + T_e\nu_{ei}(\omega_{ce}x_i + \nu_{ei} + \nu_{eN})\right\}\times$$
$$\times(\mu_{iN}\nu_{iN}A)^{-1};$$

$$D_{ixy} = \frac{x_i}{A}\left\{\frac{T_i}{\mu_{iN}\nu_{iN}}[\omega_{ce}^2 + \nu_{eN}(\nu_{eN} + 2\nu_{ei})] - \frac{T_e\nu_{ei}}{m_e}\right\}. \tag{2.92}$$

The transverse electron and ion coefficients of thermodiffusion and mutual thermodiffusion are

$$D_{e\perp}^{(T)}=\frac{T_e}{m_eA}\left[(\nu_{eN}+\nu_{ei})\left(1+\frac{m_e\nu_{ei}}{\mu_{iN}\nu_{iN}}+x_i^2\right)-C_{eN\wedge}^{(T)}\omega_{ce}\left(1+\frac{2m_e\nu_{ei}}{\mu_{iN}\nu_{iN}}+x_i^2\right)-\right.$$

$$\left.-C_{ei\wedge}^{(T)}\omega_{ce}\left(1+\frac{m_e\nu_{ei}}{\mu_{iN}\nu_{iN}}+x_i^2\right)\right];$$

$$D_{exy}^{(T)}=-\frac{T_e\omega_{ce}}{m_eA}\left(1+\frac{2m_e\nu_{ei}}{\mu_{iN}\nu_{iN}}+x_i^2\right);$$

$$D_{ei\perp}^{(T)}=\frac{T_i\nu_{ei}}{m_eA}\left[\frac{m_e(\nu_{eN}+\nu_{ei})}{\mu_{iN}\nu_{iN}}+x_i^2\right];$$

$$D_{eixy}^{(T)}=-\frac{T_ix_i\nu_{ei}}{m_eA};$$

$$D_{i\perp}^{(T)}=\frac{T_i}{\mu_{iN}\nu_{iN}A}\left[\omega_{ce}^2\left(1+\frac{m_e\nu_{ei}}{\mu_{iN}\nu_{iN}}\right)+(\nu_{eN}+\nu_{ei})^2\right];$$

$$D_{ixy}^{(T)}=\frac{T_ix_i}{\mu_{iN}\nu_{iN}A}[\omega_{ce}^2+\nu_{eN}(\nu_{eN}+2\nu_{ei})];$$

$$D_{ie\perp}^{(T)}=\frac{T_e}{m_eA}\left[\frac{m_e\nu_{ei}}{\mu_{iN}\nu_{iN}}(\omega_{ce}x_i+\nu_{ei}+\nu_{eN})-\beta_{eN\wedge}\nu_{ei}x_i-\right.$$

$$\left.-C_{ei\wedge}^{(T)}(\nu_{ei}x_i+\omega_{ce}x_i^2)\right]$$

$$D_{iexy}^{(T)}=-\frac{T_e}{m_eA}\nu_{ei}x_i. \qquad (2.93)$$

For the case $C^{(T)}_{eN\wedge}=C^{(T)}_{eN\wedge}=0$ these expressions were derived in [30].

As was mentioned, the largest error in Eqs. (2.92) - (2.93) corresponds to the situation when two conditions are fulfilled simultaneously:

$$\nu_{eN}\sim\nu_{ei} \text{ and } \omega_{ce}\lesssim(\nu_{eN}+\nu_{ei}),$$

and when the dependence of electron-neutral collision frequency ν_{eN} on velocity strongly differs from the Rutherford law. In [4] it was demonstrated that in case of argon, for example, where $\nu_{eN}\sim V^3$, in the absence of a magnetic field an error is up to three times when two collision frequences are of the same order: $\nu_{eN}\sim\nu_{ei}$. The exact calculations were made using the Chapman-Enskog method.

In the absence of a magnetic field a more accurate result for electron mobility can be obtained by applying the Frost approximation [4]. According to this approximation the modified collision frequency $\nu^{(F)}$ is inserted into Eq. (2.60) for the parallel electron mobility $b_{e\|}$. This frequency is defined to provide smooth transition from weakly to fully ionized plasmas:

$$\nu^{(F)} = \nu_{eN}(V) + 0.476 \frac{8\pi\,\Lambda e^4 n}{\sqrt{2}(4\pi\varepsilon_0)^2 V^2 m_e^{3/2} T_e^{1/2}}. \tag{2.94}$$

A similar expression for electron parallel heat conductivity can be found in [4].

More accurate results for the transport coefficients were derived in [14] by the Grad method. We shall present here the expressions for the electron heat conductivity

$$\begin{aligned}
&\kappa_{e\|} = \frac{5nT_e}{2m_e\nu_e}; \quad \kappa_{e\perp} = \frac{\kappa_{e\|}}{1+\omega_{ce}^2/\nu_e^2}; \quad \kappa_{e\wedge} = \kappa_{e\perp}\omega_{ce}/\nu_e;\\
&\nu_e = 1.9\nu_{ei} + (2.5 - 1.2B_{eN}^*)\nu_{eN};\\
&B_{eN}^* = (5\Omega_{ee}^{(12)} - \Omega_{eN}^{(13)})/3\Omega_{eN}^{(11)};\\
&\hat{C}_{e\beta}^{(T)} = \frac{\hat{\kappa}_e m_e \nu_{e\beta}}{n_e T_e}\left(\frac{2\Omega_{e\beta}^{(12)}}{5\Omega_{e\beta}^{(11)}} - 1\right); \quad \beta = i, N.
\end{aligned} \tag{2.95}$$

The integrals $\Omega_{\alpha\beta}{}^{(l,s)}$ are defined according to Eq. (2.80).

Detailed tables of the transport coefficients were obtained by Shkarofsky [31] for the power dependency of the electron-neutral transport collision frequency on velocity. He used the two-term expansion for the distribution function of electrons while the ion motion was neglected. In a small magnetic field such an approach is satisfactory if $m_e\nu_{ei} < \mu_{iN}\nu_{iN}$ since, in accordance with Eq. (2.92), the ion fluid velocities are much smaller than the electron ones. Across the strong magnetic field the ion fluid velocities often exceed the electron velocities, hence one must be careful when employing the results of [31].

2.8 ELECTRON TRANSPORT IN HIGH-FREQUENCY ELECTRIC FIELD

If the influence of high-frequency electric field

$$\vec{E}^{(1)}(\vec{r},t)=\vec{E}_0^{(1)}(\vec{r},t)\exp(-i\omega t), \qquad (2.96)$$

on formation of the electron distribution function (EDF) f_0 (for the Maxwellian EDF - on the electron temperature) is small, the derivation is straightforward. The amplitude of the field $\vec{E}^{(1)}(\vec{r},t)$ may be spatially non-uniform on distances which exceed the mean free path λ_e and time-dependent with long time scale with respect to the field period. In the presence of high-frequency electric field the distribution function of electrons in a nonuniform plasma can be sought in the form

$$f_e(\vec{r},\vec{V},t)=f_0(\vec{r},V,t)+(\vec{f}_1^{(0)}\vec{V})/V+(\vec{f}_1^{(1)}\vec{V})\exp(-i\omega t)/V, \qquad (2.97)$$

where the second term in r.h.s. is associated with the quasistationary electric field $\vec{E}^{(0)}(\vec{r},t)$, while the last term is proportional to the oscillatory electric field. Inserting Eq. (2.97) into the Boltzmann kinetic equation we have, similarly to Eq. (2.55),

$$\begin{aligned}\vec{f}_1^{(0)}&=-\hat{M}_0\left(V\vec{\nabla}f_0-\frac{e\vec{E}^{(0)}}{m_e}\frac{\partial f_0}{\partial V}\right)\\ \vec{f}_1^{(1)}&=-\hat{M}_1\frac{e\vec{E}_0^{(1)}}{m_e}\frac{\partial f_0}{\partial V},\end{aligned} \qquad (2.98)$$

where tensor $\hat{M}_0\equiv\hat{M}$ is defined by Eq. (2.56), and

$$\hat{M}_1=\begin{Bmatrix}\dfrac{\nu_{eN}(V)+i\omega}{\omega_{ce}^2+(\nu_{eN}+i\omega)^2} & \dfrac{-\omega_{ce}}{\omega_{ce}^2+(\nu_{eN}+i\omega)^2} & 0\\ \dfrac{\omega_{ce}}{\omega_{ce}^2+(\nu_{eN}+i\omega)^2} & \dfrac{\nu_{eN}(V)+i\omega}{\omega_{ce}^2+(\nu_{eN}+i\omega)^2} & 0\\ 0 & 0 & \dfrac{1}{\nu_{eN}+i\omega}\end{Bmatrix}. \qquad (2.99)$$

This tensor coincides with $\hat{M}$ with ν_{eN} replaced by $\nu_{eN}+i\omega$.

Now it is easy to calculate the mobility of electrons in a high-frequency electric field which connects the oscillatory mean velocity with the oscillatory electric field. Implying $f_0=f^{(M)}$, similarly to Eq. (2.60), we find

$$\hat{b}_e = -\frac{4\pi e}{3m_e n}\int_0^\infty V^3 \hat{M}_1(V)\frac{\partial f^{(M)}}{\partial V}dV\,. \tag{2.100}$$

In the absence of a magnetic field it is convenient to define the frequency ν_{eff} so that

$$b_e = \frac{e}{m_e(\nu_{eff} + i\omega)}\,. \tag{2.101}$$

For the velocity independent collision frequency $\nu_{eff}=\nu_{eN}$. Generally, in the limiting cases of large and low frequencies, comparing Eqs. (2.99) - (2.101), one obtains

$$\begin{aligned} \frac{1}{\nu_{eff}} &= \frac{4\pi}{3n}\int_0^\infty \frac{d(V^3/\nu_{eN})}{dV} f_0 dV\,, \quad \omega \ll \nu_{eN} \\ \nu_{eff} &= \frac{4\pi}{3n}\int_0^\infty \frac{d(V^3\nu_{eN})}{dV} f_0 dV\,, \quad \omega \gg \nu_{eN}\,. \end{aligned} \tag{2.102}$$

Thus, for low and large frequencies we can use the mobility in the form Eq. (2.101), while for intermediate frequencies and in magnetic field the general expression Eq. (2.100) is to be employed. Detailed calculations of the high-frequency conductivity for different values of magnetic field, ionization degree and for power dependencies of the electron-neutral transport collision frequency from energy can be found in [31].

2.9 TRANSPORT OF ELECTRONS IN WEAKLY IONIZED PLASMA WITH NON-MAXWELLIAN DISTRIBUTION

When the ionization degree is very low, so that $\nu_{ee} \ll \delta_{eN}\nu_{eN}$, the electron-electron collision integral can be neglected in the r.h.s. of Eq. (2.49) with respect to the electron-neutral collision integral, Eq. (2.52), and the distribution of electrons in an electric field can strongly deviate from the

Maxwellian distribution. The most important example of such a situation corresponds to glow discharge plasma in atomic gases at low currents when the non-Maxwellian distribution results from the Joule heating by quasistationary or high-frequency current. In the absence of magnetic field equation for the isotropic part of the distribution function f_0 in quasistationary electric field $\vec{E}^{(0)}$, as follows from Eqs. (2.49) - (2.52), (2.55), reads

$$\frac{\partial f_0}{\partial t}-\frac{V^2}{3}\vec{\nabla}\cdot[\vec{\nabla}\frac{f_0}{\nu_{eN}(V)}]+\frac{eV}{3m_e}\vec{\nabla}\cdot\left[\frac{\vec{E}^{(0)}}{\nu_{eN}(V)}\frac{\partial f_0}{\partial V}\right]$$
$$+\frac{e\vec{E}^{(0)}}{3m_e V^2}\frac{\partial}{\partial V}\left[\frac{V^3\vec{\nabla} f_0}{\nu_{eN}(V)}\right]-\frac{e^2(E^{(0)})^2}{3m_e V^2}\frac{\partial}{\partial V}\left[\frac{V^2}{\nu_{eN}(V)}\frac{\partial f_0}{\partial V}\right] \quad (2.103)$$
$$=\frac{1}{2V^2}\frac{\partial}{\partial V}\left[V^3\delta_{eN}\nu_{eN}(V)\left(f_0+\frac{T_N}{m_e V}\frac{\partial f_0}{\partial V}\right)\right]$$

In uniform and stationary plasma this equation can be rewritten in the form

$$\frac{1}{\sqrt{\varepsilon}}\frac{\partial}{\partial\varepsilon}\sqrt{\varepsilon}\left(D_\varepsilon\frac{\partial f_0}{\partial\varepsilon}+W_\varepsilon f_0\right)=0, \quad (2.104)$$

where $\varepsilon=m_eV^2/2$; $D_\varepsilon(\varepsilon)=(eE^{(0)}\lambda_e)^2\nu_{eN}/3+\delta_{eN}\nu_{eN}\varepsilon T_N$; $W_\varepsilon=\delta_{eN}\nu_{eN}\varepsilon$. This means that the distribution function is determined by the diffusion on energy with diffusion coefficient D_ε and by the directed velocity in the energy space ($-W_\varepsilon$); the coefficients $\varepsilon^{-1/2}$, $\varepsilon^{1/2}$ simply correspond to transition from the phase space to the energy space.

In uniform plasma it has the well-known stationary solution [2-4]

$$f_0^{(0)}=An\exp\left[-\int_0^V\frac{V'dV'}{\frac{2e^2(E^{(0)})^2}{3\delta_{eN}m_e^2\nu_{eN}^2(V)}+\frac{T_N}{m_e}}\right]. \quad (2.105)$$

This distribution function is formed by the balance of the work of electric field and of energy transfer from electrons to neutral gas due to elastic collisions. For the transport collision frequency which decreases with velocity not too fast (slower than $V^{s/2-1}$, $s>0$ for power dependence) or increases with velocity Eq. (2.105) determines the exponential function of energy which decays with the characteristic scale

$$\Delta\varepsilon(\varepsilon)=\left(\frac{\partial \ln f_0^{(0)}}{\partial\varepsilon}\right)^{-1}=\frac{D_\varepsilon(\varepsilon)}{W_\varepsilon(\varepsilon)}\sim\frac{e^2(E^{(0)})^2}{\delta_{eN}m_e\nu_{eN}^2(\varepsilon)}+T_N\,. \qquad (2.106)$$

It corresponds to the average diffusive displacement on energy with the diffusion coefficient D_ε during energy relaxation time $(\delta_{eN}\nu_{eN})^{-1}$. Since distribution Eq. (2.105) decreases exponentially, the average electron energy ('temperature') $\bar{\varepsilon}$ satisfies $\Delta\varepsilon(\bar{\varepsilon})=\bar{\varepsilon}$. For glow discharges $\bar{\varepsilon}>>T_N$ and the mean energy is equal to the energy gained in electric field on the energy relaxation length defined by Eq. (2.20)

$$\bar{\varepsilon}\sim eE\lambda_e^{(\varepsilon)}(\bar{\varepsilon})=eE\lambda_e(\bar{\varepsilon})/\sqrt{\delta_{eN}}\,. \qquad (2.107)$$

For the constant collision frequency the function $f_0^{(0)}$ coincides with the Maxwellian distribution with the effective temperature

$$T_{eff}=T_N+\frac{2}{3}\frac{e^2(E^{(0)})^2}{\delta_{eN}m_e\nu_{eN}^2}\,. \qquad (2.108)$$

It means that all properties of electron gas in strong *dc* or RF electric field

$$E^{(0)}\geq E_{cr}^{(0)}=\frac{T_N\delta_{eN}^{1/2}}{e\lambda_{eN}}$$

depend on the field strength. Numerous phenomena which are connected with this effect are discussed in [3], [30], [33]. For example, the mobility Eq. (2.77) becomes field-dependent, and Ohm's law - nonlinear. Its character is determined by the dependencies $\nu_{eN}(V)$ and $\delta_{eN}(V)$ - effective inelastic parameter, see below Eq. (2.113). In particular for the velocity independent δ_{eN} which is considered now and power law $\nu_{eN}(V)\sim V^p$ the electron mobility is

$$b_e(E^{(0)})\sim(E^{(0)})^\alpha\,; \qquad \alpha=-[1+1/(2p)]\,. \qquad (2.109)$$

For electron scattering on rigid spheres α=-1/2.

In slightly inhomogeneous nonstationary plasma, the terms with temporal and spatial derivatives in Eq. (2.107) are small when

$$\tau_\varepsilon = (\delta_{eN}\nu_{eN})^{-1} << t \ ; \ L >> \lambda_e^{(\varepsilon)}. \qquad (2.110)$$

In such a situation the fluid description can be constructed too. However, in this case as a zero approximation, the distribution function f_0 slightly different from $f_0^{(0)}$ Eq. (2.108) is to be taken. Direct substitution of the distribution (2.108) as a zero approximation in Eq. (2.55) is sometimes incorrect. The reason is that diffusive flux which is proportional to the first term in Eq. (2.55) is $\lambda_e^{(\varepsilon)}/L$ times smaller than the flux caused by electric field. But the third and the fourth terms in the l.h.s. of Eq. (2.107) are also of the order of $\lambda_e^{(\varepsilon)}/L$ of the largest ones - of the fifth term in the l.h.s. and of the term in r.h.s. Therefore, Eq. (2.107) must be solved with accuracy up to $\lambda_e^{(\varepsilon)}/L$. The corrections $f_0^{(1)}$ are to be found from the Eq. (2.107) by successive approximation method, and the sum $f_0=f_0^{(0)}+f_0^{(1)}$ is to be inserted into the expression for the directed part f_1 of the EDF Eq. (2.55), and finally into the particle flux Eq. (2.57). The corrections $f_0^{(1)}$ result in fluxes along the electric field which are linear in $\vec{\nabla} n, \vec{\nabla}(E^{(0)}/N)$ (N is the density of neutrals). Since the 'temperature' (the mean energy) for $T_N \to 0$ is determined by E/N according to Eq. (2.107), the resulting field-driven fluxes formally resemble the diffusion and thermodiffusion. Substituting $f_0^{(0)}+f_0^{(1)}$ into Eq. (2.57) yields

$$\vec{\Gamma}_e = -b_e n_e \vec{E}^{(0)} - D_e^{(0)}\vec{\nabla} n_e - \frac{n_e}{N}\vec{\nabla}(ND_e^{(0)}) - \frac{\kappa_1 \vec{E}^{(0)}(\vec{E}^{(0)}\vec{\nabla} n_e)}{(E^{(0)})^2} -$$
$$- \frac{\kappa_2 n_e \vec{E}^{(0)}}{E^{(0)}}\left[\frac{N}{E^{(0)}}\vec{\nabla}\cdot\left(\frac{\vec{E}^{(0)}}{N}\right)\right] - \frac{\kappa_3 n_e \vec{E}^{(0)}}{(E^{(0)})^2}\left[\vec{E}^{(0)}\vec{\nabla}\ln\left(\frac{E^{(0)}}{N}\right)\right] +$$
$$+ \frac{\kappa_4 n_e}{b_e E^{(0)}}\frac{\partial(\vec{E}^{(0)}/N)}{\partial t}, \qquad (2.111)$$

where b_e, $D_e^{(0)}$ are defined in a usual manner according to Eqs. (2.60), (2.61) with $f_0^{(0)}$instead of $f^{M)}$. The first three terms represent the usual mobility, diffusion and thermodiffusion. The transport coefficients κ_1, κ_2, κ_3 correspond to the fluxes, which are parallel to $\vec{E}^{(0)}$, and are linear functions of $\vec{\nabla} n, \vec{\nabla}(E^{(0)}/N)$. It means that now the diffusion and thermodiffusion coefficients are tensors - in the presence of inhomogeneity of density or effective temperature the fluxes along the electric field are different from the fluxes across it. The coefficient κ_4 corresponds to the temporal dependence of electric field. These tensorial coefficients are of the order of the diffusion coefficient $D_e^{(0)}$ and approach zero for the velocity-independent electron-neutral collision frequency ν_{eN}.

The terms with κ_1, κ_2, κ_3 reveal themselves as effective anisotropy of a diffusion and thermodiffusion coefficients. Consider for example an inhomogeneity with a spatial scale L drifting in electric field and spreading due to the diffusion. Electrons in front of the cloud have travelled further along the electric field. It means that their energy is $\sim \lambda_e^{(\varepsilon)} \bar{\varepsilon} / L$ larger than the energy of the rear electrons. The mobilities of these groups also differ by a factor of the order of $\lambda_e^{(\varepsilon)}/L$ due to velocity dependence of the transport collision frequency, and, as a result, the initially spherical cloud is transformed into an ellipsoid. If ν_{eN}, increases with velocity than the longitudinal effective diffusion coefficient is smaller than the transversal diffusion coefficient and vice versa. For power dependence $\nu_{eN}(V)=\nu_0 V^p$, $\delta_{eN}(V)=\delta_0 V^{s-2p-2}$ for $p<3$, $s>0$ the values of κ_1, κ_2, κ_3 and κ_4 are presented in [35-36] for the relatively large electric fields when the temperature of the neutral gas can be neglected. Some other data for the real gases and for the magnetized plasma can be also found there.

The energy balance equation, as has been already mentioned in Section 2.1, is unnecessary in the considered case. When $\nu_{ee} << \delta_{eN}\nu_{eN}$ the relaxation length of the distribution function λ_{ef} Eq. (2.59) coincides with the energy relaxation length $\lambda_e^{(\varepsilon)}$ Eq. (2.20). At distances larger than $\lambda_e^{(\varepsilon)}$ the electron energy is determined by the local electric field $E^{(0)}$ and by the neutral temperature in accordance with Eq. (2.107); on the other hand, for smaller distances the fluid description is invalid, and rigorous kinetic analysis is necessary.

Numerous situations exist when the energy input in plasma is connected with the Joule heating by the high-frequency electric field. If the field frequency ω is low with respect to the energy relaxation frequency Eq. (2.110) in all the expressions (2.107) - (2.111) the local instantaneous electric field intensity is to be substituted. Since the distribution function Eq. (2.108) and the mobility Eq. (2.60) now become field-dependent, the plasma properties are nonlinear, and it is inconvenient to apply the Fourier expansion of the electric field Eq. (2.111).

However, if the frequency of the oscillatory electric field exceeds considerably the energy relaxation frequency Eq. (2.110), and this field is strong and plays a considerable role in formation of f_0, another simple approach is possible. The isotropic part of the distribution f_0 is determined by the instantaneous value of the quasistationary (slowly varying with frequency lower than Eq. (2.110)) electric field and amplitude value of the high frequency electric field (not by its instantaneous value). The $(E^{(0)})^2$ in the denominator of Eq. (2.105) is to be replaced:

$$(E^{(0)})^2 \to (E^{(0)}(\vec{r},t))^2 + \frac{(E_0^{(1)}(\vec{r},t))^2}{2} \frac{\nu_{eN}^2}{\omega^2 + \nu_{eN}^2}.$$

Due to the fact that the relaxation frequency of f_0 Eq. (2.111) is low, the high frequency modulation of f_0 is small (the non-linear effects produced by it are discussed in [30]). For calculation of the high-frequency conductivity Eq(2.105) with modified $(E^{(0)})^2$ can be used.

The account of inelastic collisions. The situations when it is possible to restrict analysis only to elastic collisions, are relatively few. In most cases it is impossible to neglect the inelastic collisions completely. Especially important are collisions of electrons with neutral molecules which are accompanied by excitation and deexcitation of molecular energy levels. In gas discharges, for example, even the existence and maintenance of the plasma itself is determined by the ionizing collisions of electrons with neutral particles, i.e., by a specific kind of inelastic collisions. The collision integral in Eq. (2.1) for the processes in which an electron loses an energy quantum ε_k is given by

$$St_{eN}^{(inel)} = \int [\frac{V'}{V} \nu_k(V', \vartheta) f(\vec{V}') - \nu_k(V, \vartheta) f(\vec{V})] d\Omega, \qquad (2.112)$$

where $\nu_k(V, \vartheta)$ represent the collision frequency for electron scattering at an angle ϑ with excitation of the k-th level of a neutral molecule. Here $(V')^2 = V^2 + 2\varepsilon_k / m_e$, and the thermal motion of the neutral particles is neglected.

The molecular energy levels can be subdivided into rotational, vibrational (these levels are absent in pure atomic gases), and electronic states. The characteristic excitation energies of these groups are greatly different. The energy scale for the rotational levels corresponds to energies of the order of $10^{-3} eV$; the vibrational states are typically characterized by energies of the order of $10^{-1} eV$, and the characteristic energies which are lost by plasma electrons for the electronic excitation (and for the ionization) correspond to values from several eV up to $20eV$.

The influence of the inelastic collisions on the EDF formation depends crucially on the magnitude of the energy quantum ε_k with respect to the characteristic energy scale $\Delta\varepsilon(\varepsilon)$ of the electron distribution function (2.105). If $\Delta\varepsilon(\varepsilon) << \varepsilon_k$, the difference in (2.112) can be expanded over $(\varepsilon_k/\varepsilon)$. It follows (see, for example, [3]) that in Eqs. (2.100) - (2.105) it is necessary to introduce instead of $\delta_{eN} = 2m_e/m_N$ the effective mean energy fraction (which becomes now energy-dependent) which is lost by an electron with energy ε in one collision. If the excited molecules have Boltzmann's distribution, we have

$$\delta_{eff}(\varepsilon)=\frac{2m_e}{m_N}+[1+\exp(-\frac{\varepsilon_k}{T_N})]\frac{\varepsilon_k}{\varepsilon}\frac{\nu_k}{\nu_{eN}}. \tag{2.113}$$

This means that such collisions can be treated as quasielastic. Since the electron distribution, as a rule, decreases exponentially, and for plasma maintenance an appreciable ionization rate is necessary, the typical energy scale of the distribution function in the discharges is of the order of $1eV$. Consequently, such an approach is applied usually to the electron collisions which are accompanied by excitation of the rotational and vibrational molecular levels. But the situation with the electronic excitations in gas discharges, both in the atomic and the molecular gases, is quite different. At relatively low field intensities, when $\bar{\varepsilon}$ with δ_{eff} given by Eq. (2.113) is small with respect to the energy of the first electronic level ε_1, the distribution of the majority of electrons is determined by Eq. (2.105). The collisions with excitation of the electronic states which are accompanied by loss of practically all energy of the plasma electrons determine only the exponentially small tail of the EDF at $\varepsilon \geq \varepsilon_1$. The electron energy in this case is lost mainly in the quasielastic collisions. Since at energies which do not significantly exceed ε_1 the transport collision frequency (2.51) and the frequency of the electronic excitation ν_k usually satisfy the inequalities

$$\nu_{eN}(\varepsilon) >> \nu_k(\varepsilon), \tag{2.114}$$

$$\nu_k(\varepsilon) >> \delta_{eff}\nu_{eN}(\varepsilon), \tag{2.115}$$

the tail of EDF is also described by the two-term approximation (2.48).
From the other side, from the Eq. (2.115) it follows that the electrons of the distribution tail lose their energy mainly not in the quasielastic collisions, but to excitation of the electronic levels. It means that the distribution tail decreases with energy considerably faster than the function Eq. (2.105). The situation is reduced to the problem of diffusion over energy in the vicinity of absorbing boundary at $\varepsilon=\varepsilon_1$ (see lower, Section 3.2), and in rough approximation the zero boundary condition at $\varepsilon=\varepsilon_1$ can be imposed on the distribution function f_0 in the energy diffusion equation (2.104). The relaxation of the tail electrons with $\varepsilon \geq \varepsilon_1$ is fast, and the fluid description is valid, if the criteria Eq. (2.110) (of course, with $\delta_{eff}(\varepsilon)$ instead of δ_{eN}) are fulfilled. The source terms in the continuity (2.7) and energy balance (2.16) equations, which are determined, as a rule, by the tail of EDF, depend only

on intensity of the local electric field. Thus, the system of transport equations combined with the qusineutrality condition becomes closed.

The value of $\bar{\varepsilon}$, Eq. (2.107), which corresponds to the scale of the exponential decrease of the EDF Eq. (2.105) due to the quasielastic collisions, increases with the field intensity, and at moderate fields becomes of the order of ε_1. At such fields the transition occurs to the regime in which the electron energy balance is controlled by the 'genuine' inelastic collisions with large energy losses. Since Eq. (2.115) is fulfilled usually with significant margin, in a considerable range of the electric fields the characteristic energy scale of the EDF tail remains small with respect to ε_1, which corresponds now to the energy scale of the "distribution body" at $\varepsilon<\varepsilon_1$. It is possible to neglect the quasielastic (the second) term $W_\varepsilon f_0$ in Eq. (2.104) and to impose the approximate "absorbing wall" boundary condition $f_0=0$ at $\varepsilon=\varepsilon_1$. This situation is illustrated Fig. 2.4 [35], where the average electron energy $\bar{\varepsilon}$ versus electric field intensity is presented for several noble gases (*He, Ne, Ar*).

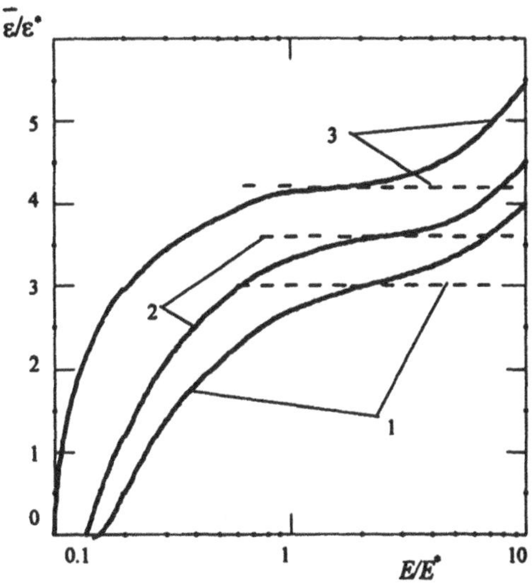

Figure 2.4. Dependencies of mean electron energy $\bar{\varepsilon}$ on electric field for *He* (1); *Ne* (2); *Ar* (3). At low fields the curves correspond to the EDF Eq. (2.104) with the values of ε_1=21.2 *eV* for *He*, 16.7 *eV* for *Ne* and 11.2 *eV* for *Ar*. The field is plotted as E/E^*, where E^*=6.4p for *He*, 2.0p for *Ne*, and 9.0p for *Ar* (the field E^* is in V/cm, the pressure p *is* in torr). The horizontal dashed lines correspond to the distribution function in the absorbing wall model; the energy loss in elastic collisions is ignored. Here $\bar{\varepsilon}/\varepsilon_1$=0.30 for *He*, 0.36 for *Ne* and 0.43 for *Ar*. In low-current discharges ($j/R\leq 100$ mA cm^{-3}) such EDFs occur at $10^{-2}\leq Rp\leq 10$ torr·cm.

The ascending branches of the curves represent the case when the energy balance is determined by the elastic collisions (Eq. (2.104)). This case corresponds to the EDF, which up to energy $\sim\varepsilon_1$ coincides with Eq. (2.105). The plateau when $\bar{\varepsilon}$ does not depend on the field and remains of the order of ε_1, corresponds to the situation when the electron energy is lost mainly in inelastic collisions (in excitations of the electronic states), but the distribution tail decreases fast, and the contribution of the electrons with $\varepsilon>\varepsilon_1$ to the total density and energy remains small. The EDF in this case at $\varepsilon<\varepsilon_1$ is practically independent on the field. According to Eq. (2.104) the energy diffusion flux is conserved up to $\varepsilon=\varepsilon_1$, and is absorbed quickly at $\varepsilon>\varepsilon_1$.

The problem of derivation and applicability of the fluid equations in this important case is rather complicated and up to now seems not completely clear. The main difficulty stems from the fact that due to the discreteness of the inelastic energy losses the processes of energy relaxation and of distribution relaxation become essentially different. The energy relaxation time, for example, is roughly equal to

$$\tau_e^{(\varepsilon)}=\varepsilon_1^2/D_\varepsilon,$$

- since an electron with $\varepsilon\geq\varepsilon_l$ loses its energy fast, - it coincides with the time of energy diffusion over 'distance' ε_1 The length of the energy relaxation is of the order of $\lambda_e^{(\varepsilon)}=\varepsilon_1/eE$. In numerous publications the expressions and calculations of effective value of δ_{eff} are presented for this case. However, the assertion, similar to Eq. (2.105), that the averaged fluid description of electrons is valid, if the characteristic temporal and spatial scales of a problem exceed $\tau_e^{(\varepsilon)}$ and $\lambda_e^{(\varepsilon)}$, is too optimistic. In real situations the corresponding criteria can be far more restrictive.

The averaged fluid description is applicable, if the distribution function in a given place and at a given moment is mainly determined by the local instantaneous values of the particles density, mean energy, fields, etc.. The plasma nonstationarity and inhomogeneity result in small corrections to this local distribution. It means that the characteristic scales of a problem must exceed the relaxation time τ_{ef} and length λ_{ef} of the distribution function f_0. If, for example, the energy losses in the quasielastic collisions are negligible, and among 'the genuine' inelastic collisions with large energy loss the excitation of only one electronic level ε_1 dominates, these characteristic scales tend to infinity. In other words, only the processes which introduce stochasticity and result in electron forgetting its 'history' determine these scales. The latter can greatly exceed the energy relaxation scales Eq. (2.105). For the case of d.c. electric field when stochasticity is connected with quasielastic collisions, and only one genuine inelastic process controls the energy loss, we have [35]

$$\tau_{ef} = (\delta_{eff} \nu_{eN})^{-1} \frac{e^2 E^2}{\delta_{eff} \nu_{eN}^2 m_e \varepsilon_1} ;$$
$$\lambda_{ef} = \lambda_e / \sqrt{\delta_{eff}} \frac{e^2 E^2}{\delta_{eff} \nu_{eN}^2 m_e \varepsilon_1} . \quad (2.116)$$

These values are $[\delta_{eff}(\varepsilon_1)\nu_{eN}(\varepsilon_1)\ \tau_e^{(\varepsilon)}]^{-1}$ times larger than Eq. (2.110). The energy relaxation time

$$\tau_e^{(\varepsilon)} = \left(\frac{\varepsilon_1}{eE}\right)^2 \left(\frac{m_e}{\varepsilon_1}\right)^{1/2} \lambda_e^{-1} . \quad (2.117)$$

corresponds to the drift time until an electron gains energy ε_1 and undergoes an inelastic collision. The examples of nonhydrodynamic behavior of electrons in situations when (2.115) is violated are presented in [35], [36].

This approach is invalid in very strong electric fields, when the characteristic decay scale of the EDF at $\varepsilon > \varepsilon_1$ becomes of the order of (and exceeds) the excitation threshold ε_1. If the energy scale of the electron distribution (which is of order of the mean energy $\bar{\varepsilon}$) significantly exceeds ε_1, the excitation of the electronic levels can also be treated as a quasielastic process. But in such extremely strong fields, generally speaking, the process of electron runaway [25]-[26] occurs. It means that the collisions are switched off, and the electron motion becomes more and more close to free-streaming regime. The electron energy depends on its history, and the fluid description of such a system becomes impossible.

If the inequality (2.114) is not fulfilled, the electron distribution becomes strongly anisotropic. In the limiting case $\nu_{eN}\tau_e^{(\varepsilon)} \ll 1$, where $\tau_e^{(\varepsilon)}$ is given by Eq. (2.117), the electron motion is practically one-dimensional. If the inelastic collision frequency ν_1 is sufficiently high, an electron undergoes the inelastic collision practically just after it gains energy ε_1 and moves ballistically until it gains it again. This situation was treated in [37]. Such a relay-race mechanism of conductivity is very close to the above-mentioned (in Section 2.5) ion motion which is controlled by the resonance charge-exchange.

2.10 QUASINEUTRALITY MAINTENANCE

Here we shall study the problem of establishing the quasineutral state for the weakly ionized plasma. To be specific, consider the pure plasma with the single ion species and constant temperatures in the absence of the magnetic field. General equations describing such plasma can be obtained by combining the continuity Eqs. (2.7) with the expressions for the particle fluxes Eq. (2.41). In the frame of reference moving with the velocity of the neutral gas the result reads

$$\begin{aligned} &\frac{\partial n_e}{\partial t} - \vec{\nabla}\cdot(D_e\vec{\nabla}n_e + b_e n_e\vec{E}) = I - S \\ &\frac{\partial n_i}{\partial t} - \vec{\nabla}\cdot(D_i\vec{\nabla}n_i - b_i n_i\vec{E}) = I - S\,. \end{aligned} \tag{2.118}$$

The self-consistent electric field is determined by the Poisson equation

$$\vec{\nabla}\cdot\vec{E} = \frac{e}{\varepsilon_0}(n_i - n_e)\,. \tag{2.119}$$

Usually in partially ionized plasma the magnetic field perturbation can be neglected, and, hence, electric field can be treated in terms of potential: $\vec{E} = -\vec{\nabla}\varphi$. The equation system (2.118) - (2.119) turns out to be nonlinear owing to the terms with electric field.

At very low densities of the charged particles when terms with electric field can be neglected, the Eqs. (2.118) are transformed to the pure diffusion equations for electrons and ions. The corresponding diffusion is called unipolar. Since in the absence of magnetic field $D_e >> D_i$, the unipolar diffusion is characterized by two time scales - $\tau_e = L^2/4D_e$ and $\tau_i = L^2/4D_i$ where $\tau_e << \tau_i$. Consequently, the process of unipolar diffusion can be divided into two stages - at the first, during $t \sim \tau_e$ electron profile is smoothed, and afterwards the ion diffusion takes place on a practically uniform electron background with the time scale $t \sim \tau_i$. Let us estimate electric field in the process of unipolar diffusion. In the absence of external electric field (external current) assuming $|n_e - n_i| \sim n$ from the Poisson equation we obtain $E \sim enL/\varepsilon_0$. One can neglect electric field in the transport Eqs. (2.118) when

$$b_\alpha EL/D_\alpha \sim eEL/T_\alpha \sim e^2nL^2/\varepsilon_0 T_\alpha \sim L^2/r_d^2 << 1.$$

Here the Debye radius with electron temperature is defined:

$$r_d = (\varepsilon_0 T_e/n_0 e^2)^{1/2}. \tag{2.120}$$

Therefore, the unipolar diffusion takes place when the scale of the inhomogeneity is smaller than the Debye radius. This is rather a rare case, characteristic for small-scale perturbations, for some processes in low-current discharges, and for phenomena of atmospheric electricity, etc..

The opposite situation is more typical. In fact Langmuir introduced the term 'plasma' specially for the systems with the spatial scales exceeding the Debye radius where the self-consistent electric fields play decisive roles. In plasmas the paradigm of quasineutrality consists in the following: the densities of the charged species are considered to be equal

$$n_e = n_i, \tag{2.121}$$

and the self-consistent electric field providing the quasineutrality is to be found from the plasma equations, in our case from the system (2.118) with $n_e = n_i$. The space charge can be calculated afterwards from the Poisson equation if necessary. In the absence of the net current flowing through the system the self-consistent potential is of the order of T/e (for $T_e \sim T_i$), and Eq. (2.119) yields $|n_e - n_i|/n \sim r_d^2/L^2 << 1$. In quasineutral approximation the Poisson equation is extraneous since it determines only the small space charge which creates the potential necessary to provide $n_e \cong n_i$.

If $r_d << L$, for arbitrary initial conditions when the quasineutrality does not hold from the very beginning, the evolution is characterized by two different time scales. At the first fast stage quasineutrality is established, and afterwards the quasineutral evolution takes place with a much longer time scale. Let us consider as an example evolution of small perturbations δn_e and δn_i on a uniform quasineutral background $n_e = n_i = n_0$ in the absence of an external current. Linearizing Eqs. (2.118) - (2.119), eliminating the potential perturbation, we have for the Fourier components

$$\begin{aligned}
&\delta n_{e\vec{k}} = C_1 \exp(-i\omega_1 t) + C_2 \exp(-i\omega_2 t); \\
&\delta n_{i\vec{k}} = -(b_i C_1 / b_e) \exp(-i\omega_1 t) + C_2 \exp(-i\omega_2 t), \\
&\text{where} \qquad\qquad\qquad\qquad\qquad\qquad\qquad\qquad (2.122) \\
&\omega_1 = -iD_e / r_d^2; \ \omega_2 = -ik^2 D_i (1 + T_e / T_i); \\
&C_1 = \delta n_{e\vec{k}}^{(0)} - \delta n_{i\vec{k}}^{(0)}; \ \ C_2 = \delta n_{i\vec{k}}^{(0)} - b_i C_1 / b_e \approx \delta n_{i\vec{k}}^{(0)}.
\end{aligned}$$

Here $\vec{k}$ is the wave vector. To obtain Eq. (2.122) the inequalities $D_e >> D_i$, $kr_d << 1$ have been used. Since $|\omega_1| >> |\omega_2|$ the quasineutrality is established at times $|\omega_1|^{-1} << t << |\omega_2|^{-1}$ when the first terms in Eq. (2.122) vanish while the ion density profile remains almost constant during this fast stage. The corresponding time scale

$$\tau_M = r_d^2 / D_e = \varepsilon_0 / \sigma_e, \tag{2.123}$$

where $\sigma_e=n_0eb_e$ is the electron conductivity, is known as a Maxwellian time. This time scale represents the time of diffusive displacement on Debye radius. The Maxwellian time does not depend on the spatial scale of the perturbation. This fact can be understood from the Poisson equation - increase of the spatial scale k^{-1} leads to the simultaneous growth of the electric field while the time of the field-driven displacement of electrons on k^{-1} remains the same. The second frequency ω_2 corresponds to the quasineutral ambipolar diffusion of electrons and ions which is discussed in detail in Chapter 4. The process of relaxation to the quasineutral state in magnetic field can be treated in a similar way; the corresponding expressions can be found in [37].

The approach discussed above is justified if $\tau_M\nu_{eN}>>1$. In plasma with rare collisions electron plasma waves are exited by the initial space charge with the time decay of the order of ν_{eN}^{-1}. Thus, in general, plasmas can be treated in quasineutral approximation on a time scale which exceeds $\max(\tau_M, \nu_{eN}^{-1})$.

REFERENCES

1. S. Chapman, T. G. Cowling, *The Mathemathical Theory of Non-Uniform Gases*, 2nd ed. (Cambr.: University Press, 1952): 350.
2. E. M. Lifshits, L. Pitaevsky, *Course of Theoretical Physics* **vol. 10**, *Physical Kineticics* (Oxford: Pergamon Press, 1981): 452.
3. I. Shkarofsky, T. Johnston, M. P. Bachinski, *The Particle Kinetics of Plasmas* (Reading: Addison-Wesley Publ., 1966): 518.
4. M. Mitchner, C. H. Kruger, *Partially Ionized Gases* (New York-London-Sydney-Toronto: Wiley, 1973): 518.
5. F.F.Chen, *Introduction to Plasma Physics and Controlled Fusion*, v.1, 4th ed. (New York: Plenum, 1990): 421.
6. V. E. Golant, A.P. Zhilinski, I.E. Sakharov, *Fundamentals of Plasma Physics* (New York: Wiley, 1980): 405.
7. F. Hinton, in *Basic Plasma Physics* **vol.1**, Editors: A. A. Galeev, R. N. Sudan (New York: North Holland, 1983) 147-197.
8. L. Landau, *Phys. Zeitsch. d. Sovietunion* **10**: 154-203 (1936).
9. J. O. Hirschfelder, C. F. Curtiss, R. B. Bird, *Theory of Gases and Liquids*, corrected print (New York: Wiley, 1967): 1249.
10. A. V. Eletskii, L. A. Palkina, B. M. Smirnov, *Transport Phenomena in Weakly Ionized Plasmas* (Moscow: Atomizdat, 1975): 333 (in Russian).
11. S. I. Braginskii, in *Rewievs of Plasma Physics*, **vol.1**, Editor: M.A. Leotovich (New York: Consultants Bureau, 1963): 183-271.
12. L. D. Tsendin, *Sov. Phys. JETP* 29: 502-506 (1969).

13. I. I. Prigogine, *Non-Equilibrium Statistical Mechanics*, (New York: Interscience, 1966): 319.
14. V. M. Zhdanov, *Transport Phenomena in Multispecies Plasma* (Moscow: Energoizdat, 1982): 176 (in Russian).
15. H. Grad, *Comm. Pure Appl. Math.* **2**: 331-369 (1949).
16. L. Herdan, B. S. Liley, *Rev. Mod. Phys.* **32**: 731-741 (1960).
17. M. N. Kogan, V. S. Galkin, O. G. Fridlender, *Sov. Phys. Uspekhi* **19**: 420-428 (1976).
18. A. Mihailovskii, *Instabilities in Inhomogeneous Plasma*, **vol.2** (New York: Consultants Bureau, 1974): 314.
19. V. N. Oraevski, Yu. B. Konikov, G. V. Khasanov, *Transport Processes in Anisotropic Space Plasma* (Moscow: Nauka, 1986): 172 (in Russian).
20. B. I. Davydov, *Phys. Z. Sovjetunion* **8**: 59-82 (1935).
21. W. Allis, in *Handbuch der Physik*, **vol. 21**, Editor: S. Flugge (Berlin:Springer, 1956): 383-503.
22. M. Druyvesteyn, *Physica* **10**: 69-103 (1930).
23. E. A. Mason, E. W. McDaniel, *Transport Properties of Ions in Gases* (New York: Wiley, 1988): 560.
24. T. Holstein, *Phys. Rev.* **70**: 367-384 (1946).
25. H. Dricer, *Phys. Rev.* **115**: 238-249 (1959).
26. A. V. Gurevich, *Sov. Phys. JETP* **12**: 904-912, (1961).
27. S. T. Beljaev in *Plasma Physics and the Problems of Controlled Thermonuclear Reactions*, **vol.3** (Pergamon Press: Oxford, 1961) 56-75.
28. V. E. Golant, *Sov. Phys. Uspekhi* **6**: 161-197 (1963).
29. Yu. L. Klimontovich, *Kinetic Theory of Nonideal Gases and Plasmas* (New York: Pergamon Press, 1982): 318.
30. A. V. Gurevich, *Nonlinear Phenomena in the Magnetosphere* (New York: Springer, 1978): 370.
31. I. Shkarovsky, *Canad. J. Phys.* **39**: 1619-1703 (1961).
32. A. A. Radzig, B. M. Smirnov, in *Plasma Chemistry*, **vol.11**, Editor: B. M. Smirnov (Moscow: Energoatomizdat, 1984): 170-207 (in Russian).
33. H. W. Ellis, *Atom. Nucl. Data Tabl.* **17**. 177-205 (1976); **22**: 179-201 (1976).
34. N. L. Aleksandrov, A. M. Ohrimovskii, A. P. Napartovich, *JETP* **82**: 214-220 (1996).
35. L. D. Tsendin, *Sov. J. Plasma Phys.* **8**: 96-109; 228-233 (1982).
36. T. Ruzicka, K. Rohlena, *Czech. J. Phys.* **B22**: 906-919 (1972).
37. A. P. Dmitriev, L. D. Tsendin, *Sov. Phys. JETP* **54**:1071-1075 (1981).
38. A. V. Gurevich, E. E. Tsedilina, *Sov.Phys. Uspekhi* **10**: 214-236 (1967).

Chapter 3

Sheath Structure And Effective Boundary Conditions For Plasma Equations

3.1 SEPARATION INTO QUASINEUTRAL PLASMA AND SPACE CHARGE SHEATH

Equations (2.7), (2.10), (2.15) for the particle, momentum and energy transport, in combination with the Maxwell equations for electromagnetic field, describe evolution of inhomogeneous profiles of plasma density, composition and temperature. Transport problems are typically characterized by relatively slow temporal and large spatial scales, and the arising electric field is, as a rule, electrostatic. If we assume, that plasma is quasineutral, neglect its influence on the magnetic field, and substitute the expressions for the particle fluxes from the momentum conservation equations (2.10) into the Eqs. (2.7), (2.15), we obtain p equations for densities of the charged particles (electrons and (p-1) species of ions), and two equations for the electron temperature and temperature of heavy particles. When supplemented by the quasineutrality condition Eq. (2.121), this system of (p+3) equations determines (p+3) unknown functions: p partial densities, two temperatures and potential. Usually the transport processes are strongly influenced by phenomena at solid boundary surfaces - vessel walls, electrodes, etc.. It means, that to solve any given problem the system of the transport equations is to be supplemented by appropriate boundary and initial conditions.

The necessary conditions can easily be formulated, if we notice, that the quasineutral equations are of the first order in temporal derivatives of densities and temperatures, and of the second order in spatial derivatives of all the unknown functions. Such so-called parabolic problems, similar to ordinary problems of diffusion and heat transport of neutral particles, demand one set of initial conditions (for the partial densities and temperatures), and one set of boundary conditions for all the unknown functions, including potential. Since the quasineutrality equation (2.121) is algebraic and allows to express one of the partial densities in terms of other ones, it follows, that it is necessary to have $(p+1)$ initial and $(p+2)$ boundary conditions. The formulation of the initial conditions can be performed in the traditional way, but the formulation of the boundary conditions is more complicated.

The boundary conditions imposed on the variables - the particle densities and temperatures - are known as the Dirichlet conditions, and the conditions on their normal derivatives - as the von Neuman's conditions. Mixed boundary conditions are also possible, which connect the boundary values of the variables with their normal derivatives. A well-known example of such conditions is presented by the radiation (Stephan's) boundary condition in the theory of the heat transfer, which connects the thermal flux to the radiating surface with its temperature (see e.g. [1]).

More complicated is the case (see Section 2.8), when an external high frequency field is applied to plasma. But in such a situation it is, as a rule, possible to separate unambigously the slowly varying (quasistationary) potential electric field, which is associated with the plasma inhomogeneity, and the high frequency non-potential field. For the latter the special boundary condition is required.

The boundary conditions, consequently, can be imposed on the densities and the temperatures of the plasma particles, their normal derivatives (particles and energy fluxes), and on the quasistationary potential or its normal derivative. At a sectioned conducting boundary, for example, the potential profile or current density distribution on the surface are given.

The main problem in formulating the boundary conditions for the plasma transport equations consists in the fact, that for the transport processes the quasineutrality fails only in the space charge sheaths in the vicinity of the boundaries, where the electric field and potential differences are large. It is unreasonable to seek for the solution of such a system in the whole volume, because the space charge density which generates the electrostatic electric field is determined in the plasma bulk by the small difference in the electron and ion densities. Such an approach demands extremely accurate calculations in the whole domain. From the other side, such a complication greatly masks the underlying physics. It seems desirable, therefore, to use a more traditional scheme. It demands to solve the complete system only in the space charge sheaths, and to use these solutions as the boundary conditions for the quasineutral plasma equations. Such a procedure allows considerable simplification of the solution of the complete problem. For analysis of the relatively thin sheaths, as a rule, a 1D stationary problem arises. Only if curvature of the boundary surface is comparable with the sheath thickness, as in the case of a small probe, is necessary the more detailed analysis.

Additional simplification is connected with the fact that typically the processes of plasma generation and removal take place in the whole plasma volume. The resulting fluxes of the charged particles recombine on the vessel walls. Since the sheath volume is negligible with respect to the bulk plasma, in most cases it is possible to neglect the generation and losses of the charged particles in the sheaths, and to assume that the normal to the wall components (along the ξ axis) of the particle fluxes $\Gamma_{\alpha\xi}$ are conserved here. Due to the fact that the thermal capacity of the sheaths is also negligible, the normal components of the chaotic energy fluxes $q_{\alpha\xi}$ are also conserved here. In other words, if the boundary conditions at the vessel walls are imposed on the particle or energy fluxes, they can be applied directly to the quasineutral plasma equations, and a transport problem can be solved without detailed analysis of the sheath structure. Of course, such an analysis is sometimes very important in applications. However, it can be performed independently from the solution of the transport problem. The situation is different when the potential or the current density profile on the boundary surface is given. In this case the solution of the 1D sheath problem, which expresses the potential difference $\Delta\varphi$ in the sheath through the conserved values of the normal fluxes $\Gamma_{\alpha\xi}$, can be used as the boundary condition for the quasineutral system. An important exception is the case of the cathode and anode sheaths in the dc discharge, and of the sheaths in the capacitively coupled radio frequency (RF) discharge. In these cases the generation terms in the sheaths are important, and an accurate account of these processes and of the sheath structure is necessary.

The processes in the sheaths are substantially different depending on whether the mean free path exceeds or is less than the sheath width L_{sh} (which is, as a rule, of the order of the local Debye radius, Eq. (2.120)). Additional difficulty is connected with the fact that the energy relaxation length Eq. (2.20), especially for the energetic electrons, which are responsible for the ionization processes in gas discharges, is often comparable with the sheath width, or even exceeds it. The fluid description adopted here is invalid in such sheaths, and rigorous nonlocal kinetic analysis is necessary, in order to gain even qualitative description of these processes [2], [3]. We shall restrict ourselves here only to the cases, when the fluid approximation is applicable. If the sheath is collisional, it implies that its width considerably exceeds the relaxation length of the distribution functions Eq. (2.59). In the case of a collisionless sheath such an approach is valid for electrons, if the ionization degree is sufficiently high, and the Maxwellian distribution is quickly restored due to the electron-electron collisions on a small spatial scale, which is comparable with the mean free path.

In typical cases the plasma is far from the thermodynamic equilibrium - the densities of the charged particles far exceed the equilibrium values, corresponding to the temperature of an environment. Accordingly, the simplest boundary conditions correspond to total absorption of the charged particles at the solid surfaces with subsequent recombination. We shall restrict ourselves mainly to this case. Even in this situation the problem turns out to be non-trivial. From a vast field of more complicated surface processes we shall discuss briefly only the absorption, thermoelectronic (or secondary electron-electron) and ion-electron emission.

In this Chapter we shall start with the collisionless sheath without magnetic field in a pure plasma, which consists of electrons and one species of positive ions. We shall derive the relation between the potential difference $\Delta\varphi$ in the sheath, and the particles fluxes $\Gamma_{\alpha\xi}$ through it at the nonemitting surface. Afterwards we proceed to the formation of double sheath at the surface with intense emission (thermal or secondary) of cold electrons. Of course, the separation into plasma and sheath is a rough simplification. In reality, a relatively strong electric field arises in plasma at distances of the order of the mean free path from the boundary. According to the well-known Bohm criterion [4] the potential difference over this presheath is approximately $0.5T/e$, see Section 3.3. In Section 3.4 the sheath near the emitting surface will be considered, while in Section 3.6 the collisional sheath at the nonemitting surface will be analyzed.

These results are also valid if the magnetic field is normal to the sheath surface. If the strong magnetic field is parallel to the boundary the sheath is always to be treated as collisional. If its width exceeds the Larmor radius, the situation is in principle the same as in the diffusion-dominated sheath in absence of the magnetic field. In Section 3.5 the case of a thin (with respect to the Larmor radius) sheath at the surface parallel to the magnetic field will be analyzed. In Sections 3.7-3.8 the situations will be briefly discussed (the cathode and anode sheaths of the dc glow discharge and sheath in the capacitively coupled RF discharge), in which the ionization processes are important.

3.2 BOUNDARY CONDITIONS FOR PARTICLE DENSITIES

Consider a situation, which arises in absence of the electric field (or in a weak one) in a plasma of low density, when the local Debye radius is large with respect to the mean free path. Even in this simplest case the macroscopic transport equations are applicable only at comparatively large scale, where the fluid description of the plasma particles is valid. In other words, they are valid if the behavior of an average particle with mean directed velocity and mean chaotic energy represents the behavior of the whole distribution in a given place. This approximation is violated near an absorbing boundary, since in the absence of a magnetic field, for example, at distances of the order of the mean free path λ from the boundary there are practically no ingoing (moving into the plasma) particles. In the presence of an absorbing wall such particles arise only due to collisions in this thin layer. It means that the distribution here is strongly anisotropic, and is formed mainly by the outgoing particles. In other words, an average particle is not representative here. In order to characterize adequately a particle, it is necessary to specify to which one of two populations it belongs - to the ingoing, or to the outgoing particles.

The crude boundary condition for the partial density at the absorbing surface corresponds to the zero density value at $\xi=0$. But in order to be absorbed, some non-zero density needs to be present here. The fluid approximation fails in the vicinity of the boundary, and in order to find the density profile here, it is necessary to solve a rather complicated kinetic problem (see for example [5]). The anisotropy of the distribution function is considerable here, and the two-term approximation given by equation. (2.48) fails. If, nevertheless, we shall extrapolate the usual two term expansion

$$f(\xi,\vec{V},t)=f_0(\xi,V,t)+f_1(\xi,V,t)\cos\vartheta, \tag{3.1}$$

where $f_1 \equiv \left|\vec{f}_1\right|$, ϑ represents the angle between the velocity and the ξ axis, up to the absorbing surface, from Eq. (2.50) we obtain

$$f_1=-\frac{V}{\nu}\frac{df_0}{d\xi}. \tag{3.2}$$

Neglecting in Eq. (2.103) the inelastic collisions we have

$$\frac{d}{d\xi}\left(\frac{V^2}{3\nu}\frac{df_0}{d\xi}\right)=0. \tag{3.3}$$

Its solution corresponds to the linear profile of $f_0(\xi)$ (and of the particle density) in the surface vicinity. In the rough approximation of equation. (3.1) it is impossible to satisfy the boundary condition at the absorbing surface at all angles in the interval $0\le\vartheta\le\pi/2$

$$f(\xi=0,\vec{V})=0\,, \qquad (3.4)$$

which corresponds to the absence of ingoing particles at the surface. Let us impose the less restricting boundary condition that only total influx equals zero:

$$\int_0^{\pi/2} f\cos\vartheta\sin\vartheta d\vartheta=0\,. \qquad (3.5)$$

We then can obtain the approximate boundary condition in the following form:

$$\frac{2}{3}\frac{\lambda df_0}{d\xi}=f_0,\quad at\ \ \xi=0\,, \qquad (3.6)$$

where $\lambda=V/\nu$ is the mean free path of the particles. The directed velocity at $\xi=0$ for the Maxwellian distribution function f_0 is

$$u_\xi(\xi=0)=V_T/2=\sqrt{\frac{2T}{\pi m}}\,. \qquad (3.7)$$

It is of the order of the sound velocity and is equal to one half of ion thermal velocity.

If the mean free path λ is velocity-independent, this means that the zero boundary condition on the density which is determined by f_0 in the approximation Eq. (3.1) is to be imposed at distance $2\lambda/3$ behind the real boundary.

In spite of the fact that the two term approximation leads to the physically senseless result that in the vicinity of the boundary at some inward-directed angles (for the ingoing particles) the distribution function Eq. (3.1) is negative, comparison with the rigorous kinetic solution (for the case of isotropic velocity-independent scattering) [5] demonstrates surprisingly good agreement. The rigorous kinetic result corresponds to the value of the coefficient 0.7104 instead of 2/3 in Eq. (3.6). The density profile, calculated according to Eqs. (3.1), (3.3) and (3.6), is shown in Fig. 3.1 by the dashed line. The strict kinetic result corresponds to the solid line. One can see that the fitting is surprisingly satisfactory, and, therefore, simplified approach can be used.

Angular dependence of the distribution of outgoing particles is shown in Fig. 3.2.

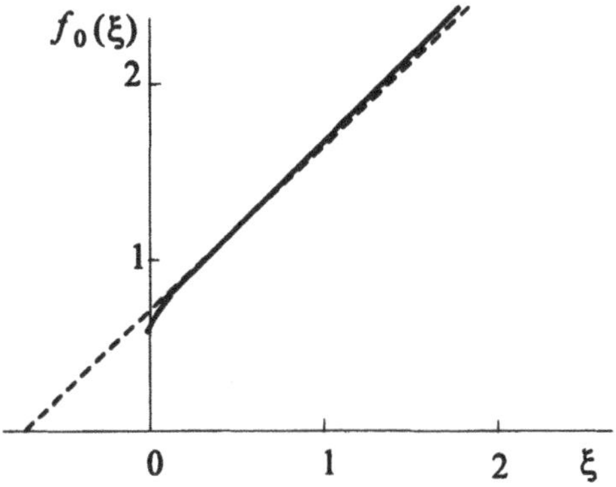

Figure 3.1. Distribution function $f_0(\xi)$ in the vicinity of absorbing wall for isotropic elastic scattering. Solid line is the kinetic solution; dashed line is the linear approximation for f_0 with 0.7104 instead of 2/3 in Eq. (3.6).

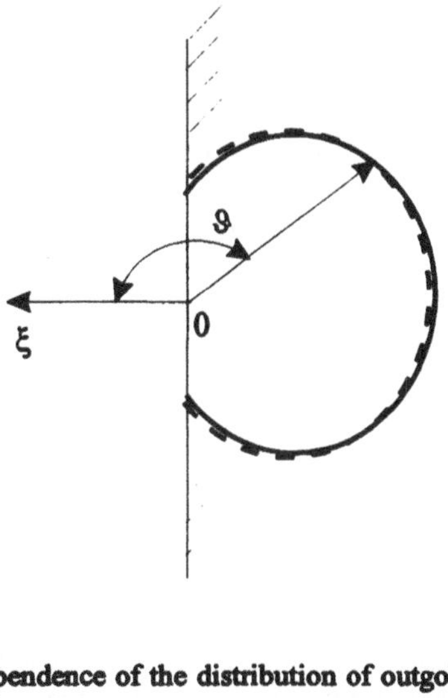

Figure 3.2. Angular dependence of the distribution of outgoing particles. The solid line - kinetic result; dashed line - approximation Eqs. (3.1), (3.2) at $\xi=0$: $f(\vartheta)=0.7104-\cos\vartheta$.

From Fig. 3.2 it follows, that the two-term approximation Eq. (3.1) also gives a rather accurate description of angular distribution of the outgoing particles.

From Eq. (3.6) the mixed boundary condition

$$n|_{\xi=0} = 0.7104\lambda dn / d\xi|_{\xi=0}. \tag{3.8}$$

for boundary density and its normal derivative follows. If $dn/d\xi \sim n(0)/L$ where $n(0)$ is the central plasma density and L is the characteristic scale of the problem, it follows that $n/n(0) \sim \xi/L << 1$. In other words, a rough zero boundary condition can be imposed on the densities at the plasma-sheath boundary. More detailed analysis of the sheath structure is necessary if boundary conditions are imposed on the potential or current density distribution on the boundary surface.

The flux of the chaotic (thermal) energy q_ξ at the boundary for the Maxwellian f_0 and for the directed part of the distribution function Eqs. (3.1), (3.2), (3.6) is given by

$$q_\xi = mnV_T^3 / 16. \tag{3.9}$$

3.3 COLLISIONLESS SHEATH IN THE ABSENCE OF MAGNETIC FIELD

The processes are significantly different, when electric field is taken into account. We consider here the case when the sheath width which is determined by the local Debye radius r_d, Eq. (2.120), is small with respect to the mean free path. Such sheath is called collisionless sheath. In such a situation inertial motion dominates in the sheath. In the absence of magnetic field (or when the direction of ξ axis is parallel to the field) the space charge in the sheath is typically determined by the ions. The reason is that the random flux of electrons,

$$\Gamma_{eT} = \frac{1}{4} n_e V_{Te} = n_e \sqrt{\frac{T_e}{2\pi m_e}}, \tag{3.10}$$

is extremely high. In other words, in the absence of electric field, considerable electron current flows from the plasma to the electrode. In order to reduce it, the surface needs to be biased negatively with respect to the plasma, with the potential difference at least of the order of several T_e/e. The plasma density profile for negative $e\varphi$ satisfies the Boltzmann law for electrons:

$$n_e = n_0 \exp(e\varphi / T_e), \tag{3.11}$$

where n_0 corresponds to the plasma density at $\varphi=0$. Flux of the Maxwellian electrons is equal to

$$\Gamma_{e\xi} = \frac{1}{\sqrt{2\pi}} n_s \sqrt{\frac{T_e}{m_e}} \exp(-e\Delta\varphi / T_e) \qquad (3.12)$$
$$= \Gamma_{eT}^{(s)} \exp(-e\Delta\varphi / T_e) = j_{e\xi} / e,$$

where n_s is the plasma density at the sheath edge and $\Delta\varphi=\varphi_s-\varphi_w$ is the potential drop in the sheath, Fig. 3.3.

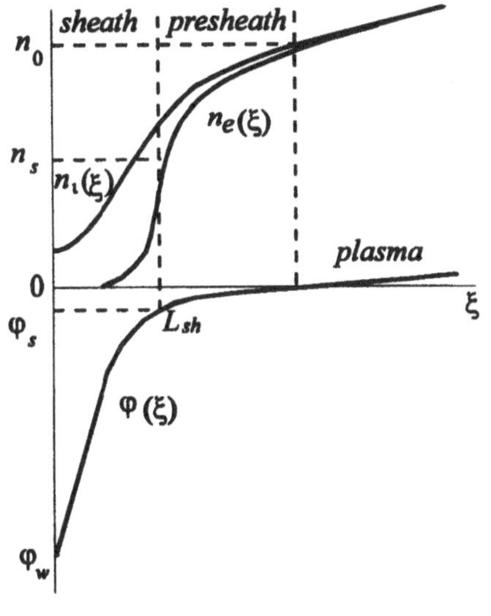

Figure 3.3. Qualitative behavior of potential and particle densities in vicinity of negatively biased surface.

The electrons, due to their high mobility, escape quickly from plasma and charge the surface negatively. This negative surface charge repels the plasma electrons and attracts ions, thus forming the layer of positive space charge in the vicinity of the boundary surface and equalizing the electron and ion currents along ξ. The ion current in this case is referred to as the ion saturation current. It corresponds to the sum of the diffusion and conductivity fluxes, and at uniform partial temperatures equals

$$j_i^{(sat)} = eD_i(1+T_e / T_i)dn / d\xi . \qquad (3.13)$$

The normal to the boundary density gradient is to be calculated in the region where plasma is quasineutral and collisional, but not too far from the wall.

This second condition is necessary in order to consider the particle fluxes as conserving. On the contrary, at positive values of φ the electron saturation current flows to the surface, and the ion distribution is the Boltzmann one.

Accordingly, in the subsequent Sections (with the exception of Section 3.5) we shall mainly focus our attention on case of strongly negatively biased surface $|\Delta\varphi| >> T_e / e$. In this situation the ion current practically equals to its saturation value and does not depend on $\Delta\varphi$. In the surface vicinity the electrons are repelled, and the Boltzmann distribution is invariant with respect to the choice of the reference point where $\varphi=0$, $n_e=n_0$. It means that in the expression (3.12) it is possible to replace n_s by n_0 and $\Delta\varphi$ by φ.

The simplest situation corresponds to the case when the field influence on the ion thermal motion is negligible (it is possible only if $T_i>>T_e$). The plasma density decreases according to a linear law towards the wall, and the potential diverges logarithmically, according to Eq. (3.11). The necessary connection between $\Delta\varphi$ and the partial fluxes can easily be obtained from Eqs. (3.8), (3.12), (3.13). The value of $\Delta\varphi$ in this case is determined only by the electron flux according to

$$e\Delta\varphi=T_e \ln(\Gamma_{eT}/\Gamma_{e\xi}),$$

and value of the boundary plasma density n_s is determined by the Eqs. (3.8), (3.13). The boundary plasma (ion) velocity is given by Eq. (3.7). Since at $T_i>>e\Delta\varphi>> T_e$ the ion density in the sheath is uniform, and the electron density is zero, the potential profile, according to the Poisson equation, is parabolic, and the sheath width is equal to

$$L_{sh} = \sqrt{\frac{2\varepsilon_0\Delta\varphi}{e n_s}} = \sqrt{2e\Delta\varphi / T_e} r_d(n_s) \qquad (3.14)$$

The separation into plasma and sheath is, of course, conventional - at considerable distance from the boundary surface the charge separation (and the field strength) smoothly increases towards the surface; this region is often referred to as presheath. The characteristic presheath scale equals the mean free path of the attracted particles (for the negatively biased surface - λ_i). It means that this problem demands rigorous kinetic treatment. In order to obtain a qualitative understanding of the presheath formation we shall restrict ourselves here to a simple fluid model, in which the influence of the ion collisions will be accounted for by introducing the effective friction force (fully equivalent results follow from the model [6], in which the ion momentum dissipation is described by introducing a source term in Eq. (3.15)). Neglecting the electron inertia, using the quasineutrality and the ion flux conservation, we obtain from Eqs. (2.10), (2.22) in the 1D stationary case

$$m_i n u_i du_i/d\xi=-d(p_i+p_e)/d\xi-\mu_{iN} n \nu_{iN} ui, \qquad (3.15)$$

or

$$(1-c_s^2/u_i^2)du_i/d\xi=-\mu_{iN}\nu_{iN}/m_i, \qquad (3.16)$$

where

$$c_s=(dp/d\rho)^{1/2} \qquad (3.17)$$

is the ion sound velocity (compare to Eq. (3.7); here $\rho=m_i n$).

From Eq. (3.16) it follows that the directed plasma velocity (which practically coincides with the ion velocity) continuously increases towards the wall from $u_i=0$ up to

$$u_i=u_0=c_s. \qquad (3.18)$$

At this point the plasma acceleration $du_i/d\xi$ (and $dn/d\xi$) tends to infinity. If the Debye radius is small with respect to the ion mean free path it means that the quasineutral approximation becomes invalid in the vicinity of this point: according to Eq. (3.11) the electric field diverges here. The transition to the sheath occurs, where the field intensity greatly exceeds the plasma field, and the density of the repulsed particles (in our case - of electrons) decreases fast toward the wall. This point, where singularity in the quasineutral solution occurs, is adopted usually as a position of the plasma-sheath boundary, and the condition Eq. (3.18) is known as the Bohm criterion. According to it, the directed plasma (i.e. the ion) velocity at the entrance of the sheath equals the sound velocity. In the case $T_e>>T_i$ instead of Eq. (3.18) we have simply

$$c_s=(T_e/m_i)^{1/2}. \qquad (3.19)$$

In other words, the plasma density decreases by the factor of the order of $e^{1/2}$ over the last ion mean free path, and according to Eq. (3.11), the potential difference over the presheath is $\sim T_e/2$ [4]. The sheath structure, if necessary, can easily be found from the Poisson equation

$$d^2\varphi/d\xi^2=\varepsilon_0 e(n_e-n_i) \qquad (3.20)$$

with account of the electron space charge according to Eq. (3.11) and collisionless expression for the ion density. The latter is obtained by neglecting the thermal ion energy

$$n_i = \frac{u_0 n_s}{u_i} = \frac{n_s}{\sqrt{1 - \frac{2e\varphi}{m_i u_0^2}}}. \tag{3.21}$$

The Eq. (3.20) describes the motion of 'particle' in 'potential'

$$W(\Phi = -e\varphi / T_e) = 1 - \exp(-\Phi) - 2\Phi_0(\Phi / \Phi_0 + 1)^{1/2}, \tag{3.22}$$

where n_s and $\Phi_0 = -m_i u_0^2/2T_e$ correspond to the ion density and energy at the plasma-sheath boundary. This so-called Sagdeev's 'potential' $W(\Phi)$ at

$$\Phi_0 \geq 1/2, \tag{3.23}$$

has its maximum at low values of Φ. In the opposite case it decreases monotonously, and in a low electric field the space charge density diverges. This means that the physically reasonable potential profile $\Phi(\xi)$ which describes the transition to the plasma where the electric field and the space charge density tend to zero simultaneously is possible only if Eq. (3.23) is satisfied. This condition is also sometimes called the Bohm criterion. If the local Debye radius, which corresponds to spatial scale of the quasineutrality violation, exceeds the ion mean free path λ_{iN}, the quasineutral equation (3.15) fails in the point where $u_i \ll c_s$. This situation corresponds to collisional sheath, Section 3.6.

If the potential difference over the sheath $\Delta\varphi$ significantly exceeds T_e/e, we can neglect n_e and Φ_0 in Eqs. (3.20), (3.22), and obtain according to the Child-Langmuir law:

$$|\varphi|^{3/4} = \frac{3}{2}\left(\frac{j_i^{(sat)}}{\varepsilon_0}\right)^{1/2}\left(\frac{m_i}{2e}\right)^{1/4}\xi; \qquad L_{sh} = \frac{\sqrt{2}}{3} r_d \left(\frac{2e\Delta\varphi}{T_e}\right)^{3/4} \tag{3.24}$$

the potential profile in it and the sheath width. The local Debye radius at the sheath edge is given by $r_d = (\varepsilon_0 T_e / e n_s)^{1/2}$.

The floating potential which corresponds to the absence of the net current through the surface is equal to

$$\Delta\varphi_{fl} = \frac{T_e}{e}\ln\frac{\sqrt{T_e/(2\pi m_e)}}{c_s} = \frac{T_e}{2e}\ln\frac{m_i}{2\pi m_e}. \tag{3.25}$$

Rigorously speaking, the traditional analysis which results in the Bohm criterion Eqs. (3.18) and (3.19) is rather crude. It is based on existence of the singularity in the fluid Eqs. (3.15) and (3.16) which arises at small spatial scale of the order of the Debye radius, where these equations are invalid. This means that more rigorous kinetic analysis is necessary. The electric field in the plasma periphery is high and usually its influence on the ion motion in the plasma and presheath is significant, and cannot be neglected.

Solution of the corresponding kinetic problem for the case of the ion motion in parent gas, when resonant charge-exchange collisions (Section 2.3) dominates, was obtained in [7]. At $T_e >> T_i$ and $\lambda_i = const$ the plasma density profile is determined by the ion flux conservation (with the field-dependent mobility Eq. (2.82)) and by Boltzmann's law for electrons given by Eq. (3.11): $n \sim (\xi)^{1/2}$, instead of the linear law for the case of constant mobility. In this case, instead of Eq. (3.18), the ion velocity at the sheath boundary according to the rigorous kinetic solution is given by the expression:

$$u_0 = 1.07(T_e/m_i)^{1/2}. \tag{3.26}$$

In other words, the precise position of the sheath boundary and the form of the solution in this region depend on rather subtle kinetic properties of the presheath. For example, for a collisionless slab model of a positive column the coefficient in Eq. (3.26) is 1.14 [8], and for the cylindrical case it equals 1.21 [9,10]. The problems of the collisionless sheath and presheath formation and structure are discussed in more detail in [11 - 13]. Peculiarities of the sheath structure in multispieces plasmas are discussed in [14] - [17].

An interesting effect is described in [14-16] for multispecies electronegative plasmas with $T_e >> T_i$. If the electron density is not too low (see Chapter 10), the floating vessel wall is negatively biased with respect to the plasma. The electrostatic potential of the plasma with respect to the wall is of the order of T_e. The corresponding strong electric field piles up the negative ions into the plasma bulk, and in the edge region they are practically absent. Hence in the perypheric pure plasma, which consists of electrons and positive ions, the boundary condition for the plasma equations has the standard form of the Bohm criterion Eqs. (3.18) and (3.19).

On the other hand, if the density of the negative ions in the edge plasma is high, the potential scale at the wall-adjacent plasma decreases from the electron temperature T_e to the ion temperature T_i. Therefore, as a boundary condition we practically have the Bohm criterion with the ion temperature. Such nonlinear phenomena as hysteresis, which arise in this situation, are described in Section 3.9 [14-16].

3.4 COLLISIONLESS SHEATH NEAR EMITTING SURFACE. DOUBLE SHEATH

As an example of a more complicated boundary condition we shall consider a surface which emits electrons with the Maxwellian distribution. Their temperature T_{ec} is assumed to be much lower than the temperature of plasma electrons T_{eh}. Let us restrict ourselves to the surface at floating potential. The emitting electrons partially neutralize the positive ion space charge near the surface thus reducing the potential difference in the sheath and increasing the flux of hot electrons from plasma to the wall. For a small flux of cold electrons Γ_{ec} we have the zero current condition

$$\Gamma_{eh} - \Gamma_{ec} = \Gamma_i, \tag{3.27}$$

where Γ_{eh} and Γ_i are defined according to Eqs. (3.12), (3.18). The Eq. (3.27) determines the potential drop in the sheath as a function of the cold electron flux Γ_{ec}. Further rise of this flux results in the formation of a double sheath with the minimum of the potential ("virtual cathode") [10] near the electrode, Fig. 3.4.

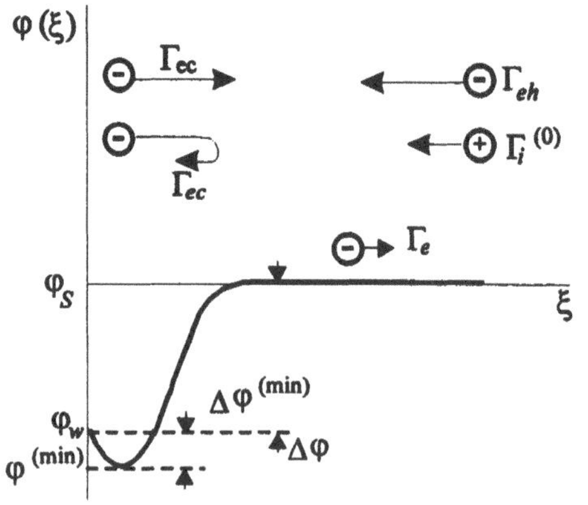

Figure 3.4. Potential distribution in the double sheath.

The negative space charge and the negative potential drop near the surface restricts the influx of cold electrons, so that only part of their flux $\Gamma_{ec}=\Gamma_{ec}^{(0)}$ is injected through the double layer into the plasma while the rest are reflected back to the electrode. The potential difference $\Delta\varphi^{(\min)}$ between the surface and the minimum of the potential profile can now be calculated from the equation $\Gamma_{ec}^{(0)} = \Gamma_{ec}\exp(-e\Delta\varphi^{(\min)}/T_{ec})$, hence $\Delta\varphi^{(\min)}$ is of the order of several T_{ec}/e. We shall restrict ourselves to the simplest case $T_{ec} << T_{eh}$ .Since the potential drop in the double layer is of the order of T_{eh}/e we shall neglect $\Delta\varphi^{(\min)}$ thus making no difference between the potential drop in the double layer and the potential difference between plasma and electrode $\Delta\varphi$.

In the double sheath the Poisson equation has to be solved:

$$d^2\varphi / d\xi^2 = e(n_{eh} + n_{ec} - n_i)/\varepsilon_0 . \tag{3.28}$$

The particle densities need to be expressed as functions of the potential. Ion density is given by the Eq. (3.21) where n_s is the ion density at the plasma side of the double sheath. The density of hot electrons must be described more accurately than by the Boltzmann distribution (Eq. (3.11)), since $\Delta\varphi$ in the double layer is significantly lower than the floating potential Eq. (3.25). Taking into account the number of passing and reflected hot electrons, we have

$$n_{eh}(\Phi) = \frac{1}{2} n_{ehs} \exp(-\Phi)[1 - erf(\sqrt{\Phi^{(\min)} - \Phi})], \tag{3.29}$$

where $\Phi = -e\varphi/T_{eh}$, n_{ehs} is the hot electron density at the plasma side. The cold electrons are accelerated in the double sheath, hence (neglecting their initial energy) [10]

$$n_{ec}(\Phi) = \frac{1}{2}\exp\left(\frac{T_{eh}(\Phi^{(\min)} - \Phi)}{T_{ec}}\right)\left(1 - erf\left(\sqrt{\frac{T_{eh}(\Phi^{(\min)} - \Phi)}{T_{ec}}}\right)\right) \approx$$
$$\approx \frac{1}{2\sqrt{\pi}} n_{ec}(\Phi^{(\min)})\sqrt{\frac{T_{ec}}{T_{eh}(\Phi^{(\min)} - \Phi)}}, \tag{3.30}$$

where $n_{ec}(\Phi^{(\min)})$ represents the cold electron density at the potential minimum. It is connected to the flux of cold electrons according to

$$\Gamma_{ec}^{(0)} = \frac{1}{\sqrt{2\pi}} n_{ec}(\Phi^{(\min)})\sqrt{\frac{T_{ec}}{m_e}}, \tag{3.31}$$

Now integrating the Poisson equation, taking into account the fact that electric field is zero both at the plasma side of the double sheath and at the potential minimum, we obtain the condition of double sheath formation

$$\int_0^{\Phi^{(min)}} (n_{eh} + n_{ec} - n_i)\, d\Phi = 0 . \tag{3.32}$$

Combining Eq. (3.32) with Eqs. (3.21), (3.29), (3.30), employing the zero current condition Eq. (3.27) with Γ_{ec} replaced by $\Gamma_{ec}^{(0)}$, and taking into account the quasineutrality condition at the plasma side $n_{ehs}+n_{ecs}=n_s$, we obtain the single equation for the potential difference $\Delta\Phi_{fl}=e\Delta\varphi_{fl}/T_{eh}$:

$$\exp(-\Delta\Phi_{fl}) = \frac{1+erf(\sqrt{\Delta\Phi_{fl}})(1-a)}{1+\dfrac{a}{\sqrt{\pi\Delta\Phi_{fl}}}} ;$$

$$a = \frac{m_i u_0^2}{T_{eh}}\left(\sqrt{1+\frac{2T_{eh}}{m_i u_0^2}\Delta\Phi_{fl}} - 1\right) . \tag{3.33}$$

Here the terms of the order of $(m_e/m_i)^{1/2}$ are neglected. For $T_{eh}>>T_i$, according to the Bohm criterion Eq. (3.19), $m_i u_0^2=T_{eh}$, and we obtain from Eq. (3.33) $\Delta\Phi_{fl}=0.95$. In the case of equal temperatures $T_{eh}=T_i$ substituting $m_i u_0^2=2T_{eh}$, we have $\Delta\Phi_{fl}=0.687$.

3.5 SHEATH IN THE MAGNETIC FIELD

The sheath structure for a magnetic field normal or almost normal to the surface is completely the same as without the magnetic field, while for the magnetic field oblique or parallel to the surface the sheath structure turns out to be quite different. This difference is caused by the fact that for the case when the Debye radius is smaller than the particle's gyroradii, the magnetic presheath of the order of gyroradius is formed with the distribution function far from Maxwellian. The particle motion in the magnetic presheath can be described only by the kinetic equation, and the potential drop in it is typically of the order of several temperatures. Therefore, expressions for the particle fluxes to the surface as functions of the potential difference are more complicated. Moreover, the cases of the magnetic field parallel to the surface and of the oblique magnetic field are essentially different. In the first situation the particles reach the wall only owing to collisions in the magnetic presheath, and the final result depends on the type of the collisions, while for an oblique magnetic field the fluxes to the wall are governed mainly by the motion parallel to the magnetic field. Therefore, these two cases should be considered separately. We shall also assume that the sheath width is much less than the particle's gyroradii.

Magnetic field parallel to the surface. Fluxes produced by Coulomb collisions. We shall start from the case when the Coulomb collisions dominate. The orbits inside the magnetic presheath can be subdivided into two groups. At a given coordinate ξ there are particles with the velocities directed so that their orbits intersect the wall. The reflected particles must also have sufficient energy to overcome the potential difference and to reach the surface. Since the time of motion along the orbit is of the order of the inverse gyrofrequency $\omega_{c\alpha}^{-1}$ these orbits in strong magnetic field are almost empty. They are filled only by the relatively rare Coulomb collisions. Hence, the distribution becomes strongly anisotropic, and the so-called loss regions are created. The occupied and empty orbits are shown in Fig. 3.5.

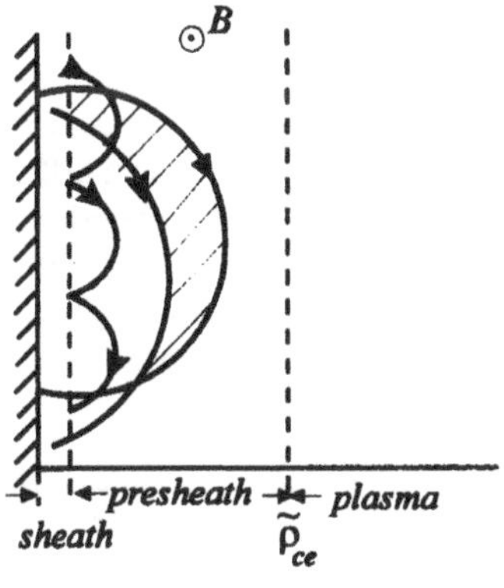

Figure 3.5. Sketch of the electron orbits for negatively biased surface. Hatched area are the loss regions.

The volume of the loss regions in the velocity space strongly depends on ξ. At ξ=0 loss regions for attracted particles occupy almost all the velocity space, while at a distance of the order of thermal gyroradius at the plasma side of the magnetic presheath their volume tends to zero. Thus, the density and the potential drop in the presheath could be very large. Consequently, the problem is of an essentially kinetic nature. The distributions are subdivided into two populations - the outgoing particles in the loss regions and trapped ones. In contrast to the case without magnetic field (Section 3.2), when the escape time from the loss region is of order of transition time from the trapped orbits to the outgoing ones, and kinetic analysis results only in an insignificant factor (0.7104 instead of 2/3 in Eq. (3.8)), in magnetized plasma the "narrow bottleneck" corresponds to transition from the trapped orbits to the outgoing ones. It means that the loss rate is limited by this process and is considerably reduced with respect to the fluid estimate $\sim nV_T$.

Let us consider the case of the surface at large negative potential with respect to the plasma, so that the electrons are trapped. At distance ξ^* of the order of electron gyroradius the electron magnetic presheath is formed. The diffusive electron flux to the surface transforms into the electron flux from the occupied to the almost empty loss regions in the velocity space. The latter can be estimated [18-19] (see also [20]) as

$$\Gamma_{e\xi} \sim n^* \tilde{\rho}_{ce} \nu_e \exp(-e\Delta\varphi / T_e), \qquad (3.34)$$

where n^* is the density at the edge of the electron magnetic presheath, $\Delta\varphi$ is the potential difference between ξ^* and the surface, and $\tilde{\rho}_{ce}$ is the gyroradius of the fast electron with energy $e\Delta\varphi$. A rigorous solution has been obtained only for $e\Delta\varphi >> T_e$ [19]. In this case the electric field in the magnetic presheath does not significantly change the shape of the orbits of the electrons which have enough energy to reach the wall. The motion of fast electrons over orbits and over energy has a diffusive character while the loss region boundary corresponds to the absorbing wall, similarly to the problem of the particle losses in the open magnetic traps [21]. The kinetic equation for fast electrons which have sufficient energy $p^2/2m_e \sim e\Delta\varphi >> T_e$ to overcome the potential drop has the form [22]

$$\frac{\partial}{\partial \vec{p}}(\vec{A}f - D_{\parallel}^{(p)}\frac{\partial f}{\partial \vec{p}_{\parallel}} - D_{\perp}^{(p)}\frac{\partial f}{\partial \vec{p}_{\perp}}) + \frac{\vec{p}}{m_e}\frac{\partial f}{\partial \vec{r}} + e[\vec{p} \times \vec{B}]\frac{\partial f}{\partial \vec{p}} = 0, \qquad (3.35)$$

$$\vec{A} = -\nu_e(p)\vec{p}\,;\; D_{\parallel}^{(p)} = \nu_e(p) m_e T_e\,;\; D_{\perp}^{(p)} = \nu_e(p) p^2 \;.$$

Here $\nu_e(p) = 4\pi e^2 \Lambda n m_e p^{-3}$ is the collision frequency of the fast electron with thermal electrons and ions. For the z-axis parallel to $\vec{B}$ we take the motion integrals as new variables: $\varepsilon = p^2/2m_e$, p_z and $Q = p_y - m_e\omega_{ce}\xi$. The loss region boundary with these new coordinates is described by theparaboloid

$$\varepsilon = e\Delta\varphi + (p_z^2 + Q^2)/2m_e.$$

The main contribution to the flux is provided by electrons with energies exceeding $e\Delta\varphi$ by few T_e. Their energy is changed by the collisions at time scale$\sim m_eT_e/[p^2\nu_e(p)]\sim T_e[e\Delta\varphi\nu_e(2m_ee\Delta\varphi)^{1/2}]$. The corresponding variation of their velocity direction is characterized by a small angle $\Delta\theta\sim(T_e/e\Delta\varphi)^{1/2}<<1$. Thus, the distribution function is strongly perturbed only near the loss regions, while at angles larger than $\Delta\theta$ from the loss region boundary the distribution function is close to the Maxwellian one. The solution can be sought in the form

$$f(\varepsilon, p_z, Q, \xi)=f_0(\varepsilon, p_z, Q)+f_1(\varepsilon, p_z, Q, \xi),$$

where f_1 is a small addition of the order of $\nu_e(2m_ee\Delta\varphi)^{1/2}/\omega_{ce}$. The kinetic equation in new variables yields

$$D_\perp^{(p)}\frac{\partial^2 f_0}{\partial p_z^2}+2e\Delta\varphi\frac{\partial}{\partial\varepsilon}[\frac{D_\parallel^{(p)}}{m_e}\frac{\partial f_0}{\partial\varepsilon}+\frac{|\vec{A}|f_0}{p}]+\frac{p_\xi}{m_e}\frac{\partial f_1}{\partial\xi}+$$
$$+D_\perp^{(p)}(1-\xi^2/\rho_{ce}^2)\frac{\partial^2 f_0}{\partial Q^2}=0, \tag{3.36}$$

where $\tilde{\rho}_{ce}=(m_e/eB)(2e\Delta\varphi/m_e)^{1/2}$. For the particles outside the loss region the boundary condition at $\xi=0$ reads $\partial f_1/\partial\xi(\xi=0)=0$. Thus, putting $\xi=0$ and employing this condition, Eq. (3.36) is reduced to the single equation for f_0 After substitution $f_0=\exp(-\varepsilon/2T_e)F$ it is transformed to

$$\frac{\partial^2 F}{\partial\tilde{p}_z^2}+\frac{\partial^2 F}{\partial\tilde{\varepsilon}^2}+\frac{\partial^2 F}{\partial\tilde{Q}^2}=F/4, \tag{3.37}$$

where

$$\tilde{p}_z=p_z/(m_eT_e)^{1/2};\ \tilde{Q}=Q/(m_eT_e)^{1/2};\ \tilde{\varepsilon}=\varepsilon/T_e.$$

The boundary condition at the edge of the loss region is reduced to $F(\tilde{\varepsilon}=\Phi_c+\tilde{p}_z^2/2+\tilde{Q}^2/2)=0$, where $\Phi_c=e\Delta\varphi/T_e$. Let us introduce the parabolic coordinates

$$(\tilde{p}_z^2+\tilde{Q}_z^2)/\sigma^2=2\tilde{\varepsilon}-2\Phi_c-1+\sigma^2;$$
$$(\tilde{p}_z^2+\tilde{Q}_z^2)/\tau^2=2\tilde{\varepsilon}-2\Phi_c-1+\tau^2.$$

The boundary of the loss region corresponds now to $\sigma=1$. The solution of Eq. (3.37) can be factorized: $F=F_1(\sigma)F_2(\tau)$, and at $\sigma\to\infty$ the distribution function tends to the Maxwellian one. The solution is given by

$$F = C\int_1^\sigma \frac{d\sigma'}{\sigma'}\exp[\sigma^2/4 - \tau^2/4 - (\sigma')^2/4];$$

$$C = nm_e[\pi T_e \int_1^\infty \frac{d\sigma'}{\sigma'}\exp(-(\sigma')^2]^{-1}.$$

The flux of electrons to the loss region boundary is calculated by integrating Eq. (3.37) over velocity space and employing the Gauss theorem. The result reads

$$\Gamma_{e\xi} = 1.85(\pi e\Delta\varphi / T_e)^{1/2} n^* \tilde{\rho}_{ce} v_e(\sqrt{2m_e e\Delta\varphi})\exp(-e\Delta\varphi / T_e), \quad (3.38)$$

where n^* is the density at $\xi^* = \tilde{\rho}_{ce}$.

The flux of attracted particles (of ions) can be estimated as

$$\Gamma_{i\xi} \cong n^{**}\rho_{ci}v_i. \quad (3.39)$$

Here n^{**} is the density at the plasma side of the ion magnetic presheath ξ^{**} - of the order of the ion gyroradius. The potential and density drop in the region between ξ^{**} and ξ^* depends on the ratio of the particle fluxes. For

$$v_i/\omega_{ci} << \Gamma_{e\xi}/\Gamma_{i\xi} < 1 \quad (3.40)$$

the ions should be trapped between ξ^{**} and ξ^*. If one assumes the opposite, then one can immediately conclude that n^*/n^{**} is very small since at ξ^* almost all ion orbits intersect the wall, and density on these orbits corresponds to pure free streaming: $\Gamma_{i\xi} \sim n^*(T_i/m_i)^{1/2}$. Now comparing this with Eq. (3.39) we find $n^*/n^{**} \sim v_i/\omega_{ci}$. The latter relation comes to a contradiction with Eqs. (3.38) - (3.40). Thus, if inequality (3.40) is valid, the ions should be trapped between ξ^{**} and ξ^*, and the potential profile turns out to be nonmonotonic. The potential drop in the ion presheath of the order of T_i/e is sufficient to trap ions. Furthermore, since the ions are trapped, the density does not change significantly, and $n^* \sim n^{**}$. Only if the electron flux to the wall is so small that $v_i/\omega_{ci} > \Gamma_{e\xi}/\Gamma_{i\xi}$, electrons are trapped everywhere, and the relation $n^*/n^{**} \sim v_i/\omega_{ci}$ takes place.

The expression for the ion flux to a wall which is biased to a large positive potential with respect to plasma, has the same form as Eq. (3.38):

$$\Gamma_{i\xi} = 1.85(\pi e\Delta\varphi / T_i)^{1/2} n^{**} \tilde{\rho}_{ci} v_i(\sqrt{2m_e e\Delta\varphi})\exp(-e\Delta\varphi / T_i), \quad (3.41)$$

where $\nu_i(p) = 2\pi e^2 \Lambda n m_i p^{-3}$. The electron flux is estimated by $\Gamma_{e\xi} \sim n^* \tilde{\rho}_{ce} \nu_e$ with $n^* \sim n^{**}$ since ions are trapped. In the case of a floating wall the fluxes given by Eqs. (3.38), (3.41) are of the same order for $\Delta\varphi \sim T_{e,i}/e$, hence, the floating wall is biased to the potential of the order of T/e while in the ion and electron magnetic presheathes the potential is nonmonotonic.

Magnetic field parallel to the surface. Fluxes caused by collisions with neutrals. Let us calculate the flux to the negatively biased wall caused by the electron-neutral collisions which scatter particles to the open trajectories. In contrast to the Coulomb collisions, scattering through large angles, i.e. from orbits on which the distribution is approximately Maxwellian, is important here. Taking this circumstance into account, extending the integration over scattering angle to zero, and adopting the integrals of motion as variables, we find [19]

$$\Gamma_{e\xi} = \frac{N}{\pi m_e T_e} \int_0^\infty d\xi n \int_0^\pi d\vartheta \int_{-\infty}^{\infty} dQ \int_{Q^2/2m_e + e\Delta\varphi}^{\infty} d\varepsilon (2\varepsilon / m_e)^{1/2} \exp(-\varepsilon / T_e)$$
$$\times \sigma(\varepsilon, \vartheta) \left[\frac{2\varepsilon}{m_e} - \left(\frac{Q}{m_e} + \omega_{ce} \xi \right)^2 \right]^{-1/2}, \tag{3.42}$$

where N is the neutral density. Since the integral over ε is governed primarily by the interval near the lower limit, we evaluate the integral over Q by the method of steepest descent, finding

$$\Gamma_{e\xi} = 1/2(\pi T_e / e\Delta\varphi)^{1/2} n^* \tilde{\rho}_{ce} \nu_{eN}(\sqrt{2m_e e\Delta\varphi}) \exp(-e\Delta\varphi / T_e), \tag{3.43}$$

where $\nu_{eN}(\sqrt{2m_e e\Delta\varphi}) = N\sqrt{2m_e e\Delta\varphi}\,\sigma_0(e\Delta\varphi)$, σ_0 is the total scattering cross section. The same result has been obtained in [23] in a different way. A similar expression is valid for the case of trapped ions:

$$\Gamma_{i\xi} = 1/2(\pi T_i / e\Delta\varphi)^{1/2} n^{**} \tilde{\rho}_{ci} \nu_{iN}(\sqrt{2m_i e\Delta\varphi}) \exp(-e\Delta\varphi / T_i). \tag{3.44}$$

Magnetic field oblique to the surface. As the thermal particle fluxes along magnetic field usually strongly exceed the collisional fluxes Eqs. (3.38), (3.41), (3.43), (3.44), the situation changes for very shallow angles Θ between the magnetic field line and the wall. For

$$\nu_e / \omega_{ce} << \Theta << \nu_i / \omega_{ci} \tag{3.45}$$

(here $\nu_{e,i}$ represent both collisions with neutrals and with the charged particles) the ion flux to the wall is still given by Eq. (3.41),(3.44), while the electron flux is determined by the parallel flux. Hence, for the case of trapped electrons we have

$$\Gamma_{e\xi} = \frac{\Theta}{\sqrt{2\pi}} n^{**} \sqrt{\frac{T_e}{m_e}} \exp(-e\Delta\varphi / T_e), \qquad (3.46))$$

where n^{**} is the density at the plasma side of the ion magnetic presheath and $\Delta\varphi$ is the corresponding potential difference. The ion flux is governed by the parallel motion when

$$\Theta >> \omega_{ci}/\nu_i. \qquad (3.47))$$

Here the ion flux can be obtained using a similar approach to the case of normal magnetic field [24]. Due to the Boltzmann distribution of electrons, the density in the ion presheath strongly decreases towards the wall as the loss region volume increases. This fact results in the existence of singularity in the fluid equations describing the plasma far from the wall, and hence the ξ component of the ion fluid velocity should be equal to Θc_s, where c_s is the sound speed. The ion flux is then

$$\Gamma_{i\xi} = \Theta n^{**} c_s. \tag{3.48}$$

It should be noted that the density and potential distribution, as well as the particle motion in the magnetic presheath, is rather complicated and was investigated by many authors, e.g. [24] - [29]. For example, the potential drop in the magnetic presheath and in the sheath can be comparable, since density in the magnetic presheath decreases strongly towards the wall. However, to obtain boundary conditions one needs only expressions for the fluxes Eqs. (3.46)) and (3.48)). As it follows from Eqs. (3.46)), (3.48)) the floating potential of the wall for the case Eq. (3.47)) is independent of the oblique angle and coincides with Eq. (3.25) for normal electric field.

The problem of stability of the magnetic presheath has been studied in [28]. It was shown that the Kelvin-Helmholtz instability often develops within the presheath. The properties of the resulting turbulent magnetic presheath and relation between the particles fluxes and the averaged sheath voltage were modelled there. The problem of the turbulent magnetic presheath is still to be investigated in more detail.

3.6 COLLISIONAL SHEATH IN THE ABSENCE OF MAGNETIC FIELD

At higher pressures and/or at lower plasma densities the sheath thickness L_{sh} exceeds the mean free path of the particles. Since $L_{sh} << L$, the zero boundary conditions still represent reasonable approximation for the particle densities. In contrast, the condition for the potential $\Delta\varphi(\Gamma_{e\xi}, \Gamma_{i\xi})$ is more complicated than Eq. (3.14). In order to derive this condition, let us consider in the boundary region, where the normal fluxes $\Gamma_{\alpha\xi}$ are conserved, the fluid equations for electrons and ions in combination with the Poisson equation

$$D_{e,i}\frac{dn_{e,i}}{d\xi} \mp b_{e,i}n_{e,i}\frac{d\varphi}{d\xi} = \Gamma_{e,i} \geq 0\ ;$$
$$\frac{d^2\varphi}{d\xi^2} = \frac{e(n_e - n_i)}{\varepsilon_0}. \qquad (3.49)$$

This equation system with zero boundary conditions was investigated in detail numerically in [30-31], see also [54]. If $\Gamma_i D_e = \Gamma_e D_i$, the electric field in plasma is absent, and the boundary surface is at the plasma potential. But in a general case the field in the plasma and in sheath is considerable. As follows from Eq. (3.49), it suppresses the electron motion, if $\Gamma_i D_e > \Gamma_e D_i$, and hinders ions, if $\Gamma_i D_e < \Gamma_e D_i$. In the quasineutral plasma (the region I in Fig. 3.6) the density profile is linear:

$$n = \frac{\Gamma_{e\xi}b_i + \Gamma_{i\xi}b_e}{D_e b_i + D_i b_e}(\xi - L_{sh}), \qquad (3.50)$$

and the field is given by:

$$\frac{d\varphi}{d\xi} = \frac{\Gamma_{i\xi}D_e - \Gamma_{e\xi}D_i}{(D_e b_i + D_i b_e)n}\ . \qquad (3.51)$$

According to Eqs. (3.50),(3.51), at $\xi \to L_{sh}$ the plasma density decreases, and the field strength (as well as the space charge density $e(n_i - n_e)$, which is the source of the field) tends to infinity. The quasineutral approximation thus becomes invalid starting from some distance $L_{sh}+\delta$ from the wall, where L_{sh} can be chosen as the sheath boundary, and δ - as a presheath width. We shall consider the case of weakly ionized plasma and a negatively charged wall with large potential drop $e|\Delta\varphi|>>T_e \geq T_i$, when a relatively simple analytic solution exists [19].

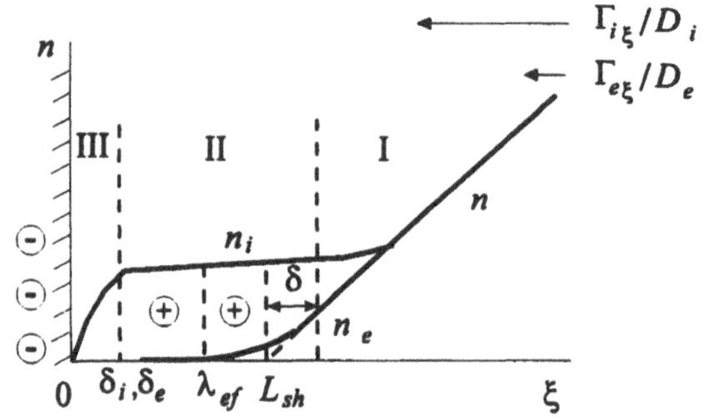

Figure 3.6. Collision dominated sheath adjacent to the negatively biased surface.

In plasma and in the main part of the sheath (regions I and II in Fig. 3.6) the repulsed particles (electrons) are distributed according to the Boltzmann law Eq. (3.11). Moreover, in the sheath their density is much smaller than the ion density, and the space charge there is generated by ions.

In order to estimate the presheath width δ we notice, using for diffusion and mobility coefficients the approximation of elementary theory Eqs. (2.71), (2.77) and the Einstein relation Eq. (2.63), that in plasma for $\Gamma_{e\xi}/D_e<<\Gamma_{i\xi}/D_i$ the electric field is given by

$$\frac{d\varphi}{d\xi}=\frac{T_e}{e(\xi-L_{sh})},$$

and the charge separation is

$$|n_e-n_i|=\frac{\varepsilon_0 T_e}{e^2(\xi-L_{sh})^2}. \tag{3.52}$$

As the position of the plasma sheath boundary we shall choose the point where deviation from quasineutrality is of the order of unity. Combining Eq. (3.50) with Eq. (3.52), we obtain (see Fig. 3.6)

$$\delta=\left(\frac{\varepsilon_0 T_e(T_e+T_i b_i)}{e^3\Gamma_{i\xi}}\right)^{1/3}. \tag{3.53}$$

Plasma density here at $\xi=\xi_{sh}+\delta$ is given by

$$n_s = \left[\frac{\varepsilon_0 T_e \Gamma_{i\xi}^2}{b_i^2 (T_e + T_i)^2}\right]^{1/3} . \tag{3.54}$$

The presheath scale δ is thus of the order of the local Debye radius, calculated using this density. The plasma density at the sheath boundary n_s is determined only by the flux of attracted particles (ions), and the electron density in plasma and in the main region II of the sheath satisfies the Boltzmann law, Eq. (3.11). The field-driven and diffusive electron fluxes practically compensate each other here, and electron density decreases exponentially towards the wall.

In the region II, both the ion diffusive term and the electron density are negligible. Accordingly, the Poisson equation here reduces to ($E<0$)

$$dE / d\xi = -en_i / \varepsilon_0 = -e\Gamma_{i\xi} / e\varepsilon_0 b_i E . \tag{3.55}$$

Since the plasma field is small with respect to the field in the sheath, integrating Eq. (3.55), we obtain

$$|E(\xi)| = \sqrt{2e\Gamma_{i\xi}(L_{sh} - \xi) / \varepsilon_0 b_i} . \tag{3.56}$$

After the second integration

$$L_{sh} - \xi = \left[\frac{9\varepsilon_0 b_i}{8e\Gamma_{i\xi}} \left(\varphi(L_{sh}) - \varphi(\xi)\right)^2\right]^{1/3} . \tag{3.57}$$

The expressions given by Eqs. (3.56), (3.57) are violated in region III, adjacent to the surface. Here the particle densities tend to zero (to be more precise, to the small values Eq. (3.6)) but the electric field remains finite. Consequently, the field-driven fluxes in Eq. (3.49) at $\xi \to L_{sh}$ become small with respect to the diffusion-driven ones. So the ion flux is transformed here from field-driven to diffusion-driven. The characteristic length of this region (see Fig. 3.6)

$$\delta_i = D_i / (b_i |E(\delta_i)|)$$

is determined by the ratio of ion diffusion coefficient to ion convective velocity in the local electric field. The corresponding electron length is slightly larger (since, as a rule $T_e > T_i$) but both δ_e and δ_i are small with respect to the scales of the Eq. (3.57). The potential drop in region III is also small - of the order of T_e, T_i. If we neglect now the potential differences over the presheath and region III, the potential drop in the sheath $\Delta\varphi$ coincides with the potential drop in region II. The latter can be estimated as follows. Up to the boundary between regions II and III δ_e the electron diffusive flux is compensated by the field driven one and at $\xi < \delta_e$ the electron flux is transformed into the pure diffusive form. At the point of transformation we have

$$\Gamma_{e\xi} \cong b_e E(\delta_e) n_s \exp(-e\Delta\varphi / T_e). \tag{3.58}$$

Substituting expression for n_s, Eq. (3.54), and for $E(\delta_e)$, Eq. (3.56) we obtain the estimate for the potential drop in the sheath

$$\frac{e\Delta\varphi}{T_e} = \ln\left(\frac{\Gamma_{i\xi} b_e}{\Gamma_{e\xi} b_i}\right) + \frac{1}{3}\ln\left[\frac{3e\Delta\varphi T_e}{(T_e + T_i)^2}\right] \tag{3.59}$$

and for the sheath width

$$L_{sh} = \left[\frac{9\varepsilon_0 b_i (\Delta\varphi)^2}{8\Gamma_{i\xi}}\right]^{1/3} \tag{3.60}$$

The important case of the floating potential at $\Gamma_{e\xi} = \Gamma_{i\xi}$ corresponds to

$$\frac{e\Delta\varphi_{fl}}{T_e} = \ln\left(\frac{b_e}{b_i}\right) + \frac{1}{3}\ln\left[\frac{3e\Delta\varphi_{fl} T_e}{(T_e + T_i)^2}\right]. \tag{3.61}$$

In spite of several rough approximations which were made during evaluation, comparison with the numerical simulation [30] - [31] demonstrates surprisingly good accuracy: the error is considerable only at $e\Delta\varphi \leq T_e$. Since the Boltzmann law Eq. (3.11) holds also in the plasma region adjacent to the surface, the wall potential with respect to an arbitrary plasma point ξ close to the wall is

$$\varphi_w = \varphi(\xi) - \frac{T_e}{e}\ln\left[\frac{b_e n(\xi)}{\Gamma_{e\xi}}\left(\frac{3\Gamma_{i\xi} T_e}{\varepsilon_0 b_i}\right)^{1/3}\right]. \tag{3.62}$$

For simplicity we have omitted here the second logarithm in Eq. (3.59).

It is possible to interpret these results as follows. The plasma density profile Eq. (3.50), its boundary value n_s Eq. (3.54), as well as the potential profile in the sheath, Eq. (2.57), are determined by the ion saturation flux Eq. (3.13). The electron flux $\Gamma_{e\xi}$ determines only the sheath thickness and the potential drop $\Delta\varphi$. The wall potential corresponds, roughly speaking, to the point where $\xi \sim \delta_e$,

$$\delta_e = D_e / [b_e |E(\delta_e)|], \tag{3.63}$$

where the free electron diffusion (Eq. (3.63)) starts.

The results of this Section are also valid for the collisional sheath across magnetic field when the gyroradii are small with respect to the sheath thickness L_{sh}, if one inserts the cross-field transport coefficients into the corresponding expressions.

Since the field strength the sheath is large, the ion mobility in the collisional sheath often becomes field dependent. In the case of constant λ_{iN} and T_N=0 the expressions for the potential and density profiles and the sheath thickness (when the ion mobility is given by Eq. (2.82)) were obtained in [54], [55]. At $T_e >> T_i$ the ion mobility in the presheath is also field dependent. In this situation the plasma density at the entrance of the collisional sheath is $n_s = (\varepsilon_0 m_i^2 \Gamma_{i\xi}^4 / e\lambda_{iN}^2 T_e)^{1/5}$ instead of Eq. (3.54). In [55] the interpolation expression for the plasma velocity at the sheath entrance $u_s = \Gamma_{i\xi} / n_s$, $u_s / u_0 = (1 + \pi r_d / (2\lambda_{iN}))^{-1/2}$ has been proposed

Influence of depletion of the electron distribution. Such an approach depends crucially on the Maxwellian character of electron and ion distribution functions with spatially uniform T_e, T_i. For ions in weakly ionized plasma it is the case, as a rule, since the Maxwellian distribution with $T_i = T_N$ is quickly established practically in one collision. The corresponding spatial scale is small - of the order of λ_{iN}. On the contrary, for electrons the validity range of all the results derived in the previous Sections is far more restricted. The reason is connected with the fact that the electron current (Joule heating) and absorption of fast electrons by the surface both strongly disturb the electron distribution function. The relaxation of distribution anisotropy (which is responsible for the coefficient 0.71 in Eq. (3.8) and $\tilde{\rho}_{ce} \nu_e$ in Eq. (3.34)) occurs relatively fast: over the time scale $\sim \nu_e^{-1}$. The corresponding spatial scale is small: of the order of λ_e ($\sim \rho_{ce}$ across magnetic field). On the contrary, the energy relaxation mechanisms, which restore the distribution equilibrium form, are relatively weak. As a result, the isotropic part of the electron distribution is perturbed on more considerable scale $\lambda_e^{(\varepsilon)}$, λ_{ef}, Eqs. (2.20), (2.59).

The effects of Joule heating are more pronounced at positive values of φ_w when electron saturation current flows to the surface. In this case T_e is non-uniform and considerably exceeds T_N [3],[32]. This phenomenon can be neglected if the sheath width exceeds the electron energy relaxation length $\lambda_e^{(\varepsilon)}$, Eq. (2.20).

At negative values of φ_w the electron current is exponentially small and cooling in electric field is, as a rule, insignificant. In contrast, the kinetic effects of the electron distribution depletion due to the escape of fast electrons to the wall are typically dominant [33], [34]. In such a situation, which is more characteristic for weakly ionized plasma, the kinetic effects of nonlocality determine the relation between $\Delta\varphi$ and $\Gamma_{e\xi}$. The distinction with the results of the fluid approach is caused by the fact that at distances less than λ_{ef} Eq. (2.59) the concept of mean directed velocity fails. The electron flux $\Gamma_{e\xi}$ to the negatively biased surface is transported only by the high energy part of electron population. Electrons with total energy ε (kinetic plus potential in electric field) lower than $e\Delta\varphi$ are trapped and do not contribute to electron current. The current in the vicinity of a surface is transported in the form of free diffusive flux of fast untrapped electrons with $\varepsilon > e\Delta\varphi$ at constant value of ε. Their density at $\xi \sim \lambda_{ef}$ is

$$n_{eh}(\lambda_{ef}) \sim n_e(\lambda_{ef}) \exp\left[\frac{-e(\varphi(\lambda_{ef}) - \varphi_w))}{T_e}\right]. \tag{3.64}$$

The density of fast electrons decreases practically to zero at $\xi=0$. The diffusive flux of fast electrons (which coincides with net electron flux) equals to

$$\Gamma_{e\xi} \sim D_e n_{eh}(\lambda_{ef}) / \lambda_{ef}. \tag{3.65}$$

The fluid results for the collisional sheaths are valid, if $\lambda_{ef} < \delta_e$. In the opposite case, due to the fact that up to the point $\xi = \lambda_{ef}$, the electron density approximately satisfies the Boltzmann law Eq. (3.11), the density of hot electrons is

$$n_{eh}(\lambda_{ef}) \sim n_e(\delta_e). \tag{3.66}$$

This means that at the same value of the potential drop $\Delta\varphi$ the slow relaxation of the distribution function reduces the electron flux from the value given by Eq. (3.58) to Eq. (3.65). In other words, the free diffusion of the fast electrons occurs from distance λ_{ef} instead of δ_e. In the case $L_{sh} > \lambda_{ef} > \delta_e$ the factor δ_e / λ_{ef} has to be inserted into the Eq. (3.58) and under the logarithm sign in the Eq. (3.59). Using the equations (3.60) and (3.63), we obtain

$$\frac{\delta_e}{\lambda_{ef}} \sim \frac{T_e}{e\lambda_{ef}}\left(\frac{9\varepsilon_0 b_i}{8e\Delta\varphi\Gamma_{i\xi}}\right)^{1/3}. \tag{3.67}$$

If λ_{ef} exceeds L_{sh}, the fast electron diffusion starts in the quasineutral plasma. According to Eqs. (3.11),(3.64),(3.65), the potential drop in the sheath is

$$\frac{e\Delta\varphi}{T_e} = \ln\left[\frac{D_e n(\lambda_{ef})}{\lambda_{ef}\Gamma_{e\xi}}\right] - \ln\frac{n(\lambda_{ef})}{n_s}, \tag{3.68}$$

where n_s is given by Eq. (3.54).

The results of Sections 3.2 - 3.5 for the collisionless sheath are also valid only if the ionization degree is sufficiently high. Indeed, if electron-electron collision frequency exceeds electron-neutral collision frequency, $\nu_{ee} \geq \nu_{eN}$, the Maxwellian distribution function is restored practically during one collision time. In the opposite case of small electron-electron collision frequency $\nu_{ee} << \nu_{eN}$ the reduction factor to be inserted under the logarithm sign into Eq. (3.14) is of the order of the ratio $(\lambda_{eN}/\lambda_{ef})$. For the thin sheath across magnetic field (Section 3.5) another factor of the order of $(\rho_{ce}/\lambda_{ef\perp})$ is to be introduced in the right hand side of Eqs. (3.54) and (3.58).

The reduction of the sheath potential due to kinetic effects is sometimes very significant. As a first example we consider the weakly ionized plasma where Coulomb collisions are completely negligible: $\nu_{ee} << \delta_{eN}\nu_{eN}$. In this case the ratio $(\lambda_{eN}/\lambda_{ef})$ is of the order of $(\delta_{eN})^{1/2}$. This means that in the case of collisionless sheath the additional kinetic factor in Eq. (3.15) for the atomic gases equals to $(m_e/m_i)^{1/2}$, and Eq (3.25) formally gives $\Delta\varphi_{fl}=0$. In other words, the assumption $\Delta\varphi_{fl} >> T_e/e$ is violated, and in contrast to the earlier considered situation the floating potential $|\Delta\varphi_{fl}|$ is of the order of T_e/e.

As a second example consider the situation in low pressure discharges when the value of λ_{ef} exceeds the total plasma scale L. In this case the electron distribution function (EDF) in the whole discharge volume is strongly depleted and falls exponentially for values of the full energy ε above the wall potential (which is equal to the floating potential) due to the escape of the fast electrons to the vessel walls. On the other hand, the electron distribution decays exponentially due to the influence of energy losses connected with excitation of electronic states and ionization. In a stationary plasma the ionization rate is to be balanced by electron loss rate (which is controlled by the escape of the fast electrons). Since both these rates depend exponentially on the full energy ε, their thresholds are to be situated rather close. Consequently, in contrast to the result of the fluid approximation Eq. (3.25), the floating potential is to be situated not far from the effective ionization potential.

An example of such a situation is presented in Fig. 3.7. One can see that in this second example, as in the previous situation, the kinetic effects result in considerable reduction of the floating potential with respect to the Langmuir expression.

For detailed quantitative analysis of these important problems, rather complicated self-consistent numerical calculations are necessary to solve the kinetic equation. Up to now, as we know, such results are practically absent

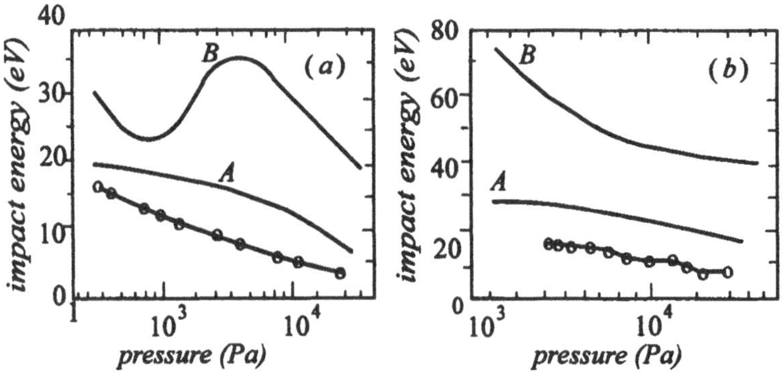

Figure 3.7 Maximal energies of ions of *Ar* (*a*) and *He* (*b*) bombarding the vessel wall in the inductively coupled RF discharge (experimental points [35]).These energies correspond to $(e\Delta\varphi_{fl}+T_e/2)$, where the second (small) term approximates the potential drop in the presheath. Curves *A* are calculated with the account of EDF depletion; curves *B* have been calculated by the Langmuir formula Eq. (3.25) [35]. Due to the non-Maxwellian character of EDF, the so-called screening temperature (see [14] and [17]) $T_e^{(sc)} = -ed \ln n_e / d\varphi \,|_{\varphi=\varphi_{sh}}$ has been used in the calculations. The ionization potentials are equal to 15.76 *eV* (Ar) and 21.58 *eV* (*He*).

3.7 IONIZATION PROCESSES IN SHEATHS

We shall consider here the influence of ionization by electron impact on the structure of the collisional sheath, Section 3.6. In all the preceding Sections we have treated the fluxes $\vec{\Gamma}_e, \vec{\Gamma}_i$ in the sheaths as outward directed vectors. They were generated in the plasma volume and conserved in relatively thin sheaths. Nevertheless, numerous important situations exist when such an approach is invalid, and the processes of generation (and sometimes of removal and transformation) of charged particles in sheaths are important, and can even dominate. The most important of these processes is electron impact ionization. The reason is that the tail of the electron distribution function, which is responsible for the ionization, strongly depends on the electric field strength. As a rule, this dependence is exponential (see e.g. Section 2.9), and the electric field in the sheaths is very high. As a result, in spite of the fact that the sheath volume is negligible with respect to the plasma size, the plasma generation here can be comparable with the generation in the whole volume.

In the absence of magnetic field the floating potential is negative. Since the electron flux is directed outward, electrons are losing energy when performing work against the electric field. Therefore, the impact ionization in the sheaths is suppressed with respect to the plasma volume. Let us consider the simplest situation, when we have a bounded plasma which is generated by some external source. A *dc* voltage U is applied to it, using parts of a vessel wall as electrodes (Fig. 3.8).

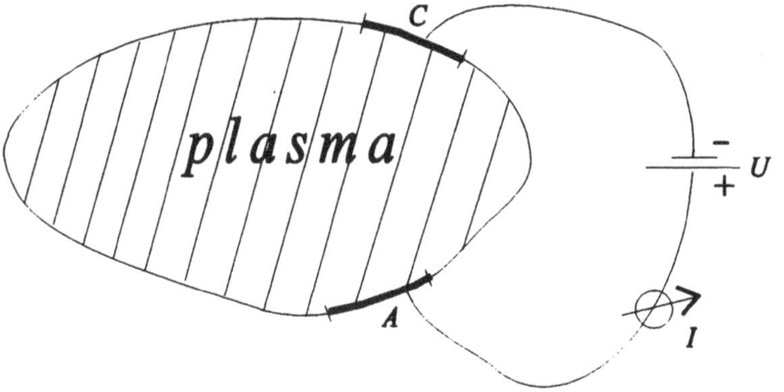

Figure 3.8

This voltage is divided between the cathode and anode sheaths in such a way as to increase the electron current to the anode and to reduce it to the cathode. At small $U \ll T_e/e$ the electron current dependence on the sheath voltage is linear. It means that the ratio of the rise of the cathode voltage to the drop of the anode's is to be equal to S_A/S_C. At moderate U values both cathode and anode remain negatively biased. The potential drop in the cathode and anode sheaths $\Delta\varphi_C \sim \Delta\varphi_A \sim \Delta\varphi_{fl}$, and ion current density everywhere coincides with its saturation value Eq. (3.13). If the anode area is not too small

$$S_A/S_C > j_i^{(sat)}/j_e^{(sat)}, \qquad (3.69)$$

the net current is restricted by the value $I_0 = S_C j_i^{(sat)}$. Such a current is reached at the values of U of the order of several T_e/e, when the electron current to the cathode becomes negligible. Further increasing the voltage practically does not change the current I_0, potential and density profiles in plasma, and the potential drop in the anode sheath $\Delta\varphi_A$. The voltage rise results only in increase of the cathode potential fall $\Delta\varphi_C$ and of the cathode sheath width. It leads to further decrease of the negligible exponentially small electron current to the cathode up to the moment when the ionization processes in the cathode fall are switched on.

Electron flux reversal and formation of the cathode potential fall. We consider the situation at large $\Delta\varphi_C >> T_e/e$. In the absence of ionization the electron density decreases exponentially towards the wall according to the Boltzmann law Eq. (3.11) from the value $n_s(\Gamma_{i\xi})$, Eq. (3.54). The density profile is determined only by the ion flux which is equal to the saturation value and does not depend on the voltage $\Delta\varphi_C$. The potential profile $\varphi(\xi)$ in the sheath, Eq. (3.57), which is also determined by the ion flux and the sheath voltage (or sheath width), also does not depend on the negligible electron density in the sheath. Therefore, in such a situation it is preferable to express the sheath characteristics in terms of ion flux $\Gamma_{i\xi}$ and voltage $\Delta\varphi_C$.

The problem for electrons becomes linear. In the absence of the ionization we have

$$\frac{d}{d\xi}\left(D_e \frac{dn_e}{d\xi} - b_e n_e \frac{d\varphi}{d\xi}\right) = 0 \qquad (3.70)$$

with boundary conditions (see Eq. (3.7)),

$$n_e(\xi = L_{sh}) = n_s(\Gamma_{i\xi}), \qquad (3.71)$$

$$n_0 V_{Te}/4 = \Gamma_{e\xi}\,;\; n_0 = n_e(\xi = 0). \qquad (3.72)$$

The solution of the Eq. (3.70) is straightforward:

$$n_e(\xi)=\Gamma_{e\xi}\int_{\xi_0}^{\xi}\exp(\Phi-\Phi'(\xi'))d\xi'/D_e=$$
$$=\Gamma_{e\xi}\exp(\Phi)[\int_{0}^{L_{sh}}\exp(-\Phi')d\Phi'/u_e(\Phi')+ \tag{3.73}$$
$$+4/V_{Te}-\int_{\xi}^{L_{sh}}\exp(-\Phi')d\Phi'/u_e(\Phi')].$$

Here we used the Einstein relation and also substituted the electron velocity $u_e=b_e d\varphi/d\xi$ driven by electric field. Here dimensionless potential is introduced as $\Phi=e\varphi/T_e$. Outside the region III (Fig. 3.6) we have an inequality: $e(\varphi-\varphi_w)>>T_e$. The terms in the square parentheses in the right hand side of Eq. (3.73) are negligible, and the electron profile in this region is the Boltzmann one. The connection between n_s, L_{sh} and $\Gamma_{e\xi}$ can be found from

$$n_s=\Gamma_{e\xi}\exp(\Delta\Phi_C)\left[\int_{0}^{L_{sh}}\exp(-\Phi')d\Phi'/u_e(\Phi')\right] \tag{3.74}$$
$$\approx\Gamma_{e\xi}\exp(\Delta\Phi_C)/u_e(\xi=0)\approx n_0\exp(\Delta\Phi_C).$$

The condition Eq. (3.72) implies total absence of electron emission by the cathode surface and outward directed electron flux $\Gamma_{e\xi}$ which decreases exponentially with voltage $\Delta\varphi_C$. The work of the electric field Ej_e over the electron gas is negative. The electrons are cooled in the sheath and impact ionization is absent here. From the other side, the ion flux $\Gamma_{i\xi}$ does not vary with $\Delta\varphi_C$. If follows that at sufficiently high voltage the ionization by a small amount of fast electrons which are born in the sheath close to the cathode becomes significant. The main source of such fast electrons is usually assumed to be the process of secondary ion-electron emission by the cathode surface. The typical values of the ion-electron emission coefficient $\gamma<<1$ (which equals the ratio of emitted electron flux to the impinging ion one) is of the order of several per cent. If the emission is sufficient for the reversal of the electron flux $\Gamma_{e\xi}$, the ingoing electrons gain energy from the electric field, and some of them become capable of impact ionization. In order to roughly describe this transition we replace the boundary condition Eq. (3.72) at the cathode surface by expression similar to Eq. (3.27):

$$\Gamma_{e\xi}(\xi=0)=\Gamma_{e\xi}{}^{(out)}+\Gamma_{e\xi}{}^{(in)}=n_s V_{Te}/4-\gamma\Gamma_{i\xi}\big|_{\xi=0}\,. \qquad (3.75)$$

According to Eqs. (3.18), (3.54) the plasma density at the sheath entrance n_s does not depend on the sheath potential. As follows from Eq. (3.74), the value of n_0 decreases exponentially with the sheath voltage $\Delta\varphi_C$, and at

$$\Delta\varphi_C=\frac{T_e}{e}\ln\frac{n_s V_{Te}}{4\gamma\Gamma_{i\xi}} \qquad (3.76)$$

the reversal of the electron flux $\Gamma_{e\xi}$ occurs. In order to describe the increase of the inverse inward electron flux due to ionization we apply the rough fluid approximation. Usually the electron density in sheaths is sufficiently low - $\nu_{ee}<<\delta_{eN}\nu_{eN}$, and the scales $\lambda_e^{(\varepsilon)}$ and λ_{ef}, Eqs. (2.20), (2.59), coincide. We shall consider the case of relatively thick sheaths $L_{sh}>>\lambda_e^{(\varepsilon)}$. In such a situation the electron distribution is non-Maxwellian and is determined by local values of the electric field. Traditionally this process is described by the field-dependent Townsend coefficient $\alpha(E/p)$ (see e.g. [36], [37]):

$$\frac{d\Gamma_{e\xi}}{d\xi}=\alpha(\mathrm{E}(\xi)/p)p\Gamma_{e\xi}, \qquad (3.77)$$

where p is the neutral gas pressure. Since the field profile in the sheath is independent of the electron density, it is given by Eq. (3.56). The electron multiplication can be thus calculated using this field profile:

$$\Gamma_{e\xi}(\xi)=\Gamma_{e\xi}(\xi=0)\exp[\int_0^{\xi}\alpha\left(\frac{E(\xi')}{p}\right)pd\xi']. \qquad (3.78)$$

This relation at $\xi=L_{sh}$ presents the necessary expression of the outgoing flux $\Gamma_{e\xi}(\xi=L_{sh})$ in terms of the ingoing flux $\Gamma_{i\xi}(\xi=L_{sh})$ and sheath voltage $\Delta\varphi_C$. In calculating the ion space charge it is necessary to account for ionization too:

$$\frac{d\Gamma_{i\xi}}{d\xi}=-\alpha(\xi)p\Gamma_{e\xi}\;. \qquad (3.79)$$

At $\gamma\Gamma_{i\xi}>>n_s V_{Te}\exp(-e\Delta\varphi_C/T_e)$ it is possible to replace $\Gamma_{e\xi}(\xi=0)$ in Eq. (3.76) by $\gamma\Gamma_{i\xi}(\xi=0)$. Further increase of $\Delta\varphi_C$ leads to extremely fast rise of the electron current since it exponentially depends on Townsend coefficient α

according to Eq. (3.78). In a medium electric field the α value in its turn exponentially depends on E/p [37]. It follows that the Boltzmann term in the expression for electron density Eq. (3.73) becomes dominant only in the small vicinity of plasma-sheath transition $\xi = L_{sh}$. The boundary condition Eq. (3.75) transforms into

$$(\Gamma_{e\xi} = \gamma \Gamma_{i\xi})\Big|_{\xi=0}. \tag{3.80}$$

The electron multiplication factor

$$M(\Delta\varphi_C) = \int_0^{L_{sh}} \exp(\alpha(\xi'))d\xi'$$

increases quickly. At some critical value $\Delta\varphi_{C0}$

$$M(\Delta\varphi_{C0}) = 1/\gamma + 1 \tag{3.81}$$

the 'breakdown' of the cathode sheath occurs - it becomes practically independent from the plasma region. In such a situation one electron which is emitted from the cathode generates $(1/\gamma)$ ions in the sheath. When impinging on the cathode these ions give rise to the next generation of secondary electrons, and the ionization processes in the sheath become independent from the bulk plasma. This case corresponds to short *dc* glow discharge with a cold cathode, when the plasma region is absent or small. The situation in standard dc discharge is very similar. In homogeneous plasma of positive column the ion current is negligible ($\sim b_i/b_e$ of the total one), and at the cathode surface practically all current (more precisely $(\gamma+1)^{-1}$ part of it) is transported by ions. It means that, as in the case of sheath breakdown, almost all ions falling on the cathode have been generated in the sheath. The quantitative calculation of cathode sheath structure in this local (i.e. based on the assumption that ionization rate is determined by the local field, Eq. (3.77)) approximation can be found in numerous publications (see, e.g., [38], [39]).

However, such a simple scenario is practically never met in practice. The increase of the current density results in proportional increase of the ion density n_i in the sheath and to decrease of its width L_{sh} according to the Poisson equation. From the other side, since the value of $M(\Delta\varphi_{C0})$, Eq. (3.81), is extremely sensitive to the field strength, especially in the vicinity of the cathode, here the value of the field remains practically unchanged. Therefore, such a 1D sheath has a falling *I-V* characteristic. It is unstable with respect to formation of cathode spots which occupy only part of the cathode surface. The current density in it - the normal current density - is current-independent, and the area of the cathode spot is proportional to the total current. The normal current density corresponds to the situation when the field strength is so high that exponential dependence of the Townsend coefficient $\alpha(E/p)$ is switched out. But in such strong fields, as a rule, electron runaway takes place - the electron distribution depends not on the local field strength, but on the potential difference between generation and observation points - the local fluid approach becomes invalid, and self-consistent kinetic description is necessary (see, e.g.[40], [41]).

Anode region. The anode phenomena in the dc glow discharges are also rather complicated. In an unmagnetized plasma current is transported mainly by electrons. Due to the boundary conditions Eq. (3.80) practically all the current is to be transformed in the cathode region from the electron form into the ion one. In the anode region, from the other side, only small ion current is to be generated. Thus, the ionization rate in the cathode region, its voltage $\Delta\varphi_C$, luminosity, etc. far exceed the corresponding values at the anode. The cathode part of dc glow discharges is considered as the most important for discharge maintenance.

In a sufficiently strong magnetic field the situation reverses - since according to Eqs. (2.77), the plasma current across $\vec{B}$ is transported mainly by ions, the anode fall $\Delta\varphi_A$ increases with B and $\Delta\varphi_C$-decreases [42]. Across the strong magnetic field the main ionization phenomena occur in the anode sheath, Fig. 3.9.

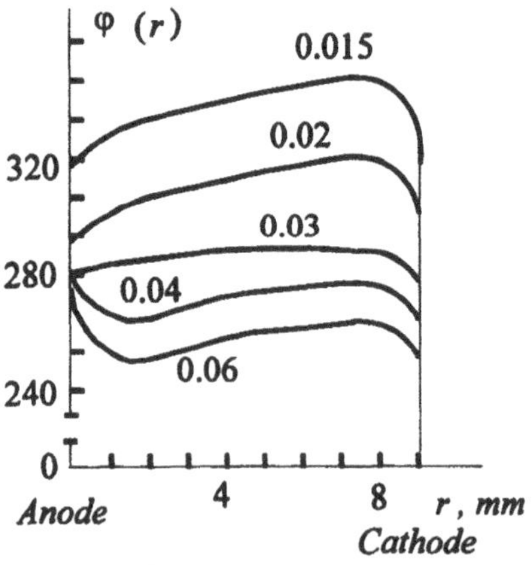

Figure 3.9. The potential profile in glow discharge in *He* across magnetic field [42]. The inner of two coaxial cylinders - anode, the outer - cathode. The discharge conditions are: I=5A; p=133 *Pa*; numbers at curves denote the magnetic field induction in Tesla.

As we have seen before, in the absence of a magnetic field with small current in plasma generated by an external source, the anode sheath is practically the same as in the case of the floating surface. The anode potential fall $\Delta\varphi_A$ reflects electrons and is slightly less than $\Delta\varphi_{fl}$. We also have the same situation in self-sustained discharges at high current density, when the electron distribution function is close to the Maxwellian one. The anode fall is also negative (reflects electrons) at low current density, if the anode area S_A is small. From the other side, at moderate values of S_A the anode fall changes its sign with the current decrease [43].

At low currents and high neutral gas pressures, when the anode sheath is thick and the nonlocality of the electron distribution is unimportant, for the case of positive $\Delta\varphi_A$ the detailed measurements were performed in [44], [45]. If the diffusion is negligible*, the ion flux $\Gamma_{i\xi}$ is directed from the anode into the plasma and equals to zero at the anode surface. The transition to quasineutral plasma occurs when the ion flux, which is generated by the ionization in the sheath, reaches critical value. This process determines the sheath characteristics.

With accuracy up to $b_i/b_e \ll 1$ the total current, which is conserved in the sheath, coincides with the electron current $e\Gamma_{e\xi}$:

$$\Gamma_{e\xi}=b_e(E)En_e=j/e. \tag{3.82}$$

For the dependence $b_e(E)$ the power approximation is often used [46]; see Section 2.9):

$$b_e(E)=b_e^{(0)}(E_0/E)^{\alpha}, \tag{3.83}$$

where $0<\alpha<1$; the sub/superscript zero corresponds to the values at the sheath-plasma boundary. The field strength $E(\xi)$ is at a maximum at the anode, and falls practically to zero towards the plasma, where $n_e \approx n_i$. Therefore, the ion space charge in the main part of the sheath is negligible, and from the Poisson equation and Eqs. (3.82),(3.83) we obtain

$$E_{\max}^{2-\alpha} - E^{2-\alpha}(\xi) = \frac{(2-\alpha)j}{\varepsilon_0 b_e^{(0)} E_0^{\alpha}}\xi. \tag{3.84}$$

The sheath width can be expressed in terms of the maximal field strength at the anode:

---

* In the discharge current-carrying plasma the diffusion can be important only on rather short spatial scale $l_T=T_e/eE$ which is of the order of $\lambda_e^{(\varepsilon)}$, Eq. (2.20), see below Chapter 9.

$$L_{sh} = \frac{\varepsilon_0 b_e^{(0)} E_0^{\alpha}}{j(2-\alpha)} E_{\max}^{2-\alpha}. \tag{3.85}$$

The ion flux which is generated in the sheath, equals

$$\Gamma_{i\xi}(L_{sh}) = \int_0^{L_{sh}} (pj/e)\alpha(E(\xi)/p)d\xi. \tag{3.86}$$

The equations (3.84) - (3.86) present the necessary connection between j, $\Gamma_{i\xi}$ and $\Delta\varphi_A$. Similarly to the cathode case, due to the fact that the coefficient α exponentially depends on (E/p), and the sheath spatial scale decreases with current according to Eq. (3.84), the falling current-voltage characteristic of the anode sheath and phenomenon of the normal current density on anode are also possible [44]. If the sheath width $L_{sh} \leq \lambda_{ef}$, Eq. (2.59), the behavior of the anode region is determined by the kinetic effects ([47] - [49]), see also the discussion in the end of the previous Section. This sign reversal of $\Delta\varphi_A$ was investigated experimentally in [43] (see Fig. 3.10). Thorough theoretical analysis and derivation of the transition criteria between different regimes still do not exist.

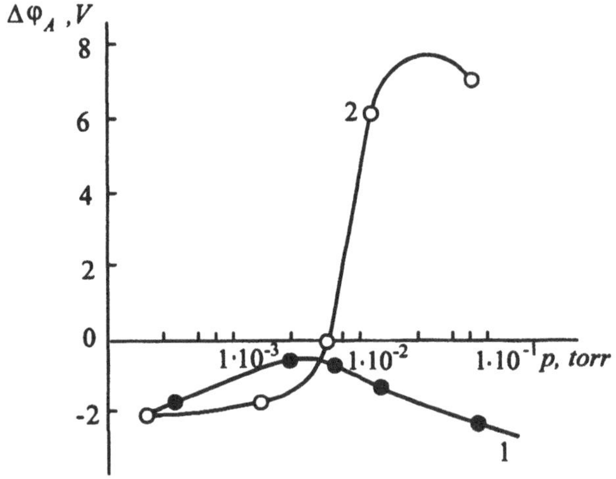

Figure 3.10. The anode potential drop at a planar anode in *Ne* versus current and pressure in *Hg*. Tube radius a=1.6 cm. Current 1 - 2 *A*; 2 - 0.1 *A*.

3.8 SHEATHS IN CAPACITIVE RADIO FREQUENCY DISCHARGE

Sheaths of a specific kind arise in capacitively coupled radio frequency discharges. These discharges are maintained by applying high frequency voltage to conducting electrodes, which can be immersed into plasma or separated from it by insulating plates (or by vessel walls). The sheath structure in radio frequency discharges attracts considerable interest nowadays (see, for example, [50]). The sheath width in such discharges considerably exceeds the Debye radius, and approximate zero boundary conditions for the plasma density are often inadequate.

The ionization in the sheaths in RF discharge is also far more important than in the absence of the RF field. To describe such phenomena, as application of an additional *dc* voltage to the electrodes or to the vessel walls, the formulation of more precise boundary conditions is also necessary.

As was discussed in Chapter 2, the characteristic frequencies for electron and ion density relaxation to the quasineutral state are equal to the inverse electron and ion Maxwellian times $\tau^{(e)}{}_M$ and $\tau^{(i)}{}_M$ defined according to Eq. (2.123), for the case of frequent collisions. These frequencies are determined by the electron and ion plasma frequencies $\omega_{pe}=(n_e e^2/\varepsilon_0 m_e)^{1/2}$, $\omega_{pi}=(n_i e^2/\varepsilon_0 m_i)^{1/2}$, if collisions are rare.

The most physically interesting and practically important is the case when the driving RF frequency ω lies between these frequencies. This being the case, the electron drift velocity is determined by the instantaneous electric field intensity $E(t)$ according to the expressions of Section 2.8, while the ion motion occurs on a far longer time scale which is determined by the time averaged over the RF oscillations electric field in the discharge.

The evolution of the ion density profile is controlled by slow processes of ionization, recombination and diffusion. On the other hand, the electron density profile oscillates quickly between two electrodes with the RF frequency ω (see Fig. 3.11). Therefore, the central part of the discharge gap is occupied by the quasineutral plasma, where $n_e \approx n_i$. At its periphery stationary positive space charge is partially compensated by the oscillating electron negative space charge contribution produced by electron density profile.

If the discharge voltage considerably exceeds T_e/e, the edge of the electron profile is quite sharp. Its structure is similar to that described above, in Sections 3.3 and 3.6 ([50] and [51]). The corresponding spatial scale, which is determined by the local Debye radius r_d, Eq. (2.120), is small with respect to the amplitude of the electron displacement in an oscillatory electric field. The potential drop over this thin boundary of the electron profile is of the order of several T_e/e, and hence remains small with respect to the dc voltage over the sheath.

At each moment the sharp oscillating boundary separates the plasma region and the region of positive space charge where electrons are practically absent. We define as the sheath boundary position the point of maximal displacement of this sharp edge of the electron profile (Fig. 3.11). In other words, in the plasma, at $\xi > L_{sh}$, the space charge is negligible, and at any point of the sheath during part of the RF period the quasineutral plasma phase lasts; the remaining time is occupied by the phase of positive space charge.

The radio frequency current in the plasma phase (as well as in the bulk plasma) is transported by electrons. On the contrary, during the space charge phase the electron and ion conductivity currents are negligible with respect to the displacement current. Considering the sheath structure, we shall restrict ourselves to the case $T_e >> T_i$, and purely sinusoidal RF current density $j(t)$.

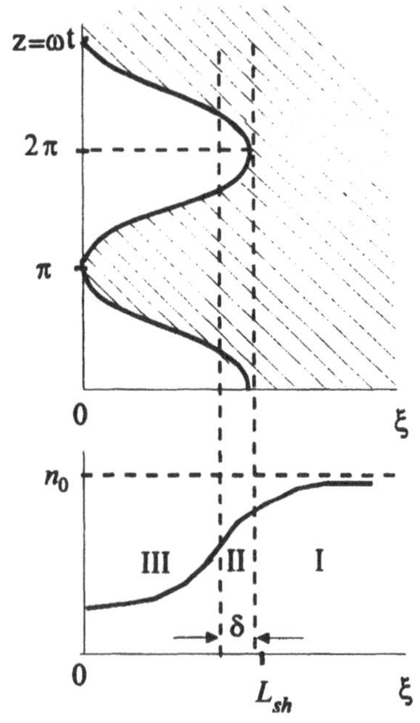

Figure 3.11 The motion of sharp boundary of the electron profile $n_e(\xi, t)$, *a*), and the ion density profile $n_i(\xi)$, *b*). I-the plasma region, II, III-the sheath. In the region III the ion motion is controlled by the convection, in the region II the transition from the diffusion-dominated to the convection-dominated ion motion occurs.

RF sheath without ionization processes. In the absence of the external *dc* current, the ion current to the boundary surface averaged over the RF period is to be balanced by electrons. It means that the sharp boundary of the electron profile 'licks' the surface, and the duration of the plasma phase tends to zero at $\xi \to 0$. On the other hand, at $\xi \to L_{sh}$ duration of the space charge phase tends to zero (Fig. 3.11).

The evolution of the plasma density profile is determined by relatively slow processes of generation and removal of ions. In other words, for the sheath analysis we can consider the ion density profile as stationary and to perform the averaging of the ion momentum equation over the fast electron motion with the RF frequency ω. In the plasma region (I in Fig. 3.11), at $\xi \geq L_{sh}$, the average field is proportional to the density gradient and corresponds to the diffusive ion motion. The ion flux is equal to

$$\Gamma_{i\xi} = D_{eff} \frac{dn_i}{d\xi}, \tag{3.87}$$

where D_{eff} is the effective diffusion coefficient. It is a sum of the ordinary ambipolar diffusion coefficient, Eq. (4.8), and of the specific RF diffusion coefficient

$$D_{HF} = \frac{b_i j_0^2 \varepsilon_0}{2e^3 b_e^2 n_i^3(\xi)} \tag{3.88}$$

where j_0 is the current amplitude. This expression is derived below, in Section 4.3, Eq. (4.53). In the sheath the field averaging needs to be performed over the plasma phase and the space charge phase separately. The resulting expression is thus given by the sum of two terms. The averaging over the plasma phase in the vicinity of the plasma-sheath boundary, $E=\langle E\rangle_2$, results in the diffusive flux, Eq. (3.87). The averaged over the space charge phase field $E=\langle E\rangle_1$ results in convective ion flux which is proportional to the local ion density. The field strength in the plasma phase is very low with respect to the space charge field. It means that in the main part of the sheath (region III in Fig. 3.11) the ion flux is driven by the space charge field. The contribution of the diffusive flux is significant only in the small transition region II (Fig. 3.11), where the ion flux is transformed from the diffusive form, Eq. (3.87), into the convective form. In this transition region the plasma phase lasts during almost all the RF period, and with acceptable accuracy we can use here the plasma expressions, Eqs. (3.87) and (3.88) for the diffusive flux.

In order to calculate the convective flux, it is necessary to derive the equation for position of the sharp boundary between the plasma and space charge phases $\xi(z)$ (Fig. 3.11). Since the ion and electron currents in the space charge phase are negligible, the current here is transported in the form of displacement current. On the other side, in the plasma phase the electron conductivity current dominates. Integrating the Poisson equation in the sheath from $\xi=0$ to the oscillating boundary of the $n_e(\xi,t)$ profile, we obtain

$$E(\xi,t) = E(0,t) + \int_0^{\xi(z)} \varepsilon_0^{-1} e n_i(\xi) d\xi = 0 . \tag{3.89}$$

Differentiating over t and accounting for the fact that at $\xi=0$ only the displacement current exists, we have (in the zero order in $\omega\tau^{(e)}_M$ or in ω/ω_{pe}) the equation for the sharp boundary in the form

$$e\omega n_i \frac{d\xi}{dz} + j(z) = 0 . \tag{3.90}$$

Since the electron profile 'licks' the floating boundary surface, $d\xi/dz=0$ at $\xi=0$. Integrating the field strength $E(\xi,t)$, Eq. (3.89), over duration of the space charge phase, we obtain the convective ion drift velocity

$$V(\xi) = b_i \langle E \rangle_1 (z(\xi)) = \varepsilon_0^{-1} b_i j_0 (\sin z - z\cos z)/(\pi\omega), \tag{3.91}$$

where $z=z(\xi)$ corresponds to the moment when the plasma phase boundary is at the given point ξ.

The profile of the ion density in the sheath is determined from

$$\Gamma_{i\xi} = D_{eff} \frac{dn_i}{d\xi} + V(z(\xi)) n_i . \tag{3.92}$$

in combination with Eqs. (3.90), (3.91). The values of the sheath dc voltage $U_{dc}^{(fl)}$, and of the sheath width L_{sh}, can be expressed in terms of $\Gamma_{i\xi}$. Neglecting the ion diffusion, for the case of constant ion mobility from Eqs. (3.90) and (3.91) we obtain

$$L_{sh} = \frac{3 j_0^2 b_i}{4\varepsilon_0 e \omega^2 \Gamma_{i\xi}}, \tag{3.93}$$

and for the negative dc self-biasing voltage, with respect to plasma of the floating surface,

$$U_{dc}^{(fl)} = \frac{j_0^3 b_i}{e\varepsilon_0^2 \pi^2 \omega^3 \Gamma_{i\xi}} \left(\frac{\pi^2}{3} - \frac{16}{27} \right). \tag{3.94}$$

Since the average field in the RF sheath is rather high, in practice the ion mobility is, as a rule, field dependent. For the case of constant ion mean free path λ_{iN} at $e\langle E\rangle\lambda_{iN}>>T_N$ (when b_i is proportional to $E^{-1/2}$, see Section 2.5) we have

$$L_{sh} = 0.78 \frac{j_0^{3/2}}{\varepsilon_0^{1/2} e^{1/2} \omega^{3/2} \Gamma_{i\xi}} \sqrt{\frac{\lambda_{iN}}{m_i}}, \tag{3.95}$$

$$U_{dc}^{(fl)} = 0.4 \frac{j_0^{5/2}}{\varepsilon_0^{3/2} e^{1/2} \omega^{5/2} \Gamma_{i\xi}} \sqrt{\frac{\lambda_{iN}}{m_i}}. \tag{3.96}$$

The time-averaged *dc* biasing voltage of the floating surface $U_{dc}^{(fl)}$ is negative and can be expressed in terms of the ion density profile from integration of the Eqs. (3.90), (3.91). If the discharge geometry is asymmetric, the *dc* self-biasing voltage arises between the RF electrodes. It is determined by the difference between the floating *dc* voltages applied to the opposite RF electrodes. Application of additional *dc* voltage results in increase of the width of the cathode sheath. The sharp boundary of the $n_e(\xi,t)$ profile in it now oscillates far from the cathode surface, and the *dc* current arises, which is equal to the ion current from the plasma to the cathode [50].

The width δ of the transition region II (Fig. 3.11), where the ion flux transformation occurs, can be estimated as

$$\delta = D_{eff}/V(\delta). \tag{3.97}$$

Expanding Eq. (3.91) at $z \to 0$, we obtain

$$\delta = \left[\frac{9\pi^2 D_{eff}^2 j_0}{8 b_i^2 \omega e n_s^3} \right]^{1/5}, \tag{3.98}$$

where n_s is the boundary plasma density at $\xi = L_{sh}$. The approximate zero boundary condition for the ambipolar diffusion equation (4.7) can be imposed at $\xi = (L_{sh} - \delta)$.

At lower pressures, the fluid approach fails to describe the ion motion, and it is necessary to account for the ion inertia, as in Section 3.3, Eq. (3.15). Using the approximate ion momentum equation, and neglecting the ion pressure, we obtain

$$m_i V_i \frac{dV_i}{d\xi} = e\langle E\rangle - \mu_{iN} \nu_{iN} V_i, \tag{3.99}$$

where the time-averaged electric field $\langle E\rangle$ consists of a diffusion term which is proportional to the density gradient, and of a convective term $\langle E\rangle_1$, Eq. (3.91). Neglecting the RF diffusion, we obtain

$$\frac{dV_i^2}{d\xi} = \frac{2V_i^2}{c_s^2 - V_i^2}\left(\mu_{iN} V_i^2 / (m_i \lambda_{iN})\right) - e\langle E\rangle_1 / m_i. \tag{3.100}$$

The difference in the parentheses in the r.h.s. of Eq. (3.100) is equal to zero at $\xi=(L_{sh}-\delta)$, Eq. (3.97). In order to describe the profile $V_i(\xi)$ which monotonously increases towards the wall, this point has to coincide with the sonic point. In other words, if δ, Eq. (3.97), is small with respect to the ion mean free path λ_{iN}, and the fluid description fails, the Bohm criterion, Eqs. (3.18), (3.19) can be imposed at the plasma-sheath boundary

Ionization processes in RF sheaths. We shall briefly sketch here the situation in the self-sustained discharges when plasma itself is maintained by the RF current. Since the width of the RF sheaths is considerable, the ionization processes here, generally speaking, play a far more important role, than in the *dc* plasmas. It means that frequently there are the situations when the variation of the electron and ion fluxes in the sheaths cannot be neglected. Strictly speaking, the flux conservation scenario described above takes place only in non-self-sustained discharges when the external ionization is more or less uniformly distributed over the plasma volume, and the ionization processes in the sheaths are negligible. Such a situation corresponds also to self-sustained discharges at low pressure, when due to nonlocality of the electron distribution function ([2]; see also Section 2.9) the ionization processes at the plasma periphery are suppressed.

On the other hand, the second specific feature of the RF plasmas is that in the zero (in $(\omega\tau^{(l)}_M)^{-1}$ or in ω_{pi}/ω) approximation, in the absence of electrons (in the space charge phase) the total current is transported not by ions, but in the form of the displacement current. The role of the ionization processes in the sheaths in some cases can be considerably reduced with respect, for example, to the *dc* discharges. The secondary ion-electron emission from the electrode (or, more generally speaking, the so-called γ-processes [36],

[37]) are now unnecessary for discharge maintenance: it can exist in so-called α-form, when the γ-processes, which generate electron flux at the surface during the cathode phase, are negligible. And even in the γ-form, when 'sheath breakdown' occurs, the ionization rate is not now obliged to transform the total current from electron form into the ion form, as it was in the case of the cathode fall of the *dc* discharge (see the previous Section). Its only function remains to maintain the stationary ion density profile.

The simplest situation corresponds to the α-form of self-sustained discharge at high and medium pressure, when the electron distribution function (and the ionization rate) are determined by the local value of oscillatory RF electric field intensity. The uniform plasma density in the discharge bulk n_0 is determined by balancing of the ionization and volume recombination processes. Due to recombination at the vessel walls, the plasma flux is directed outward, and the plasma density decreases at the plasma periphery and in the sheath.

Since the RF current in plasma and in the sheath during the plasma phase is transported in the form of electron conductivity, the oscillatory electric field and ionization rate increase towards the wall. Hence, a sort of negative biasing arises: increase of the RF field at the plasma periphery results in increase of the ionization rate which tends to restore the equilibrium plasma density n_0. The dependence of the ionization rate on the field strength is rather steep since it is determined usually by the exponential distribution tail. As a result, this density stabilization mechanism is very effective, and typical values of the ion density in the sheath are of the order of n_0, see [50] and [51]. More accurate expressions for the plasma density at the plasma-sheath boundary are presented in [51]. The ion flux $\Gamma_{i\xi}$ at the surface can be estimated as follows. It is of the order of

$$\Gamma_{i\xi} \sim b_i(\langle E \rangle_1) n_0 \langle E \rangle_1, \tag{3.101}$$

where the field $\langle E \rangle_1$ averaged over the plasma phase corresponds to the wall vicinity. Since duration of the plasma phase here coincides with the RF period, and current is transported in form of the displacement current, we have

$$\langle E \rangle_1 \sim j_0/(\varepsilon_0 \omega). \tag{3.102}$$

Eqs. (3.101) and (3.102) correspond to an ionization rate in the sheath which is produced by the plasma electrons (they are sometimes referred to as the α-electrons) during the part of the RF period when they are present in the sheath.

The flux of the secondary γ-electrons which are ejected by the surface is small: of the order of $\gamma\Gamma_{i\xi}$. But in contrast to the α-electrons, the γ-electrons move in the sheath mainly during the space charge phase when the electric field strength is very high, Eq. (3.102). This electric field (and the additional ionization rate produced by the γ-electrons) increases with current, and at some critical current the transition to the γ-form occurs ([50] and [51]). In this regime the main ionization in the sheath is determined only by the γ-electrons. In some sense this transition corresponds to the sheath 'breakdown' - the ionization processes in the sheath become practically independent from the bulk plasma. It is to be noted that this regime differs substantially from the case of *dc* cathode (Section 3.7).

In the *dc* case the total current is mainly transported at the plasma-sheath boundary by the multiplicated γ-electrons, and at the surface by ions. In contrast, both in the α- and γ-regimes, the total current in plasma (and in sheath during the plasma phase) is transported mainly by the plasma electrons, and the contribution of ions and γ-electrons is small [51]. In Figs.3.12 and 3.13 the typical calculated ion density profiles in the RF sheath for the α- and γ-regimes [51] are presented.

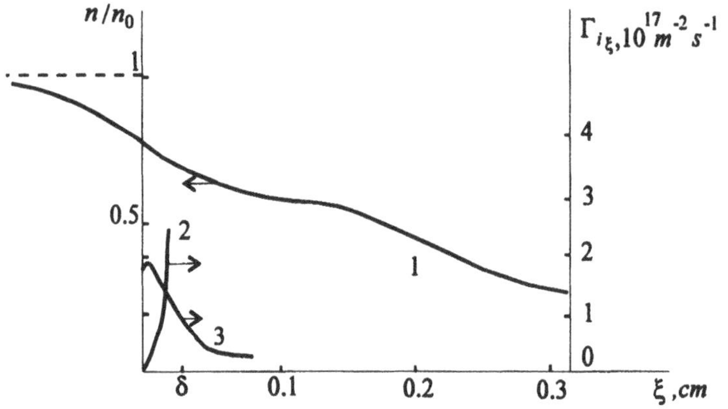

Figure 3.12. The ion density profile (1) in the sheath in N_2 at p=200 Pa, $\omega/2\pi$=13.56 MHz; j_0=150 A/m^2 . Curves 2, 3 - the convective and diffusive terms in Eq. (3.92) for Γ_i; the value of δ is calculated according to Eq. (3.98).

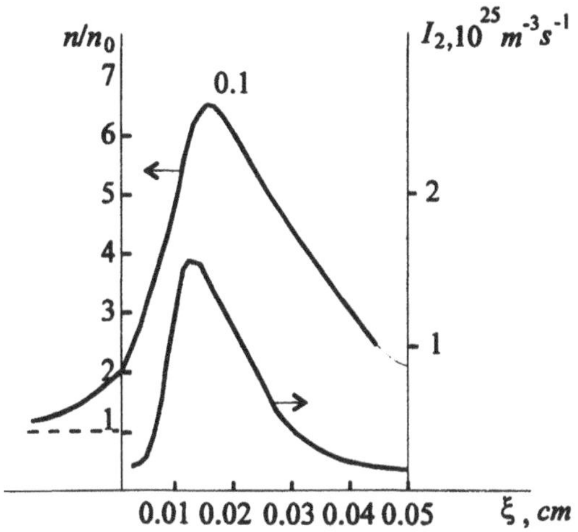

Figure 3.13 The ion density profile in the sheath in N_2 at 200 *Pa*; $\omega/2\pi$=13.56 MHz; γ=0.01; j_0=2.4 kA/m^2. 1-numerical solution for density; 2 - ionization density by the γ - electrons I_2.

3.9 THE SHEATH FORMATION IN PLASMAS WITH MORE COMPLICATED KINETICS. SHEATH IN ELECTRONEGATIVE PLASMAS

As we have seen in the preceding Sections, the sheath processes are rather complicated and depend crucially on subtle processes of the space charge formation. In numerous situations the fluid approach, which was used traditionally for analysis of the sheath dynamics, is sometimes oversimplified, especially in the case of the collisionless sheath. This fact is not surprising, since in this situation we are trying to apply the fluid equations, which are valid for description of large-scale phenomena, on short spatial scale of the Debye radius. Up to now the rigorous kinetic analysis of the transition from plasma into the collisionless sheath is practically absent. The existing examples (e.g., see Section 10.1 of the book [3]) demonstrate that this problem is complicated enough. It remains unclear even, whether an arbitrary system of charged particles without collisions tends to the quasineutral plasma state, and which scenaria of this transition are possible. In order to illustrate the situation, we consider in this Section the plane 1D collisionless problem of currentless plasma consisting of one species of positive ions with mass M and temperature T, and of two species of trapped negative particles of masses m_e and m_2 with temperatures T_e and

T_2. In [14] such an approach was applied to the analysis of the Bohm criterion in electronegative plasmas.

On one hand, in the gas discharges, for example, the Maxwell distribution is rather exotic. The electron distribution function is determined by various collisions, by complex interaction with fields, and with electrodes, and thus often represents a superposition of populations of cold and fast electrons. The most widely known example of this kind corresponds to the Faraday dark space of the *dc* glow discharge. Such phenomena occur also in RF and UHF discharges of various kinds. Such a situation can often be approximated in terms of bi-Maxwellian distribution function. In this case the two types of the negative particles correspond to the cold and hot electron populations (m_e=m_2; T_e>>T_2, T).

One of the reasons for such a phenomenon arises in low-pressure discharges, if the electron heating occurs at the plasma periphery. The electrons in the discharge center are trapped by the plasma ambipolar electric field and cannot reach the heating region. As a result, the majority of the electrons are cold. Only a small portion of electrons of high energy can reach the energy source, forming a distribution tail greatly enriched by energetic particles. This strongly non-Maxwellian electron distribution function can also be roughly described by such a two-temperature approximation [52].

We shall consider the simplest problem of this kind which corresponds to the positive column of the gas discharge. Pairs of charged particles appear due to ionization of neutrals by the fast electrons, and recombine at the vessel walls, $x=\pm L$. The ionization rate is approximated as $n_e\nu^{(ion)}(T_e)$.

Since we consider the negative particles as trapped by the ambipolar field, they can be considered to be in thermodynamic equilibrium, and their densities obey Boltzmann's law

$$\begin{aligned} n_e &= n_e^0 \exp(e\varphi / T_e); \\ n_2 &= n_2^0 \exp(e\varphi / T_2), \end{aligned} \tag{3.103}$$

where $\varphi \leq 0$ is the ambipolar potential. The density of the positive ions in an observation point x equals the sum of partial contributions of the ions which were born everywhere between x'=0 and x'=x

$$p(x) = \int_0^x \frac{n_e(x')\nu^{(ion)}(T_e)dx'}{\sqrt{2e(\varphi(x') - \varphi(x))/M}}. \tag{3.104}$$

Since the positive ions start their motion after ionization with negligible energies, which are of the order of the room temperature of neutrals T, we have neglected their initial velocity.

In pure plasma, which contains only the electrons and positive ions, $n_2=0$, the quasineutrality condition $n_e(x)=p(x)$ determines the ambipolar potential $\varphi(x)$:

$$\exp(e\varphi(x)/T_e)=\int_0^x\frac{\nu^{(ion)}\exp(e\varphi(x')/T_e)dx'}{\sqrt{2e(\varphi(x')-\varphi(x)/M}}. \quad (3.105)$$

This integral equation can easily be solved by the Abel transform [8]. Multiplying the right- and left-hand sides of Eq. (3.105) by $(\varphi_0-\varphi(x))^{-1/2}$, integrating it over φ from zero up to φ_0, and changing the order of integration in the resulting double integral in the r.h.s., we obtain

$$\frac{dx}{d\varphi}=-E^{-1}(x)=\sqrt{\frac{2e}{M}}\frac{\exp(-e\varphi/T_e)}{\pi\nu^{(ion)}}\frac{d}{d\varphi}\int_0^{\varphi}\frac{\exp(e\varphi'/T_e)d\varphi'}{\sqrt{\varphi'-\varphi}}. \quad (3.106)$$

The function

$$\Omega(\eta)=\Omega(-e\varphi/T_e)=\int_0^{\Phi}\frac{\exp(-\Phi')d\Phi'}{\sqrt{\Phi-\Phi'}}=2\int_0^{\sqrt{\Phi}}\exp(t^2-\Phi)dt \quad (3.107)$$

increases at small η, and decreases at $\eta\to\infty$. It is at a maximum at $\varphi=\varphi_{sh}=-0.855T_e/e$, and equals to $\Omega(|e\varphi_{sh}/T_e|)=1.08$. According to Eq. (3.106), the electric field in the quasineutral approximation becomes singular in this point. Hence, it can be identified as the transition point from the plasma to the sheath with the ion space charge and strong electric field. Neglecting the Debye radius with respect to the discharge spacing $2L$, it is necessary to impose the boundary condition

$$\pi\nu^{(ion)}\sqrt{\frac{M}{2T_e}}L=\int_0^{0.855}\exp(\Phi)[\Phi^{-1/2}-\sqrt{\pi}erf(\sqrt{\Phi})]d\Phi. \quad (3.108)$$

With accuracy up to the numerical factor of the order of unity, it means that in a stationary state the average lifetime of the charged particles $(L\sqrt{M/T_e})$ is equal to the inverse ionization frequency. The average plasma mass velocity (it coincides with the velocity of positive ions) at the plasma-sheath boundary is

$$u_i=0.344e^{0.855}(2T_e/M)^{1/2}=1.14(T_e/M)^{1/2}. \quad (3.109)$$

This value practically coincides with the Bohm criterion Eq. (3.18).

In the presence of the second species of the negative trapped particles, the equation for the plasma quasineutral potential is analogous to Eq. (3.106):

$$\frac{dx}{d\varphi} = \frac{\exp(-e\varphi / T_e)}{\pi \nu^{(ion)}} \sqrt{\frac{2T_e}{M}} \frac{d}{d\varphi} [\Omega(|e\varphi / T_e|) + \\ +(n_2^0 / n_e^0)\sqrt{T_2 / T_e}\, \Omega(|e\varphi / T_2|)]. \qquad (3.110)$$

In a rather wide range of the parameters T_e/T_2 and n_2^0/n_e^0 the expression in the brackets in the Eq. (3.110), which is proportional to the flux of the positive ions in the quasineutral approximation, can have two maxima, Fig. 3.14. At $T_e/T_2 >> 1$, for example, these maxima occur at $|\varphi| = |\varphi^1_{sh}| \cong 0.855 T_2/e$ and at $|\varphi| = |\varphi^2_{sh}| \cong 0.855 T_e/e$. For $n_2^0/n_e^0 > 1$ at the second maximum the flux of the positive ions Γ_p is lower, than at the first one. The quasineutral solution corresponds to monotone increase both of Γ_p and η from zero at the discharge midplane. It means that in this case the whole discharge volume up to the wall is occupied by the quasineutral plasma with low electric field. The characteristic scale of the potential profile in the plasma is determined by the temperature of the cold trapped particles T_2.

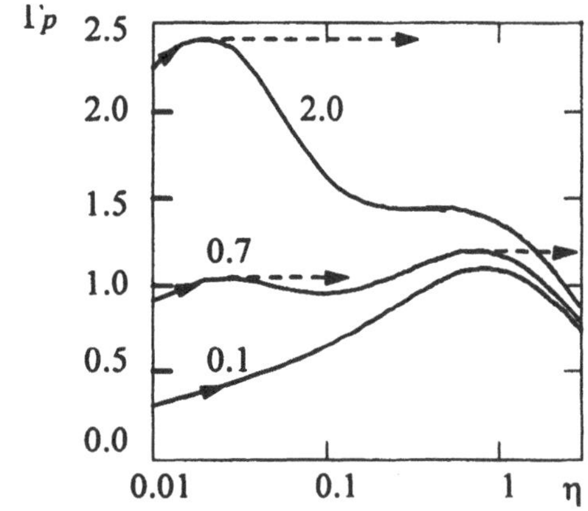

Figure 3.14. Dependence of the positive ions flux $\Gamma_p(\eta) = \Gamma_p(-e\varphi/T_e) = \Omega(\eta) + n_2^0/n_e^0 \Omega(\eta T_e/T_2)$ from potential η in collisionless plasma with two species of the negative trapped particles at $T_e = 10T_2$. The numbers at the curves correspond to $n_2^0/n_e^0 = 0.1; 0.7; 2$.

Depending on the plasma parameters and the type of boundary conditions on the partial fluxes, the potential drop in the sheath, which is adjacent to the wall, is also determined by T_2, or by T_e.

If the fraction of the cold trapped particles is not too high, and the second maximum of $\Gamma_p(\eta)$, which is connected with the energetic trapped particles, corresponds to a higher value of Γ_p than the first maximum, the situation is essentially different. Density profiles consisting of two quasineutral plasma sections are possible. The central region is occupied by cold plasma with the potential scale T_2/e. On the other hand, the monotone quasineutral rise both of Γ_p and η at values of η above the first maximum is impossible, and the quasineutrality is violated at this point. Since the spatial scale of a non-quasineutral solution is small, the flux Γ_p is conserved in it, and transition to the second quasineutral profile portion is possible.

The characteristic scale of the electric field in this peripheral plasma is determined by the high electron temperature T_e. The potential drops in the double sheath, which separates the cold and hot plasmas, and in the sheath adjacent to the wall, are also of the order of electron temperature T_e, see [52] and [55].

REFERENCES

1. H. S. Carslaw., J. C. Jaeger, *Conduction of Heat in Solids* (Oxford: Claderon, 1959): 488.
2. L. D. Tsendin, *Plasma Sources, Sci. & Technology* **4**: 200-211 (1995); V. A. Godyak, V. I. Kolobov, *IEEE Trans.* **PS-23**: 503-531 (1995).
3. *Thermoionic Convertors and Low-Temperature Plasma,* Editors: K. Hansen, B. Ya. Moyzhes, G. Ye. Pikus (Washington: Tech. Inf. Center/US Dept. Energy, 1978): 484; F. G. Baksht, V. G. Yurjev, *Sov. Phys. Techn. Phys.* **21**: 531-548 (1976); **24**: 535-557 (1979).
4. D. Bohm, *The Characteristics of Electrical Discharges in Magnetic Field,* Editors: A. Guthrie, R. Wakerling (New York: McGraw Hill, 1949): 346. Chaps. 1, 2, 9.
5. P. Morse, H. Feshbach, *Methods of Theoretical Physics,* v.1 (New York: McGraw Hill, 1953): 997.
6. R. Franklin, *Plasma Phenomena in Gas Discharges* (Oxford: Clarendon, 1976): 250.
7. F. G. Baksht, B. Ya. Moyzhes, V. A. Nemchinsky, *Sov. Phys. Techn. Phys.*, **14**: 417-422 (1969).
8. E. R. Harrison, W. B. Thompson, *Proc. Phys. Soc.* **72(4)**: 145-152 (1959).

9. B. M. Smirnov, *Physics of Weakly Ionized Gases* (Moscow: Mir, 1981): 430.
10. I. Langmuir, *Phys. Rev.* **23**: 954-980 (1929).
11. K.-U. Riemann, *J. Phys.D* **24**: 493-518 (1991).
12. P. C. Stangeby, in *Physics of Plasma-Wall Interaction in Controlled Fusion,* D. E. Post, R. Berish Eds., (New York: Plenum, 1986): 41-97.
13. G. A. Emmert, R. M. Wieland, T. Mense, J. N. Davidson, *Phys. Fluids* **23**: 803-812 (1980).
14. R. L. F. Boyd, J. B. Thompson, *Proc. Roy. Soc.* **A252**: 102-119 (1959).
15. N. S. J. Braithwaite, J. E. Allen, *J. Phys. D: Appl. Phys.* **21**: 1733-1737 (1988).
16. J. W. Braley, H. Amemya, *Jap. J. Appl. Phys.* **33**: 3578-3585 (1994).
17. K.-U. Riemann, *IEEE Trans.* **PS-23**: 709-706 (1995).
18. I. I. Litvinov, *J. Applied Math. Theoretical Phys.* **1**: 52-57 (1977) (in Russian).
19. V. A. Rozhansky, L. D. Tsendin, *Sov. J. Plasma Phys.* **4**: 218-221 (1978).
20. U. Daybelge, B. Bein, *Phys. Fluids* **24**: 1190-1194 (1981).
21. G. I. Budker, S. T. Belyaev in *Plasma Physics and the Problem of Controlled Thermonuclear Reactions* v.2 (Oxford: Pergamon, 1961): 431-450.
22. V. Sivykhin in *Rewievs of Plasma Physics*, **vol.4**, Editor: M.A. Leotovich (New York: Consultants Bureau, 1966): 93-241.
23. F. G. Baksht, *Sov. Phys. Techn. Phys.* 23: 1014-1019 (1978); **24**: 253-254 (1979).
24. R. Chodura, *Phys. Fluids* **25**: 1628-1639 (1982).
25. K.-U. Riemann, *Phys. Plasmas* **1**: 552-558 (1994).
26. D. L. Holland, B. D. Fried, G. L. Morales, *Phys. Fluids* **B5**: 1723-1737 (1993).
27. R. H. Cohen, D. D. Ryutov, *Phys. Plasmas* **2**: 2011-2019 (1995); D. D. Ryutov, *Contrib. Plasma Phys.* **2/3**: 207-219 (1996).
28. K. Theihaber, K. Birdsall, *Phys. Fluids* **B1**: 2244-2259 (1989); B1: 2260-2272 (1989).
29. H. A. Claaβen, H. Gerhauser, *Contrib. Plasma Phys.* **2/3**: 381-385 (1996).
30. C. H. Su, S. H. Lam, *Phys. Fluids* **6**: 1479-1491 (1963).
31. I. M. Cohen, *Phys. Fluids* **6**: 1492-1499 (1963).
32. F. G. Baksht, G. A. Djuzhev, S. M. Shkolnik, *Sov. Phys. Techn. Phys.* **22**: 940-943; 944-950 (1977).
33. L. D. Tsendin, *Sov. Phys. Solid State* 7: 864-865 (1965).
34. F. G. Baksht, B. Ya. Moyzhes, V. A. Nemchinsky *Sov. Phys. Techn. Phys.* **12**: 520-530 (1967).
35. U. Korchagen, M. Zethoff, *Plasma Sources Sci. Technol.* **4**: 541-546.

36. A. von Engel, M. Steenbeck, *Elektrische Gasentlodungen* (Berlin: Springer, 1932): **B1**, 248; **B2**, 352.
37. Yu. P. Raizer, *Gas Discharge Physics* (New York: Springer, 1991) 449.
38. A. L. Ward, *Phys. Rev.* **112**: 1852-1857 (1958).
39. V. I. Kolobov, L. D. Tsendin, *Sov. Phys. Techn.Phys.* **34**: 1239-1243 (1989).
40. V. I. Kolobov, L. D. Tsendin, *Phys. Rev.* **A46**: 7837-7852 (1992).
41. T. J. Sommerer, W. N. G. Hitchon, J. Lawler, *Phys. Rev.* **A39**, 6356-6366 (1989).
42. S. D. Vagner, O. Yu. Kotelnikova, V. P. Pyadin, *Sov. Phys. Techn. Phys.* **34**: 854-857 (1989).
43. B. N. Kliarfeld, N. A. Neretina, *Sov. Phys. Techn. Phys.*, **3**: 271-288 (1958); 4: 13-20 (1959).
44. Yu. S. Akishev, A. P. Napartovich, P. N. Peretjatko, N. I. Trushkin, *Teplofizika Vysokikh Temperatur* **18**: 831-834 (1980).
45. S. V. Pashkin, *Teplofizika Vysokikh Temperatur* **14**: 563-569 (1976); **19**: 647-650 (1981).
46. L. G. H. Huxley, R. W. Crompton, *The Diffusion and Drift of Electrons in Gases* (New York: Wiley, 1974): 669.
47. A. von Engel, *Phyl. Mag.* **32**, N214: 417-428 (1941).
48. L. D. Tsendin, *Sov. Phys. Techn. Phys.* **31**: 169-175 (1986).
49. H. al Havat, Yu. B. Golubovsky, L. D. Tsendin, *Sov. Phys. Techn. Phys.* **31**: 169-175 (1986).
50. Yu. P. Raizer, M. N. Shneider, N.A. Yatsenko, *Capacitive Discharges* (Bace Raton: CRS Press, 1995): 292.
51. A. S. Smirnov, L. D. Tsendin, *IEEE Trans.* **PS-19**: 130-140 (1995).
52. L. L. Beilinson, I. D. Kaganovich, L. D. Tsendin, *Proc. Russ. Conf. Physics Low-Temp. Plasma,* L. A. Luizova, Ed., (Petrozavodsk: University Press, 1995), p. 313-314.
53. I. D. Kouznetsov, A. J. Lichtenberg, M. A. Lieberman, *Plasma Sources, Sci.&Technol.*, **5**: 662-669 (1996).
54. M. A. Lieberman, A. J. Lichtenberg, *Principles of Plasma Discharges and Materials Processing* (New York: Wiley, 1994):572.
55. V. A. Godyak, N. Sternberg, *IEEE Trans.* **PS-18**: 159-168 (1990).

Chapter 4

Transport Processes in the Absence of Magnetic Field

4.1 AMBIPOLAR DIFFUSION IN PURE PARTIALLY IONIZED PLASMA

Let us consider the process in a pure unmagnetized plasma, with constant and uniform partial temperatures, in the reference frame bounded to the neutral gas u_N=0. Substituting the quasineutrality condition Eq. (2.121) n_e=n_i=n into Eqs. (2.7), (2.41), we obtain the basic equations which describe the evolution of density and potential:

$$\frac{\partial n}{\partial t}+\vec{\nabla}\cdot\vec{\Gamma}_e=I-S\ ,$$
$$\frac{\partial n}{\partial t}+\vec{\nabla}\cdot\vec{\Gamma}_i=I-S\ ; \tag{4.1}$$

$$\vec{\Gamma}_e=-(D_e\vec{\nabla}n-b_e n\vec{\nabla}\varphi),$$
$$\vec{\Gamma}_i=-(D_i\vec{\nabla}n+b_i n\vec{\nabla}\varphi)\ . \tag{4.2}$$

Taking the difference between the two Eqs. (4.1), we have

$$\vec{\nabla}\cdot\vec{j}=\vec{\nabla}\cdot e(\vec{\Gamma}_i-\vec{\Gamma}_e)=0\ . \tag{4.3}$$

The last condition means that in quasineutral plasmas space charge does not accumulate, and current is to be vortex. After substituting the expressions for particle fluxes from Eq. (4.2), the Eq. (4.3) yields

$$\vec{\nabla}\cdot[n(b_e+b_i)\vec{\nabla}\varphi]=\vec{\nabla}\cdot[(D_e-D_i)\vec{\nabla}n]\,. \tag{4.4}$$

Potential which satisfies the linear Eq. (4.4) can be sought in the form $\varphi=\varphi_d+\varphi_c$. Here φ_d is the diffusive part of the potential corresponding to the absence of the current in plasma

$$\vec{\nabla}\varphi_d=\frac{D_e-D_i}{b_e-b_i}\vec{\nabla}n/n\,. \tag{4.5}$$

The part φ_c satisfies Eq. (4.4) with the r.h.s. at zero,

$$\vec{\nabla}\cdot[n(b_e+b_i)\vec{\nabla}\varphi_c]=0\,, \tag{4.6}$$

and corresponds to the external current through plasma. When there is no such current $\varphi_c=0$, and the fluxes of both species are equal everywhere: $\vec{\Gamma}_e=\vec{\Gamma}_i$. Since, according to Eq. (2.89), the mobilities b_e and b_i depend on density in a similar way, even in partially ionized plasma with density dependent mobilities, the part φ_c does not contribute to the Eq. (4.1) because $\vec{\nabla}\cdot[n(b_e+b_i)\vec{\nabla}\varphi_c]=\vec{\nabla}\cdot(nb_e\vec{\nabla}\varphi_c)=\vec{\nabla}\cdot(nb_i\vec{\nabla}\varphi_c)=0$.

Substituting the value of $\vec{\nabla}\varphi_d$ into any of the Eqs. (4.1), we obtain [1]

$$\frac{\partial n}{\partial t}-\vec{\nabla}\cdot(D_a\vec{\nabla}n)=I-S \tag{4.7}$$

- the well-known ambipolar diffusion equation with the coefficient of the ambipolar diffusion

$$D_a=(D_eb_i+D_ib_e)/(b_e+b_i). \tag{4.8}$$

The ambipolar diffusion coefficient can usually be simplified by employing Eq. (2.89). With accuracy $\sim(m_e/m_i)^{1/2}$ we have

$$D_a=(T_e+T_i)/\mu_{iN}\nu_{iN}. \tag{4.9}$$

In weakly ionized plasma

$$D_a=(1+T_e/T_i)D_i. \qquad (4.10)$$

In spite of the extremely important role of the electric field which is characterized by the factor T_e/T_i in Eqs. (4.9) and (4.10) the field does not appear explicitly in the ambipolar diffusion Eq. (4.7) which determines the plasma density. Hence the problems of the determining density and potential are split. To resolve Eq. (4.7) for density the boundary condition for potential is unnecessary, and thus analysis of the sheath is extraneous too. The density evolution is determined simply by the linear Eq. (4.7) with the zero boundary condition at the surface. Effects connected with finite value of an ion mean free path (Section 3.2), which result in deviations from zero boundary condition for the plasma density, are discussed in detail in [8], [14], [15].

The potential profile can be calculated from Eq. (4.5). For $m_e\nu_{ei} \ll \mu_{iN}\nu_{iN}$

$$\varphi_d = \frac{D_e - D_i}{b_e + b_i}\ln n + const. \qquad (4.11)$$

In weakly ionized plasma $D_e \gg D_i$, so

$$\varphi_d = \frac{T_e}{e}\ln n + const \qquad (4.12)$$

corresponds to the Boltzmann distribution for electrons, the example of polarization is shown in Fig. 4.1*a*.

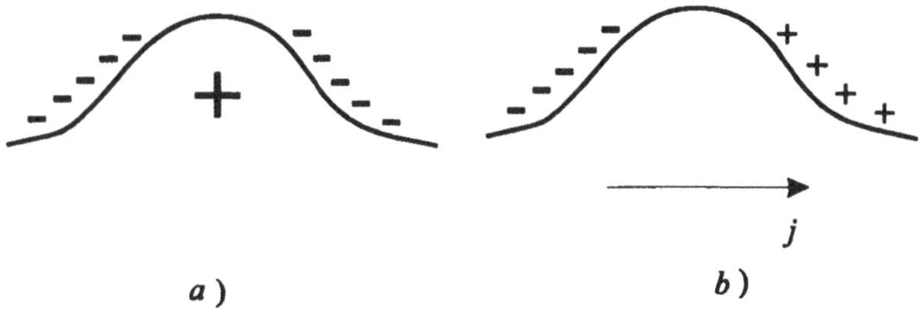

Figure 4.1. Polarization of the plasma density perturbation in the process of ambipolar diffusion: *a*) j=0, $\varphi=\varphi_d$; *b*) additional polarization φ_c caused by an external current j.

In other words in this case the electric field almost completely decelerates the more mobile particles (electrons) and accelerates ions, therefore evolution is determined by the less mobile component. In the presence of external current the additional polarization field $\vec{\nabla}\varphi_c$ according to Eq. (4.6) has inverse density dependence, see Fig. 4.2*b*:

$$\vec{\nabla}\varphi_c = \vec{j} \,/\, ne(b_e + b_i)\,. \tag{4.13}$$

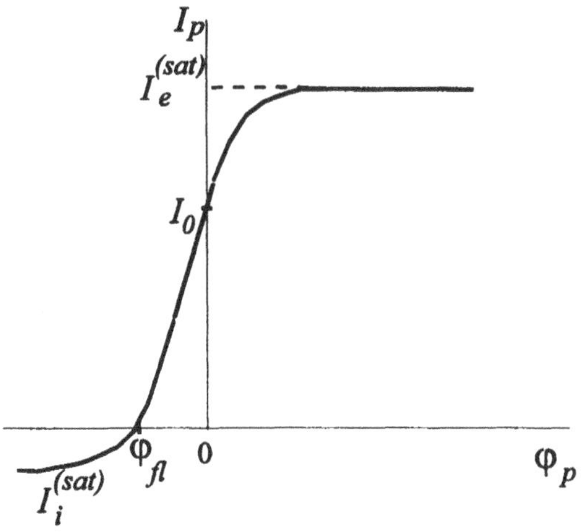

Figure 4.2. Typical current-voltage characteristic of a diffusive probe.

In the nonuniform neutral gas the problem of plasma evolution in general is not reduced to the equation (4.7) of ambipolar diffusion. Indeed, since b_e and b_i can now have different spatial dependencies, the component φ_c which is determined by the Eq. (4.6) is canceled after substituting in Eqs. (4.1), (4.2). Similar situation when the mobility of electrons depends on electric field will be discussed in Chapter 9.

The term ‘ambipolar’ diffusion is used by some authors as a synonym of diffusion of quasineutral plasma. In contrast, we shall use this term in a more narrow sense, defining as ambipolar the diffusion which is described by the equation similar to Eq. (4.7) when the diffusion is determined by the slow plasma component. Below we present the simplest examples of its solution.

Diffusive decay. This type of diffusion takes place in the absence of sources and sinks in Eq. (4.7), e.g. after switching off the external ionization source or external current in gas discharge. For time scales larger than $(\delta_{eN}\nu_{eN})^{-1}$ the plasma becomes almost isothermal. If the recombination in the volume is negligible, the 1D solution of Eq. (4.7) in a cylindrical geometry with the boundary condition $n(r=a)=0$ can be obtained by the standard method of separation of variables. It has the form

$$n(r,t)=\sum_{k=1}^{\infty}A_k \exp(-t/\tau_k)J_0(\zeta_k r/a), \tag{4.14}$$

where $\tau_k=a^2/(\zeta_k^2 D_a)$, J_0 is the zero-order Bessel function, ζ_k are the roots of the Bessel function, $\zeta_1=2.405$. Since the sequence ζ_k increases quickly with k, modes with $k>1$ decay faster than the fundamental mode $k=1$ which becomes the dominant mode at $t>\tau_1$. The solution in a slab geometry ($n(0,L)=0$) has the form

$$n(x,t)=\sum_{k=1}^{\infty}A_k \exp(-t/\tau_k)\sin(\pi kx/L), \tag{4.15}$$

where $\tau_k=L^2/(k^2\pi^2 D_a)$.

For a cylinder of a finite length L:

$$n(r,t)=\sum_{j,k=1}^{\infty}A_{j,k} \exp(-t/\tau_{jk})J_0(\zeta_k r/a)\sin(\pi jz/L), \tag{4.16}$$

where

$$1/\tau_{jk}=D_a(\zeta_k^2/a^2+\pi^2 j^2/L^2). \tag{4.17}$$

One-dimensional steady state solution with ionization. An important example of such a diffusion problem is the positive column of glow discharge in a long cylindrical tube or radius a. The density profile is determined by

$$-\frac{1}{r}\frac{d}{dr}(rD_a\frac{dn}{dr})=I-S=n\nu^{(ion)}-K^{(rec)}n^2. \tag{4.18}$$

Here the average ionization rate $\nu^{(ion)}=K^{(ion)}N$ (N is the neutral density) is integrated over the electron distribution function frequency of ionizing collisions. It is a function of the reduced axial electric field E_z/N in the discharge; $K^{(rec)}$ is the recombination rate coefficient. In the case $K^{(rec)}n(0)<<\tau_1^{-1}=D_a\zeta_1^2/a^2$ the volume recombination is negligible. Then after imposing the boundary condition $n(r=a)=0$, the Eq. (4.18) corresponds to the Sturm-Liouville problem of determining the eigenfunction $n(r)$ and the eigenvalue $\nu^{(ion)}$. The solution of the Bessel equation (4.18) which is positive everywhere is given by

$$n(r) = n(0)J_0(\zeta_1 r / a);$$
$$\nu^{(ion)} = \zeta_1^2 D_a / a^2. \tag{4.19}$$

According to Eq. (4.19) the axial electric field is determined only by the tube radius a, and is independent from the discharge current, while the plasma density varies proportionally to the discharge current. Since the radial current is equal to zero, the radial potential profile is determined by the Eq. (4.12):

$$\varphi(r) = (T_e / e)\ln J_0(\zeta_1 r / a) + const. \tag{4.20}$$

The potential strongly decreases towards the wall. Its drop over the plasma between the center and the sheath edge is $(T_e/e)\ln(n(0)/n_s) >> T_e/e$. To obtain the potential drop in the sheath one has to know the ion flux to the wall. The latter according to Eq. (4.19) is

$$\Gamma_{ir} = -D_a \frac{dn}{dr}(r = a) = D_a n(0)\zeta_1 J_1(\zeta_1) / a. \tag{4.21}$$

In the case of the collisionless sheath, for example, the potential drop over plasma is equal to $(T_e/e)\ln(a/\lambda_{iN})$, and the potential drop in the sheath is given by Eq. (3.25).

When the volume recombination dominates over the diffusive losses $K^{(rec)}n(0) >> \tau_1^{-1}$ the profile with plateau $K^{(rec)}n = \nu^{(ion)}$ is formed. Only near the wall at distances of the order of the diffusive length $l_D = (D_a/\nu^{(ion)})^{1/2}$ does the density fall to zero.

Probe in the diffusive regime. A probe (a small electrode, inserted into plasma and biased negatively or positively with respect to it) is one of the most useful tools for plasma diagnostics. Its current-voltage characteristic contains information on the plasma density and temperatures of the species. When the sheath width and species mean free paths are both less than the probe dimensions, the probe is said to operate in the diffusive regime (the opposite case of large mean free path is usually referred to as the Langmuir probe [2]). In the pure plasma the density profile in the probe vicinity is determined by Eq. (4.7) of ambipolar diffusion. After neglecting ionization and recombination processes in the small region near the probe where the density is perturbed, Eq. (4.7) is reduced to the Laplace equation

$$\Delta n = 0. \tag{4.22}$$

It has to be solved with the boundary conditions (subscript s corresponds to the probe surface)

$$n|_s = 0\,; \qquad n(\vec{r} \to \infty) = n_0\,.$$

Approximating the probe by the ellipsoid of rotation with semiaxes (a, a, b), one can solve the Laplace equation in the elliptic coordinates

$$\frac{x^2}{a^2+\xi} + \frac{y^2}{a^2+\xi} + \frac{z^2}{b^2+\xi} = 1\,.$$

Here ξ is the elliptic coordinate, $\xi=0$ corresponds to the probe surface. The solution of Eq. (4.2) is (the equivalent electrostatic problem is solved in [3])

$$n/n_0 = \begin{cases} 1 - \dfrac{Arth\sqrt{(b^2-a^2)/(b^2+\xi)}}{Arth\sqrt{1-a^2/b^2}}\,; & b>a\,; \\ 1 - \dfrac{arctg\sqrt{(b^2-a^2)/(b^2+\xi)}}{arctg\sqrt{1-a^2/b^2}}\,; & b<a \end{cases} \,. \tag{4.23}$$

For a spherical probe

$$n/n_0 = 1 - a/r. \tag{4.24}$$

For large negative potential electron current to the probe is negligible. Electrons are trapped in the plasma by electric field and have a Boltzmann distribution, and the ion flux, according to Eq. (4.2), is determined by the density profile: $\vec{\Gamma}_i = -D_a \vec{\nabla} n$. The net ion current to the probe is independent of its potential and is known as the ion saturation current. It can be obtained by integrating the normal component of the ion current, calculated in accordance with Eq. (4.23) at $\xi=0$, over the probe surface. The result is proportional to the electric capacity of the probe C [3] :

$$I_i^{(sat)} = 4\pi e D_a n_0 C\,, \tag{4.25}$$

where

$$C=\begin{cases}\dfrac{\sqrt{b^2-a^2}}{Arch(b/a)}; & b>a\\ \dfrac{\sqrt{a^2-b^2}}{\arccos(b/a)}; & a>b\\ a; & a=b.\end{cases} \tag{4.26}$$

For a large negative probe potential the ions are trapped in plasma, the potential corresponds to the Boltzmann distribution for ions, and electron saturation current is thus given by

$$I_e^{(sat)}=4\pi eD_e(1+T_i/T_e)n_0C. \tag{4.27}$$

Consider the intermediate part of I-V characteristics for spherical probes surrounded by a thin collisional sheath $L_{sh}<<a$ (the sheath width L_{sh} is determined by Eqs. (3.59), (3.60)). Since $I_e^{(sat)}>>I_i^{(sat)}$ the considerable part of the I-V characteristics corresponds to the negative probe potentials φ_p (plasma potential far from the probe is assumed to be zero), Fig. 4.2. The ion current to the probe for large negative φ_p is equal to the ion saturation current, and the ion flux to the probe is

$$\Gamma_i^{(sat)}=I_i^{(sat)}/(4\pi ea^2)=D_a n_0/a.$$

The electron current to the probe is $I_p+I_i^{(sat)}$. For $I_p<<I_e^{(sat)}$ the electric field in the plasma and the main part of the sheath corresponds to the Boltzmann distribution for electrons. Employing the results of Section 3.6 (in particular, the expression for the potential drop in the sheath as a function of electron current, Eq. (3.59)), after adding the potential drop in plasma $(T_e/e)\ln(n_0/n_s)$, we obtain

$$\frac{e\varphi_p}{T_e}=\ln\left(\frac{I_p+I_i^{(sat)}}{I_e^{(sat)}}\right)-\ln\left(\frac{3a^2}{r_d^2(1+T_i/T_e)^2}\right), \tag{4.28}$$

where the Debye radius r_d with the electron temperature is given by Eq. (2.120). Expression Eq. (4.28) determines the I-V characteristics for large negative probe potentials $|\varphi_p|>>T_e/e$. The floating potential of the probe φ_{fl} is negative and can be calculated by putting $I_p=0$ in Eq. (4.28):

$$\frac{e\varphi_{fl}}{T_e} = \ln\left(\frac{D_a}{D_e(1+T_i/T_e)}\right) - \ln\left(\frac{3a^2}{r_d^2(1+T_i/T_e)^2}\right). \qquad (4.29)$$

For the probe at plasma potential φ_p=0 there is no electric field in the plasma and the sheath vanishes. The density profile still obeys Eq. (4.24), while the current to the probe is given by the difference of two unipolar diffusive fluxes:

$$I_0 = 4\pi e(D_e - D_i)n_0 a. \qquad (4.30)$$

For positive probe potentials the *I-V* characteristics can be obtained in a way similar to Eq. (4.28). Probe diagnostics at the collisional limit is discussed in more detail in [4]-[6].

4.2 DIFFUSION OF WEAKLY IONIZED MULTISPECIES PLASMA

The ambipolar diffusion equation Eq. (4.7) for the pure plasma which was investigated in the previous Section has several striking peculiarities.

First of all, it is formulated solely in terms of the plasma density. The self-consistent electric field which is always present in the inhomogeneous plasma remains 'hidden' in the ambipolar diffusion coefficient Eqs. (4.8) - (4.10). The self-consistent problem of calculating the plasma density and potential profiles is reduced to two separate problems. The plasma density profile can be found from Eq. (4.7) with the zero boundary condition (or, more accurately, using the Bohm criterion Eqs. (3.18) and (3.19)) independently of the potential profile. And the latter can be found, if necessary, from Eq. (4.4). In other words, the evolution of the density profile does not depend on the presence of the *dc* current through the plasma inhomogeneity. The rate of the diffusion process is determined by the less mobile particles, i.e. by ions. The electrons are hindered by the arising electric field which drags the ions. This drag force enhances the ion diffusion and causes the factor $(1+T_e/T_i)$ in the expression for the ambipolar diffusion coefficient Eq. (4.10).

The second remarkable property of the ambipolar diffusion process consists in the fact that in spite of the nonlinearity of the initial problem, which is described by Eqs. (4.1), (4.2), both Eq. (4.4) for potential and Eq. (4.7) for density are linear.

These properties, and the striking simplicity of the ambipolar diffusion equation, stimulated numerous attempts to apply such an approach to more complicated situations - to multispecies plasmas, to plasmas in magnetic field, etc.. These attempts have not attained serious success during several decades. The reason is that such an approach is principially wrong. It follows from the fact that the ambipolar diffusion problem, which is described by Eqs. (4.4) and (4.7), corresponds to a degenerate case. It means that such a reduction of the non-linear problem to a system of two linear equations for plasma density and potential is possible only for the case of pure plasma with scalar constant diffusion and mobility coefficients. In this situation several essentially nonlinear effects strictly compensate each other. But in the general case such compensation is impossible, and the problem Eqs. (4.1) (4.2) cannot be reduced to the ambipolar diffusion equation (4.7), and remains essentially nonlinear and current-dependent. Even in the case of diffusion in currentless plasma [7] the electric field is determined by the density gradients of all charged particles. Hence, the problem becomes essentially nonlinear, since every partial field-driven flux is proportional to product of density of a given species and the field strength which is determined by linear combination of density gradients of all species.

In order to understand these nonlinear mechanisms, we consider the simplest 1D example, when N test ions are injected at t=0 into a small region of the boundless and uniform ambient weakly ionized plasma with density n_0 and $T_e = T_i$. We neglect the ionization-recombination processes, and assume that the mobility and diffusion coefficients of the injected and ambient ions coincide. The only distinction with the ambipolar problem Eq. (4.7) consists in the fact that we shall now distinguish between the injected and the ambient particles.

The evolution process for the injected n_1 and for the ambient n_2 ion densities is described by the same Eqs. (2.7), (2.41), as the ambipolar diffusion process. The quasineutrality condition states $n_e = n_1 + n_2$. For the electron profile $n_e(x,t)$ equation (4.7) is valid, with the ambipolar diffusion coefficient $D_a = 2D_1$. Its solution for the case of the point source is

$$n_e(x,t) = n_0 + n_a(x,t) = n_0 + \frac{N}{\sqrt{4\pi D_a t}} \exp\left(-\frac{x^2}{4D_a t}\right). \qquad (4.31)$$

The electric field $e\varphi = T_e \ln n_e$ corresponds to the Boltzmann distribution for electrons. The total Gaussian profile n_e with the effective width of the order of $(4D_a t)^{1/2}$ is, nevertheless, in a rather complex way formed by the diffusive and field-driven fluxes of the injected and ambient particles. In the linear case $n_1 \ll n_0$ the electric field perturbation is small, and its influence on the motion of the injected ions is negligible. Their profile is thus determined by the unipolar ion diffusion:

$$n_1(x,t) = \frac{N}{\sqrt{4\pi D_1 t}} \exp\left(-\frac{x^2}{4D_1 t}\right). \tag{4.32}$$

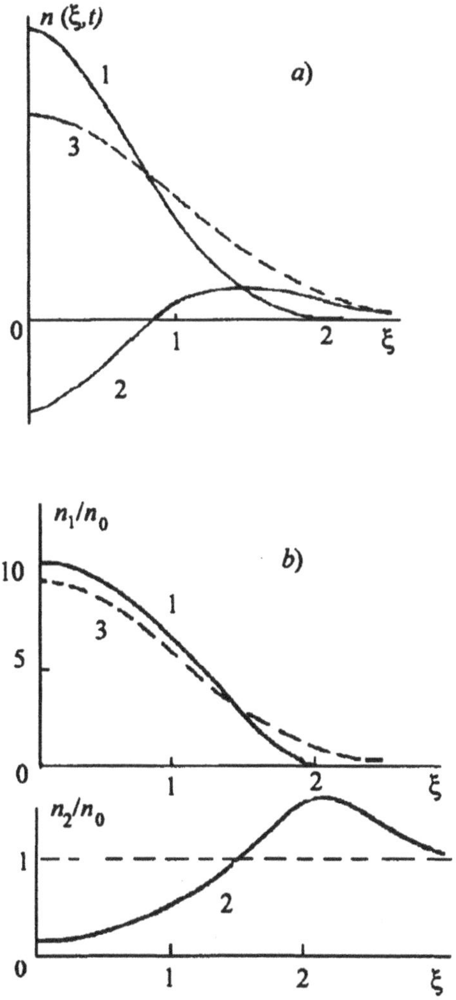

Figure 4.3. Density profiles of injected (1) and ambient (2) ions of equal mobilities for point instantaneous source. The curve 3 is the ambipolar net plasma density profile Eq. (4.31). The spatial coordinate ξ is measured in units of $(4D_i t)^{1/2}$. The figure *a*) corresponds to small disturbance; $n_2^{(0)} \gg N(4D_i t)^{-1/2}$; *b*) is a numerical calculation for strong disturbance; the densities are in units of n_0.

The perturbation of the ambient ion density is caused by the electric field. It can be found as the difference between Eqs. (4.31) and (4.32). The ambient ions are extracted by the field from the neighboring zone $x \leq (4D_1 t)^{1/2}$, where the injected ones dominate, and are piled up outside of it (Fig. 4.3*a*).

If $n_0 \leq n_1$, the problem becomes nonlinear. Since the total number of the ambient ions in the perturbed region is of the order of $n_0(4D_a t)^{1/2}$, the net profile Eq. (4.31) consists mainly of the injected ions. Their diffusion is practically ambipolar: the field here simply doubles their diffusive flux. The ambient ions are almost absent in the neighboring zone, as they are forced out by the electric field. In Fig. 4.3b an example of the numerical solution for such a situation is displayed.

Diffusion in multispecies plasmas. General consideration. The previous analysis was greatly simplified by the fact that the potential and total density profiles were known beforehand - they were determined by the ambipolar diffusion Eq. (4.31). If the ion mobility and diffusion coefficients are different, the situation becomes more complicated, essentially nonlinear and current-dependent.

Even in the absence of global current through the plasma inhomogeneity, vortex current in the plasma bulk arises. It is evident that the zero current condition

$$\vec{j}/e = -\sum_{j=1}^{p}(n_j b_j \vec{\nabla}\varphi + Z_j D_j \vec{\nabla} n_j) = 0 \tag{4.33}$$

is too restrictive. It is fulfilled only if the potential satisfies the special relation

$$\vec{\nabla}\varphi = -\sum_{j=1}^{p} Z_j D_j \vec{\nabla} n_j \Big/ \sum_{j=1}^{p} b_j n_j \,. \tag{4.34}$$

At constant b_j, D_j Eq. (4.34) is equivalent to collinearity condition of the vectors $\vec{\nabla}(\sum_{j=1}^{p} b_j n_j)$, and $\vec{\nabla}(\sum_{j=1}^{p} Z_j D_j n_j)$. Generally this is not the case. It means that even if at plasma boundaries current is absent, the evolution of 2D or 3D plasma inhomogeneities generates vortex currents in the plasma volume. Since the plasma conductivity is usually determined by electrons, this current is small - it is proportional to the ion's contribution to the conductivity.

For the 1D inhomogeneity the vortex current is absent. Nevertheless, even in the absence of global current, the problem of evolution of a plasma inhomogeneity remains rather complicated - due to the influence of the electric field each partial flux is determined by the density gradients of all the plasma species. Substituting Eq. (4.34) into Eqs. (4.1), (4.2), we obtain

$$\frac{\partial n_j}{\partial t}+\frac{\partial}{\partial x}[-D_j\frac{\partial n_j}{\partial x}+Z_j b_j n_j\sum_{k=1}^{p}(Z_k D_k\frac{\partial n_k}{\partial x})/\sum_{k=1}^{p}b_k n_k]=I_j-S_j. \tag{4.35}$$

In the simplest case of evolution of small density perturbation in uniform unbounded background plasma, the system Eq. (4.35), together with the quasineutrality condition $\sum_{j=1}^{p}Z_j n_j=0$, can be linearized, and the Fourier expansion can be used. The resulting dispersion equation determines (p-1) branches $\omega_q(k)$=$-iD_qk^2$ which correspond to different decaying modes of the Fourier harmonics $\exp(ikx-i\omega_q(k)t)$, where the coefficients ω_q depend on all the mobilities and diffusion coefficients of the plasma components. It is to be noted that during plasma evolution corresponding to every diffusive mode, all the partial densities are perturbed.

Two species of positive ions. We consider mechanisms of formation of the nonlinear evolution patterns for the case of spreading plasma inhomogeneity with two species of positive ions with strongly different mobilities. For simplicity we restrict ourselves to the 1D boundless situation on a uniform stationary background $n_{1,2}(t,x\rightarrow\infty)=n_{1,2}^{(0)}$; $b_e>>b_2>>b_1$; $I_j=S_j=0$. From Eq. (4.34) and the quasineutrality condition we find

$$e\varphi=T_e\ln[(n_1+n_2)/(n_1^{(0)}+n_2^{(0)})]. \tag{4.36}$$

The density of the light ions can be found from

$$\frac{\partial n_2}{\partial t}=-\vec{\nabla}\cdot\vec{\Gamma}_2=\vec{\nabla}\cdot[D_2\vec{\nabla}n_2+\frac{T_e b_2 n_2}{e(n_1+n_2)}(\vec{\nabla}n_1+\vec{\nabla}n_2)]. \tag{4.37}$$

This equation corresponds to diffusion of the light ions of the second species with effective variable diffusion coefficient $[D_2+T_eb_2n_2/e(n_1+n_2)]$, and their drift with velocity $[-T_eb_2\vec{\nabla}n_1/e(n_1+n_2)]$. This drift extracts the light ions from regions where the density of the heavy ones is high. Since the mobility ratio b_2/b_1 is large, the evolution process can be subdivided into two stages. During the first fast stage the density $n_1(x,t)$ in Eq. (4.37) is practically equal to the initial density $n_1(x,t=0)$. Eq. (4.37) describes the redistribution of light ions at a fixed density profile of the heavy component. The characteristic temporal scale of this stage for $n_1 \sim n_2$ is $L^2/[4D_2(1+T_e/T_i)]$, where L is the initial inhomogeneity scale. At the end of this stage n_2 satisfies asymptotically $\vec{\Gamma}_2(n_1(\vec{r},t=0),n_2)=0$. This condition corresponds to the collinearity of the gradients $\vec{\nabla} n_1$ and $\vec{\nabla} n_2$. Therefore, at the end of the fast stage

$$-dn_1/dn_2=(1+T_i/T_e)+T_i n_1/(T_e n_2). \tag{4.38}$$

Hence, the minima of the asymptotic $n_2(x)$ profile are formed in the points where the maxima of the initial profile $n_1(x,t=0)$ are situated. In other words, during the fast stage the initial inhomogeneity of the plasma composition increases. This tendency is especially stressed in the case of strong nonlinearity when the perturbed density strongly differs from the background one. In Fig. 4.4 the asymptotic profile of the ambient ions $n_2(x)$ is shown for the case $T_e = T_i$ and for the Gaussian initial profile $n_1(x,t=0)=An_2^{(0)}\exp(-x^2/a^2)$.

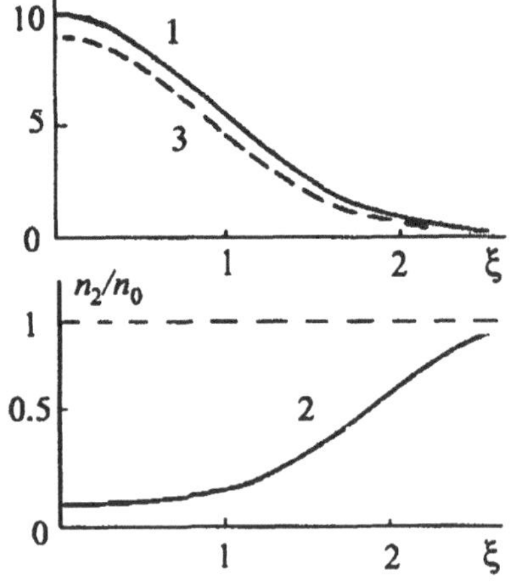

Figure 4.4. Density profiles of heavy injected (1) and light ambient (2) ions in the end of the first fast stage for $A=10$; the curve 3 is (n_e/n_0-1) profile.

From Eq. (4.38) we have

$$2n_2 = \sqrt{n_1^2 + 4(n_2^{(0)})^2} - n_1.$$

The light ion density at $x=0$ drops almost tenfold with respect to its unperturbed value. The region width from which they are pushed out considerably exceeds its width in the linear case; (see Fig. 4.3).

The subsequent slow stage is characterized by the far larger temporal scale $L^2/[4D_1(1+T_e/T_i)]$. During this stage the light ion density adiabatically follows $n_1(x,t)$ evolution according to Eq. (4.38). Substituting Eq. (4.38) into Eqs. (4.1), (4.2) for $n_1(x,t)$, and using the Einstein relation Eq. (2.63) we obtain the nonlinear diffusion equation

$$\partial n_1 / \partial t = \vec{\nabla} \cdot D_{eff} \vec{\nabla} n_1, \tag{4.39}$$

where the effective diffusion coefficient is

$$D_{eff} = D_1(1+T_e/T_i)/[1+n_2T_e/(T_i(n_1+n_2))]. \tag{4.40}$$

Plasma decay in isothermal electronegative gas. As the second example of plasma separation into regions with different ion composition we shall consider the problem of diffusive decay of a cylindrical positive column of weakly ionized plasma which consists of electrons and two ion species - positive ions with density $p(x,t)$, and negative ones with density $n(x,t)$. The quasineutrality condition demands $n_e=p-n$. Expressing the radial electric field from the zero current condition at the tube wall $\Gamma_{er}+\Gamma_{nr}=\Gamma_{pr}$, we obtain the flux of the negative ions

$$\vec{\Gamma}_n = \frac{n\vec{\nabla} n_e(D_e b_n - b_n D_p) - n_e \vec{\nabla} n(D_n b_e + D_n b_p) - 2n\vec{\nabla} n D_n b_p}{b_p p + b_n n + b_e n_e}$$
$$\approx \frac{D_e b_n n\vec{\nabla} n_e - D_n b_e n_e \vec{\nabla} n - 2nD_n b_p \vec{\nabla} n}{b_p p + b_n n + b_e n_e}. \tag{4.41}$$

Imposing the zero boundary condition on all the three partial densities, and neglecting all the plasmachemical processes, we find that the problem of plasma decay has a solution which corresponds to the fundamental diffusive mode Eq. (4.14) [8]. For the case of cylindrical geometry it means that all the densities are proportional to the Bessel function $J_0(\zeta_1 r/a)$ and to time-dependent factors $N(t)$, $P(t)$, $N_e(t)$. Eqs. (4.1), (4.2) yield the nonlinear system

$$\frac{dN}{dt}=\frac{ND_n}{N_eD_e}\frac{dN_e}{dt}=-\frac{2NPD_nb_p}{Nb_n+Pb_p+N_eb_e}\left(\frac{\zeta_1}{a}\right)^2. \tag{4.42}$$

The radial field strength is

$$E_r=-\frac{D_eN_e+D_nN-D_pP}{N_eb_e+Pb_p+Nb_n}\frac{d}{dr}\left(\ln J_0(\zeta_1 r/a)\right). \tag{4.43}$$

At $T_e=T_i$ the electric field Eq. (4.43) corresponds to the Boltzmann law both for electrons and for negative ions, until the Bessel partial profiles satisfy the condition

$$N_e>>Pb_p/b_e. \tag{4.44}$$

Since relaxation of the partial temperatures occurs usually considerably faster than that of the particle densities, the plasma is isothermal, during the main part of the decay, $T_e=T_i=T_N$. In this case the flux Γ_{nr}, according to Eq. (4.41), is small, and the decay process at this stage is determined by the electron-ion ambipolar diffusion, when the radial positive ions flux is practically equal to the electron flux. During this stage plasma is continuously enriched by the negative ions, and when the relation (4.44) is violated, the field, Eq. (4.43), ceases to hinder electrons. The rate of electron loss increases drastically up to a value which corresponds to the electron free unipolar diffusion, and the whole tube volume is filled with the ion-ion plasma. The second stage of the decay process is determined by the ion-ion ambipolar diffusion.

The situation is essentially different, if the electron temperature even slightly exceeds T_i [9]. It follows from the expression for the radial flux Eq. (4.41) that the solution Eq. (4.42) fails provided

$$T_e/T_i-1>2pb_p/(n_eb_e). \tag{4.45}$$

This means that the assumption of similar density profiles demands in this case the physically senseless condition of inward directed flux Γ_{nr} at the tube wall. Therefore, at the first decay stage the partial density profiles are not similar. If the initial electron density is considerable (satisfies to Eq. (4.44)), during this stage the electric field, which corresponds to the Boltzmann law for electrons, piles up negative ions into the plasma bulk. The boundary condition at the tube wall for the negative ions demands

Γ_{nr}=0, and the $n(r)$ profile is formed, which is separated from the wall by the region of the electron-ion plasma. In other words, if the number of negative ions is considerable, the nonisothermal plasma is divided into regions of different ion composition. In the central part an ion-ion plasma is formed, where the electron fraction is small, and the peripheric region is occupied by the electron-ion plasma. The resulting plasma density profiles will be discussed in more detail in Chapter 10.

The sharp change of the plasma decay regime consistent with the above-developed pattern was observed experimentally in [10], [11]. In Fig. 4.5 the fluxes Γ_{pr} (curve 1), and Γ_{nr} (curve 2) at the tube wall, and the electron probe current in the tube center (3), which is proportional to the electron density, are presented [11]. At the first stage (up to 3.8ms after the start of the plasma decay) the flux Γ_{nr} remains negligible. The electron density practically disappears at this moment.

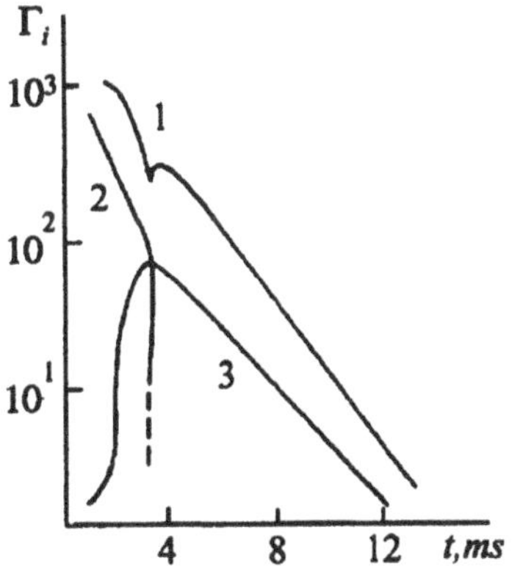

Figure 4.5. Temporal variation of the flux densities of ions O_2^+ (1) and O^- (2) to tube wall which comprise the dominant positive and negative ions in discharge afterglow in mixture of *Kr* with O_2 (1:1) [11]. Pressure is 4MPa; tube radius is a=5.5cm. The curve 3 is the electron saturation current to a probe in the tube center.

Decrease of the decay rate at the second stage is connected, evidently, with decrease of the electric field caused by transition from electron-ion to ion-ion ambipolar diffusion. It seems natural to identify the short, abrupt fall of Γ_{pr} at this moment as the beginning of the second stage, when ion-ion plasma separated from the wall is formed. The potential profile Fig. 4.6 [11] is also correlated with this simple explanation.

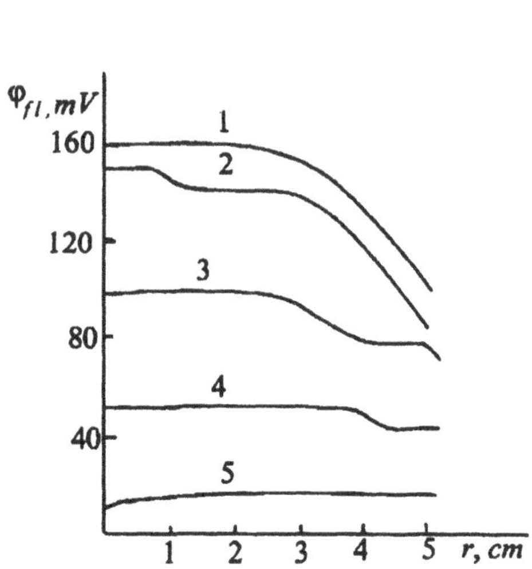

Figure 4.6. Radial dependencies of the probe floating potential φ_{fl} during discharge afterglow (numbers correspond to the time in ms after the start of afterglow) [11]. The plasma parameters are the same as in Fig. 4.5.

The weak field region which appears in the tube center, and expands up to $t \approx 3.8$ms, probably corresponds to the ion-ion plasma

Positive column in an extremely electronegative gas. As it follows from the preceding example, the situation in a positive column, where $T_e >> T_i$, is rather complicated. Here we consider only the case when all the partial densities have similar Bessel profiles. The balance equations for the partial densities we shall take in the simplest form:

$$-\frac{1}{r}\frac{d}{dr}[r(D_p\frac{dp}{dr} - b_p E_r p)] = \nu^{(ion)} n_e \, ,$$
$$-\frac{1}{r}\frac{d}{dr}[r(D_n\frac{dn}{dr} + b_n E_r n)] = K_{att} N_n n_e - K_{dt} n, \qquad (4.46)$$

where the coefficients K_{att}, K_{dt} describe phenomenologically the electron attachment and detachment processes, $p=(n_e+n)$. The electric field Eq. (4.43) in this regime is weak (of the order of T_i/ea) and does not influence the free electron diffusion. Substituting the expression for field Eq. (4.43), we obtain for the ionization frequency

$$\nu^{(ion)} = \frac{1/\tau_e + 2(1/\tau_e + K_{att})/(K_{dt}\tau_n)}{1 + 2/(K_{dt}\tau_n) + D_n/D_p}, \qquad (4.47)$$

where partial unipolar life times $\tau_j=(a/2.4)^2/D_j$. This regime corresponds to practically free electron diffusion, and the discharge maintenance demands very high longitudinal electric field strength E_z to compensate the fast electron losses. But such a solution exists only at extremely high electronegativity, $n/n_e>D_e/D_n$. If the electron fraction increases and exceeds the critical value, the flux Γ_{nr}, according to Eq. (4.41), becomes directed inward. At given K_{att} the growth of the electron fraction corresponds to decrease of the detachment rate K_{dt}. Hence, at $T_e>>T_i$ this regime corresponds to

$$K_{dt}<K_{dt}^{c}=2K_{att}D_p/D_e. \tag{4.48}$$

At lower values of K_{dt} the inward directed flux of the negative ions at the tube wall is necessary for existence of this solution. In a real positive column the density profiles become non-similar, and $n(r)$ profile is formed which is separated from the tube wall by the region of the electron-ion plasma. This situation and the corresponding solutions will be discussed in Chapter 10.

4.3 AMBIPOLAR DIFFUSION IN RADIO FREQUENCY FIELD

A specific kind of ambipolar diffusion arises in RF plasmas in the presence of a potential oscillatory electric field [12]. It is caused by the additional *dc* electric field which arises in inhomogeneous collisional plasma with slowly varying profile $n(x)$ in the presence of the sinusoidal RF current $\tilde{j}(t)$. If the electron Maxwellian time τ_M Eq. (2.123) exceeds the RF period ω^{-1} and the electron collision time ν_{eN}^{-1}, the current is transported mainly by electrons, and in order to maintain the quasineutrality the oscillatory electric field arises which is equal to

$$\tilde{E}(x,t)=\tilde{j}(t)/(eb_e n(x)). \tag{4.49}$$

This field is produced by small oscillatory redistribution of the electron density

$$\delta\tilde{n}_e=-\varepsilon_0\frac{d\tilde{E}}{dx}/e. \tag{4.50}$$

The product $\delta\tilde{n}_e\tilde{E}$ corresponds to the *dc* spatially non-uniform electron current

$$\delta j_e = -\frac{1}{2}\varepsilon_0 b_e \frac{d\tilde{E}^2}{dx}. \tag{4.51}$$

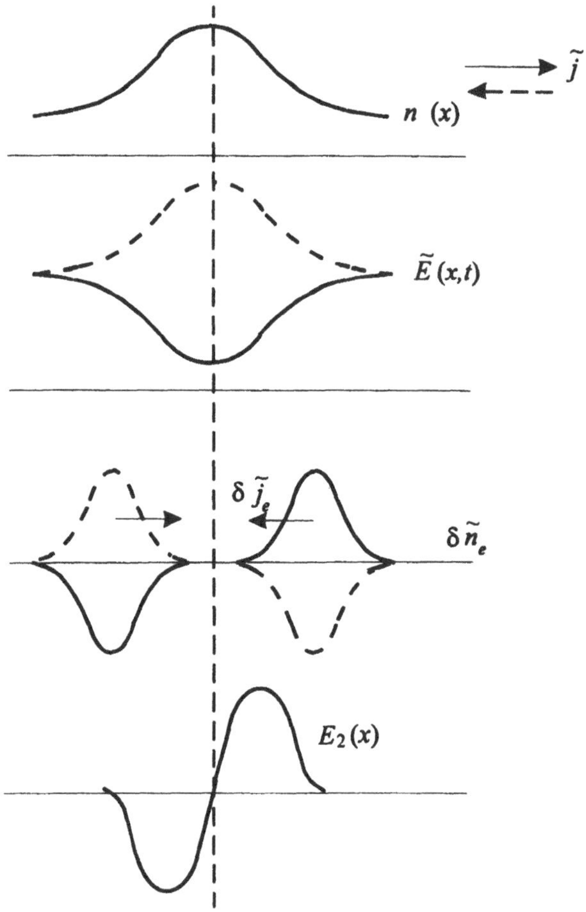

Figure 4.7. Sketch of the formation of *dc* field at RF diffusion. Solid lines correspond to the periods of positive RF current; dashed lines to the negative values of $\tilde{j}(t)$.

In order to avoid the piling up of the electron space charge and to maintain the quasineutrality, the *dc* electric field $\langle E\rangle_2(x)$ arises which hinders the electron motion:

$$\langle E\rangle_2(x) = -\frac{\delta j_e}{eb_e n} = \frac{\varepsilon_0}{2en}\frac{d\tilde{E}^2}{dx} = -\frac{\varepsilon_0 j_0^2}{2e^3 b_e^2 n^4}\frac{dn}{dx}, \tag{4.52}$$

where j_0 is the amplitude of the RF current. This field is proportional to the ion density gradient, Fig. 4.7. It enhances the ordinary ambipolar diffusion which corresponds to the ion motion under influence of the gradient of the ion pressure $T_i \vec{\nabla} n$, and of the ambipolar electric field Eq. (4.12) which compensates the gradient of the electron pressure $T_e \vec{\nabla} n$. Combining Eq. (4.52) with the expression for the ordinary ambipolar diffusion coefficient, we obtain

$$D_{eff} = D_a + D_{HF}. \tag{4.53}$$

The effective ambipolar diffusion coefficient in presence of the longitudinal RF field is

$$D_{HF} = \frac{\varepsilon_0 b_i j_0^2}{2e^3 b_e^2 n^3}. \tag{4.54}$$

It is interesting to note that the effective average force which is applied to electrons and causes the dc current δj_e is equal to the well-known gradient of the RF pressure ([13], Section 1.6), but has the opposite sign.

4.4 TRANSITION FROM FREE TO AMBIPOLAR DIFFUSION

At relatively high plasma densities the sheath at the vessel wall, according to the Sections 3.3, 3.6, is thin with respect to the vessel size so that the quasineutral plasma occupies practically all its volume. This ambipolar case was discussed in Section 4.1. As the plasma density decreases, the sheath thickness, where the field is strong and electrons are practically absent, becomes more and more significant. At very low densities, when the Debye radius exceeds the vessel size, influence of the space charge and the electric field on the particle motion becomes negligible. In this case the electric field can be neglected in Eqs. (2.7), (2.41) and the resulting free diffusion of electrons and ions occurs independently. The evolution of arbitrary initial profiles of the electron and ion densities in this situation is given by the same expressions as Eqs. (4.14) - (4.17), where the ambipolar diffusion coefficient D_a is replaced by the partial diffusion coefficients D_e and D_i correspondingly. Since the electron diffusion is several orders of magnitude faster, than that of ions, the decay process, for example, can be subdivided into two stages. During the first stage the ion motion can be neglected, their density profile practically coincides with the initial density while the electrons escape fast to the vessel walls. The characteristic time of this process for the cylinder of finite length L, analogously to Eq. (4.17), is

$$1/\tau_{11}^{(e)} = D_e(\zeta_1^2/a^2 + \pi^2/L^2) = D_e/\Lambda^2\,, \tag{4.55}$$

where Λ corresponds to the characteristic diffusive length. The second, slow, stage is connected with a decay of the ion density profile with the characteristic time $\tau_{11}^{(i)} = D_i/\Lambda^2$.

If we consider the stationary positive column at such a low plasma density (the volume recombination in such conditions can, of course, be neglected; we shall consider for simplicity by the plane-parallel 1D case with spacing $2L$), it is described by the equations

$$\begin{aligned} -D_e\frac{d^2n_e}{dx^2} &= \nu^{(ion)}n_e \\ -D_i\frac{d^2n_i}{dx^2} &= \nu^{(ion)}n_e\,, \end{aligned} \tag{4.56}$$

with the solution

$$\begin{aligned} &n_e(x) = n_e(0)\cos(x/\Lambda)\,; \\ &n_i(x) = n_i(0)\cos(x/\Lambda)\,; \\ &n_i(0) = n_e(0)D_e/D_i\,; \\ &\nu_0^{(ion)} = D_e/\Lambda^2 = D_e/(2L/\pi)^2. \end{aligned} \tag{4.57}$$

The relative value of electron density is extremely low and in order to compensate the fast electron losses and to maintain the stationary regime a very high ionization rate is necessary.

The rise of the plasma density is accompanied by an increase of the electric field which hinders the electron diffusion and enhances the ion diffusion. The field-driven electron and ion fluxes become more and more important, and the Poisson equation, which determines the electric field, becomes now necessary. The equations system for this transition from free to ambipolar diffusion becomes nonlinear; its numerical solution has been obtained in [16] and discussed in [14], [17]. The electron mean lifetime increases drastically during this transition from the free diffusion value, Eq. (4.55), up to the ambipolar lifetime, Eq. (4.19). In contrast, the ion lifetime considerably ($(1+T_e/T_i)^{-1}$ times) decreases.

The interesting peculiarity of this transition process consists in the fact, that dependence of the mean ion lifetime on the plasma density turns out to be non-monotonous. For illustration of this phenomenon let us consider the simplest case of the stationary plane positive column in the approximation of zero mean free path. According to [14], we shall define the average partial lifetimes as

$$\tau^{(e,i)} = \nu^{(ion)} n_e(0) / n_{e,i}(0), \tag{4.58}$$

and the partial effective diffusion coefficients as

$$D_{e,i}^{(eff)} \nu^{(ion)} \Lambda^2 n_e(0) / n_{e,i}(0). \tag{4.59}$$

The results of numerical calculations [14], [16] are presented in Fig. 4.8.

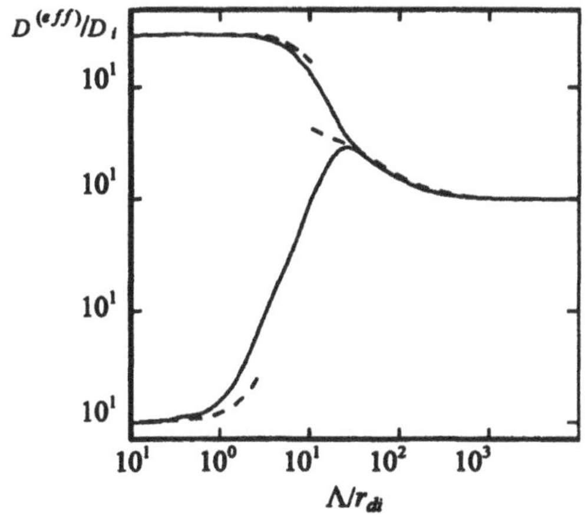

Figure 4.8. Transition from free to ambipolar diffusion. The values $D_e^{(eff)} / D_i$ (upper curves), $D_i^{(eff)} / D_i$ (lower curves) are presented as functions of the parameter κ for b_e/b_i=32, T_e/T_i=100. These values roughly correspond to an active discharge in H_2. The solid lines correspond to the simulations [14], [16]; the dashed lines to the approximations Eqs. (4.63), (4.64).

At low plasma density (we shall characterise it by a central ion density $n_i(0)$ or by a dimensionless ratio $\kappa = e^2 n_i(0)\Lambda^2 / \varepsilon_0 T_i = \Lambda^2 / r_{di}^2$ of the squared ratio of the characteristic vessel size to the ion Debye radius in the tube center) the negative charge in the volume is practically absent. It is concentrated at the vessel wall. On the other hand, the positive space charge and the electric field at $\kappa \ll 1$ can be calculated from the unperturbed ion density Eq. (4.57). This field is given by

$$E = (en_i(0)\Lambda / \varepsilon_0)\sin(x / \Lambda). \tag{4.60}$$

If we refer the electron $N(x)$ and the ion $n(x)$ partial densities to the unperturbed ones, Eqs. (4.57), distance x - to the characteristic scale Λ, and introduce the dimensionless ionization rate Z in units of the unperturbed field-free one $\nu_0^{(ion)}$, Eq. (4.57), we obtain the system

$$\begin{aligned} &N'' + ZN = -\kappa T_i / T_e \cos(2x), \\ &n'' + ZN = \kappa \cos(2x) \end{aligned} \tag{4.61}$$

with the boundary conditions $N'(0) = n'(0) = N(\pi/2) = n(\pi/2) = 0$. Correspondingly,

$$\begin{aligned} &Z = 1 - 4T_i\kappa / (3\pi T_e); \\ &N(x) = \cos x(1 - 2T_i\kappa / (3\pi T_e)) + T_i\kappa / 3T_e \cos 2x; \\ &n(x) = \cos x(1 - 2T_i\kappa / (3\pi T_e)) + T_i\kappa / 3T_e + \\ &\kappa T_i / 12T_e(1 + 2\cos x) - \kappa T_i / 4T_e(1 + \cos 2x). \end{aligned} \tag{4.62}$$

In the simplest case $T_e \gg T_i$ the influence of the space charge field on the ion motion is far more pronounced than on the electron motion, and

$$\begin{aligned} &D_i^{(eff)} = D_i(1 + \kappa / 2), \\ &D_e^{(eff)} = D_e(1 - 4T_i\kappa / (3\pi T_e)). \end{aligned} \tag{4.63}$$

At large $\kappa \gg 1$ the main part of the discharge gap is occupied by the quasineutral plasma. The sheath width L_{sh} at the floating wall is given by Eq. (3.62). The plasma density at the plasma-sheath entrance n_s, which is given by Eq. (3.54), is small with respect to the central plasma density. The ion mean lifetime consists of two terms: it is determined by the ion drift caused by the electric field in the plasma and in the sheath. Since the field strength in the sheath is high, in the lowest approximation we can neglect the second term, and consider the ion lifetime as the time of the ambipolar diffusion over distance $(L - L_{sh})$. Finally

$$D_i^{(eff)} = D_a(1 + 2L_{sh} / L). \tag{4.64}$$

The approximations Eqs. (4.63), (4.64) are presented by the dashed lines in Fig. 4.8.

REFERENCES

1. W. Schottky, *Phys. Zs.* 23: 635-640; **25**: 342-348 (1924).
2. .F. F.Chen, *Introduction to Plasma Physics and Controlled Fusion*, v.1, 4th ed. (New York: Plenum, 1990): 421.
3. L. D. Landau, E. M. Lifshits, *Electrodynamics of Continuous Media* (Oxford: Pergamon Press, 1965) 418.
4. C. H. Su, S.H. Lam, *Phys. Fluids* **6**: 1479-1491 (1963).
5. I. M. Cohen, *Phys. Fluids* **6**: 1492-1499 (1963).
6. P. M. Chung, L. Talbot, K. J. Touryan, *Electric Probes in Stationary and Flowing Plasmas: Theory and Application* (Berlin: Springer, 1975) 238.
7. J. M. Ramshow, C. H. Cheng, *Plasma Chem. Plasma Processing* **11**: (1991); 12: 299-309 (1992); **13**: 189-195 (1993); **13**: 489-498 (1993); J. M. Ramshow, *J. Non-Equilib. Thermodyn.* **18**: 121-132 (1993).
8. H. J. Oskam, *Philips Res. Rept.* **13**: 335-400 (1958).
9. L. D. Tsendin, *Sov. Phys. Techn. Phys.* **30**: 169-175 (1985).
10. L. J. Puckett, C. Lieberger, *Phys. Rev.* **A1**: 1635-1641 (1970).
11. D. Smith, I. C. Plumb, *J. Phys.* **D6**: 1431-1436 (1973); D. Smith, A. G. Dean, N. G. Adams, *J. Phys. D7*: 1944-1962 (1962).
12. G. I. Shapiro, A. M. Soroka, *Sov. Phys. Techn. Phys. Lett.* **5**: 51-52 (1979).
13. L. A. Artsimovich, R. Z. Sagdeev, *Plasma Physics for Physicists* (Moscow: Atomizdat, 1979): 320 (in Russian); Plasmaphysik fuer Physiker (Stuttgart: Teubner, 1983), 368.
14. A. V. Phelps, *J. Res. Natl. Inst. Standards Technol.* **95**: 407-431 (1990).
15. H. Chantry, *J. Appl. Phys.* **62**: 1141-1151 (1987).
16. W. P. Allis, R. J. Rose, *Phys. Rev.* **93**: 84-92 (1954).
17. T. Holstein, *Phys. Rev.* **75**: A 1323-1332 (1949).

Chapter 5

One-Dimensional Diffusion in a Magnetized Plasma

5.1 AMBIPOLAR DIFFUSION IN PURE PARTIALLY IONIZED PLASMA ACROSS MAGNETIC FIELD

Let us consider pure magnetized plasma with constant and uniform partial temperatures in the reference frame which moves with the neutral gas: $u_N=0$. The basic equations are given by Eqs. (2.7), (2.41) and the quasineutrality condition Eq. (2.121):

$$\frac{\partial n}{\partial t}-\vec{\nabla}\cdot(\hat{D}_e\vec{\nabla}n-\hat{b}_e n\vec{\nabla}\varphi)=I-S \ ;$$
$$\frac{\partial n}{\partial t}-\vec{\nabla}\cdot(\hat{D}_i\vec{\nabla}n+\hat{b}_i n\vec{\nabla}\varphi)=I-S \ , \tag{5.1}$$

where tensors $\hat{D}_\alpha$, $\hat{b}_\alpha$ are defined according to Eqs. (2.43), (2.91), (2.92). The particle fluxes are

$$\vec{\Gamma}_e=-\hat{D}_e\vec{\nabla}n+\hat{b}_e n\vec{\nabla}\varphi \ ;$$
$$\vec{\Gamma}_i=-\hat{D}_i\vec{\nabla}n-\hat{b}_i n\vec{\nabla}\varphi \ . \tag{5.2}$$

The quasineutral state across the magnetic field is established in a way similar to the case B=0, Section 2.10. If the gyroradius or mean free path is small with respect to the Debye radius, in the absence of external current through the plasma, the space charge dissipates monotonously, and the quasineutral state is established on a Maxwellian time scale [1]

$$\tau_M = r_d^2/(D_{e\perp}+D_{i\perp}). \tag{5.3}$$

In a collisionless situation oscillatory relaxation to the quasineutral state occurs. The oscillations with wave lengths larger than (or of the order of) r_d are damped during the collision time.

It is easy to notice that in the uniform magnetic field the nondiagonal components of the diffusion tensor do not contribute to the transport equations:

$$\vec{\nabla}_\perp \cdot (\hat{D}_\alpha \vec{\nabla}_\perp n) = \vec{\nabla}_\perp \cdot (D_{\alpha\perp} \vec{\nabla}_\perp n). \tag{5.4}$$

Mathematically Eq. (5.4) follows from the antisymmetric character of the nondiagonal elements of the diffusion tensor. Physically it means that the diamagnetic fluxes, which are represented by these nondiagonal terms, are directed along the equidensities. Therefore, they do not result in particle redistribution. The nondiagonal (Hall) components of the mobility tensor can be rewritten in a similar way:

$$\vec{\nabla}_\perp \cdot (\hat{b}_\alpha \vec{\nabla}_\perp n) = \vec{\nabla}_\perp \cdot (b_{\alpha\perp} \vec{\nabla}_\perp n) + [\vec{\nabla}_\perp \varphi \times \vec{\nabla}_\perp (b_{\alpha xy} n)]\vec{B}/B. \tag{5.5}$$

In 1D geometry the second term in the r.h.s. vanishes in the absence of external current parallel to the equidensities (the situation with external current will be considered in Chapter 11). Employing Eqs. (5.4), (5.5), similar to the case B=0, Section 4.1, in the absence of external current through the plasma, the initial system can be reduced to a single equation of ambipolar diffusion

$$\frac{\partial n}{\partial t} - \vec{\nabla}_\perp \cdot (D_{a\perp} \vec{\nabla}_\perp n) = I - S, \tag{5.6}$$

where the ambipolar diffusion coefficient is given by

$$D_{a\perp} = \frac{D_{e\perp} b_{i\perp} + D_{i\perp} b_{e\perp}}{b_{i\perp} + b_{e\perp}}. \tag{5.7}$$

As in the case B=0, the equation of ambipolar diffusion corresponds to $\Gamma_{e\perp}=\Gamma_{i\perp}$. However, the fluxes in the Hall direction are different $\Gamma_{e\wedge}\neq\Gamma_{i\wedge}$, and the diamagnetic current flows. Potential is also obtained in the same manner as for B=0, and, in the absence of the external current, has the form

$$\varphi_d = -\frac{D_{i\perp} - D_{e\perp}}{b_{e\perp} + b_{i\perp}} \ln n + const\,. \tag{5.8}$$

In 2D case ($n=n(x,y)$) the initial system Eq. (5.1) is also reduced to the Eq. (5.6) of ambipolar diffusion, since the potential Eq. (5.8) turns the second term in Eq. (5.5) to zero.

Substituting expressions for diffusion and mobility coefficients from Eqs. (2.91), (2.92) for arbitrary ionization degree [2], we obtain

$$D_{a\perp} = \frac{(T_e + T_i)(\nu_{eN} + \nu_{ei})}{m_e\omega_{ce}^2 + (\nu_{eN} + \nu_{ei})\mu_{iN}\nu_{iN}}\,. \tag{5.9}$$

In the weakly ionized plasma two limiting cases correspond to different relations between perpendicular mobilities of electrons and ions. For small magnetic fields when $b_{e\perp} >> b_{i\perp}$, as in the case $B=0$ Eq. (4.10), the ambipolar diffusion coefficient is determined by ions:

$$D_{a\perp} \approx (1 + T_e / T_i) D_{i\perp} \approx (1 + T_e / T_i) D_{i\parallel},$$

and the potential corresponds to the Boltzmann distribution for electrons, Eq. (4.12), $\varphi_d \approx (T_e/e)\ln n + const$. In strong magnetic field when $b_{e\perp} << b_{i\perp}$, the ambibolar diffusion coefficient is determined by the less mobile particles, i.e. electrons:

$$D_{a\perp} \approx (1+T_i/T_e) D_{e\perp}, \tag{5.10}$$

and the potential corresponds to the Boltzmann distribution for ions

$$\varphi_d \approx -(T_i/e)\ln n + const. \tag{5.11}$$

Plasmas where $b_{e\perp} << b_{i\perp}$ are usually referred to as strongly magnetized. This condition is equivalent to the condition

$$x_e x_i >> 1, \tag{5.12}$$

where $x_e = \omega_{ce}/(\nu_{eN}+\nu_{ei})$, $x_i = eB/(\mu_{iN}\nu_{iN})$. It is usually said that electrons are strongly magnetized if $x_e >> 1$, and when $x_i >> 1$, that ions are strongly magnetized. The ratio x_e/x_i is usually a large parameter. For example, for weakly ionized plasma $x_e/x_i = \mu_{iN}\nu_{iN}/(m_e\nu_{eN}) \sim (m_i/m_e)^{1/2}$. Therefore, electrons become strongly magnetized at small magnetic fields when $x_e x_i << 1$. In contrast, ions are strongly magnetized at much larger magnetic fields. An example of the dependencies of the perpendicular diffusion coefficients on the magnetic field in weakly ionized plasma [3] are presented in Fig. 5.1.

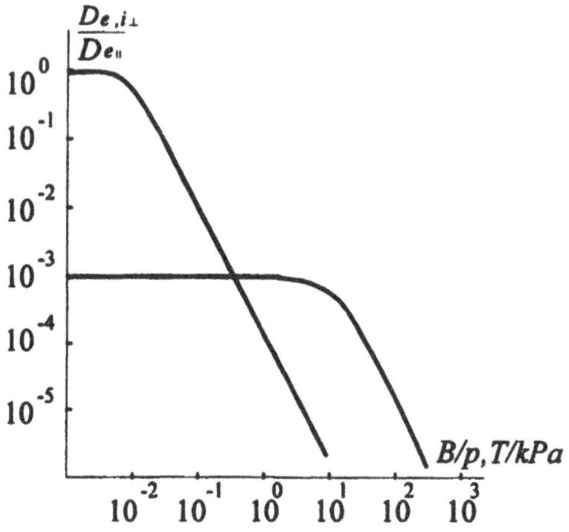

Figure 5.1. Dependence of the perpendicular diffusion coefficients $D_{e\perp}$ (1) and $D_{i\perp}$ (2) on magnetic field for *He*, $T_e=T_i=300K$ [3].

Both the density gradient and self-consistent electric field generate diamagnetic and Hall fluxes respectively along the equidensities. Their difference results in the diamagnetic and Hall currents. In strongly magnetized plasma where ions are distributed according to the Boltzmann law, these currents are carried by electrons, while ions remain at rest. Substituting Eqs. (2.91), (2.92) into Eq. (5.2), in accordance with Eq. (2.46), we obtain

$$\vec{j}_{\wedge} = \frac{[\vec{B}\times\vec{\nabla}p]}{B^2}\ ;\ p = n(T_e + T_i)\,. \tag{5.13}$$

The diamagnetic effect caused by this current is characterized by the parameter β

$$\Delta B_z/B_z \sim \beta = 8\pi n(T_e+T_i)/B^2. \tag{5.14}$$

In partially ionized plasma, as a rule, β<<1, consequently the magnetic field can be considered unperturbed, and the electric field can be described in terms of potential.

The solutions of the 1D ambipolar diffusion equation in weakly ionized plasma across a magnetic field are identical to those described in Section 4.1 for B=0. The situation is different for strongly magnetized ($x_e x_i$>>1) partially ionized plasma, when ν_{ei}>>ν_{eN}. In this case the coefficient of ambipolar diffusion, according to Eq. (5.9), is itself proportional to the plasma density:

$$D_{a\perp} = \frac{(T_e + T_i)\nu_{ei}}{m_e \omega_{ce}^2} = 2\varepsilon n \,, \tag{5.15}$$

where

$$\varepsilon = \frac{\sqrt{2\pi}}{3(4\pi\varepsilon_0)^2} \frac{e^2 m_e^{1/2}(T_e + T_i)\Lambda}{B^2 T_e^{3/2}} .$$

The equation of ambipolar diffusion thus becomes nonlinear:

$$\frac{\partial n}{\partial t} = \varepsilon \Delta_\perp n^2 \,, \tag{5.16}$$

where $\Delta_\perp$ is the Laplace operator.

Let us obtain its self-similar solution for unbounded plasma [4]. It has the form

$$n(x,t) = t^\gamma \tilde{n}(\xi)/\varepsilon \;; \qquad \xi = x/t^\delta . \tag{5.17}$$

Inserting Eq. (5.17) into Eq. (5.16), equating exponents in t, we have γ=2δ-1. One more link results from the particle conservation

$$\int_{-\infty}^{\infty} n dx = const .$$

Substituting Eq. (5.17) yields $\gamma+\delta=1$, and thus in a slab geometry γ=-1/3, δ=1/3. The function $\tilde{n}(\xi)$ satisfies the ordinary differential equation

$$\frac{1}{3}\tilde{n} - \frac{1}{3}\xi \frac{d\tilde{n}}{d\xi} = \frac{d^2 \tilde{n}^2}{d\xi^2} . \tag{5.18}$$

Its solution is

$$\tilde{n}(\xi)=\frac{\xi_0^2}{12}(1-\xi^2/\xi_0^2),$$

where the constant ξ_0 corresponds to the places ($\xi=\pm\xi_0$) where density is equal to zero. Denoting this constant as $\xi_0=a/t_0^{1/3}$, in the initial variables we have

$$n(x,t)=\frac{a^2}{12\varepsilon t_0^{2/3}t^{1/3}}\left[1-\frac{x^2}{a^2}(t/t_0)^{2/3}\right]. \tag{5.19}$$

According to the choice of constants, at the moment $t=t_0$ density turns to zero at finite distance $x=\pm a$, and this distance is increasing with time as $t^{2/3}$. For the cylindrical geometry $\gamma=-1/2$, $\delta=1/4$, the self-similar solution is, analogously,

$$n(r,t)=\frac{a^2}{16\varepsilon t_0^{1/2}t^{1/2}}\left[1-\frac{r^2}{a^2}(t/t_0)^{1/2}\right]. \tag{5.20}$$

Another self-similar solution of the Eq. (5.16) corresponds to the case of diffusive decay. In the slab geometry with the material boundaries at $x=\pm a$, we have $n(x=\pm a)=0$. The total number of particles now decreases with time. The solution of Eq. (5.16) can also be sought in the form of Eq. (5.17) with $\delta=0$ to satisfy boundary conditions. Then $\gamma=-1$, and instead of Eq. (5.18) we have

$$\frac{d^2\tilde{n}^2}{d\tilde{\xi}^2}+\tilde{n}=0, \tag{5.21}$$

where $\tilde{\xi}=x/a$. Its solution is

$$\tilde{\xi}=\int_{\tilde{n}}^{\tilde{n}(0)}\frac{\tilde{n}}{\sqrt{\tilde{n}^3(0)-\tilde{n}^3/3}}d\tilde{n}. \tag{5.22}$$

The maximum density is determined by the boundary conditions $\tilde{n}(\tilde{\xi}=\pm 1)=0$:

$$\tilde{n}(0)=2^{4/3}\pi^2/[9\Gamma^4(2/3)]=0.448.$$

The profile Eq. (5.22) is shown in Fig. 5.2. In the initial variables $n(x,t) = a^2\tilde{n}(x/a)/(\varepsilon t)$. For cylindrical geometry $\tilde{n}(0) = 0.208$, the corresponding density profile is also shown in Fig. 5.2.

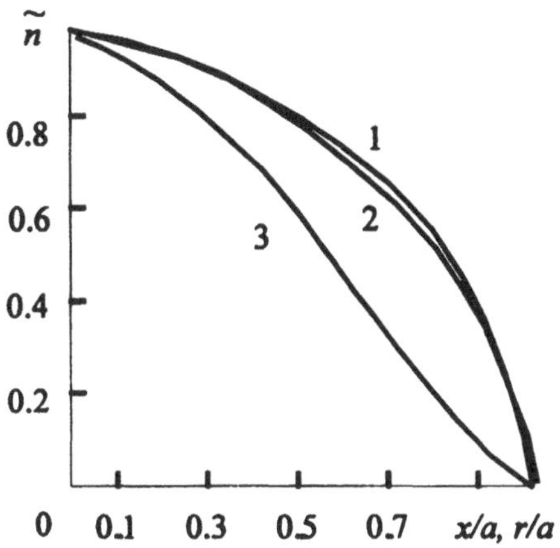

Figure 5.2. Density decay profile for a partially ionized plasma when $\nu_{ei} >> \nu_{eN}$. 1 is the slab geometry, the solution is normalized to 0.448; 2 is the cylindrical case, the solution is normalized to 0.208; 3 is the decay profile for a weakly ionized plasma $\tilde{n} = J_0(2.4r/a)$.

For additional examples of the solution of the nonlinear Eq. (5.16) see [2].

In weakly ionized plasma the equation of ambipolar diffusion in 1D case (as for B=0) also remains valid in the presence of external current $j_\perp$. But, in contrast to Eq. (5.8), the electric field is determined both by density profile and current:

$$\vec{E} = \frac{\vec{j}_\perp}{en(b_{e\perp} + b_{i\perp})} - \vec{\nabla}\varphi_d . \tag{5.23}$$

In two-dimensonal case the second term in the right hand side of Eq. (5.5) is not canceled, and, therefore, the initial system is not reduced to Eq. (5.6) of the ambipolar diffusion.

The transport of partially ionized plasma cannot be described by Eq. (5.6) in the presence of external current even in the 1D case, since the first term in the r.h.s. of Eq. (5.53) is not canceled after substitution into the transport equations. Furthermore, even in pure plasma inhomogeneity moves in the direction of the current, and during its nonlinear evolution shocks are formed. These phenomena are described in Chapter 11.

5.2 DIFFUSION OF STRONGLY IONIZED PLASMA ACROSS MAGNETIC FIELD

In strongly magnetized plasma ($x_e x_i$>>1) the coefficient of ambipolar diffusion, given by Eq. (5.9), becomes independent of the collisions with neutrals for ν_{ei}>>ν_{eN}:

$$D_{a\perp} = \frac{(T_e + T_i)\nu_{ei}}{m_e \omega_{ce}^2}. \tag{5.24}$$

However, the perpendicular diffusion coefficient for ions exceeds that of electrons up to much higher ionization degrees $m_e \nu_{ei} \le \mu_{iN} \nu_{iN}$, see Eqs. (2.92). When $m_e \nu_{ei}$>>$\mu_{iN} \nu_{iN}$ both electron and ion transport coefficients coincide with the values for fully ionized plasma. After neglecting collisions with neutrals, for magnetized particles ($x_{e,i}$>>1) the general expressions Eqs. (2.91), (2.92) in this case yield

$$\begin{aligned} &b_{e\perp} = b_{i\perp} = 0 \ ; \\ &D_{e\perp} = D_{i\perp} = \frac{(T_e + T_i)\nu_{ei}}{m_e \omega_{ce}^2} \ ; \\ &D_{e\perp}^{(T)} = -\frac{1}{2}\frac{T_e \nu_{ei}}{m_e \omega_{ce}^2} = D_{ie\perp}^{(T)} \ ; \\ &D_{ei\perp}^{(T)} = \frac{T_i \nu_{ei}}{m_e \omega_{ce}^2} = D_{i\perp}^{(T)} \ . \end{aligned} \tag{5.25}$$

The particle fluxes are then

$$n\vec{u}_{e\perp} = n\vec{u}_{i\perp} = -\frac{n\nu_{ei}}{m_e \omega_{ce}^2}[(T_e + T_i)\vec{\nabla}_\perp n / n - \vec{\nabla}_\perp T_e / 2 + \vec{\nabla}_\perp T_i]. \tag{5.26}$$

Therefore, in fully ionized plasma both electron and ion mobilities are equal to zero, while the diffusion coefficients of electrons and ions are equal to each other. In other words, diffusion is automatically ambipolar. The physical reason consists in the following: electron and ion partial pressure gradients cause diamagnetic fluxes of the species in the magnetized plasma with velocities

$$\vec{u}_{pe,i} = \pm \frac{[\vec{\nabla} p_{e,i} \times \vec{B}]}{eB^2 n}. \tag{5.27}$$

The difference in these velocities results in friction between electrons and ions

$$\vec{R}_{ei}^{(\vec{u})} = -nm_e \nu_{ei} (\vec{u}_{pe} - \vec{u}_{pi}).$$

Together with thermal force $\vec{R}_{ei}^{(T)}$ it causes drift of the species in the direction of density and temperature gradients:

$$\vec{u}_{e\perp} = \vec{u}_{i\perp} = \frac{[(\vec{R}_{ei}^{(\vec{u})} + \vec{R}_{ei}^{(T)}) \times \vec{B}]}{eB^2 n}.$$

Since the momentum exchange terms in the momentum balance equations for ions and electrons have different signs, this drift is independent of the species charge and is a linear function of the density and temperature gradients. Electric field causes drift of the particles with the same velocity $\vec{u}_{\vec{E}} = [\vec{E} \times \vec{B}] / B^2$, and thus does not result in the corresponding friction force. Therefore, the flux in the direction of electric field is absent, and perpendicular mobility $b_{\alpha\perp}$=0. One can find discussion of the transport coefficients in fully ionized plasma based on the analysis of single particle collisions in [2].

To determine the self-consistent electric field it is necessary to take into account either collisions with neutrals or perpendicular ion viscosity and inertia, which are usually neglected in partially ionized plasma. In fully ionized plasma viscosity and inertia are connected with the inhomogeneity of the ion drift velocity. These forces cause additional (with respect to Eq. (5.26)) unipolar flux of ions in the direction of the pressure gradient and electric field. As a result, effective perpendicular conductivity arises [5], [6], proportional to the small parameter ρ_{ci}/L, where L is the spatial scale of the density or electric field inhomogeneity. Therefore, this mechanism of the perpendicular conductivity is effective on small scales. However, in the 1D case in the absence of current in plasma, even small conductivity leads to the creation of a large self-consistent electric field.

For the constant temperatures T_e and T_i this electric field can be obtained without knowledge of explicit expressions for viscosity in magnetic field. Since the shear of ion Hall flux is the source of the perpendicular inertia, viscosity and, therefore, corresponding perpendicular current of ions, the ambipolarity condition $j_\perp$=0 is equivalent to the condition $\vec{u}_{pi} + \vec{u}_{\vec{E}} = 0$. The latter means that the ion diamagnetic flux is balanced by the drift in the ambipolar electric field. The ambipolar electric field thus corresponds to the Boltzmann distribution for ions [7] $\varphi=-(T_i/e)\ln n+const$.

5.3 ONE-DIMENSIONAL DIFFUSION OF PARTIALLY IONIZED PLASMA AT ARBITRARY ANGLE WITH MAGNETIC FIELD

Here we shall study the situation when plasma density depends only on the coordinate ζ which forms an angle β with the magnetic field. It is convenient to introduce the new coordinate system (x, η, ζ), Fig. 5.3, instead of the old one (x, y, z), where the z axis was parallel to the magnetic field.

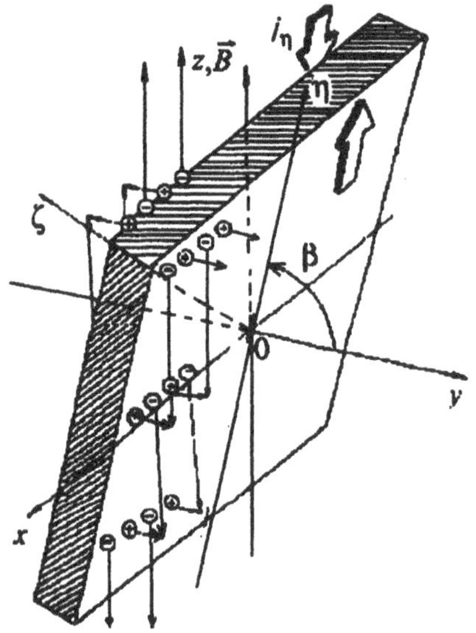

Figure 5.3. 1D diffusion $\vec{\nabla}n||\vec{\zeta}$ for $\mu_e<<\mu<<\mu_i$; the species fluxes are shown by arrows; the hatched region corresponds to enhanced density.

The new system is obtained by rotation of the old one over the x axis. The diffusion and mobility tensors in the new system are

$$\hat{D}' = \hat{A}\hat{D}\hat{A}' \ ; \ \hat{b}' = \hat{A}\hat{b}\hat{A}' \ , \tag{5.28}$$

where

$$\hat{A}=\begin{pmatrix}1 & 0 & 0\\ 0 & \cos\beta & -\sin\beta\\ 0 & \sin\beta & \cos\beta\end{pmatrix}$$

is the operator of rotation over the x axis. In the new coordinate system the diffusion tensor has the form

$$\hat{D}'=\begin{pmatrix}D_\perp & D_{xy}\cos\beta & -D_{xy}\sin\beta\\ -D_{xy}\cos\beta & D_\perp\cos^2\beta+D_\parallel\sin^2\beta & (D_\parallel-D_\perp)\cos\beta\sin\beta\\ D_{xy}\sin\beta & (D_\parallel-D_\perp)\cos\beta\sin\beta & D_\parallel\cos^2\beta+D_\perp\sin^2\beta\end{pmatrix}. \tag{5.29}$$

Here components $D_\perp$, $D_\parallel$ and D_{xy} are defined by Eqs. (2.91), (2.92). The mobility tensor has the similar form. The initial system Eq. (5.1) in these new coordinates is given by

$$\frac{\partial n}{\partial t}-\frac{\partial}{\partial\zeta}\left(D_{\alpha\zeta\zeta}\frac{\partial n}{\partial\zeta}\pm b_{\alpha\zeta\zeta}n\frac{\partial\varphi}{\partial\zeta}\right)=I-S\,, \tag{5.30}$$

where $\alpha=i,e$. In the absence of external current in ζ direction this system, as in the previous Sections, can be reduced to the form [1], [8], [9]:

$$\frac{\partial n}{\partial t}-\frac{\partial}{\partial\zeta}[D(\mu^2)\frac{\partial n}{\partial\zeta}]=I-S\,, \tag{5.31}$$

where $\mu=\cos\beta$;

$$\begin{aligned}D(\mu^2)&=\frac{D_{e\zeta\zeta}b_{i\zeta\zeta}+D_{i\zeta\zeta}b_{e\zeta\zeta}}{b_{i\zeta\zeta}+b_{e\zeta\zeta}}\\ &=\{[D_{e\parallel}\mu^2+D_{e\perp}(1-\mu^2)][b_{i\parallel}\mu^2+b_{i\perp}(1-\mu^2)]\\ &+[D_{i\parallel}\mu^2+D_{i\perp}(1-\mu^2)][b_{e\parallel}\mu^2+b_{e\perp}(1-\mu^2)]\}\\ &\times[(b_{e\parallel}+b_{i\parallel})\mu^2+(b_{e\perp}+b_{i\perp})(1-\mu^2)]^{-1}.\end{aligned} \tag{5.32}$$

The potential profile is

$$\varphi = \frac{(D_{e\parallel} - D_{i\parallel})\mu^2 + (D_{e\perp} - D_{i\perp})(1-\mu^2)}{(b_{e\parallel} + b_{i\parallel})\mu^2 + (b_{e\perp} + b_{i\perp})(1-\mu^2)} \ln n + const. \qquad (5.33)$$

The Eq. (5.31) has the form of the equation of ambipolar diffusion, however, it cannot be derived by equating the species fluxes, $\Gamma_{e\parallel} \neq \Gamma_{i\parallel}$; $\Gamma_{e\perp} \neq \Gamma_{i\perp}$. In other words, even in the absence of external current along ζ, current is generated in the η direction. It decays at $\zeta = \pm\infty$ in homogeneous background plasma. Let us discuss the corresponding physical mechanism for the most simple case of weakly ionized plasma in strong magnetic field when $b_{e\parallel} >> b_{i\parallel}$, $b_{i\perp} >> b_{e\perp}$(the last inequality is equivalent to $x_e x_i >> 1$). If the density gradient is almost perpendicular to magnetic field, so that $\mu << \mu_0$, where

$$\mu_0 = (b_{i\perp}/b_{e\parallel})^{1/2} << 1, \qquad (5.34)$$

we have

$$D(\mu^2) = D_e(\mu^2) = (1 + T_e/T_i)[D_{e\parallel}\mu^2 + D_{e\perp}(1-\mu^2)]. \qquad (5.35)$$

This coefficient is $(1+T_e/T_i)$ times larger than the electron unipolar diffusion coefficient $D_{e\zeta\zeta}$. For $\mu << \mu_0$ electrons become less mobile particles, and hence, according to Eq. (5.33), for $T_e \sim T_i$ the plasma perturbation is biased negatively. The electric field accelerates electrons and corresponds to the Boltzmann distribution for ions. As follows from Eq. (5.35), there exists a critical angle

$$\mu_e = (b_{e\perp}/b_{e\parallel})^{1/2} << 1. \qquad (5.36)$$

For $\mu << \mu_e$ electrons move almost perpendicular to the magnetic field, and the diffusion coefficient Eq. (5.35) coincides with the perpendicular ambipolar diffusion coefficient Eq. (5.10). On the other hand, at angles $\mu_e << \mu << \mu_0$ electrons move mostly parallel to $\vec{B}$, and the diffusion coefficient $D(\mu^2)$ exceeds the perpendicular ambipolar diffusion coefficient. Ions, in contrast, diffuse mostly perpendicular to magnetic field in the direction of the density gradient since their unipolar coefficient $D_{i\zeta\zeta} \approx D_{i\perp}$. As a result, the species fluxes along and across magnetic field remain completely different.

For larger angles $\mu >> \mu_0$ between the density gradient and perpendicular to $\vec{B}$ plane, Eq. (5.32) is reduced to

$$D(\mu^2) = D_i(\mu^2) = (1 + T_i/T_e)[D_{i\parallel}\mu^2 + D_{i\perp}(1-\mu^2)], \qquad (5.37)$$

and the diffusion coefficient is controlled by ions. The potential for $T_e \sim T_i$ corresponds to the Boltzmann distribution for electrons. Now electrons move mostly in the parallel direction, while for ions the flux direction depends on the relation between μ and the parameter

$$\mu_i = (b_{i\perp}/b_{i\parallel})^{1/2}. \tag{5.38}$$

For $\mu \ll \mu_i$ the ion flux is determined mostly by the perpendicular motion, and in the opposite case by the parallel one.

At the special angle $\mu_c = (D_{i\perp}/D_{e\parallel})^{1/2} = \mu_0 (T_i/T_e)^{1/2}$ the unipolar diffusion coefficients equal $D_{e\zeta\zeta} = D_{i\zeta\zeta} = D_{i\perp}$, the diffusive fluxes in the ζ direction are also equal, and hence the electric field vanishes. Note that in spite of the fact that the density gradient is almost perpendicular to the magnetic field, the coefficient $D(\mu^2)$ coincides now with $D_{i\perp}$, i.e. it is much larger than the perpendicular ambipolar diffusion coefficient Eq. (5.10). The general dependence $D(\mu^2)$ is shown in Fig. 5.4.

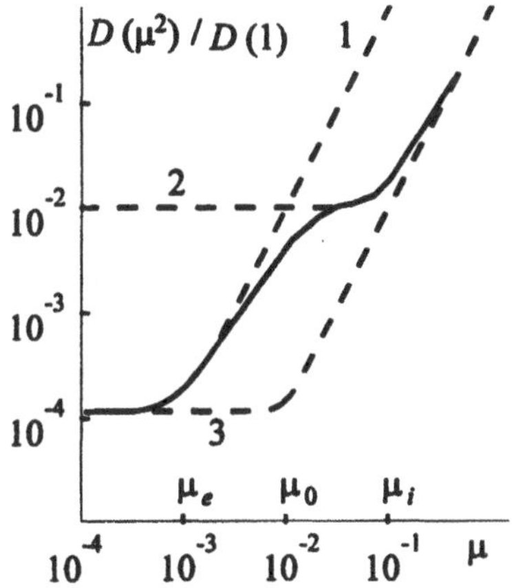

Figure 5.4. Function $D(\mu^2)$ (solid line) for $T_e = T_i$, $x_e = 10^3$, $x_i = 10$ calculated in the elementary approximation. Dashed lines are the calculations using different approximate expressions: 1 - Eq.(5.35); 2 - Eq.(5.37); 3 - Eq.(5.40).

Since in two limiting cases μ=0,1 (Sections 5.1, 5.2) the initial system can be reduced to the single equation of ambipolar diffusion, one might ask whether a simple equation of anisotropic ambipolar diffusion

$$\frac{\partial n}{\partial t} - \vec{\nabla}_\parallel (D_{a\parallel} \vec{\nabla}_\parallel n) - \vec{\nabla}_\perp \cdot (D_{a\perp} \vec{\nabla}_\perp n) = I - S \tag{5.39}$$

might be adequate to describe the evolution process. Here the coefficients of ambipolar diffusion $D_{a\|}$, $D_{a\perp}$ are defined by Eqs. (4.10), (5.10). However, Eq. (5.39) can be obtained only by assuming simultaneously $\Gamma_{e\|}=\Gamma_{i\perp}$ and $\Gamma_{e\perp}=\Gamma_{i\perp}$ which requires nonpotential (vortex) electric field. For our 1D problem Eq. (5.39) yields

$$D_a(\mu^2) = D_{a\|}\mu^2 + D_{a\perp}(1-\mu^2). \tag{5.40}$$

Its value strongly deviates from the correct result Eq. (5.32) especially at $\mu\sim\mu_0$, see Fig. 5.4.

Current along equidensities is given by

$$j_\eta = en[-(D_{i\eta\zeta} - D_{e\eta\zeta})\frac{\partial n}{\partial\zeta} - n(b_{i\eta\zeta} + b_{e\eta\zeta})\frac{\partial\varphi}{\partial\zeta}], \tag{5.41}$$

where components of the mobility and diffusion tensors are defined according to Eq. (5.32). For $\beta=0$; $\pi/2$ the current is equal to zero. For weakly ionized plasma, after substituting the potential Eq. (5.33) we have

$$j_\eta = \frac{enD_{e\|}(1+T_i/T_e)b_{i\perp}\sin\beta\cos\beta}{b_{e\|}\cos^2\beta + b_{i\perp}\sin^2\beta}\frac{\partial n}{\partial\zeta}. \tag{5.42}$$

The behavior of the diffusion coefficient $D(\mu^2)$ is connected with the conditions $D_{e\|}>>D_{i\|}$ and $D_{e\perp}>>D_{i\perp}$. Therefore, the character of the dependence $D(\mu^2)$ remains the same for partially ionized plasma provided $m_e\nu_{ei}<<\mu_{iN}\nu_{iN}$. Only in the opposite case for large ionization degree, when $D_{e\perp}=D_{i\perp}=D_{a\perp}$; $D_{e\|}=D_{i\|}=D_{a\|}$, the coefficient $D(\mu^2)$ coincides with Eq. (5.40).

5.4 DIFFUSION ACROSS MAGNETIC FIELD IN MULTISPECIES PLASMA

Equation (5.6) of ambipolar diffusion only describes the density evolution in pure plasma. In multispecies plasma, similar to the case $\vec{B}=0$, the ion flux of a given species depends on the density profiles of all other species. To illustrate the complexity of the arising problems we shall analyze two examples of such evolution for the case of two positive ion species.

Partially ionized plasma. Consider uniform background plasma with the ion density $n_2^{(0)}$ in a uniform magnetic field parallel to the z axis. At a moment t=0 plasma inhomogeneity which consists of ions of different species with density n_1 is created. We consider slab geometry, and to be specific, we choose the initial profile of injected particles

$$n_{10}(y)=An_2^{(0)}\exp(-y^2/a^2). \tag{5.43}$$

The magnetic field is assumed to be strong enough: $D_{i\perp}>>D_{e\perp}$, and the ionization degree to be not very high $m_e\nu_{ei}<<\mu_{iN}\nu_{iN}$. The diffusion in multispecies plasma is of great importance in the ionosphere where the ratio $D_{i\perp}/D_{e\perp}$ reaches two-three orders of magnitude. Therefore, for typical parameters we can completely neglect the perpendicular electron diffusion coefficient. Assuming $T_1=T_2=T$, using Einstein relation, in the absence of sources and sinks we have the continuity equations for ions

$$\begin{aligned}\frac{\partial n_1}{\partial t}&=\frac{\partial}{\partial y}[D_{1\perp}\frac{\partial n_1}{\partial y}+(T/e)n_1\frac{\partial\varphi}{\partial y}] \;; \\ \frac{\partial n_2}{\partial t}&=\frac{\partial}{\partial y}[D_{2\perp}\frac{\partial n_2}{\partial y}+(T/e)n_2\frac{\partial\varphi}{\partial y}] \;,\end{aligned} \tag{5.44}$$

where $D_{1,2\perp}$ are the ion perpendicular diffusion coefficients. For electrons in this approximation

$$\frac{\partial n_e}{\partial t}=0 \;;\; n_e\equiv n=n_1+n_2=n_{10}(y)+n_2^{(0)}. \tag{5.45}$$

The system of Eqs. (5.44), (5.45) has been solved in [10]. Summing up Eqs. (5.44), accounting for Eq. (5.45), using boundary condition $\varphi(|y|\to\infty)=0$, after integrating over y, we obtain

$$\varphi=-\frac{T}{e}\ln\frac{D_{1\perp}n_1+D_{2\perp}n_2}{D_{2\perp}n_2^{(0)}}. \tag{5.46}$$

Inserting potential into Eqs. (5.44) yields

$$\frac{\partial n_{1,2}}{\partial t}=D_{1,2\perp}\frac{\partial}{\partial y}\left[\frac{\partial n_{1,2}}{\partial y}-\frac{n_{1,2}}{D_{1\perp}n_1+D_{2\perp}n_2}\frac{\partial(D_{1\perp}n_1+D_{2\perp}n_2)}{\partial y}\right]. \tag{5.47}$$

In the case of small perturbation $n_1<<n_2$ ($A<<1$) it is possible to neglect the second term in the r.h.s. of Eq. (5.47) for n_1. The density evolution is thus governed by the unipolar diffusion equation with the coefficient $D_{1\perp}$. In other words, injected ions diffuse with their unipolar coefficient, and the perturbed electric field redistributes only the ambient ions. The profile $n_2(y)$ presents the difference between the electron density profile Eq. (5.45) and n_1. Finally

$$\frac{n_1}{n_2^{(0)}}=\frac{Aa\exp[-y^2/(a^2+4D_{1\perp}t)]}{(a^2+4D_{1\perp}t)^{1/2}}\ ;$$
$$\frac{n_2}{n_2^{(0)}}=A\{\exp(-y^2/a^2)-\frac{a\exp[-y^2/(a^2+4D_{1\perp}t)]}{(a^2+4D_{1\perp}t)^{1/2}}\}+1. \tag{5.48}$$

Therefore, the injected ions are quickly, on a time scale $t\sim a^2/(4D_{1\perp})$, spread, and are replaced by the background ions. The latter are gathered by electric field from a distance $|y|\sim a$. Here depletion regions - regions where density of the ambient ions decreases, are formed. In the approximation considered when the perpendicular electron mobility is neglected, the electron density remains unchanged and $n_2(y,t\to\infty)=n_{10}(y)+n_2^{(0)}$.

Let us now consider strongly nonlinear perturbation $A>>1$ for the special case $D_{1\perp}=D_{2\perp}=D$. We introduce the dimensionless variables: $\tilde{y}=y/a, \tilde{t}=t=tD/a^2, \tilde{n}_{1,2}=n_{1,2}/n_2^{(0)}, \tilde{n}=n/n_2^{(0)}$. In these variables Eq. (5.47) for n_1 has the form

$$\frac{\partial \tilde{n}_1}{\partial t}=\frac{\partial}{\partial \tilde{y}}\left(\frac{\partial \tilde{n}_1}{\partial \tilde{y}}-\tilde{n}_1\frac{\partial \ln \tilde{n}_1}{\partial \tilde{y}}\right), \tag{5.49}$$

where $\tilde{n}=A\exp(-\tilde{y}^2)+1$. We shall obtain the solution for $\tilde{t}>>1$. In the region $|\tilde{y}|>>1$ it is still possible to neglect the second term in the r.h.s. of Eq. (5.47) since here the perturbation is small $\tilde{n}_1<<1$ and $\partial\tilde{n}/\partial\tilde{y}<<1$. The solution of the diffusion equation in this region can be expressed through the Green function $G(\tilde{y},\tilde{t})$

$$\tilde{n}_1(\tilde{y},\tilde{t})=\int_0^{\tilde{t}} B(\tilde{t}\,')G(\tilde{y},\tilde{t}-\tilde{t}\,')d\tilde{t}\,'\ ;$$
$$G(\tilde{y},\tilde{t}-\tilde{t}\,')=\frac{1}{\sqrt{4\pi(\tilde{t}-\tilde{t}\,')}}\exp\left[-\frac{\tilde{y}^2}{4(\tilde{t}-\tilde{t}\,')}\right]. \tag{5.50}$$

The 'source' which is situated at $|\tilde{y}|\le 1$ can be considered as pointlike. It is also easy to notice that at $|\tilde{y}|\le 1$ the l.h.s. of Eq. (5.49) is small with respect to each term in the r.h.s. Hence the solution here can be obtained by equating the r.h.s. to zero

$$\tilde{n}_1 = C_1(\tilde{t})\tilde{n} \ ; \ C_1(\tilde{t}) = 1 - p(\tilde{t}) \ ; \ p << 1. \tag{5.51}$$

In other words, profiles n_1, n_2 and n are proportional to each other, and the ion densities of both species correspond to the Boltzmann distribution in the electric field. The boundary between the inner and the outer parts of the profile can be defined with logarithmic accuracy by $|\tilde{y}| = |\tilde{y}_0| \approx (\ln A)^{1/2}$. We obtain the functions $B(\tilde{t}), C_1(\tilde{t})$ using the total particle conservation and matching two solutions at the boundary

$$\int_{-\infty}^{\infty} (\tilde{n} - 1) d\tilde{y} = \sqrt{\pi} A C_1(\tilde{t}) + \int_0^{\tilde{t}} B(\tilde{t}') d\tilde{t}' \ ; \tag{5.52}$$

$$\int_0^{\tilde{t}} \frac{B(\tilde{t}')}{\sqrt{4\pi(\tilde{t} - \tilde{t}')}} \exp\left[-\frac{\tilde{y}_0^2}{4(\tilde{t} - \tilde{t}')}\right] d\tilde{t}' = C_1(\tilde{t}) \ . \tag{5.53}$$

The accuracy of such a procedure is of the order of $|\tilde{y}_0| / \tilde{t}^{1/2}$. In the outer region one can put $\tilde{y}_0 = 0$.

The character of the solution depends on the parameter t_0. Until this moment most of the injected ions remain inside the inner region $|\tilde{y}| \leq 1$, and $C_1(\tilde{t}) \approx 1$. By putting $\tilde{y}_0 = 0$ we obtain from Eq. (5.53)

$$B(\tilde{t}) = 2 / \sqrt{\pi \tilde{t}} \ ,$$

and then from Eq. (5.52)

$$p(\tilde{t}) = 4\tilde{t}^{1/2} / (\pi A) .$$

The characteristic time scale of this fast stage is determined from the condition p=1, so $\tilde{t}_0 = (\pi A)^2 / 16$. Later, at $t >> t_0$, the main part of the injected particles is spread over the outer region. Now from Eq. (5.52) we have $B(\tilde{t}) = A\sqrt{\pi}\delta(\tilde{t})$, and in accordance with Eq. (5.53) $C_1(\tilde{t}) \approx A / \sqrt{4\tilde{t}}$.

In the initial variables for A>>1 solution has the form

$$\frac{n_1}{n_2^{(0)}} = A\exp(-\frac{y^2}{a^2})\left(1-\frac{4\sqrt{Dt}}{\pi aA}\right) + \int_0^t \frac{\exp[-y^2/4D(t-t')}{\sqrt{(t-t')t'}}dt' =$$

$$A\exp(-\frac{y^2}{a^2})\left(1-\frac{4\sqrt{Dt}}{\pi aA}\right) + 1 - erf(\sqrt{y^2/4Dt})\ ; \qquad (5.54)$$

$$\frac{a^2}{D} \ll t \ll \frac{(\pi Aa)^2}{16D}\ ,$$

where $erf(z) = (2/\sqrt{\pi})\int_0^z \exp(-t^2)dt$;

$$\frac{n_1}{n_2^{(0)}} = \frac{Aa}{\sqrt{4Dt}}[A\exp(-y^2/a^2) + \exp(-y^2/4Dt)]\ ;$$

$$t \gg \frac{(\pi aA)^2}{16D}\ . \qquad (5.55)$$

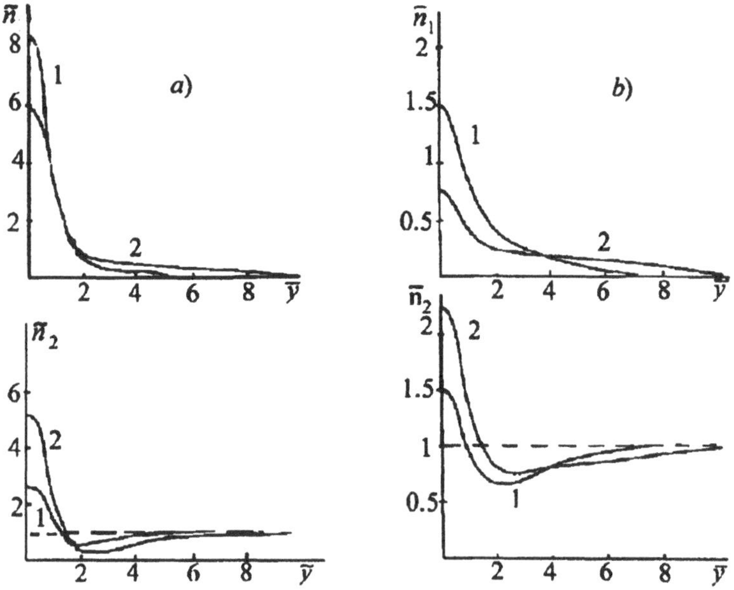

Figure 5.5. Density profiles of the injected and ambient ions for two different moments: *a*- ambipolar stage, *A*=10, calculations are according to Eqs. (5.54), (5.56). *b* - the case of intermediate nonlinearity, *A*=2, calculations are according to Eqs. (5.55) (5.56); 1 - $\tilde{t}$ =4; 2 - $\tilde{t}$ =16.

The density of ambient ions is

$$n_2 = n_{10}(y) - n_1 + n_2^{(0)}. \tag{5.56}$$

Two stages of the diffusion are characterized by the following features: at the first ambipolar stage the most injected ions remain at rest (with account for $D_{e\perp}\neq 0$ slowly diffuse with ambipolar coefficient $D_{a\perp}$ Eq. (5.10)). Only a small fraction of injected ions can diffuse with unipolar coefficient D, thus creating a pedestal with the maximum density $n_2^{(0)}$.

The background ions, in contrast, leave this outer region, forced by the electric field, and are gathered in the central peak. Their depletion is formed according to Eqs. (5.54), (5.56) with the minimum density $n_2 \sim n_2^{(0)}(Dt)^{1/2}/(Aa)$. The size of the depleted region increases with time as $t^{1/2}$, and at the same time the fraction of the background ions increases in the inner region. Electron density does not change in this process.

At later stages $t > t_0 = (\pi a A)^2/16D$ the initial plasma profile consists to a large extent of background ions. On the other hand, the most of the injected ions (the second term in the r.h.s. of Eq. (5.55)) diffuse with a unipolar diffusion coefficient similar to the linear case $A \ll 1$. We shall call this regime 'the stage of intermediate nonlinearity' since a small fraction of injected ions with large amplitude still remains inside the central peak (the first term in the r.h.s. of Eq. (5.55)). The depletion regions are already shallow at this stage. This picture will remain for a long time until electrons in the central peak spread owing to ambipolar diffusion. Examples of the density evolution are shown in Fig. 5.5.

Impurities in fully ionized plasma. Transport of impurities across a magnetic field, as the particles transport in pure fully ionized plasma, Section 5.2, is caused by the friction forces which arise due to the difference in diamagnetic species fluxes. For clarity we consider here the linear case $n_1 \ll n_2$ (subscript 1 corresponds to the impurities, 2 - to the ambient ions), restricting ourselves to heavy impurities $m_1 \gg m_2$ with large charge number Z at $T_1 = T_2$. In this situation it is possible to neglect friction between impurities and electrons. Indeed, this force is proportional to $m_1\nu_{1e} = m_e\nu_{e1}$, and the ion-ion friction force - to $m_1\nu_{12} = m_2\nu_{21}$, where the collisions frequencies are [7]

$$\nu_{e1} = \frac{4\sqrt{2\pi}e^4 Z^2 \Lambda n_1}{3(4\pi\varepsilon_0)^2\sqrt{m_e}T_e^{3/2}}\ ;\ \nu_{21} = \frac{4\sqrt{2\pi}e^4 Z^2 \Lambda n_1}{3(4\pi\varepsilon_0)^2\sqrt{m_2}T_2^{3/2}}. \tag{5.57}$$

Hence, the ion-impurity friction force is $(m_2/m_1)^{1/2}T_e^{3/2}/T_2^{3/2}$ times larger. It is proportional to the difference between the diamagnetic velocities of the impurities and main ions:

$$\vec{R}_{21}^{(\vec{u})} = -n_2 m_2 \nu_{21}(\vec{u}_{p2} - \vec{u}_{p1}). \tag{5.58}$$

The diamagnetic velocity of the main ions is given by Eq. (5.27), and for impurities we have a similar expression: $\vec{u}_{p2} = [\vec{B}/B \times \vec{\nabla} p_1]/(eZBn_1)$. In the presence of the ion temperature gradient it is necessary to add the expression for ion-impurity thermal force, which is fully analogous to the expression for the electron-ion thermal force Eqs. (2.26), (2.27), (2.85):

$$\vec{R}_{21}^{(T)} = -(3/2)n_2(\nu_{21}/\omega_{c2})[\vec{B}/B \times \vec{\nabla} T_2]. \tag{5.59}$$

Similar to the case of the pure plasma, the flux of impurities is proportional to the total friction force $\vec{R}_{21} = \vec{R}_{21}^{(\vec{u})} + \vec{R}_{21}^{(T)}$ [7], [11]:

$$\vec{\Gamma}_1 = n_1\vec{u}_1 = \frac{[\vec{B}/B \times \vec{R}_{21}]}{eZB} = \frac{n_2 m_2 \nu_{21} T_2}{Ze^2B^2}\left[\frac{\vec{\nabla} n_2}{n_2} - \frac{\vec{\nabla} n_1}{Zn_1} - \frac{\vec{\nabla} T_2}{2T_2}\right]. \tag{5.60}$$

Since $Z \gg 1$, when the impurities density has a shallow gradient $|\vec{\nabla} n_1|/n_1 \sim |\vec{\nabla} n_2|/n_2$ the impurity flux has a convective character. In the absence of the temperature gradient it is directed along the density gradient of the main ions. In other words, impurity ions are dragged to the maximum of the density profile of the main ions. The stationary distribution which corresponds to the absence of the impurity flux Eq. (5.60) thus has the form

$$n_1 = A n_2^Z T_2^{-(1/2)Z}. \tag{5.61}$$

Therefore, in the absence of a temperature gradient the resulting profile is very sharply peaked near the maxima of the main ions.

The flux of the main ions also arises due to the friction with impurities. Since $\vec{R}_{21} = -\vec{R}_{12}$, their flux has the opposite direction:

$$Z\vec{\Gamma}_2 = -\vec{\Gamma}_1. \tag{5.62}$$

As a result, there is no transport of the space charge, and hence no additional electric field is necessary to balance the impurity transport.

REFERENCES

1. A. V. Gurevich, E. E. Tsedilina, *Sov.Phys. Uspekhi* **10**: 214-236 (1967).
2. V. E. Golant, *Sov. Phys. Uspekhi* **5**: 161-197 (1963).
3. A. P. Zhilinsky, L. D. Tsendin, *Sov.Phys. Uspekhi* **23**: 331-355 (1980).
4. C. L. Longmire, *Elementary Plasma Physics* (New York: Wiley, 1963): 296.
5. V. Rozhanskij, A. Ushakov, S. Voskoboynikov, *Nuclear Fusion* **36**: 337-340 (1996).
6. V. Rozhansky, *Contrib. Plasma Phys.* **36**: 366-370 (1996).
7. S. I. Braginskii, in *Rewievs of Plasma Physics*, vol.**1**, Editor: M.A. Leotovich (New York: Consultants Bureau, 1963): 183-271.
8. A. V. Gurevich, *Sov. Phys. JETP* **17**: 878-881 (1963).
9. A. A. Ganichev, V. E. Golant, A. P. Zhilinsky, *Sov. Phys. Techn. Phys.* **19**: 58-67 (1964).
10. V. A. Rozhansky, L. D. Tsendin, Geomagnetism and Aeronomia **24**: 755-759 (1984).
11. V. M. Zhdanov, *Transport Phenomena in Multispecies Plasma* (Moscow: Energoizdat, 1982): **176** (in Russian).

Chapter 6

Diffusion of Unbounded Plasma in Magnetic Field

In the general 3D case the diffusion process in the absence of sources and sinks is described by Eqs.(5.1)

$$\frac{\partial n}{\partial t} - \vec{\nabla}\cdot(\hat{D}_e\vec{\nabla}n - \hat{b}_e n\vec{\nabla}\varphi) = 0 ;$$
$$\frac{\partial n}{\partial t} - \vec{\nabla}\cdot(\hat{D}_i\vec{\nabla}n + \hat{b}_i n\vec{\nabla}\varphi) = 0 \quad . \tag{6.1}$$

We shall consider in this Chapter the evolution of inhomogeneities of the uniform background plasma with constant uniform temperatures $T_{e,i}$=*const.* The magnetic field is parallel to the z axis. The boundary conditions for Eqs. (6.1) are

$$n(\vec{r} \to \infty) = n_0 \ ; \quad \varphi(\vec{r} \to \infty) = 0 \, . \tag{6.2}$$

To focus on the main physical effects we shall study pure, weakly ionized and strongly magnetized ($x_e x_i$>>1) plasma. These assumptions makes the final results more compact since it is possible to use the Einstein relations. However, the main results of this Chapter can easily be generalized to the case of partially ionized plasma provided $D_{e\parallel}$>>$D_{i\parallel}$, $D_{i\perp}$>>$D_{e\perp}$.

In principle there exist two extreme cases for 3D diffusion in magnetic field. In plasma of extremely low density, when the Debye radius exceeds the density perturbation scale, the influence of the electric field is negligible. Under these conditions the independent anisotropic diffusion of each species occurs, which is known as unipolar diffusion. In unbounded plasma the equidensities for unipolar diffusion are of the form of ellipsoids of revolution with semiaxes determined by the coefficients $D_{\alpha\|}$, $D_{\alpha\perp}$ of the given species. Another extreme situation is described by Eq. (5.39) of the anisotropic ambipolar diffusion, when the species fluxes are supposed to be equal at each space point and the diamagnetic current flows only along the equidensities. The equidensities now are also ellipsoids but with much smaller semiaxes, which are determined by the ambipolar diffusion coefficients $D_{a\|}$, $D_{a\perp}$, Eqs.(4.10), (5.10), or, in other words, by the diffusion coefficients of the less mobile particles. We shall show in this Chapter that in a magnetic field such ambipolar diffusion takes place only in degenerate situations (for extremely strong inhomogeneity, in 1D cases with density gradient parallel or perpendicular to magnetic field, inside some special spatially restricted regions). In general, multidimensional diffusion is accompanied by flowing of short-circuiting currents which provide quasineutrality. The spreading of the initial inhomogeneity is usually determined by the diffusion of electrons along $\vec{B}$, and of ions across $\vec{B}$, and occurs considerably faster than the diffusion decay even with the unipolar diffusion coefficients.

6.1 SPREADING OF A SMALL INHOMOGENEITY. THE GREEN'S FUNCTION

For small density perturbation $\delta n \ll n_0$ solution of Eqs.(6.1) was investigated in [1-9]. It can be sought in the form of Fourier integrals

$$\delta n(\vec{r},t) = \frac{1}{(2\pi)^3}\int \delta n_{\vec{k}} \exp(i\vec{k}\vec{r})d\vec{k} \; ;$$
$$\varphi(\vec{r},t) = \frac{1}{(2\pi)^3}\int \varphi_{\vec{k}} \exp(i\vec{k}\vec{r})d\vec{k} \; . \qquad (6.3)$$

Inserting Eq. (6.3) into the linearized Eqs.(6.1), we find

$$\delta n_{\vec{k}}(t) = \delta n_{\vec{k}}(0)\exp[-D(\mu^2)k^2 t] \; ;$$
$$\varphi_{\vec{k}} = \frac{D_{e\|}\mu^2 - D_{i\perp}(1-\mu^2)}{b_{e\|}\mu^2 + b_{i\perp}(1-\mu^2)} \frac{\delta n_{\vec{k}}}{n_0} \; , \qquad (6.4)$$

where diffusion coefficient $D(\mu^2)$ is defined according to Eq. (5.32), $\mu=\cos\beta=\vec{k}\vec{B}/kB$. In weakly ionized plasma for $T_e=T_i$, $x_e x_i>>1$ Eq. (5.32) can be reduced to

$$D(\mu^2)=\frac{2[D_{e\|}\mu^2+D_{e\perp}(1-\mu^2)][D_{i\|}\mu^2+D_{i\perp}(1-\mu^2)]}{[D_{e\|}\mu^2+D_{i\perp}(1-\mu^2)]}. \tag{6.5}$$

Let us analyze the dynamics of the point-like initial perturbation

$$\delta n(\vec{r},0)=N\delta(\vec{r}), \tag{6.6}$$

where N is the total number of injected particles, $\delta(\vec{r})$ is the delta-function. Here the ratio $G(\vec{r},t)=\delta n(\vec{r},t)/N$ corresponds to Green's function of the linearized system Eqs.(6.1). In the presence of an arbitrary source its solution is expressed through Green's function

$$\begin{aligned}\delta n(\vec{r},t)=&\int\delta n(\vec{r}\,',0)G(\vec{r}-\vec{r}\,',t)d\vec{r}\,'\\&+\int_0^t\int G(\vec{r}-\vec{r}\,',t-t')I(\vec{r}\,',t')d\vec{r}\,'dt'\end{aligned}. \tag{6.7}$$

According to Eqs.(6.3), (6.4), (6.6) Green's function is given by

$$G(\vec{r},t)=\frac{1}{(2\pi)^3}\int\exp[i\vec{k}\vec{r}-D(\mu^2)k^2t]d\vec{k}\ . \tag{6.8}$$

We introduce the polar coordinates β, $\tilde{\varphi}$ in $\vec{k}$ space with polar axis parallel to magnetic field, $\tilde{\varphi}$ is the azimuth angle, and α is the angle between magnetic field and the radius vector of the observation point. After integrating over k, we have [5]

$$\begin{aligned}G(\vec{r},t)=&-\frac{1}{32\pi^{5/2}t^{3/2}}\int_{-1}^{1}d\mu\int_0^{2\pi}d\tilde{\varphi}\\&\times\frac{1}{D^{3/2}(\mu^2)}\left[1-\frac{r^2F^2}{2D(\mu^2)t}\right]\exp\left[-\frac{r^2F^2}{4D(\mu^2)t}\right],\end{aligned} \tag{6.9}$$

where $F=\mu\cos\alpha+\sqrt{1-\mu^2}\sin\alpha\cos\tilde{\varphi}$. The potential perturbation is given by

$$\varphi(\vec{r},t)=-\frac{1}{32\pi^{5/2}t^{3/2}}\int_{-1}^{1}d\mu\int_{0}^{2\pi}d\tilde{\varphi}\frac{D_{e\parallel}\mu^2-D_{i\perp}(1-\mu^2)}{b_{e\parallel}\mu^2+b_{i\perp}(1-\mu^2)}$$
$$\times\frac{1}{D^{3/2}(\mu^2)}\left[1-\frac{r^2F^2}{2D(\mu^2)t}\right]\exp\left[-\frac{r^2F^2}{4D(\mu^2)t}\right]. \tag{6.10}$$

Asymptotic behavior of Green's function. To find an asymptotic expression for Green's function we convert from the integration over μ in Eq. (6.9) (with polar axis along magnetic field) to an integration over F with polar axis along the radius vector of the observation point, $\vec{r}$, and we change the order of integration [6]

$$G(\vec{r},t)=-\frac{1}{32\pi^{5/2}t^{3/2}}\int_{0}^{2\pi}d\varphi'\int_{-1}^{1}dF$$
$$\times\frac{1}{D^{3/2}(F,\varphi')}\left[1-\frac{r^2F^2}{2D(F,\varphi')t}\right]\exp\left[-\frac{r^2F^2}{4D(F,\varphi')t}\right] \tag{6.11}$$

Here

$$\mu=F\cos\alpha+(1-F^2)^{1/2}\sin\alpha\cos\varphi', \tag{6.12}$$

and φ' is the angle in $\vec{k}$ space between the planes $(\vec{B},\vec{r})$ and $(\vec{k},\vec{r})$. At large values of r the integral over F is dominated by a small region near F=0. In the inner integral we use the variable substitution $Z=rF/[4D(F,\varphi')t]^{1/2}$. The integral is dominated by the region $Z\sim1$, so that at large values of r the inner integral can be extended to infinity and the expression in the denominator can be expanded in a power series in Z. We find that terms of second order vanish and that the integrals containing Z and Z^3 do not contribute to the Green's function because the integrand is of odd parity. The integral from the zero-order terms vanishes with exponential accuracy, because the error which is introduced by the extension of integration in the inner integral over Z to infinity is exponentially small. Finally, we have

$$G(\vec{r},t)=-\frac{t}{8\pi^2r^5}\int_{0}^{2\pi}\frac{d^4D(0,\varphi')}{dF^4}d\varphi'. \tag{6.13}$$

Converting to a differentiation with respect to μ^2 with the help of Eq. (6.12), making the substitution $\sin\alpha\cos\varphi'=\mu$, and exploiting the even parity of the integrand, we find [6] (the equivalent expression has been obtained in [4])

$$G(\vec{r},t)=-\frac{2t}{\pi^2 r^5}\int_0^{\sin\alpha}\frac{d\mu}{\sqrt{\sin^2\alpha-\mu^2}}$$
$$\times[4\mu^4\cos^4\alpha\frac{d^4D(\mu^2)}{(d\mu^2)^4}+12\mu^2\cos^2\alpha(\cos^2\alpha-\mu^2)\frac{d^3D(\mu^2)}{(d\mu^2)^3}. \quad (6.14)$$
$$+3(\cos^4\alpha+\mu^4-6\mu^2\cos^2\alpha\frac{d^2D(\mu^2)}{(d\mu^2)^2}] .$$

The resulting integrals can be calculated easily using the theorem of residues in the complex μ plane, through an integration along a contour enveloping the branch point $\pm\sin\alpha$.

We finally find [6]

$$G(\vec{r},t)=\frac{D_{i\perp}t}{\pi r^5\mu_0^4}\frac{12-36\frac{\sin^2\alpha}{\mu_0^2}+\frac{9}{2}\frac{\sin^4\alpha}{\mu_0^4}}{(1+\frac{\sin^2\alpha}{\mu_0^2})^{9/2}} \quad (6.15)$$

for $\sin\alpha\sim\mu_0=(b_{i\perp}/b_{e\|})^{1/2}$;

$$G(\vec{r},t)=\frac{9}{2}\frac{D_{i\perp}t}{\pi r^5}\frac{\mu_0}{\sin^5\alpha} \quad (6.16)$$

for $\sin\alpha\gg\mu_0$.

In the same way we can derive the asymptotic expression for the potential, performing expansion in F to second order

$$\frac{e\varphi(\vec{r})}{T}=\frac{N}{2\pi r^3\mu_0^2 n_0}\frac{\frac{\sin^2\alpha}{\mu_0^2}-2}{\left(\frac{\sin^2\alpha}{\mu_0^2}+1\right)^{5/2}} \quad (6.17)$$

for $\sin\alpha\sim\mu_0$;

$$\frac{e\varphi(\vec{r})}{T} = \frac{\mu_0 N}{2\pi r^3 n_0} \tag{6.18}$$

for $\sin\alpha >> \mu_0$.

The range of applicability of the asymptotic expressions is defined by the inequalities $r>>(D_{e\parallel}t)^{1/2}$ for Eqs.(6.15), (6.17), and $r>>(D_{i\perp}t)^{1/2}$ for Eqs.(6.16), (6.18). The asymptotic expression for the potential corresponds to the time-independent quadrupole. It changes sign at $\sin\alpha=2^{1/2}\mu_0$, and decreases as r^{-3} with distance. The density perturbation decreases faster with distance, as r^{-5}, and becomes negative at $\alpha\sim\mu_0$, which means that depletion regions are formed in the background plasma [6].

The non-Gaussian asymptotic behavior of the density and potential perturbation at $r\to\infty$ is closely connected with the nonanalytical dependence $D(\vec{k})$ [5], Eq. (6.5). It depends only on the angle between $\vec{k}$ and $\vec{B}$, thus is not defined and has singularity at $\vec{k}=0$. Since the density perturbation with $\vec{k}=0$ corresponds to the injection of additional particles into the ambient plasma, the perturbation which does not change the total number of particles has Gaussian asymptotic behavior.

Solution in the neighboring zone. To evaluate the solution for smaller r we partition the integration range into three subranges: $(0, \mu^*)$, (μ^*, μ^{**}), $(\mu^{**}, 1)$, (for negative μ: $(-1, -\mu^{**})$ $(-\mu^{**}, -\mu^*)$, $(-\mu^*, 0)$, correspondingly), where

$$(b_{e\perp}/b_{e\parallel})^{1/2}=\mu_e<<\mu^*<<\mu_0;$$

$$\mu_0<<\mu^{**}<<\mu_i=(b_{i\perp}/b_{i\parallel})^{1/2}. \tag{6.19}$$

In the first ('electron') of these subranges we can use $D(\mu^2)$ in the form of $D_e(\mu^2)$ in Eq. (5.35). Inserting $D_e(\mu^2)$ into Eq. (6.9) and extending the integration over μ to unity, we obtain

$$G^{(e)} = -\frac{1}{32\pi^{5/2}t^{3/2}}\int_{-1}^{1} d\mu \int_{0}^{2\pi} d\tilde{\varphi} \times \frac{1}{D_e^{3/2}(\mu^2)}\left[1-\frac{r^2F^2}{2D_e(\mu^2)t}\right]\exp\left[-\frac{r^2F^2}{4D_e(\mu^2)t}\right]. \tag{6.20}$$

To calculate $G^{(e)}$ it is convenient to return to the form of Eq. (6.8) and to convert to Cartesian coordinates. The integration yields

$$G^{(e)} = \frac{\exp\left[-\frac{z^2}{4(1+T_i/T_e)D_{e\parallel}t} - \frac{x^2+y^2}{4(1+T_i/T_e)D_{e\perp}t}\right]}{8\pi^{3/2}(1+T_i/T_e)^{3/2}t^{3/2}D_{e\parallel}^{1/2}D_{e\perp}}. \tag{6.21}$$

This expression exponentially decreases outside the 'electron' ellipsoid

$$r\cos\alpha \le [4(1+T_i/T_e)D_{e\parallel}t]^{1/2};$$

$$r\sin\alpha \le [4(1+T_i/T_e)D_{e\perp}t]^{1/2}. \tag{6.22}$$

Analogously, from Eq. (6.10) for the potential in the electron ellipsoid we have

$$\varphi^{(e)} = -\frac{T_i}{e}\frac{N}{n_0}G^{(e)}. \tag{6.23}$$

For the 'ion' subrange (μ^{**}, 1) and (-1, μ^{**}) coefficient $D(\mu^2)$ coincides with $D_i(\mu^2)$ Eq. (5.37). Inserting Eq. (3.57) into Eq. (6.9) and extending the integration over μ to zero, after integrating we obtain

$$G^{(i)} = \frac{\exp\left[-\frac{z^2}{4(1+T_e/T_i)D_{i\parallel}t} - \frac{x^2+y^2}{4(1+T_e/T_i)D_{i\perp}t}\right]}{8\pi^{3/2}(1+T_e/T_i)^{3/2}t^{3/2}D_{i\parallel}^{1/2}D_{i\perp}}. \tag{6.24}$$

Similarly the potential

$$\varphi^{(i)} = \frac{T_e}{e}\frac{N}{n_0}G^{(i)}. \tag{6.25}$$

Eqs.(6.33), (6.34) are small outside the 'ion' ellipsoid

$$r\cos\alpha \le [4(1+T_e/T_i)D_{i\parallel}t]^{1/2};$$

$$r\sin\alpha \le [4(1+T_e/T_i)D_{i\perp}t]^{1/2}. \tag{6.26}$$

Integrals over (μ^*, μ^{**}), ($-\mu^{**}$,$-\mu^*$) for $T_e \sim T_i$ and $\mu^* \sim \mu^{**} \sim \mu_0$ can be estimated as

$$G^{(2)} \sim \frac{1}{16\sqrt{2}\pi^{3/2}t^{3/2}D_{e\|}^{1/2}D_{i\perp}}\,;$$
$$\varphi^{(2)} \sim \frac{T}{e}\frac{N}{n_0}G^{(2)}, \tag{6.27}$$

provided the radius vector corresponds to the region

$$r\cos\alpha \leq [4(1+T_i/T_e)D_{e\|}t]^{1/2};$$
$$r\sin\alpha \leq [4(1+T_e/T_i)D_{i\perp}t]^{1/2}. \tag{6.28}$$

Since $G^{(e)}$, $G^{(i)}$ decrease exponentially outside the electron, Eq. (6.22), and ion, Eq. (6.26), ellipsoids where they are smaller than $G^{(2)}$, it is possible to combine Eqs. (6.21), (6.24) and (6.27) for the neighboring zone Eq. (6.28):

$$\delta n(\vec{r},t) = n^{(e)} + n^{(i)} + O(n_0 h), \tag{6.29}$$

$$\varphi(\vec{r},t) = -\frac{T_i}{e}n^{(e)} + \frac{T_e}{e}n^{(i)} + O(\frac{T_{e,i}}{e}n_0 h), \tag{6.30}$$

where $n^{(e,i)}=NG^{(e,i)}$, and

$$h = \frac{N}{8\pi^{3/2}(1+T_e/T_i)(1+T_i/T_e)^{1/2}n_0 t^{3/2}D_{e\|}^{1/2}D_{i\perp}}. \tag{6.31}$$

Let us estimate the error which arises due to replacement of the function $D(\mu^2)$ by $D_e(\mu^2)$ and $D_i(\mu^2)$ while deriving Eqs.(6.29) - (6.30). For example, in electron ellipsoid $r^2F^2/[D_e(\mu^2)t]\leq 1$, so the terms thrown off have the order

$$\Delta G^{(e)} \sim \frac{1}{32\pi^{5/2}t^{3/2}}\int_{\mu^*}^{1}d\mu\int_{0}^{2\pi}d\tilde{\varphi}\frac{1}{D_e^{3/2}(\mu^2)} \sim G^{(e)}\left(\frac{\mu_e}{\mu_*}\right)^2 << G^{(e)}.$$

In the remote zone which can be defined by the inequalities inverse to Eq. (6.28) the asymptotic expressions Eqs. (6.16) - (6.18) are valid. Dimensionless parameter h by order of magnitude coincides with the relative density perturbation at the boundary of applicability of the asymptotic expressions. The maximum relative depth of the depletion regions is also of the same order.

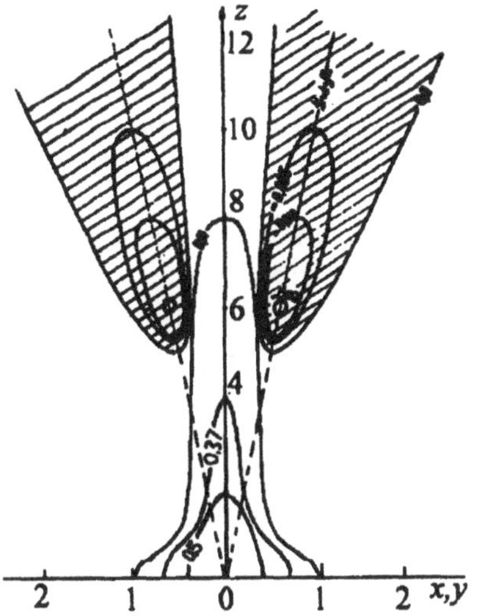

Figure 6.1. Equidensities for the point perturbation $\delta n(\vec{r},t)/\delta n(0,t)$. The calculations were carried out for isothermal weakly ionized plasma with model values x_e=30, x_i=0.3; $\vec{z}\|\vec{B}$; distances are expressed in the units of $(8D_{i\|}t)^{1/2}$. The hatching shows the depletion regions. The number of particles leaving the depletion regions is of the order of 0.4N.

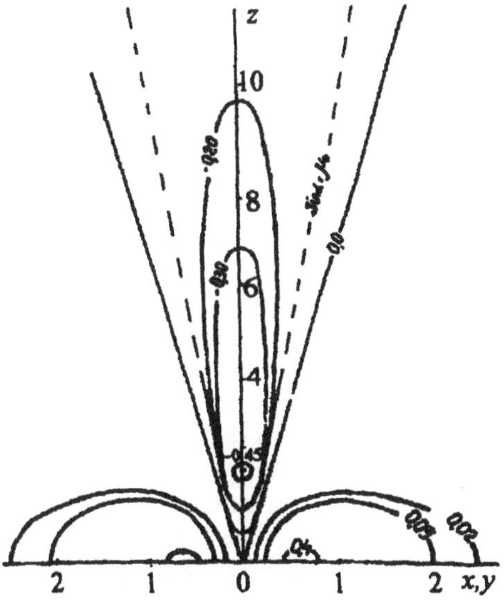

Figure 6.2. Equipotentials for the point perturbation. The potential is expressed in the units of $T\delta n(0,t)/en_0$, the other notations are the same as in Fig. 6.1

Analytic expressions for the density and potential perturbations for the neighboring zone, given by equations. (6.29) and (6.30), were compared with the numerical calculations performed in [5] and [6]. The agreement between analytical and numerical results turned out to be quite satisfactory. Some numerical examples are shown in Figs. 6.1 and 6.2.

Qualitative picture of the diffusion. The reason for the quite unusual character of the solution described above - non-Gaussian asymptotics, anomalous nonellipsoidal shape of the equidensities, large spatial scales - $[4(1+T_i/T_e)D_{e\|}t]^{1/2}$ along $\vec{B}$ and $[4(1+T_e/T_i)D_{i\|}t]^{1/2}$ across $\vec{B}$, formation of the depletion regions, and of the quadrupole potential - consists in the following: the process of spreading inhomogeneity is governed by self-consistent electric fields which are determined by the quasineutrality constraints. Along the magnetic field the electrons are the fastest particles, and the electric field E_z formed in the plasma must confine the electrons in the vicinity of the injection point. The electron mobility across the magnetic field, on the other hand, is low, so that the transverse electric field must confine ions and must be directed toward the center. Accordingly, at the origin the potential must have a saddle point. The boundary condition at infinity, $\varphi(\infty)=0$, results in the formation of two symmetric potential minima on the z axis, at the points $z=\pm z_0$. In the plane perpendicular to z and passing through the origin there is a maximum at $\rho=\rho_0$ ($\rho=(x^2+y^2)^{1/2}$). Accordingly, the electric field at large distances has a quadrupole configuration, $\varphi\sim r^{-3}$, Fig. 6.2. The fluxes which are induced by the electric field are thus proportional to $\Gamma_{e,i}\sim\nabla\varphi\sim r^{-4}$, and the density is $\delta n\sim\vec{\nabla}\cdot\vec{\Gamma}_{e,i}\sim r^{-5}$. The diffusion fluxes in this case, which are of the order of $\nabla n\sim r^{-6}$, are negligible, and the electric field for $|z|>z_0$ and $\rho>\rho_0$ is directed in such a manner that it pulls the most mobile particles in the corresponding directions (it pulls the electrons along $\vec{B}$ and the ions across $\vec{B}$). Hence, for a considerable number of the injected particles the electric field enhances their diffusive spreading in the direction of their maximal mobility.

The particles of the opposite sign must move into these regions in order to maintain quasineutrality. Since these particles are relatively immobile, and the electric field opposes their motion, they cannot arrive from the origin, and the corresponding fluxes must be created in the background plasma. An eddy current arises, which is transported both by the injected particles (in the electron and ion ellipsoids), and by the ambient ones (in the vicinity of the depletion regions). The ambient particles leave the regions, which form an angle μ_0 with the magnetic field - the Green's function in these regions is negative. The angle μ_0 can be determined by equating the time required for the background ions to move across $\vec{B}$ to the z axis ($\sim r^2\mu_0^2/(b_{i\perp}\varphi)$) to the time required for the electrons to move along the magnetic field ($r^2/(b_{e\|}\varphi)$), so that we have $\mu_0=(b_{i\perp}/b_{e\|})^{1/2}$, Fig. 6.1. As a result, the spreading occurs far more rapidly than the ambipolar spreading due to the flow of short-circuiting currents in the background plasma. The mechanism responsible for the high diffusion rate resembles the widely known short-circuiting effect. It was originally proposed by Simon [10]-[11] in order to explain the fact that the plasma diffusion process is greatly enhanced in a vessel with conducting walls. In this case the plasma electrons leave the volume along the magnetic field, plasma ions - across the field, and the plasma current is short-circuited through the conducting walls (see below, Section 7.2). In our situation the ambient plasma itself plays the role of the conducting walls. We shall call this mechanism the short-circuiting effect over the background plasma.

In the neighboring zone diffusive and field-driven fluxes are of the same order. Electrons and ions spread out, roughly speaking, independently and occupy volumes of electron and ion ellipsoids, Eqs.(6.22), (6.26). In the electron ellipsoid electric field confines ions across magnetic field. From the absence of the potential perturbation at the infinity it follows that the electric field in the electron ellipsoid also hinders ions along $\vec{B}$. In other words, the ions are distributed here according to Boltzmann's law, Eq. (6.30). In contrast, in the ion ellipsoid the electron Boltzmann distribution holds. As a result, electrons and ions in their 'own' ellipsoids spread faster than with unipolar coefficients. For electrons the acceleration factor is $(1+T_i/T_e)$, and for ions - $(1+T_e/T_i)$. Therefore, the density perturbation in the neighboring zone presents the superposition of the solutions which corresponds to almost independently spreading electrons and ions, Eq. (6.29). The ratio of electron to ion ellipsoid volumes is given by

$$\frac{V^{(e)}}{V^{(i)}}=\frac{T_i^{3/2}}{T_e^{3/2}}\frac{D_{e\|}^{1/2}D_{e\perp}}{D_{i\|}^{1/2}D_{i\perp}}=\frac{T_i^{3/2}}{T_e^{3/2}}\frac{1+x_i^2}{x_i^{3/2}x_e^{1/2}}.$$

When $V^{(e)}>V^{(i)}$ the electrons are more mobile in the 'three-dimensional sense'. In this case the potential at the origin is positive, Eq. (6.30), and along the magnetic field there exists a region of ambipolar diffusion, where the density decreases from the values $n^{(i)}$ to $n^{(e)}$. In the opposite case $V^{(i)}>V^{(e)}$ the corresponding region exists in the direction across $\vec{B}$, and the potential at the origin is positive. The total number of particles in each ellipsoid $\int n^{(e,\,i)}dzdxdy$ according to Eq. (6.29) is equal to the total number of injected particles N. Since the deficient particles come from the background plasma, the integral over the depletion regions should be also of the order of N. Fig. 6.3 shows the electron and ion diffusive and field-driven fluxes, which lead to the formation and spreading of the depletion regions in the far zone, and extinction of these regions in the neighboring zone; instead the electron and ion ellipsoids are created there.

One should keep in mind that all features of the solution obtained are connected with the redistribution of the ambient plasma in the self-consistent electric field. The spreading of the injected particles, similar to 1D problems, is quite different. If, for example, the small initial perturbation of the test ion density has the form $\delta n_I(\vec{r},t) = N\delta(\vec{r})$, then, neglecting the influence of the perturbed electric field on the injected particles, we have

$$\delta n_I(\vec{r},t) = \frac{N \exp\left[-\frac{z^2}{4D_{I\parallel}t} - \frac{x^2+y^2}{4D_{I\perp}t} \right]}{4\pi^{3/2} t^{3/2} D_{I\parallel}^{1/2} D_{I\perp}}. \tag{6.32}$$

This unipolar solution by a factor $(1+T_e/T_i)$ deviates from the density distribution in the ion ellipsoid.

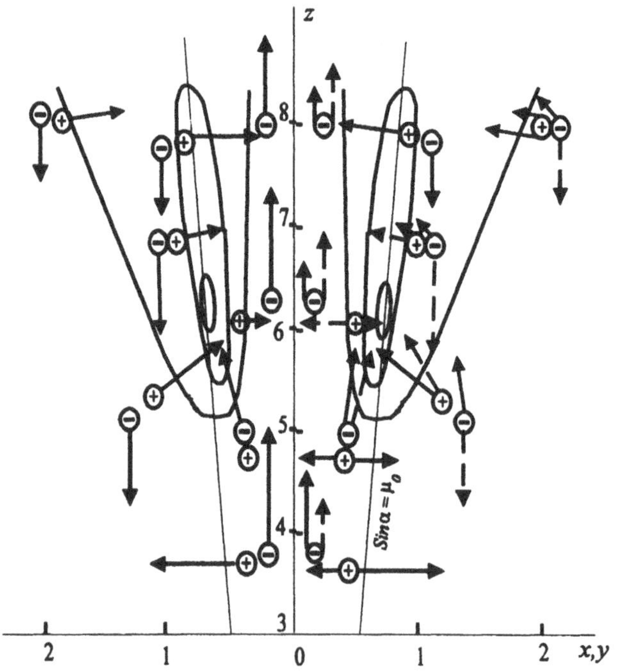

Figure 6.3. Electron and ion fluxes near the depletion region. The dashed arrows at the right show the field fluxes, the solid arrows show the gradient-driven fluxes; shown at the left are the net fluxes. Notations are the same as in Fig. 6.1.

General asymptotic expressions. Here we shall obtain the asymptotic expressions which are valid for arbitrary temperatures and arbitrary initial profiles of injected particles. From the analysis carried out in the previous Sections we can conclude that in the far zone it is possible to neglect the ion motion along the magnetic field and the electron motion across the magnetic field. We start from the condition, which is obtained by equating the second terms in Eq. (6.1),

$$\vec{\nabla}\cdot\vec{\Gamma}_e = \vec{\nabla}\cdot\vec{\Gamma}_i , \tag{6.33}$$

where we shall neglect the small terms of the order of $D_{e\perp}/D_{i\perp}$ and $D_{i\|}/D_{e\|}$. After introducing a new variable ζ and a new function Ψ, according to

$$\zeta = z(b_{i\perp}/b_{e\|})^{1/2}; \qquad \Psi = \varphi - (T_e/e)\ln(n/n_0), \tag{6.34}$$

we have

$$\vec{\nabla}\cdot(n\vec{\nabla}\Psi) = -\frac{T_e + T_i}{e}\Delta_\perp n . \tag{6.35}$$

In the linear approximation this is the Poisson equation for Ψ. In the far zone we can neglect the density perturbation $\delta n/n_0$ with respect to potential perturbation $e\varphi/T_e$, while the main density perturbation is localized in the neighboring zone. Therefore, the function Ψ can be sought in the form of expansion in multipole series. The net 'charge'

$$Q = \int \frac{\varepsilon_0(T_e + T_i)}{en_0}\Delta_\perp n d\zeta dx dy = 0 ,$$

since at $r\rightarrow\infty$ the spatial density derivatives vanish. The dipole moment is also zero due to the symmetry of the problem. Therefore, the potential has the quadrupole character [12]

$$\Psi = M\frac{2\zeta^2 - x^2 - y^2}{16\pi\varepsilon_0(\zeta^2 + x^2 + y^2)^5} . \tag{6.36}$$

Here the quadrupole moment

$$M = \int (2\zeta^2 - x^2 - y^2)\frac{\varepsilon_0(T_e + T_i)}{en_0}\Delta_\perp n d\zeta dx dy = -\frac{4\varepsilon_0(T_e + T_i)\mu_0 N}{en_0} . \tag{6.37}$$

Substituting Eq. (6.37) into Eq. (6.36), returning to the initial variables, taking into account that in the far zone $\Psi\approx\varphi$, we obtain

$$\varphi(\vec{r}) = \frac{N(T_e + T_i)}{4\pi e n_0 r^3} \frac{\mu_0(2\mu_0^2\cos^2\alpha - \sin^2\alpha)}{(\mu_0^2\cos^2\alpha + \sin^2\alpha)^{5/2}} . \tag{6.38}$$

The density perturbation can be calculated from each of the Eqs.(6.1), for example, from the ion equation. Keeping only the field-driven flux across magnetic field, we have

$$\frac{\partial n}{\partial t} = \vec{\nabla}_\perp \cdot (b_{i\perp} n \vec{\nabla}_\perp \Psi) = b_{i\perp} n_0 \Delta_\perp \Psi . \tag{6.39}$$

Substituting Eq. (6.38) into Eq. (6.39) yields

$$\delta n(\vec{r},t) = \frac{N(1 + T_e / T_i) D_{i\perp} t}{2\pi\mu_0^4 r^5} \times \frac{12\cos^2\alpha - 36\cos^2\alpha\sin^2\alpha / \mu_0^2 + (9/2)\sin^4\alpha / \mu_0^4}{(\cos^2\alpha + \sin^2\alpha / \mu_0^2)^{9/2}} . \tag{6.40}$$

The Eqs.(6.38) and (6.40) generalize the Eqs.(6.15) - (6.18). The asymptotic expressions are determined only by the total number of injected particles N, and, therefore, are valid for small perturbation of an arbitrary initial shape.

6.2 INTERMEDIATE AND STRONG PERTURBATIONS

Let us now analyze diffusion of a nonlinear point perturbation [7]. We shall assume that enough time has passed, so that the initial scales of the perturbation $a_{\|,\perp}$ are unimportant: $(D_{e,i\|,\perp}t)^{1/2} >> a_{\|,\perp}$. In other words, we shall obtain the self-similar solution when density and potential depend only on t and combination $\vec{r}/\sqrt{t}$. However, the density perturbation $\delta n(\vec{r},t)$ remains much larger than the density of the ambient plasma n_0.

Regime of intermediate nonlinearity. We seek a solution of Eq. (6.1) as an expansion

$$\delta n = \delta n^{(1)} + \delta n^{(2)} + ..., \qquad \varphi = \varphi^{(1)} + \varphi^{(2)} + ..., \tag{6.41}$$

where $\delta n^{(1)}$, $\varphi^{(1)}$ are the solutions of linearized system Eq. (6.1). Let us consider for simplicity the case of weakly ionized and isothermal plasma $T_e=T_i=T$. For the second correction we find

$$\delta n^{(2)}(\vec{r},t)=\frac{1}{(2\pi)^4}\int \exp[-i(\omega t-\vec{k}\vec{r})]\delta n_{\vec{k}}^{(2)}d\omega d\vec{k} \ ;$$

$$\delta n_{\vec{k}}^{(2)}=\frac{1}{(2\pi)^4 n_0}\int d\lambda \delta n_{\vec{k}_1}^{(1)}\Phi_{\vec{k}_2}^{(1)} \tag{6.42}$$

$$\times\frac{(\vec{k}\hat{D}_i\vec{k})(\vec{k}\hat{D}_e\vec{k}_2)-(\vec{k}\hat{D}_e\vec{k})(\vec{k}\hat{D}_i\vec{k}_2)}{-i\omega[(\vec{k}\hat{D}_e\vec{k})+(\vec{k}\hat{D}_i\vec{k})]+2(\vec{k}\hat{D}_e\vec{k})(\vec{k}\hat{D}_i\vec{k})},$$

where

$$\Phi=\frac{e\varphi}{T} \ ;$$

$$\delta n_{\vec{k}_1}^{(1)}=2\pi\delta[\omega-\omega(\vec{k}_1)] \ ;$$

$$\Phi_{\vec{k}_2}^{(1)}=\frac{2\pi}{n_0}\frac{\mu_2^2-\mu_0^2}{\mu_2^2+\mu_0^2}\delta[\omega-\omega(\vec{k}_2)] \ ;$$

$$\omega(\vec{k})=-iD(\vec{k})k^2 \ ;$$

$$d\lambda=d\vec{k}_1 d\vec{k}_2 d\omega_1 d\omega_2\delta(\vec{k}-\vec{k}_1-\vec{k}_2)\delta(\omega-\omega_1-\omega_2) \ .$$

The integrand in Eq. (6.42) has a first-order pole at $\omega=\omega(\vec{k})$. The integration contour passes below this pole, therefore, the integral splits into the principal value and the term $-i\pi\delta(\omega-\omega(\vec{k}))$ [13]. Enclosing the pole makes a small correction to $\delta n^{(1)}$, so that we consider only the principal value of the integral. Substituting into Eq. (6.42) the expressions for the linear approximation $\delta n^{(1)}$, $\varphi^{(1)}$ obtained in the previous Section, Eqs.(6.3), (6.4), and carrying out the integration, we find

$$\delta n^{(2)}(\vec{r},t)=V.P.\int d\vec{k}_1 d\vec{k}_2 \exp[i(\vec{k}_1+\vec{k}_2)\vec{r} \tag{6.43}$$

$$-D(\mu_1^2)k_1^2 t-D(\mu_2^2)k_2^2 t]\frac{\mu_2^2-\mu_0^2}{\mu_2^2+\mu_0^2}\frac{A(\vec{k}_1,\vec{k}_2)}{B(\vec{k}_1,\vec{k}_2)},$$

where

$$\begin{aligned}
&A(\vec{k}_1,\vec{k}_2)=[(\vec{k}_1+\vec{k}_2)\hat{D}_i(\vec{k}_1+\vec{k}_2)][(\vec{k}_1+\vec{k}_2)\hat{D}_e\vec{k}_2)]\\
&-[(\vec{k}_1+\vec{k}_2)\hat{D}_e(\vec{k}_1+\vec{k}_2)][(\vec{k}_1+\vec{k}_2)\hat{D}_i\vec{k}_2)]\ ;\\
&B(\vec{k}_1,\vec{k}_2)=-[D(\mu_1^2)k_1^2+D(\mu_2^2)k_2^2][(\vec{k}_1+\vec{k}_2)\hat{D}_e(\vec{k}_1+\vec{k}_2)\,.\qquad (6.44)\\
&+(\vec{k}_1+\vec{k}_2)\hat{D}_i(\vec{k}_1+\vec{k}_2)]+2[(\vec{k}_1+\vec{k}_2)\hat{D}_e(\vec{k}_1+\vec{k}_2)\\
&\times(\vec{k}_1+\vec{k}_2)\hat{D}_i(\vec{k}_1+\vec{k}_2)].
\end{aligned}$$

In the previous Section it was shown that inside the electron and ion ellipsoids, where the density perturbation is the most pronounced, the main contribution to the integral over the $\vec{k}$ space is connected with small and large values of $|\mu|$ correspondingly. The contribution from the interval $|\mu|\sim\mu_0$ is small. Therefore, to evaluate the integrals in Eq. (6.42) we account for the pairs of contributions from the regions in the $\vec{k}_1,\vec{k}_2$ spaces which determine these integrals. The dominant terms are those with $0\le|\mu_1|\le\mu^*$, $\mu^{**}\le\mu_2\le 1$ (and vice versa). Here the ratio $A(\vec{k}_1,\vec{k}_2)/B(\vec{k}_1,\vec{k}_2)$ is 1/2, and Eq. (6.42) reduces simply to the product of the corresponding integrals over $d\vec{k}_1$ and $d\vec{k}_2$ spaces:

$$\delta n^{(2)}(\vec{r},t)=n^{(e)}n^{(i)}/n_0\ , \qquad (6.45)$$

where $n^{(e,i)}=(N/n_0)G^{(e,i)}$, Eqs.(6.21), (6.24). All other regions in the $\vec{k}_1,\vec{k}_2$ spaces make small contribution in comparison with Eq. (6.45), of the order of $(n^{(e)}+n^{(i)})O(h)$, where h is given by Eq. (6.31). The other corrections are calculated analogously:

$$\delta n^{(m)}(\vec{r},t)=(n^{(e)}+n^{(i)})O(h^{m-1})+\frac{n^{(e)}n^{(i)}}{n_0}O(h^{m-2})\ ;\ m>2.$$

Similarly, the dominant corrections to the potential are

$$\Phi^{(m)}(\vec{r},t)=\left[\frac{(-1)^m}{m}\left(\frac{n^{(e)}}{n_0}\right)^m-\frac{(-1)^m}{m}\left(\frac{n^{(i)}}{n_0}\right)^m\right][1+O(h)]. \qquad (6.46)$$

Summing the series for the potential and density for the regions, defined by Eqs.(6.22), (6.26), we obtain

$$n \approx n_0 + n^{(e)} + n^{(i)} + n^{(e)} n^{(i)} + O(n_0 h); \tag{6.47}$$

$$\Phi \approx -\ln(1+n^{(e)}/n_0) + \ln(1+n^{(i)}/n_0) + O(Tn_0 h/e). \tag{6.48}$$

The applicability of these equations is limited by the condition

$$h \ll 1, \tag{6.49}$$

so that they give the correct description of not only small perturbations $\delta n \ll n_0$ but also strong plasma perturbations when $h \ll 1$ and $\delta n \gg n_0$. We shall refer to this situation as to the regime of intermediate nonlinearity. In this case, according to Eq. (6.47), the maximum density in the electron ellipsoid is of the order of $(D_{i\perp}/D_{e\perp})n_0 h$, while in the ion ellipsoid it is $\sim(D_{e\parallel}/D_{i\parallel})^{1/2} n_0 h$. Both values can considerably exceed n_0. In the region of intersection of these ellipsoids with the characteristic scales $(8D_{i\parallel}t)^{1/2}$ along $\vec{B}$, and $(8D_{e\perp}t)^{1/2}$ across $\vec{B}$ (which we shall call the ambipolar ellipsoid), the maximum density is even larger - of the order of $(D_{i\perp}D_{e\parallel}^{1/2}/D_{e\perp}D_{i\parallel}^{1/2})n_0 h^2$, Fig. 6.4.

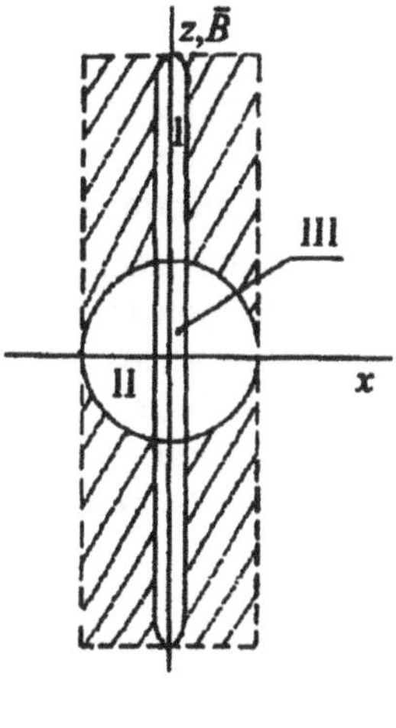

Figure 6.4. Characteristic regions for the case of strong perturbation: I is the electron ellipsoid; II is the ion ellipsoid for $x_i > 1$; III is the ambipolar ellipsoid; hatched is the region from where the ambient particles can escape, outside this region the asymptotic expressions are applicable.

Outside the ellipsoids the density perturbation $\sim n_0 h$ - is small and the linear solution is valid. The generalization of Eqs.(6.47), (6.48) to the case of different temperatures has the form

$$\begin{aligned} n &= n_0 + n^{(e)} + n^{(i)} + n^{(e)} n^{(i)} + O(n_0 h) \\ &= n_0 + n^{(e)} + n^{(i)} + n_a h + O(n_0 h) \ ; \\ \varphi &= -\frac{T_i}{e} \ln\left(1 + \frac{n^{(e)}}{n_0}\right) + \frac{T_e}{e} \ln\left(1 + \frac{n^{(i)}}{n_0}\right) + O\left(\frac{T_{e,i}}{e} n_0 h\right), \end{aligned} \tag{6.50}$$

where

$$n_a = \frac{N \exp\left[-\dfrac{z^2}{4(1+T_e/T_i)D_{i\parallel}t} - \dfrac{x^2+y^2}{4(1+T_i/T_e)D_{e\perp}t} \right]}{8\pi^{3/2}(1+T_e/T_i)^{1/2}(1+T_i/T_e)D_{i\parallel}^{1/2} D_{e\perp} t^{3/2}} \tag{6.51}$$

satisfies Eq. (5.39) of anisotropic ambipolar diffusion. In other words, in the case $h \ll 1$, in which the relative depth of the depletion regions is small, the mechanism of short-circuiting the currents through the background plasma remains effective. The majority of the injected particles can thus spread even faster than with unipolar velocity. In the region of intersection of the electron and ion ellipsoids there is an ambipolar peak (the third term in Eq. (6.50)), which contains a small fraction of the order of h of the total number of injected particles. In this region the diffusion is governed by the ambipolar mechanism, so that the electron and ion currents are equal at each point. In fact, the electron density gradient along $\vec{B}$ is balanced by the second term of the expression for potential, Eq. (6.50), while the first term balances the ion density gradient across the magnetic field. Outside this region the potential profile in the electron and ion ellipsoids, Eq. (6.50), is quite different from that in the linear problem, Eq. (6.30). However, the expressions for the fluxes and thus the density profile remain the same as in the linear case.

Ambipolar regime. When the background plasma is strongly depleted ($h>1$), the nature of the solution should change substantially. In this case the fraction of particles in the central peak tends to the total number of particles N injected into the plasma, and the density profile is given by

$$\delta n(\vec{r}, t) = n_a \,. \tag{6.52}$$

It is easy to see that Eq. (6.52) satisfies the initial Eq. (6.1) for $n_0 \to 0$, and the potential is

$$\varphi_a \approx -\frac{T_e}{e}\frac{z^2}{4(1+T_e/T_i)D_{i\parallel}t}+\frac{T_i}{e}\frac{x^2+y^2}{4(1+T_i/T_e)D_{e\perp}t}+\tilde{\varphi}(t), \qquad (6.53)$$

where $\tilde{\varphi}(t)$ is an arbitrary function of time. However, the potential, Eq. (6.53), diverges at infinity instead of falling to zero. Therefore, the ambipolar solution, Eq. (6.52), (6.53), in the presence of the ambient plasma is valid only in the vicinity of the ambipolar ellipsoid, Fig. 6.4. Outside the ambipolar ellipsoid the short-circuiting mechanism is switched on. As a result, a small fraction of the particles spreads into the electron and ion ellipsoids due to this short-circuiting mechanism. In this region the potential falls to zero. The density in the electron and ion ellipsoids can be estimated on the basis of the characteristic dimensions of the region from which the particles required for the quasineutrality can be extracted. Ions of the background plasma are driven by the electric field from the depletion region across the magnetic field. This region, in its turn, is filled owing to the ion diffusion from the unperturbed plasma. The transverse dimension of the depletion region is thus $(4D_{i\perp}t)^{1/2}$. The longitudinal scale of the depletion region is determined by the electron diffusion, and is of the order of $(4D_{e\parallel}t)^{1/2}$. The amount of particles which reach the ellipsoids is limited by the amount of ambient plasma which can be extracted from the depletion region. It is given by the product of the volume of the depletion region and n_0, see Fig. 6.4. This number determines the density in the electron and ion ellipsoids:

$$\delta n^{(e)}=(D_{i\perp}/D_{e\perp})n_0 \quad ; \delta n^{(i)}=(D_{e\parallel}/D_{i\parallel})^{1/2}n_0. \qquad (6.54)$$

The Eqs. (6.52), (6.54) are reduced to Eq. (6.50) when $h\approx 1$. Outside the region with the scales $(8D_{e\parallel}t)^{1/2}$ and $(8D_{i\perp}t)^{1/2}$ (hatched in Fig. 6.4) the asymptotic dependencies are applicable. Introducing the 'compressed' coordinates Eq. (6.34), and using the fact that potential decreases with distance faster than δn, from Eq. (6.35) we obtain the potential asymptotics in the form of Eq. (6.36). Yet the quadrupole moment is smaller than in the linear case since the nonlinear term $(\vec{\nabla}n)(\vec{\nabla}\Psi)$ is important now. The spatial dependence of the potential, and hence the spatial dependence of the density perturbation therefore coincide with the linear case. The time dependent factor must satisfy the condition $\delta n \sim n_0$ at $h\sim 1$, therefore, at $h>>1$ the asymptotic expressions must have the form

$$\begin{aligned}\delta n(\vec{r},t) &= \delta n_{lin}(\vec{r},t)2K_1/5h\ ;\\ \varphi(\vec{r},t) &= \varphi_{lin}(\vec{r},t)K_1/h,\end{aligned} \qquad (6.55)$$

where K_1 is the numerical factor, and δn_{lin}, φ_{lin} are the linear solutions Eqs.(6.38), (6.40).

We have been treating a positive perturbation of the plasma density; for negative perturbations, the value of $|\delta n|$ is always smaller than n_0, so that the short-circuiting mechanism is effective.

6.3 DIFFUSION OF PERTURBATIONS OF FINITE DIMENSIONS.

We assume that the initial inhomogeneity of the plasma density has a Gaussian distribution

$$\delta n(\vec{r},t) = An_0 \exp[-z^2 / a_\parallel^2 - (x^2 + y^2) / a_\perp^2], \qquad (6.56)$$

where $A = N/(\pi^{3/2} a_\parallel a_\perp^2)$.

Small perturbation. For $A \ll 1$ one can use the Fourier transform Eqs.(6.3), (6.4). Fourier component of the Gaussian initial profile Eq. (6.56) is

$$\delta n_{\vec{k}}(0) = Nn_0 \exp(-a_\parallel^2 k_z^2 / 4 - a_\perp^2 k_\perp^2 / 4). \qquad (6.57)$$

With the initial perturbation chosen in this manner, the equation for δn is the same as Eqs.(6.3), (6.4) for Green's function, with $D(\mu^2)$ replaced by $\widetilde{D}(\mu^2)$, where

$$\widetilde{D}(\mu^2) = D(\mu^2) + a_\parallel^2 \mu^2 / 4t + a_\perp^2 (1 - \mu^2) / 4t. \qquad (6.58)$$

As has been already mentioned in Section 6.1, the asymptotic expression does not depend on the shape of the initial perturbation, and coincides with Eqs.(6.38), (6.40). Therefore, let us analyze the solution in the neighboring zone [6]. As in the case of point perturbation, we partition the integration range in Eq. (6.9) into three subranges.

Let us consider inhomogeneities stretched out along $\vec{B}$ with

$$a_\perp / a_\parallel \ll \mu_0 = (b_{i\perp} / b_{e\parallel})^{1/2}. \qquad (6.59)$$

At small times

$$t \ll a_{\parallel}^2/4D_{e\parallel} \tag{6.60}$$

the main contribution to the integral in the $\vec{k}$ space is connected with the 'electron' region $0 \le |\mu| \ll \mu^*$, where $\mu_e \ll |\mu^*| \ll a_\perp/a_\parallel$. In contrast, the corresponding contribution from the 'ion' region $|\mu| \gg \mu_0$, where $D(\mu^2) = (1 + T_e/T_i)[D_{i\parallel}\mu^2 + D_{i\perp}(1-\mu^2)]$ is exponentially small. Therefore, we have

$$\delta n = n^{(e)} = \frac{N/\pi^{3/2}}{[a_\parallel^2 + 4(1+T_i/T_e)D_{e\parallel}t]^{1/2}[a_\perp^2 + 4(1+T_i/T_e)D_{e\perp}t]} \times \exp\left[-\frac{z^2}{a_\parallel^2 + 4(1+T_i/T_e)D_{e\parallel}t} - \frac{x^2+y^2}{a_\perp^2 + 4(1+T_i/T_e)D_{e\perp}t}\right]. \tag{6.61}$$

The contribution from the region $\mu \sim \mu_0$ is important near the ends of the inhomogeneity $z \sim \pm a_\parallel$. Here the short-circuiting mechanism becomes effective, so that the transverse spreading scale becomes large: of the order of $(a_\perp^2 + 4D_{i\perp}t)^{1/2}$. The parallel spatial scale of these end regions $\Delta a_\parallel \sim (4D_{e\parallel}t)^{1/2}$. As a result, inhomogeneity acquires a 'dumbbell' shape. When the scale $\Delta a_\parallel$ reaches $a_\parallel$ (at the moment $t \sim a_\parallel^2/4D_{e\parallel}$) the contribution from the 'ion' region in the $\vec{k}$ space becomes important. Starting from this moment

$$\delta n = n^{(e)} + n^{(i)}, \tag{6.62}$$

where

$$n^{(i)} = \frac{N}{\pi^{3/2}[a_\parallel^2 + 4(1+T_e/T_i)D_{i\parallel}t]^{1/2}[a_\perp^2 + 4(1+T_e/T_i)D_{i\perp}t]} \times \exp\left[-\frac{z^2}{a_\parallel^2 + 4(1+T_e/T_i)D_{i\parallel}t} - \frac{x^2+y^2}{a_\perp^2 + 4(1+T_e/T_i)D_{i\perp}t}\right]. \tag{6.63}$$

Equation (6.62) corresponds to the short-circuiting in the ambient plasma. Furthermore, the 1D diffusion across the magnetic field discussed in the previous Chapter holds only for very stretched out inhomogeneties

$$a_\parallel/a_\perp \gg (b_{e\parallel}/b_{e\perp})^{1/2} \tag{6.64}$$

at times less than given by Eq. (6.60) (note that (6.64) is much more severe than Eq. (6.59)). In the opposite situation the longitudinal diffusion will become important before electrons can spread in the transverse direction (at time scale $t\sim a_\perp^2/4D_{e\perp}$). The evolution of an inhomogeneity which satisfies Eq. (6.64) is shown in Fig. 6.5. According to Eq. (6.62), at $t>a_\parallel^2/4D_{i\parallel}$ the particles 'forget' the shape of the initial inhomogeneity, and the diffusion reaches the self-similar stage described by Green's function, Eq. (6.29).

For the inhomogeneity

$$a_\parallel/a_\perp << \mu_0^{-1} \tag{6.65}$$

we find analogously, that in the neighboring zone

$$\delta n = n^{(i)}, \qquad t << a_\perp^2/4D_{i\perp}t \ ; \tag{6.66}$$

$$\delta n = n^{(e)} + n^{(i)}, \qquad t >> a_\perp^2/4D_{i\perp}t. \tag{6.67}$$

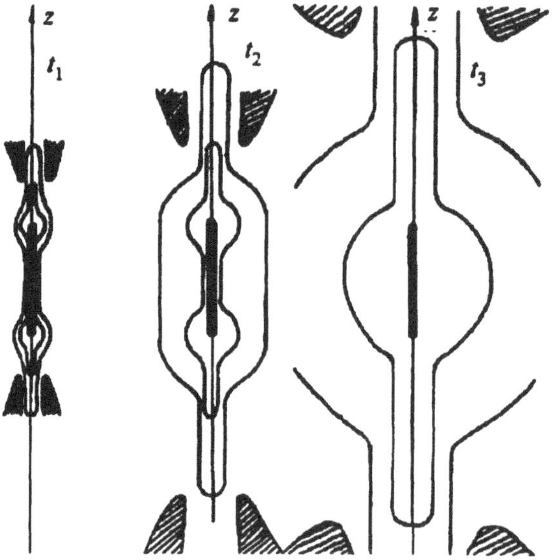

Figure 6.5. Spreading of a linear plasma perturbation, highly stretched along the magnetic field $a_\parallel/a_\perp >> (b_{e\parallel}/b_{i\perp})^{1/2}$: moment $t_1 << a_\parallel^2/(4D_{e\parallel})$ corresponds to 1D diffusion across the magnetic field; moment t_2 to the intermediate case, $t_3 >> a_\parallel^2/(4D_{i\parallel})$ to the self-similar regime.

The ends of the inhomogeneity with the scale $\Delta a_\perp \sim (4D_{i\perp}t)^{1/2}$ from the beginning spread along magnetic field much faster - with the diffusion coefficient of electrons. The 1D diffusion along $\vec{B}$ is possible if

$$a_\parallel/a_\perp \ll (b_{i\parallel}/b_{i\perp})^{1/2}. \quad (6.68)$$

Combining Eq. (6.67) with Eq. (6.62), we finally obtain

$$\delta n = n^{(e)} + n^{(i)}; \qquad t \gg t_0 = \max(a_\parallel^2/4D_{e\parallel},\ a_\perp^2/4D_{i\perp}t). \quad (6.69)$$

The expressions for potential can be obtained in the same manner. The potential Eq. (6.30) corresponds to the density profile Eq. (6.69), and the potential profiles $(-T_i/e)\ln(n^{(e)}/n_0)$ and $(T_e/e)\ln(n^{(i)}/n_0)$ - to the solutions Eqs.(6.61), (6.66) respectively. It is worthwhile to repeat that Eq. (6.69) is not valid for very short times. If one formally takes this expression at $t \to 0$, the limit will be two times larger than the initial perturbation.

Strong perturbation of finite dimensions. The results obtained above can be generalized for the case of a strong perturbation in the following way. If $h(t_0) \ll 1$, then instead of Eq. (6.69), the more general Eq. (6.50) is valid for $t \gg t_0$, where $n^{(e,i)}$ are defined according to Eqs.(6.61), (6.63). This solution corresponds to the case of intermediate nonlinearity. At times smaller than t_0, as in the linear case, the 1D ambipolar diffusion across the magnetic field takes place if Eq. (6.64) is satisfied, and along $\vec{B}$ if Eq. (6.68) is valid. In the intermediate situation for $t \le t_0$ the approximating expression can be used which describe both 1D diffusion along and across a magnetic field in the situations when either Eq. (6.64) or Eq. (6.68) are valid. The approximating ambipolar expression has the form

$$n = n_a = \frac{N}{\pi^{3/2}[a_\parallel^2 + 4(1 + T_e/T_i)D_{i\parallel}t]^{1/2}[a_\perp^2 + 4(1 + T_i/T_e)D_{e\perp}t]}$$

$$\times \exp\left[-\frac{z^2}{a_\parallel^2 + 4(1 + T_e/T_i)D_{i\parallel}t} - \frac{x^2 + y^2}{a_\perp^2 + 4(1 + T_i/T_e)D_{e\perp}t}\right];$$

$$\varphi_a = -\frac{(T_e/e)z^2}{a_\parallel^2 + 4(1 + T_e/T_i)D_{i\parallel}t} + \frac{(T_i/e)(x^2 + y^2)}{a_\perp^2 + 4(1 + T_i/T_e)D_{e\perp}t} + \tilde{\varphi}(t).$$

(6.70)

If $h(t_0) \gg 1$ then the central part of the perturbation spreads according to the ambipolar scenario, and its evolution is described by Eq. (6.70). The ambipolar diffusion takes place until $t \sim t_1 \gg t_0$, where the moment t_1 corresponds to $h(t_1) = 1$. At larger times (as well as in the remote zone) the short-circuiting mechanism becomes effective, and the evolution of the main part of the plasma clot is described by Eq. (6.50).

The Eqs.(6.50), (6.70) were compared with the results of the numerical simulation of the initial system Eq. (6.1) in [14]. The following parameters were chosen: $a_{\|}=a_{\perp}=a$, $D_{e\|}=10D_{i\|}$, $D_{i\|}=D_{i\perp}$, $D_{e\perp}=0.1D_{i\perp}$, $1\le A\le 100$, $T_e=T_i$. Results are shown in Figs.6.6-6.10. Here the dimensionless variables are: $\tilde{n}=n/n_0$, $\tau=D_{i\|}t/a^2$, $\zeta=z/a$, $\tilde{\rho}=(x^2+y^2)^{1/2}/a$. For A=10 the critical values are : $h(\tau_0)$=10, τ_0=0.25, $\tau_1\approx1.2$. It is seen from Fig. 6.6 that the transformation to the case of intermediate nonlinearity occurs at $\tau\sim0.4\tau_1$. Already at τ=0.7, as follows from Figs.6.7, 6.8, the density profiles in the neighboring zone correspond to Eq. (6.50). In the case of the more pronounced inhomogeneity A=100 the calculated profiles in the vicinity of ambipolar ellipsoid are satisfactory described by the ambipolar solution Eq. (6.70), see Figs.6.9, 6.10. Outside the ambipolar region the character of the solution is quite different, since some of the particles spread in a unipolar manner due to the field-driven fluxes from the depletion regions. To provide the necessary fluxes the background plasma is strongly depleted with respect to the case A=10, the maximum depth of the depletion regions for A=100 reaches 20%.

As a result of the comparison we can conclude that the simple expressions Eqs.(6.50), (6.70) can describe satisfactorily the diffusion of the main part of strong inhomogeneity in unbounded plasma in magnetic field.

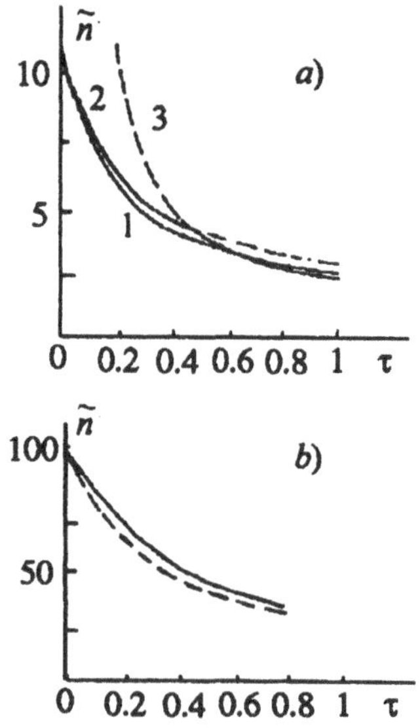

Figure 6.6. Time dependence of the density at the origin. *a*): A=10; 1 is the numerical simulation; 2 corresponds to the approximation Eq. (6.66); 3 is the approximation Eq. (6.50); *b*): A=100; solid line is the numerical simulation, dashed curve corresponds to the approximation Eq. (6.70).

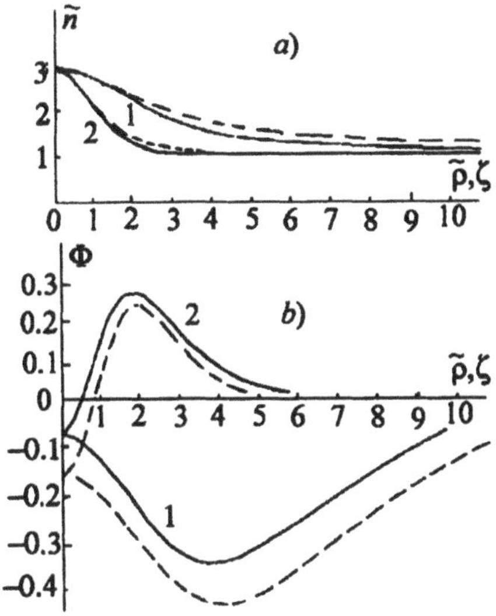

Figure 6.7. *a*): density profiles, *b*): potential profiles. 1 -along, 2 - across the magnetic field. Solid curves are the numerical simulation; dashed curves correspond to the approximation Eq. (6.50). τ=0.7; *A*=10.

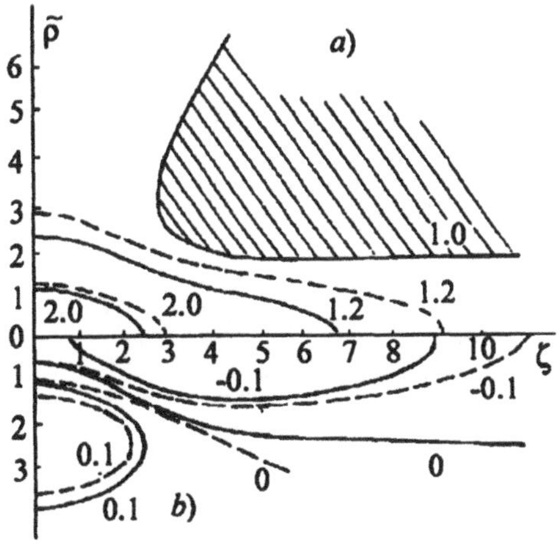

Figure 6.8. *a*): equidensities; *b*): equipotentials. Solid curves are the numerical simulation; dashed curves are the approximation Eq. (6.50). τ=0.7; *A*=10.

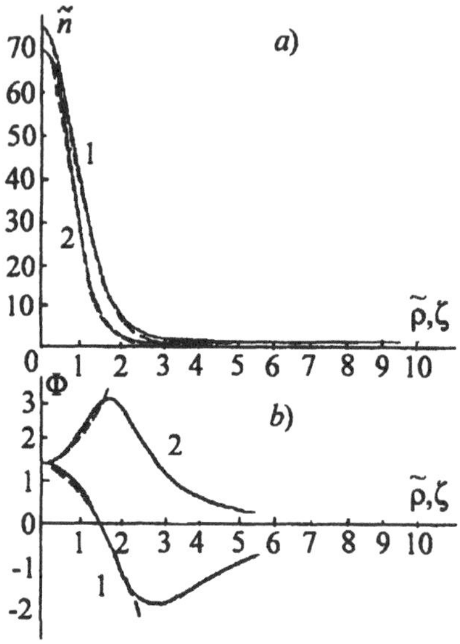

Figure 6.9. *a*): density profiles; *b*): potential profiles. 1 -along, 2 - across the magnetic field. Solid curves are the numerical simulation; dashed curves are the approximation Eq. (6.66). τ=0.1; A=100.

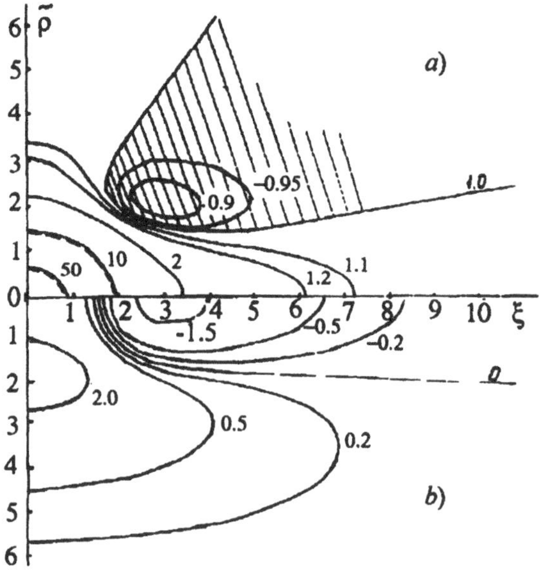

Figure 6.10. *a*): equidensities; *b*): equipotentials. Solid curves are the numerical simulation; dashed curves are the approximation Eq. (6.66). τ=0.1; A=100.

Ionospheric inhomogeneities. The results obtained above were used in [15]-[18] to describe the evolution of the weak ($\delta n \ll n_0$), small scale inhomogeneities in the ionosphere. Numerous perturbations of such a type are generated in the ionosphere by inhomogeneous motions of the neutral gas, currents and fluxes of energetic particles from the magnetosphere. Different inhomogeneities can emerge as a result of the development of the various instabilities in the ambient plasma [19]. Plasma inhomogeneities can be also created artificially due to plasma injection from space vehicles or by a powerful electromagnetic wave from the Earth [20]. In these latter experiments a wide spectrum of inhomogeneities stretching along magnetic field is generated. The inhomogeneities of this artificial turbulence are characterized practically by a single scale along the magnetic field, $a_{\|} \sim 10 Km$, while their perpendicular sizes vary in the range 1 - 100*m*. Their relaxation times have been measured experimentally and are displayed in Fig. 6.11.

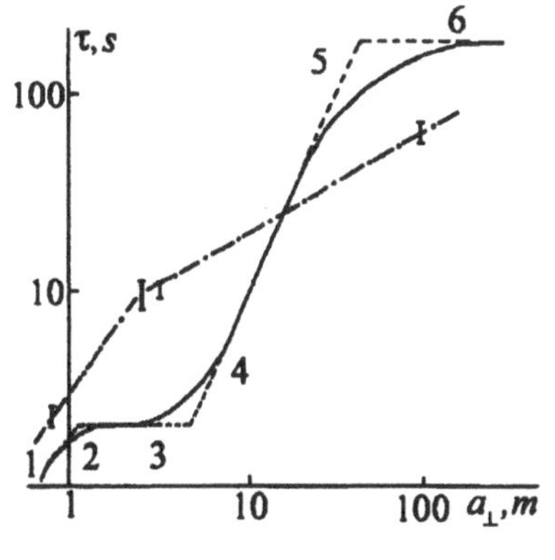

Figure 6.11. The lifetime of a small perturbation versus its transverse scale for a given longitudinal scale. The parameters chosen are: $T_e=T_i$, $a_{\|}=10$ km, $p=2.7$, $D_{e\|}=10^7 m^2/s$, $D_{e\perp}=0.2 m^2/s$, $D_{i\perp}=50 m^2/s$, $D_{i\|}=6.\ 10^4 m^2/s$. Since in the experiment the wave number $k_\perp$ has been measured, for comparison the Gaussian profile with $a_\perp=2/k_\perp$ was taken. Dashed line is the calculation, solid line is the smoothed curve, dashed-dotted line is the approximation of experiment.

Calculations according to Eqs.(6.69), (6.61) and (6.63) lead to the following decay time dependence on the transverse scale. We define the decay time τ as the time when the central density of an isolated inhomogeneity falls p times with respect to the initial value. When $a_{\|}/a_\perp \gg (b_{e\|}/b_{e\perp})^{1/2}$, the inhomogeneity lifetime is determined by the perpendicular diffusion of electrons (region 1 in Fig. 6.11)

$$\tau=a_\perp^2(p-1)/4D_{e\perp}. \tag{6.71}$$

The ionospheric plasma of the F-region, where these measurements were performed, is partially ionized, and the Einstein relation for the perpendicular electron transport coefficients is not valid. Since for $\nu_{ei}>\nu_{eN}$, $m_e\nu_{ei}>\mu_{iN}\nu_{iN}$ and $x_i>>1$, according to Eq. (2.91), (2.92), $D_{e\perp}>(T/e)b_{e\perp}$, the factor $(1+T_i/T_e)D_{e\perp}$ in Eq. (6.61) should be replaced by $D_{e\perp}$.

For $a_{\|}/a_{\perp}\sim(b_{e\|}/b_{e\perp})^{1/2}$ the decay is determined by 3D diffusion of electrons, and from Eq. (6.61) we have (region 2 in Fig. 6.11)

$$\tau=\frac{a_{\perp}^{4/3}a_{\|}^{2/3}p^{2/3}}{4D_{e\|}^{1/3}D_{e\perp}^{2/3}(1+T_i/T_e)^{1/3}}. \tag{6.72}$$

When $a_{\perp}/a_{\|}>>(b_{e\perp}/b_{e\|})^{1/2}$, the electron contribution in Eq. (6.61) results in the saturation of the decay time due to the 1D longitudinal diffusion (region 3 in Fig. 6.11)

$$\tau=\frac{a_{\|}^{2}(p^2-1)}{4(1+T_i/T_e)D_{e\|}}. \tag{6.73}$$

For $a_{\perp}/a_{\|}\sim\mu_0$ the ion and electron contributions to Eq. (6.69) are of the same order, while for $a_{\perp}/a_{\|}>>\mu_0$ the lifetime is determined by the perpendicular ion diffusion (region 4)

$$\tau=\frac{a_{\perp}^{2}(p^2-1)}{4(1+T_e/T_i)D_{i\perp}}. \tag{6.74}$$

For $a_{\perp}/a_{\|}\sim(b_{i\perp}/b_{i\|})^{1/2}$, $p>>1$ we have (region 5)

$$\tau=\frac{a_{\perp}^{4/3}a_{\|}^{2/3}p^{2/3}}{4D_{i\|}^{1/3}D_{i\perp}^{2/3}(1+T_e/T_i)}. \tag{6.75}$$

For $a_{\perp}/a_{\|}>>(b_{i\perp}/b_{i\|})^{1/2}$ (region 6)

$$\tau=\frac{a_{\perp}^{2}(p^2-1)}{4(1+T_e/T_i)D_{i\|}}. \tag{6.76}$$

The results of the calculations are compared with experimental data in Fig. 6.11.

6.4 DIFFUSION OF PLASMA PINCH

This configuration is typical, for example, for the meteor or rocket traces in the ionosphere. The difference with the previously analyzed problems is connected with the absence of the axial symmetry with respect to magnetic field. Therefore, the Hall fluxes, which are driven by self-consistent electric field, generated in the process of diffusion, are not parallel to equidensities, as in the previous Sections. The Hall fluxes can, in principle, strongly influence the inhomogeneity evolution. Furthermore, since in a strong magnetic field the Hall components of the diffusion and mobility tensors are inversely proportional to the magnetic field, while the perpendicular coefficients $D_{e,i\perp}$~B^{-2}, the impact of Hall components can dramatically enhance the inhomogeneity spreading [21]. However, we shall show that such an enhancement only takes place for small angles between the pinch axis and magnetic field. More frequent is the situation when divergence of the Hall fluxes is compensated by divergence of the corresponding parallel or perpendicular fluxes.

We choose the reference frame with the y' axis parallel to the pinch axis, z' belonging to the plane $\vec{B}\vec{y}'$, and the x axis parallel to $[\vec{y}' \times \vec{B}]$, Fig. 6.12.

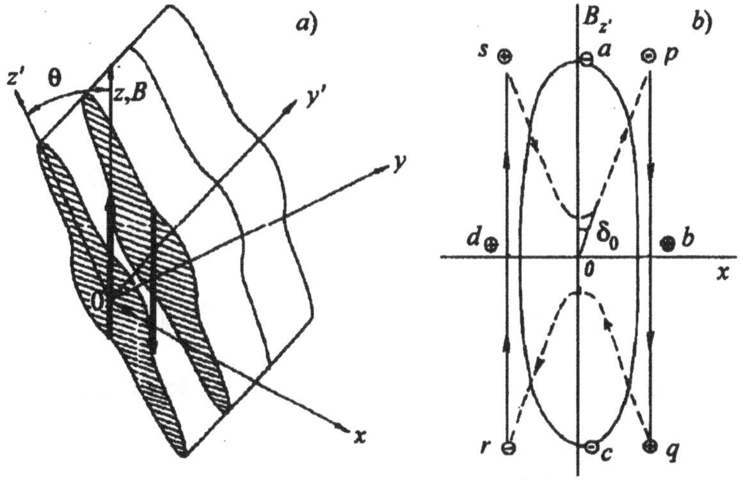

Figure 6.12. Diffusion of a plasma pinch for $\theta >> \mu_0^{1/2}$. a) is the sketch of the pinch profile, *b*) is the ambipolar ellips; *s*, *q* correspond to an additional positive space charge and *p*, *r* to an additional negative space charge. Solid arrows indicate electron fluxes along the magnetic field, dashed arrows indicate short-circuiting Hall electron fluxes.

In these coordinates the problem reduces to 2D. The diffusion tensors Eqs.(2.43), (2.45) in this system are

$$\hat{D}_{e,i}(\theta) = \begin{pmatrix} D_{e,i\perp} & \mp D_{e,i\wedge}\cos\theta \\ \pm D_{e,i\wedge}\cos\theta & \tilde{D}_{e,i\parallel} \end{pmatrix} \qquad (6.77)$$

$$\tilde{D}_{e,i\parallel} = D_{e,i\parallel}\sin^2\theta + D_{e,i\perp}\cos^2\theta \ ,$$

where θ is the angle between $\vec{y}'$ and $\vec{B}$.

In the linear approximation the Hall fluxes are divergence free, and the solution of Eqs.(6.1) with the boundary conditions Eq. (6.2), is simply the 2D analog of the 3D results previously obtained in Section 6.1 [22]. At $\tilde{D}_{e\parallel} >> \tilde{D}_{i\parallel}$ it is valid for not too small inclination angles $\theta >> \mu_0 = (D_{i\perp}/D_{e\parallel})^{1/2}$. The solution in the neighboring zone has the form similar to Eqs.(6.61), (6.63):

$$n^{(e,i)} = \frac{N}{\pi[a_\parallel^2 + 4(1+T_{i,e}/T_{e,i})\tilde{D}_{e,i\parallel}t]^{1/2}[a_\perp^2 + 4(1+T_{i,e}/T_{e,i})D_{e,i\perp}t]^{1/2}}$$
$$\times \exp\left[-\frac{(z')^2}{a_\parallel^2 + 4(1+T_{i,e}/T_{e,i})\tilde{D}_{e,i\parallel}t} - \frac{x^2}{a_\perp^2 + 4(1+T_{i,e}/T_{e,i})D_{e,i\perp}t}\right], \qquad (6.78)$$

where N is the total number of injected particles per unit length of the pinch. The relative depth of the depletion regions is given by

$$h = \frac{N}{4\pi n_0 t(1+T_i/T_e)^{1/2}(1+T_e/T_i)^{1/2}\tilde{D}_{e\parallel}^{1/2}D_{i\perp}^{1/2}} . \qquad (6.79)$$

The potential decreases with distance as r^{-2}, and the density perturbation as r^{-4}.

In the case of intermediate nonlinearity the solution may be constructed in a way similar to that described in Section 6.2 - by the successive approximation method. In contrast to the 3D case, there are two parameters of expansion. Besides the condition $h<<1$, the inequality

$$D_{e\perp}^{1/2}x_e\tilde{D}_{i\parallel}^{1/2} / \tilde{D}_{e\parallel}^{1/2} = \mu_0^2/\theta^2 << 1$$

must be fulfilled. Therefore, Eq. (6.47) is valid when $h \ll 1$ and $\theta \gg (\mu_0)^{1/2}$.

The condition

$$\theta \gg (\mu_0)^{1/2} \tag{6.80}$$

corresponds to the following physical picture. Inside the electron and ion ellipses the equidensities coincide with the equipotentials, so that the Hall fluxes fail to cause any change of the plasma density there. Outside the ellipses, however, the density and potential are only slightly perturbed. Therefore, the Hall fluxes may have an appreciable effect only in a region where the ellipses overlap, i.e., in the ambipolar ellipse. If these currents were not taken into account, the plasma would be polarized as a quadrupole with a negative potential along the z' axis, and positive across the magnetic field, see *abcd* in Fig. 6.12*b*. The maximum divergence of the Hall fluxes in this field occurs in the directions which form an angle $\delta_0 = (D_{e\perp} / \widetilde{D}_{i\parallel})^{1/2}$ with the z' axis (the angle between the fluxes and equidensities is almost 90^0 in this case). The divergence of the electron Hall flux in the field *abcd*, which is of the order of $D_{e\perp} x_e n / (D_{e\perp}^{1/2} \widetilde{D}_{i\parallel}^{1/2} t)$, causes additional polarization *pqrs* in Fig. 6.12*b*. The divergence of the longitudinal electron flux of the order of $\widetilde{D}_{e\parallel} n / (\widetilde{D}_{i\parallel} t)$ is θ^2/μ_0 times larger, if the condition Eq. (6.80) holds. Therefore, a small ($\sim\mu_0/\theta^2$) potential and density perturbations are sufficient to create a small correction to the longitudinal electron current which compensates the electron $\vec{E} \times \vec{B}$ drifts. In other words, the perturbation of the parallel electron fluxes short-circuits the Hall electron fluxes. The ion Hall flux is similarly short-circuited across the magnetic field.

For the strong density perturbation $h \gg 1$ this compensation mechanism remains the same. Hence, if Eq. (6.80) is satisfied, the majority of the particles are described by ambipolar diffusion

$$\delta n = n_a = \frac{N}{\pi [a_\parallel^2 + 4(1 + T_e / T_i)\widetilde{D}_{i\parallel} t]^{1/2} [a_\perp^2 + 4(1 + T_i / T_e) D_{e\perp} t]^{1/2}}$$
$$\times \exp\left[-\frac{z'^2}{a_\parallel^2 + 4(1 + T_e / T_i)\widetilde{D}_{i\parallel} t} - \frac{x^2}{a_\perp^2 + 4(1 + T_i / T_e) D_{e\perp} t} \right];$$
$$\varphi_a = -\frac{(T_e / e) z'^2}{a_\parallel^2 + 4(1 + T_e / T_i)\widetilde{D}_{i\parallel} t} + \frac{(T_i / e) x^2}{a_\perp^2 + 4(1 + T_i / T_e) D_{e\perp} t} + \widetilde{\varphi}(t).$$
$$(6.81)$$

At angles

$$\mu_0<\theta<\mu_0^{1/2} \tag{6.82}$$

for $\delta n>n_0$, the divergence of the Hall fluxes cannot be balanced by the electron flux along the magnetic field and the ion flux across $\vec{B}$. Let us consider the strong perturbation $\delta n>>n_0$, so that the ambient density can be neglected inside the pinch, and the case when the ions are unmagnetized, $D_{i\|}=D_{i\perp}=D_i$, but the plasma is magnetized, $x_e x_i>>1$. This situation is typical for ionospheric meteor traces. The initial Hall fluxes driven by the polarization *abcd* (Fig. 6.12*b*) flow in the direction which forms the angle δ_0 with the magnetic field projection. As in the previous case, they cause polarization *pqrs* (Fig. 6.12*b*). However, now this charge separation cannot be short-circuited along the magnetic field. Therefore, the new polarization *pqrs* causes new electron Hall fluxes which are shown in Fig. 6.13 with dashed lines. These new fluxes change the polarization radically, the resulting polarization *fghk* is shown in Fig. 6.13. The pinch periphery is thus biased negatively with respect to its center. Finally, the Hall fluxes flow almost along the equidensities and, hence, are almost divergence free. Electric field no longer confines ions across the magnetic field, so that the pinch transverse scale increases up to $(4D_i t)^{1/2}$. The small remaining polarization *pqrs* causes the additional Hall fluxes which are not divergence free to provide quasineutrality.

The solution in this case can be constructed by successive approximation method. The zero order approximation has the form ($T_e=T_i=T$, $a_\perp=a_\|=a$) [22]:

$$\delta n^{(0)}=\frac{N}{\pi(a^2+8D_i t)}\exp\left(-\frac{z'^2+x^2}{a^2+8D_i t}\right);$$
$$\frac{e\varphi^{(0)}}{T}=-\frac{z'^2+x^2}{a^2+8D_i t}\frac{1-2D_i/\tilde{D}_{e\|}}{1+2D_i/\tilde{D}_{e\|}}. \tag{6.83}$$

The first order terms

$$\delta n^{(1)}=\frac{xz'}{a^2+8D_i t}\frac{2D_i}{D_{e\wedge}}<<\delta n^{(0)};$$
$$\frac{e\varphi^{(1)}}{T}=\frac{xz'}{a^2+8D_i t}\frac{2D_i}{D_{e\wedge}}<<\frac{e\varphi^{(0)}}{T} \tag{6.84}$$

determine the additional electron Hall fluxes. It can easily be seen that the solutions Eqs.(6.83), (6.84) satisfy the initial Eq. (6.1) when $n_0\to 0$. At large distances $|x|$, $|z'|>>(4D_i t)^{1/2}$ this solution fails since here the density perturbation becomes small, and the short-circuiting mechanism through the ambient plasma is effective. The Hall fluxes become unimportant here, and the solution transforms into a linear one.

For very small angles $\theta<\mu_0$ when $\tilde{D}_{e\parallel} < \tilde{D}_{i\parallel}$ the spreading of the pinch is determined by the ambipolar diffusion. The spatial scale of the pinch drastically decreases up to $(4D_{e\perp}t)^{1/2}$ across $\vec{B}$ and up to $(4\tilde{D}_{i\parallel}t)^{1/2}$ along $\vec{B}$.

A numerical calculation of the spreading of an oblique to $\vec{B}$ pinch has been carried out for a broad range of angles θ and of the ionospheric parameters in [23]. The amplitude chosen $A=N/\pi=2$ corresponds to the case of intermediate nonlinearity. In accordance with the discussed pattern, the potential formed a quadrupole; at large θ its extrema were located practically on the x and z' axis and the density profile was almost symmetrical with respect to x and z. This confirms the minor role of the Hall fluxes. However, a considerable 'bending' of the density and potential profiles was observed at small θ. In the peripheral regions the profile 'bending' was small. The value of μ_0 was close to 1^0 and $\mu_0^{1/2}\sim 7^0$. The rate of density decay in the center increased sharply when θ changed from 0 to 1^0. It was slowed down when $1^0<\theta<5^0$, and variation of θ from 5^0 to 90^0 had practically no effect on the decay rate. This is in agreement with the model presented above. A similar picture was observed in the simulation [24].

Diffusion of meteor traces in the ionosphere has been observed in numerous experiments. For example, in [25] and [26] it was observed that at $\theta>\mu_0$ the spreading was determined by the anisotropic ambipolar diffusion, Eq. (6.81).

REFERENCES

1. A. V. Gurevich, *Sov. Phys. JETP* **17**: 878-881 (1963).
2. A. V. Gurevich, E. E. Tsedilina, *Geomagnetism I Aeronomia* **5**: 251-259 (1965) (in Russian)
3. E. E. Tsedilina, *Geomagnetism I Aeronomia* **5**: 679-687 (1965) (in Russian).
4. A. V. Gurevich, E. E. Tsedilina, *Geomagnetism I Aeronomia* **6**: 255-265 (1966) (in Russian).
5. A. V. Gurevich, E. E. Tsedilina, *Sov.Phys. Uspekhi* **10**: 214-236 (1967).
6. V. A. Rozhanskii, L. D. Tsendin, *Sov. J. Plasma Phys.* **1**: 516-521 (1975).
7. V. A. Rozhanskii, L. D. Tsendin, *Sov. J. Plasma Phys.* **3**: 217-220 (1977).

8. A. P. Zhilinsky, L. D. Tsendin, *Sov.Phys. Uspekhi* **23**: 331-355 (1980).
9. V. A. Rozhansky, *Geomagnetism and Aeronomia* **25**: 762-765 (1985) (in Russian).
10. A. Simon, *Phys. Rev.* **98**: 317-318 (1955).
11. A. Simon, *Phys. Rev.* **100**: 1557-1559 (1955).
12. L. D. Landau, E. M. Lifshits, *Electrodynamics of Continuous Media* (Oxford: Pergamon Press, 1965): 418.
13. E. M. Lifshits, L. Pitaevsky, *Course of Theoretical Physics* vol. **10** *Physical Kineticics* (Oxford: Pergamon Press, 1981): 452.
14. S. P. Voskoboynikov, Yu. V. Rakitskii, V. A. Rozhanskii, Yu. B. Senichenkov, L. D. Tsendin, *Sov. J. Plasma Phys.* **6**: 751-754 (1980).
15. N. Sh. Blaunshtein, E. E. Tsedilina, *Geomagnetism I Aeronomia* **25**: 51-57 (1985) (in Russian).
16. E. E. Tsedilina, *Geomagnetism I Aeronomia* **26**: 401-405 (1986) (in Russian).
17. V. N. Oraevski, N. Sh. Blaunshtein, Yu. Ya. Ruzhin, N. D. Filip, *Artificial Plasma Clouds in the Ionosphere of the Earth* (Springer: Heidelberg, 1990): 318.
18. N. D. Filip, N. Sh. Blaunshtein, L. M. Erukhimov, V. I. Ivanov, *Modern Methods of Investigation of the Dynamic Processes in the Ionosphere,* (Shtiinitsa: Kishinev, 1990) 287 (in Russian).
19. H. Rishbath, O. K. Garriot, *Introduction to Ionospheric Physics* (Academic Press: New York, London, 1969): 374.
20. B. M. Gershman, L. M. Eruhimov, Yu. Ya. Yashin, *Wave Effects in the Ionospheric and Space Plasmas* (Moscow: Nauka, 1984) 538. (in Russian).
21. K. H. Geissler, *Phys. Rev.* **171**: 179-180 (1968).
22. V. A. Rozhansky, L. D. Tsendin, *Geomagnetism I Aeronomia*, **17**: 1002-1007 (1977) (in Russian).
23. T. R. Kaiser, W. M. Pickering, C. D. Watkins, *Planet Space Sci.* **17**:519-552 (1969).
24. A. M. Lyatskaya, M. P. Klimov, *J. Atm. Terr. Phys.* **32** 56-60 (1988).
25. W. J. Baggaley, T. H. Webb, *Planet Space Sci.* **28**: 997-1001 (1980).
26. S. M. Levitsky, Abdrackhmanov, V. P. Timchenko *Izvestia Vuzov, Radiophysika* **25**: 1240-1243 (1982).

Chapter 7

Diffusion of a Bounded Weakly Ionized Magnetized Plasma

The character of 2D or 3D diffusion in a strong magnetic field critically depends on the type of boundary conditions on the vessel walls. In different situations the distinction in the particle lifetimes can reach orders of magnitude. Moreover, inserting special electrodes and applying voltage to them one can control the density profile in different parts of the vessel volume. The corresponding phenomena are very diverse and rather complicated, so in this Chapter we shall discuss only the simplest problem of plasma decay in basic configurations of a cylindrical vessel with dielectric or conducting walls.

7.1 DIFFUSION IN DIELECTRIC TUBE

We consider the decay of the arbitrary initial profile of pure, weakly ionized plasma in a dielectric tube of radius a and length L with the axis parallel to the magnetic field [1]. We introduce four characteristic times:

$$\tau_{\alpha\|}=L^2/(\pi^2 D_{\alpha\|}) \quad ; \quad \tau_{\alpha\perp}=a^2/(\zeta_1^2 D_{\alpha\perp}), \tag{7.1}$$

where ζ_1=2.405. In strongly magnetized plasma ($D_{e\perp}<<D_{i\perp}$) with not very high ionization degree ($m_e\nu_{ei}<<\mu_{iN}\nu_{iN}$) we have

$$\tau_{e\perp}>>\tau_{i\perp} \quad , \quad \tau_{e\|}<<\tau_{i\|}. \tag{7.2}$$

We are interested in the case in which the two fast times $\tau_{e\parallel}$ and $\tau_{i\perp}$ are smaller than the two slow times $\tau_{i\parallel}$ and $\tau_{e\perp}$. The opposite situation is of less interest; in this case practically ambipolar diffusion takes place, since the short-circuiting processes (see below), which occur on the longest of the fast times, do not manage to reveal themselves during the ambipolar diffusion time (the shortest of the slow times). So, while considering the fast processes it is possible to neglect the smallest diffusion coefficients $D_{e\perp}$ and $D_{i\parallel}$ in the basic equations (6.1):

$$\frac{\partial n}{\partial t} = \frac{\partial}{\partial z}\left(D_{e\parallel}\frac{\partial n}{\partial z} - nb_{e\parallel}\frac{\partial \varphi}{\partial z}\right) - \vec{\nabla}_\perp \cdot \left(n\hat{b}_e \vec{\nabla}_\perp \varphi\right) ; \tag{7.3}$$

$$\frac{\partial n}{\partial t} = \vec{\nabla}_\perp \cdot \left(\hat{D}_{i\perp}\vec{\nabla}_\perp n + n\hat{b}_i \vec{\nabla}_\perp \varphi\right) . \tag{7.4}$$

Let us assume that the characteristic dimensions of the initial perturbation $n_0(r, \theta, z)$ are comparable to the dimensions of the tube (evolution of small-scale perturbations is similar to evolution in the unbounded plasma described in the previous Chapter and in Chapter 11; the current distribution is specified by the large-scale profile).

The boundary conditions at the dielectric walls are

$$n|_S=0 \quad ; \quad \Gamma_{e\xi}|_S=\Gamma_{i\xi}|_S , \tag{7.5}$$

where $\Gamma_{\alpha\xi}|_S$ is the flux component normal to the wall. It was stated in Chapter 3 that in the thin sheaths adjacent to the walls the fluxes are conserved (within the ratio of the sheath dimensions to the dimensions of the tube), and the plasma density is low, the boundary conditions Eq. (7.5) may be imposed at a surface near the wall but actually inside the plasma, where quasineutrality conditions holds. This circumstance allows us to solve the problem on the basis of plasma equations, without resorting to an analysis of the processes in the sheaths. According to the division of the characteristic time scales, Eq. (7.2), the decay process can also be subdivided into two stages. The first, fast, stage is controlled by the fast pair of times, Eq. (7.2). Since $\Gamma_{e\perp}$ and $\Gamma_{i\parallel}$ are determined by the slow processes which are neglected during the fast stage, conditions Eq. (7.5) are reduced to the vanishing of $\Gamma_{e\parallel}$ at the ends and the vanishing of $\Gamma_{i\perp}$ at the side walls. Hence, the total amount of plasma particles remains unchanged during this stage.

If $\tau_{e\parallel}<\tau_{i\perp}$ (a 'short device'), the terms on the r.h.s. of Eq. (7.3) are large in comparison with the corresponding terms in Eq. (7.4), so the terms on the r.h.s. of Eq. (7.3) almost compensate each other, which means $\Gamma_{e\parallel}\approx 0$ in the whole plasma volume. The potential is thus

$$\varphi(r,\theta,z,t) = (T_e / e)\ln n(r,\theta,z,t) + \Psi(r,\theta,t) . \tag{7.6}$$

Small corrections to Eq. (7.6) of the order of $\tau_{e\parallel}/\tau_{i\perp}$ ensure the equality of the right sides of Eqs.(7.3), (7.4). To find $\Psi(r,\theta,t)$, we substitute the potential from Eq. (7.6) into Eqs.(7.3), (7.4) and integrate over z from 0 to L (over the length of the device):

$$\begin{aligned} \frac{\partial N_1}{\partial t} &= \vec{\nabla}\cdot(N_1\hat{b}_e\vec{\nabla}\Psi) \ ; \\ \frac{\partial N_1}{\partial t} &= \vec{\nabla}_\perp\cdot[\hat{D}_i(1+T_e / T_i)\vec{\nabla}_\perp N_1 + \hat{b}_i N_1\vec{\nabla}_\perp\Psi] , \end{aligned} \tag{7.7}$$

where

$$N_1(r,\theta,t) = \int_0^L n(r,\theta,z,t)dz$$

is the number of plasma particles in a column with bases at the ends of the device. In the tensor $\hat{b}_e$ only Hall components are to be kept. The potential which reduces to zero the radial current to the wall integrated over z (expression in square brackets on the r.h.s. of second equation (7.7)) is

$$\Psi(r,\theta,t) = -(T_e + T_i)\ln N_1 / e. \tag{7.8}$$

The electron Hall fluxes in this field are divergence free. The longitudinal escape of electrons is blocked by the boundary condition $\Gamma_{e\parallel}=0$, while the transverse escape is blocked by the condition $D_{e\perp}=0$. Hence, the integral N_1, according to Eq. (7.7), is independent of time in the fast stage:

$$N_1(r,\theta,t) = \int_0^L n_0(r,\theta,z)dz = N_1(r,\theta,0) , \tag{7.9}$$

where $n_0(r,\theta,z)$ is the initial density profile. Substituting Eqs.(7.6), (7.8) into Eq. (7.4), we find an equation which is linear in n:

$$\frac{\partial n}{\partial t} = \vec{\nabla}_\perp\cdot[\hat{D}_i(1+T_e / T_i)\vec{\nabla}_\perp n - \hat{b}_i(T_e + T_i)n_\perp] . \tag{7.10}$$

This equation describes diffusion in a medium which moves with a spatially nonuniform velocity proportional to $\vec{\nabla}\ln N_1$. The ion flux

$$\vec{\Gamma}_{i\perp} = (1 + T_e / T_i)\hat{D}_i[n\vec{\nabla}_\perp \ln N_1 - \vec{\nabla}_\perp n]$$

turns to zero at r=0, a. As a result of particle redistribution due to their diffusion, the ion flux $\Gamma_{i\perp}$ becomes zero throughout the volume. The steady state solution of the Eq. (7.10) which satisfies the boundary conditions is thus $n(r,\theta,z)$=$N_1(r,\theta,)f(z)$. Since the number of particles in the band running perpendicular to magnetic field,

$$N_2(z,t) = \int_0^{2\pi} d\theta \int_0^a r n_0(r,\theta,z)dr = N_2(z,0), \tag{7.11}$$

is also conserved in the fast stage, the steady-state solution of the Eq. (7.10) is

$$n(r,\theta,z) = \frac{N_1(r,\theta,0)N_2(z,0)}{\int r n_0 d\theta dr dz}. \tag{7.12}$$

Eq. (7.10) describes the fast relaxation of the initial density profile $n_0(r,\theta,z)$ to the steady-state profile Eq. (7.12). This relaxation occurs over the fast time, of the order of $\tau_{i\perp}$, because of short-circuiting through the plasma. Figure 7.1 shows a simplified diagram of this first fast stage of the decay.

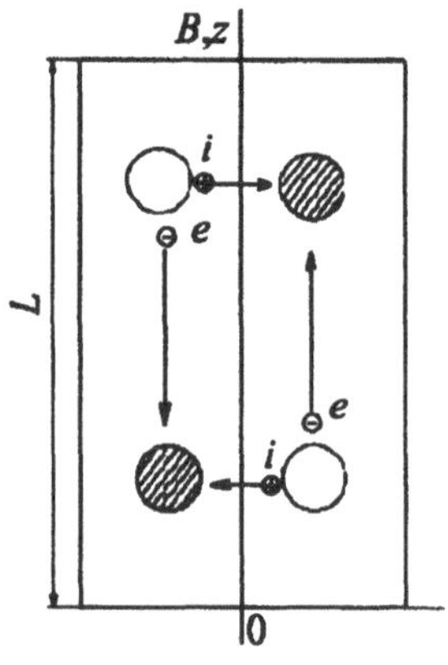

Figure 7.1. Particle fluxes in the case of short circuiting through the plasma.

At the end of the fast stage the density profile Eq. (7.12) corresponds to the potential

$$e\varphi(r,\theta,z) = T_e \ln N_2 - T_i \ln N_1 . \tag{7.13}$$

The solutions Eqs. (7.12), (7.13) can be considered as the initial conditions for the second, slow stage. In this stage the density profile can be sought as a product of two functions on z and r,θ, and the potential as a sum of the two corresponding functions. Since the initial conditions for this stage Eqs. (7.12), (7.13) has the same form, the solution of the initial system Eqs. (6.1) can be obtained by the variable splitting method [2]. In other words, the Eqs.(6.1) system reduces to the ordinary equation of anisotropic ambipolar diffusion Eq. (5.39). Its solution for the decay of the axially symmetric profile is given by (compare with 1D decay profile Eq. (4.16))

$$n = \sum_{j,k=1} A_j B_k J_0(\zeta_k r / a)\sin(\pi j z / L)\exp(-t / \tau_{jk}) ; \tag{7.14}$$

$$\frac{1}{\tau_{jk}} = (1 + T_e / T_i) D_{i\parallel}(\pi j / L)^2 + (1 + T_i / T_e) D_{e\perp}(\zeta_k / a)^2 . \tag{7.15}$$

In the case of 'long' device, $\tau_{e\parallel} > \tau_{i\perp}$, the fast stage occurs in a time scale $\tau_{e\parallel}$, while the slow stage again occurs over the ambipolar time. It is natural to expect that the fast stage leads to the ambipolar profiles Eqs. (7.12), (7.13) also in the case $\tau_{e\parallel} \sim \tau_{i\perp}$.

The observation of the ambipolar diffusion in strong magnetic field, $D_{e\perp} < D_{i\perp}$, is often masked by the development of different types of instabilities which result in the anomalous diffusion perpendicular to magnetic field. Therefore, the effective plasma lifetime becomes considerably shorter than the classical value of Eq. (7.15). Analysis of numerous experiments can be found in the reviews [2] and [3]. Classical ambipolar diffusion has been observed, e.g., in decaying plasma [4] where instabilities have been stabilized by the metal electrode perpendicular to magnetic field. In Fig. 7.2 the perpendicular ambipolar diffusion coefficient which was calculated from the measured plasma lifetimes [4], [5] are displayed versus magnetic field strength. The result is in reasonable agreement with the classical values given by Eq. (7.15). However, the situation when one of the tube's walls, or part of them, is replaced by a conductor, is more complicated. The presence of the conductor may reduce the component of the electric field parallel to the electrode (see the next Section). This effect cancels the factors $(1+T_e/T_i)$ or $(1+T_i/T_e)$ in the corresponding terms in Eq. (7.15).

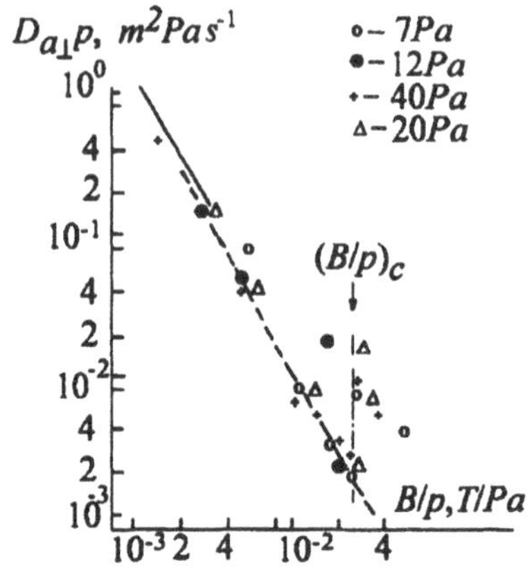

Figure 7.2. Dependence of the transverse ambipolar diffusion coefficient in a helium plasma on the magnetic field; sharp increase at $B=B_c$ is due to transition to the turbulent state. Dashed line corresponds to the classical theory, a=2cm, L=75-80cm, solid line are data of [5].

If the dielectric side walls of a chamber are inclined to the magnetic field, the diffusion occurs much faster. For the slab geometry plasma decay is described by Eq. (5.31). The plasma lifetime for the fundamental diffusive mode is

$$\tau^{-1}=D(\mu^2)(\pi/d)^2, \tag{7.16}$$

where d is the distance between the walls, coefficient $D(\mu^2)$ is defined by Eq. (5.32), $\mu=\cos\beta$; β is the angle between the normal to the wall and magnetic field. In accordance with analysis in Section 5.3, when $\mu_i=(b_{i\perp}/b_{i\|})^{1/2}>>\mu>>\mu_e=(b_{e\perp}/b_{e\|})^{1/2}$ electrons move mainly along the magnetic field, while ions move across the magnetic field. Since usually $\mu_e<<1$, even the small angle between the wall and magnetic field is sufficient for strong reduction of the particle lifetime. This effect has been investigated in [6], see Fig. 7.3, for the plasma decay in cylindrical dielectric tubes.

Eq. (7.16) cannot be applied directly to the cylindrical geometry. In this geometry the Hall currents accelerate diffusion for small inclination angles, see Section 6.4. In the interval $(\mu_0)^{1/2}>\mu>\mu_0=(b_{i\perp}/b_{e\|})^{1/2}$ this effect results, roughly speaking, in the replacement of $D_{i\perp}$ by $D_{i\|}$. Furthermore, for $\mu<a/L$ the influence of the tube ends is important, and diffusion must be ambipolar. Interpretation of the experiments for such small angles is also ambiguous, since plasma was unstable and the diffusion coefficients were anomalous. However, at least qualitative agreement with Eq. (7.16) has been reported.

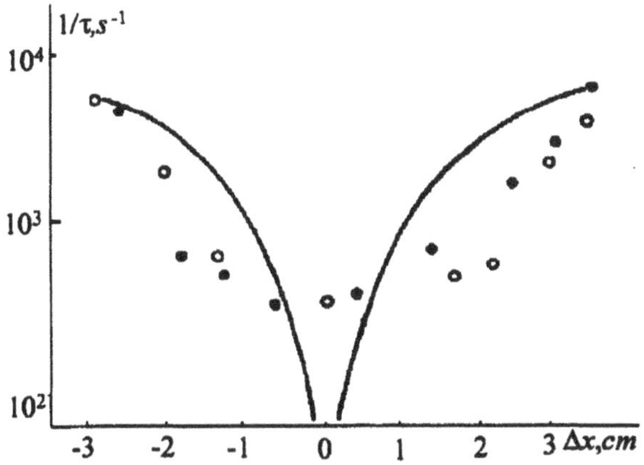

Figure 7.3. Effect of inclination of dielectric tube on the particle lifetime [6]. Experiments in *He* in a tube of length 90cm, B=0.1T. Angle between the tube axis and the magnetic field is $\alpha=\Delta x/L$. Different type of dots correspond to the inclination in two perpendicular directions. Solid line - calculations by Eq. (7.16) for $L\rightarrow\infty$ with d replaced by $\pi a/\zeta$

7.2 DIFFUSION IN A VESSEL WITH EQUIPOTENTIAL CONDUCTING WALLS

We shall consider in this Section the 2D problem of decay of axial symmetrical profile of weakly ionized plasma in the conductive tube with side walls parallel to the magnetic field [1]. The inequality $\mu_{iN}\nu_{iN}>>m_e\nu_{ei}$ is assumed to be satisfied, and the profile is considered to be symmetrical also with respect to z=0 plane. In contrast with the preceding case, the boundary conditions at the wall are imposed on the potential, which can change substantially across the sheaths, rather than on the fluxes, which vary insignificantly in the sheaths. Therefore, we should use the effective boundary conditions, formulated in Chapter 3, for the quasineutral plasma equations. For example, if the potential of the end wall is strongly negative with respect to the plasma, and the sheath adjacent to this wall can be treated as collisionless, the potential difference between the wall and a point in the plasma (close enough to the end that the problem can be assumed 1D along the magnetic field), according to Eq. (3.12), is

$$\varphi(r,z,t)=-\frac{T_e}{e}\ln\frac{\sqrt{2\pi}\Gamma_{el}(r,z=0,L,t)}{n\sqrt{T_e/m_e}}, \tag{7.17}$$

provided the Maxwellian electron distribution is not distorted by the escape of fast electrons to the wall.

In a 'short' device ($\tau_{e\parallel}<\tau_{i\perp}$) we can again, as in the case of a dielectric device, seek the potential in the form of Eq. (7.6). Hence, Eq. (7.17) holds nearly throughout the whole plasma volume, except in the immediate vicinity of the side wall. We also assume, as in the previous Section, that both slow times $\tau_{i\parallel}$ and $\tau_{e\perp}$ exceed the two fast times $\tau_{e\parallel}$ and $\tau_{i\perp}$, so that the small coefficients $D_{e\perp}$ and $D_{i\parallel}$ can be neglected in the transport equations. In this approximation the electron flux to the end walls is related to the change in the number of particles in the column $N_1(r,t)$, by

$$\frac{\partial N_1}{\partial t} = -2\Gamma_{e\parallel}(r, z = 0, t). \tag{7.18}$$

Substituting Eq. (7.18) into Eq. (7.17) we find

$$e\varphi(r,z,t) = -T_e \ln \frac{\sqrt{\pi/2}\,|\partial N_1/\partial t|}{n\sqrt{T_e/m_e}}. \tag{7.19}$$

The substitution of the potential into Eq. (7.4) leads to

$$\frac{\partial n}{\partial t} = (1+\frac{T_e}{T_i})\frac{1}{r}\frac{\partial}{\partial r}(rD_{i\perp}\frac{\partial n}{\partial r}) - \frac{T_e}{T_i}\frac{1}{r}\frac{\partial}{\partial r}\left[rD_{i\perp}\left(\frac{n\partial^2 N_1/\partial r\partial t}{\partial N_1/\partial t}\right)\right]. \tag{7.20}$$

If the potential of the side wall, $\varphi_c \le 0$, is negative with respect to the potential of the end wall or equal to it, the width of the region δ, in which Eq. (7.20) does not hold, can be estimated as follows. Near the side wall the electrons are trapped both along and across magnetic field. Potential with respect to the end wall is given by the Eq. (7.17), while that with respect to the side wall is described by analogous equation (3.43) provided the electron gyroradius exceeds the sheath width. Near the side wall the perpendicular electron flux can be expressed through the potential from Eq. (3.43):

$$\Gamma_{e\perp} = 1/2(\pi T_e/e\Delta\varphi)^{1/2} n^* \tilde{\rho}_{ce} \nu_{eN} (\sqrt{2m_e e\Delta\varphi}) \exp(-e\Delta\varphi/T_e), \tag{7.21}$$

where $\Delta\varphi=\varphi+\varphi_c$. Combining Eq. (7.21) with Eq. (7.17), we obtain

$$\frac{\sqrt{\pi/2}\,\tilde{\rho}_{ce}\nu_{eN}}{\sqrt{T_e/m_e}} \exp\left[-\frac{e|\varphi_c|}{T_e}\right] = \frac{\Gamma_{e\perp}(r=a,z,t)}{\Gamma_{e\parallel}(r=a-\delta, z=0, t)}. \tag{7.22}$$

The ratio $\Gamma_{e\perp}/\Gamma_{e\parallel}$ is thus very small, and only in the thin layer of width δ near the side wall, where the divergence of parallel and perpendicular fluxes are comparable, the electron motion becomes two-dimensional. Therefore, the width of this layer is

$$\delta \sim L\Gamma_{e\perp}/\Gamma_{e\parallel} \sim L\exp(-e|\varphi_c|/T_e)\nu_{eN}/\omega_{ce}. \qquad (7.23)$$

Equation (7.20) has the particular solution

$$n(r,z,t) = AJ_0(\zeta_1 r/a)f(z)\exp(-t/\tau_{i\perp}) \qquad (7.24)$$

which corresponds to the decay with the 'short-circuiting' time $\tau_{sc}=\tau_{i\perp}$, where $\tau_{i\perp}$ is defined by Eq. (7.1). Due to the proper account for the potential drop in the sheaths adjacent to the end walls, this decay time differs by the factor $(1+T_e/T_i)$ from the original value derived by Simon [7] from qualitative consideration. The difference becomes particularly important in the case $T_e>>T_i$ which is typical for devices with gas discharge plasmas [8]. The potential drop between a point in the plasma and the end wall is determined by Eq. (7.19). For the profile Eq. (7.24), the radial dependencies of the density and of particle escape rate in the column, $\partial N_1 / \partial t$, coincide, and, as a result, the potential is independent from the radius.

In other words, radial electric field is absent almost in the whole volume of the tube, with exception of a thin layer of the order of δ, Eq. (7.23), near the side walls. When $\varphi_c<0$ (the end walls are biased positively with respect to the side walls), a further change in φ_c affects only the potential drop in the side sheath and the plasma profile on the scale δ. The plasma decay in the case of positive ends is completely analogous to the short-circuiting case $\varphi_c=0$.

The solution for the case of a 'long' device, $\tau_{e\parallel}<\tau_{i\perp}$, at $\varphi_c\geq 0$ can be obtained analogously. The ions are trapped with respect to the side walls, and potential corresponds to the ion Boltzmann distribution. The potential is bounded to the side wall, the longitudinal electric field is absent in the tube volume, and variation of φ_c for negative ends changes only the potential drop within the sheaths adjacent to the end walls, and the decay time is $\tau_{sc}=\tau_{e\parallel}$.

Finally, we can conclude that for an arbitrary device the decay occurs on a time scale which is the largest among the fast times:

$$\tau_{sc}=\max(\tau_{i\perp}, \tau_{e\parallel}). \qquad (7.25)$$

While obtaining the solution Eq. (7.24) for 'short' device, we used the assumption about the collisionless character of the sheaths adjacent to the end walls. In the opposite case of collisional sheaths the potential in plasma with respect to the end wall is the sum of the potential difference over the plasma and the potential drop in the sheath, Eq. (3.59):

$$\frac{e\varphi(r,z,t)}{T_e} = \ln\frac{n}{n_s[\Gamma_{i\parallel}(r,z=0,t)]} - \ln\frac{D_{i\parallel}\Gamma_{e\parallel}(r,z=0,t)}{D_{e\parallel}\Gamma_{i\parallel}(r,z=0,t)}. \quad (7.26)$$

Here we neglected the second logarithmic correction in Eq. (3.59). The plasma density at the sheath edge is given by Eq. (3.54):

$$n_s=[T_e\Gamma_{i\parallel}{}^2/(8\pi e^2(T_e+T_i)^2 b_{i\parallel}{}^2)]^{1/3}. \quad (7.27)$$

Substituting Eq. (7.18) into Eq. (7.26), (7.4), we find

$$\frac{\partial n}{\partial t} = (1+\frac{T_e}{T_i})\frac{1}{r}\frac{\partial}{\partial r}(rD_{i\perp}\frac{\partial n}{\partial r}) - \frac{T_e}{T_i}\frac{1}{r}\frac{\partial}{\partial r}\left[rD_{i\perp}\left(\frac{n\partial^2 N_1/\partial r\partial t}{\partial N_1/\partial t}\right)\right]$$
$$+\frac{T_e}{3T_i}\frac{1}{r}\frac{\partial}{\partial r}\left[rD_{i\perp}n\frac{\partial}{\partial r}\ln\Gamma_{i\parallel}(r,z=0,t)\right]. \quad (7.28)$$

This equation differs from Eq. (7.20) by the last term; the density profile depends explicitly on the ion flux to the end walls. However, its partial solution has the same form Eq. (7.24) as for the collisionless sheaths, but with $\tau_{i\perp}$ replaced by $\tau_{i\perp}/(1+T_e/3T_i)$. It is seen from Eq. (7.26) that in this case there is a transverse electric field in the plasma which accelerates the ions to the side walls, so that the plasma decay time is slightly shorter than in the preceding case. The potential distribution in the plasma is shown schematically in Fig. 7.4.

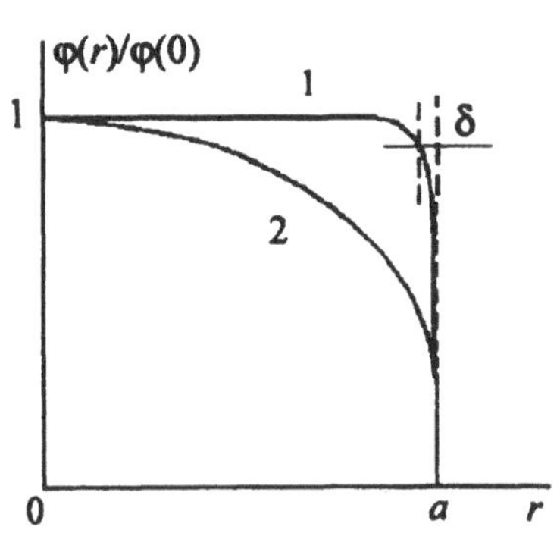

Figure 7.4. Potential profile across $\vec{B}$ during the decay in a 'short' ($\tau_{e\parallel} \ll \tau_{i\perp}$) tube with the equipotential walls. 1-collisionless end sheath; 2-collisional end sheath.

The nature of the plasma decay thus depends on the nature of the sheath adjacent to the end, in the case of a short device, or on the nature of the sheath adjacent to the side wall in the case of a long device.

The character of the diffusion in the vessel with the conducting walls strongly depends on the initial profile of the plasma. Let us examine how the initial profile affects the decay. We assume $\varphi_c \leq 0$ in a short device, and we assume that the initial distribution is Eq. (7.24) with a small additional perturbation, for which the characteristic dimensions are L along z and $\tilde{a}$ along r. If $\tilde{\tau}_{i\perp} = \tilde{a}^2 / D_{i\perp} < \tau_{e\parallel}$ the influence of the walls is unimportant and the perturbation spreads and decays with $\tau \sim \tilde{\tau}_{i\perp}$ similar to the case of unbounded plasma. In contrast, if $\tilde{\tau}_{i\perp} >> \tau_{e\parallel}$, than the situation resembles the one in the dielectric tube discussed in the preceding Section. During $\tau \sim \tilde{\tau}_{i\perp}$ a profile of the type of $N_1(r)N_2(z)$ is established. Later, in spite of the fact that the profile has two different scales in r direction, $\tilde{a}$ and a, the diffusion across the magnetic field is determined by the tube radius and occurs in a time scale $\tau_{i\perp} > \tilde{\tau}_{i\perp}$ [1]. The reason is connected with the fact that, on neglecting the electron transverse diffusion, the integral $N_1(r,t)$ can only decrease with time due to escape of electrons to the end walls. Thus, the transverse diffusion of ions of the perturbation is blocked by the electric field since no electrons can come from the end walls to compensate the ion space charge.

If, for example, the initial density perturbation is a high, narrow peak of width $\tilde{a}$, the ions cannot move across $\vec{B}$ in this perturbation (and thus electrons cannot move along the magnetic field), so that the potential with respect to the ends is high and increases in the direction across $\vec{B}$ (Fig. 7.5).

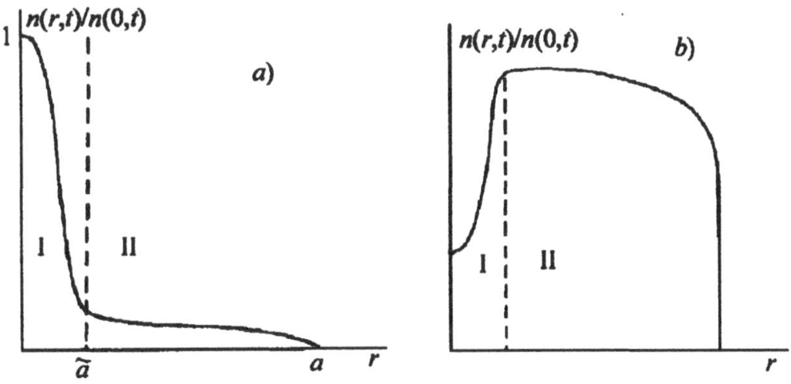

Figure 7.5. Density (a) and potential (b) distributions in the case of sharply peaked pinch in a tube with the equipotential walls.

The electrons in region II are thus blocked along the magnetic field far more strongly than in region I, and the density profile in region II does not change during the decay time of the central peak. Accordingly, the ions from the central peak, region I, pass through region II and escape to the side walls.

If the number of particles in the central peak is high, the plasma decay time is far longer than time which is necessary for ions to reach the side wall, $\tau_{i\perp}$, and we can distinguish a fast and slow stage. In the fast stage, the profile Eq. (7.12) is established in a time of the order of $\tau_{i\perp}$. In the slow stage, the ion current to the side wall

$$I_{i\perp}|_{r=a} = e\int_0^L \Gamma_{i\perp}\,dz \approx -eD_{i\perp}\partial N_1 / \partial r|_{r=a}$$

is limited, constant and formed by particles from the central peak. Consequently, the height of the peak falls off linearly with the time according to

$$\frac{\partial N_1^{(I)}}{\partial t} \approx -\frac{D_{i\perp}N_1^{(II)}}{\tilde{a}^2} \tag{7.29}$$

until $N_1^{(I)}$ becomes of the order of $N_1^{(II)}$.

The idea of acceleration of collisional diffusion in partially ionized plasma in the devices with conducting walls has been put forward by Simon [7]. Later this effect has been investigated in numerous experiments. In [9], [10] - [12] hot plasma pinch was localized in the central part of the device while the edge region was filled by low temperature plasma due to its diffusion. A diffusion coefficient much faster than the ambipolar coefficient has been reported, [9] and [7]. In [10] it was first observed that electrons diffuse along the magnetic field, while ions escape across the magnetic field.

Plasma decay in the metallic chambers has been studied in [5], [13] and [14]. Species temperatures in these experiments were equal to room temperature, and the short decay time with respect to the ambipolar one was reported. In [13] the acceleration of the decay process can be initiated in any arbitrary moment by short-circuiting of the end and the side walls. Before short-circuiting the slow ambipolar decay process, as described in the preceding Section, took place, and after the short-circuiting of the walls, the regime of fast diffusion was observed. The decay rate after short-circuiting increased drastically by more than two orders of magnitude, see Fig. 7.6.

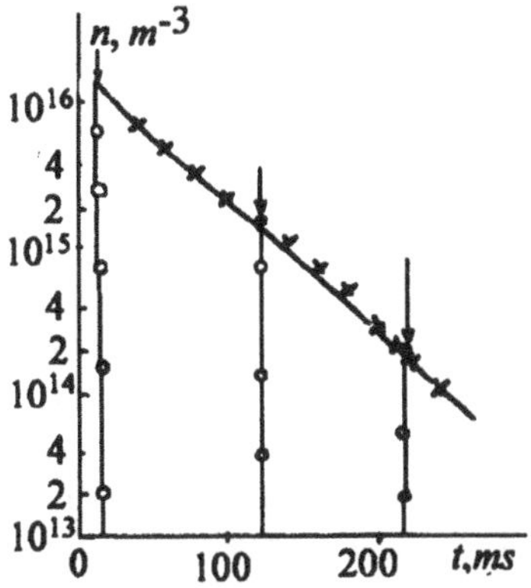

Figure 7.6. Decay of helium plasma in a metallic chamber [13]; a=2cm, p=13.3Pa, B=0.2T. The end electrode was simulated by a cylinder in a weak field. Arrow indicates moment when the end and side electrodes are connected; x - ambipolar decay, o - short circuited walls.

7.3 CONTROL OF THE PLASMA PARAMETERS BY BIASING

The application of a potential difference between the walls results in a controlled variation of the plasma parameters, which is very important for numerous applications. Moreover, if the walls are divided by several conducting sectors, the biasing of separate sectors may result in more or less local control of the plasma parameters. Local plasma control is of special significance, for example, in the task of material processing. The modern devices for electron cyclotron heated discharges, for example, (see [15] and references therein), which are widely used for processing, often contain both dielectric and conductive walls in a rather strong magnetic field. The biasing of vessel walls is a flexible and powerful tool for the control of the plasma parameters in such devices. Here we consider several examples when a voltage is applied between the end and the side walls of a vessel which contains magnetized decaying plasma, to focus on the new effects, which emerge in the various biasing schemes and result in the plasma control.

In the 'short' ($\tau_{e\parallel} \ll \tau_{i\perp}$) cylindrical tube the plasma decay is described by Eqs.(7.19) and (7.20) independently of the biasing potential difference φ_c between the end and the side walls. In the case of the collisionless sheath adjacent to the end wall, the potential difference between the central part of the plasma and the end wall is given by Eq. (7.19). For the profile Eq. (7.24) we have

$$e\varphi(z) = -T_e \ln \frac{2\int_0^L f(z)dz}{\tau_{i\perp} V_{Te} f(z)} \sim T_e \ln \frac{\tau_{i\perp} V_{Te}}{L}, \tag{7.30}$$

where $V_{Te}=[T_e/(2\pi m_e)]^{1/2}$. As discussed in the preceding Section, the transverse electric field is absent in the main part of the volume.

In the short-circuiting case, $\varphi_c=0$, the potential difference between the central part of the plasma and the side wall is located in the narrow region of the order of δ, Eq. (7.23). If the side wall is biased negatively with respect to the end wall, $\varphi_c<0$, the situation remains almost the same - the potential in the plasma is again bounded to the end wall according to Eq. (7.30), and the additional potential difference is localized in the narrow region of width δ. This value decreases with φ_c according to Eq. (7.30).

In contrast, when the side wall is biased positively with respect to the end wall, $\varphi_c>0$, the effect is much more pronounced. For a positive side wall the potential difference between the plasma center and the side wall is less than Eq. (7.30), and at

$$e\varphi_c \approx T_e \ln \frac{\tau_{i\perp} V_{Te}}{L} \tag{7.31}$$

the potential drop between the center and the side wall becomes zero. A further increase of φ_c results in the perpendicular electric field in the plasma volume which decelerates transverse diffusion of ions. Furthermore, the decay slows, and the decay time τ becomes longer than $\tau_{i\perp}$. The considerable difference in the decay times means that the ions in the plasma are distributed in accordance with the Boltzmann law. If, for example, the sheath adjacent to the side wall is also collisionless, the potential drop in the sheath is given by Eq. (3.44). Let us consider the situation when the potential drop in the sheath near the side wall is of the order of T_i/e. In this case ions are already trapped across the magnetic field, so that Boltzmann distribution for them is more or less satisfied, but the decay time is not yet reduced strongly, $\tau \sim \tau_{i\perp}$. The potential difference between the center and the edge of the sheath is, with logarithmic accuracy, of the order of $T_i / e \ln(a / \tilde{\rho}_{ci})$. Therefore, the strong increase of the decay time starts when the biasing potential exceeds the value

$$\varphi_c^* = \frac{T_e}{e} \ln \frac{\tau_{i\perp} V_{Te}}{L} + \frac{T_i}{e} \ln \frac{a}{\tilde{\rho}_{ci}}. \tag{7.32}$$

In other words, the difference φ_c-φ_c^* is dropped in the sheath near the side wall, so to reduce strongly the ion flux to the side walls the applied voltage has to be large: φ_c-φ_c^*>>T_i/e. The further increase of the potential drop in this sheath results in a sharp decrease of the perpendicular ion flux and exponential increase of the plasma lifetime.

For φ_c-φ_c^*>>T_i/e, as in the case of a tube with dielectric walls discussed in Section 7.1, it is possible to distinguish between fast and slow stages of the decay. In the fast stage the number of particles in the volume is conserved (within terms of the order of $\tau_{i\perp}/\tau$) and the profile given by Eq. (7.12) is established over a time ~ $\tau_{i\perp}$. In the slow stage we can ignore the left side of Eq. (7.20) and put $\Gamma_{i\perp}\approx 0$. After integrating the equation $\Gamma_{i\perp}\approx 0$ over z we find

$$\frac{\partial N_1}{\partial t} = C_1(t) N_1^{\gamma+1}, \tag{7.33}$$

where

$$C_1(t) = \left(\frac{\partial N_1}{\partial t} N_1^{-\gamma-1} \right)\Big|_{r=a-\delta}; \quad \gamma = T_i/T_e.$$

One can prove that Eq. (7.33) is established in the end of the fast stage. To do this it is necessary to express Ψ, Eq. (7.6), through Eq. (7.8), (7.19).

The solution of Eq. (7.33) is

$$N_1^{-\gamma}(r,t) - N_1^{-\gamma}(0,t) = N_1^{-\gamma}(r,0) - N_1^{-\gamma}(0,0). \tag{7.34}$$

It is valid in the whole plasma volume except the small region of width δ near the side walls where the motion of electrons becomes 2D. To derive an equation for $N_2(z,t)$ we note that the potential drop between a point in the plasma and the side wall is

$$e\varphi(r) - e\varphi_c = T_i \ln \frac{\Gamma_{i\perp}(r=a,z,t)}{\tilde{\rho}_{ci} \nu_{iN} n(r)}. \tag{7.35}$$

Here we assumed for simplicity ω_{ci}>>ν_{iN}, and used Eq. (3.44) with logarithmic accuracy in combination with the Boltzmann distribution for ions in the volume. The quantity $\Gamma_{i\perp}(r=a,z,t)$ is related to the escape of particles from the band perpendicular to z in accordance with

$$\left| \frac{\partial N_2}{\partial t} \right| = 2\pi a \Gamma_{i\perp}(r=a,z,t). \tag{7.36}$$

Using Eqs.(7.35), (7.36) we can rewrite the Eq. (7.3) for electrons in a form analogous to Eq. (7.33):

$$\frac{\partial N_2}{\partial t} = C_2(t) N_2^{1+1/\gamma}, \tag{7.37}$$

where

$$C_2(t) = \left(\frac{\partial N_2}{\partial t} N_2^{-1-1/\gamma} \right) \Big|_{z=0} .$$

As a result, the potential $\varphi(r)$ is determined by the Eqs.(7.35) - (7.37), and the potential drop between the center and the end wall by Eqs.(7.19), (7.33). Summing up Eqs. (7.19) and (7.35), we find the potential difference between the end and the side walls:

$$\varphi_c = \frac{T_i}{e} \ln \frac{2\pi a\, n\, \tilde{\rho}_{ci} \nu_{iN}}{|\partial N_2 / \partial t|} + \frac{T_e}{e} \ln \frac{n V_{Te}}{2|\partial N_1 / \partial t|} . \tag{7.38}$$

Since the l.h.s. of Eq. (7.34) is increasing with time exponentially (as will be shown later), at $t > \tau$ it is possible to put to zero the r.h.s. of this equation. This means that except for small regions near the walls, the function $N_1(r,t)$ is nearly independent of r. An analogous assertion for the function $N_2(z,t)$ follows from Eq. (7.37).

This density profile, which is practically independent of the coordinates, is established for the following reasons. On the one hand, since τ is longer than $\tau_{e\|}$, $\tau_{i\perp}$, the particle fluxes must be smaller than $D_{e\|}n/L$, $D_{i\perp}n/a$. On the other hand, the potential is 'tied' to the conducting walls, so both the longitudinal and transverse electric fields in the volume are weak. The fluxes can accordingly be small only if the corresponding components of the density gradients are very small. The small corrections to N_1 and N_2 which depend on r, z and which provide the necessary fluxes can be found by the method of successive approximations.

Relaxation of the initial profile of the slow stage $N_1(r,0)$ to the plateau occurs in following manner. Initially, the transverse ion diffusion is balanced by the electric field-driven fluxes. Therefore, the longitudinal electron motion near the side walls is opposed far more effectively than in the central region, so that the density decays more rapidly in the central region at small values of r. Analogously, the profile $N_2(z,t)$ becomes flatter. Thus at $t > \tau$ we have

$$N_1 \approx N/(\pi a^2) \;\; ; \;\; N_2 \approx N/L , \tag{7.39}$$

where N is the total number of particles. Finally, combining Eqs.(7.38), (7.39) we find an exponential decay. For the case $T_e=T_i=T$ the time constant is

$$\tau = \sqrt{2La/(\tilde{\rho}_{ci}\nu_{iN}V_{Te})}\exp(e\varphi_c/2T). \tag{7.40}$$

Exponential dependence on $e\varphi_c/(2T)$ is connected with the fact that, since electrons are trapped along the magnetic field, and the ions are trapped across the field, the rise of φ_c results only in an increase of the potential drop in the sheaths. In order to reduce the longitudinal electron flux and the transverse ion flux by a factor of e, each of these potential drops have to be increased by T.

This situation takes place all the way up to values τ of the order of $\tau_{i\parallel}$, when the longitudinal escape of ions becomes important, and longitudinal ambipolar diffusion begins to control the decay. In this case, at $\tau\sim\tau_{i\parallel}$, a distribution $\cos(\pi z/L)$ is established along $\vec{B}$, while the distribution across $\vec{B}$ is again independent on r. The only result of a further increase of φ_c is an increase in the potential drop in the sheath at the side wall.

In the 'long' ($\tau_{e\parallel}>>\tau_{i\perp}$) tube the solution is constructed in the same manner. For $\varphi_c\leq 0$ the density profile practically coincides with the profile in the case of short-circuiting walls. The potential is 'tied' to the side walls, so that the longitudinal electric field is absent in the whole plasma volume except the small region near the end walls. The plasma decay time is $\tau_{e\parallel}$ for $\varphi_c\leq 0$. For $\varphi_c>0$ the decay is decelerated starting with

$$\varphi_c^* = \frac{T_e}{e}\ln\frac{L}{\lambda_{eN}} + \frac{T_i}{e}\ln\frac{\tau_{e\parallel}\tilde{\rho}_{ci}\nu_{iN}}{a}, \tag{7.41}$$

where λ_{eN} is the electron mean free path. At $\varphi_c>\varphi^*$ the decay is described by the Eqs.(7.38) - (7.40). At $\tau\sim\tau_{e\perp}$ the decay is determined by transverse ambipolar diffusion, and a further increase of φ_c results in an increase of the potential drop in the sheath near the end wall.

Figure 7.7 shows the experimental behavior $\tau(\varphi_c)$ from [16]. Curves 2 and 3 are calculated from Eq. (7.40) for a 'short' device, while curve 1 is calculated from the same equation for a 'long' device. At the transition from the short-circuiting case to the case of very negative ends we see that there is agreement between the calculated and the experimental data. For the saturation region at large values of τ the difference is due to the observed instabilities, which accelerate the plasma decay. The reason for the difference at small values of τ is not clear; it may be due to the complicated geometry of the experimental device. At large values of φ_c when the decay time becomes of the order of $\tau_{i\parallel}$ or $\tau_{e\perp}$, the further increase of φ_c results only in an increase of the potential drop in the sheath near the side wall for the 'short' device, and near the end wall in the case of 'long' device.

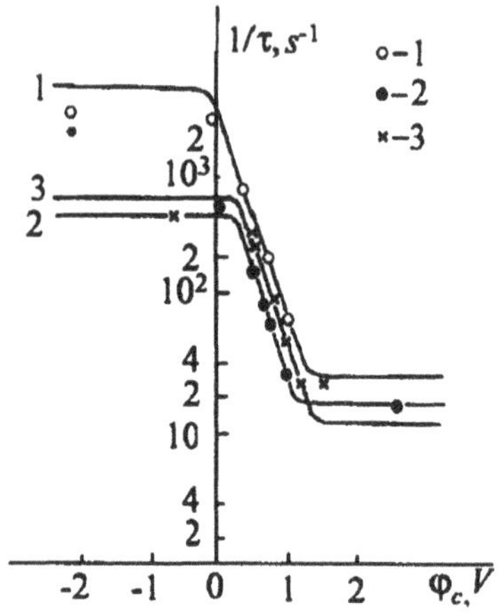

Figure 7.7. Decay constant as a function of φ_c. Experimental points correspond to helium in a cylindrical tube with a=1.9cm, L=75cm; 1: B=0.05T, p=7Pa; 2: B=0.2T, p=7Pa; 3: B=0.2T, p=11Pa.

Therefore, exponential decrease of the perpendicular ion flux $\Gamma_{i\perp}$ in the first case, or parallel electron flux $\Gamma_{e\parallel}$ in the second case, must take place. As a result, the plasma decay can be even slower than the ambipolar decay Eq. (7.15). In Fig. 7.8 an example of such a decay is displayed.

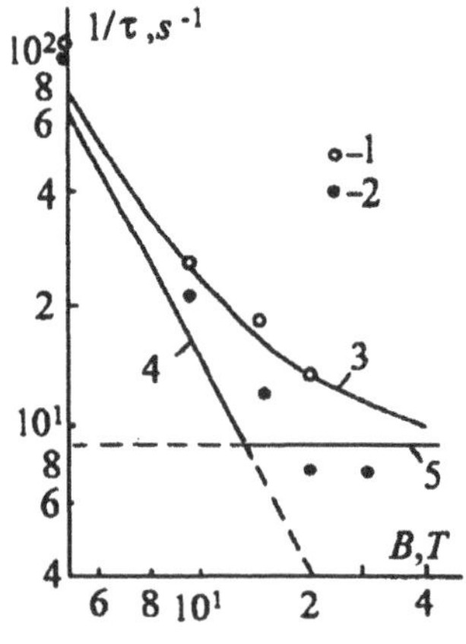

Figure 7.8. Decay time in the negative end regime in *He* [16]; a=1.9cm, L=70cm, p=16Pa. 1 - ends under floating potential (ambipolar regime); 2 - negative ends (φ_c=2.5V); 3 - $\tau_a^{-1}=\tau_{a\perp}^{-1}+\tau_{a\parallel}^{-1}$; 4 - $\tau_{a\perp}^{-1}$; 5 - $\tau_{a\parallel}^{-1}$.

In more complicated geometry it is possible to control the local plasma parameters by means of biasing. If, for example, only part of the side wall of a tube is a conductor, while the other part is dielectric, then the short-circuiting or biasing results in the acceleration of the decay only in the small cylinder with the conducting side walls. In Fig. 7.9 the time history of the plasma density in this small cylinder of length L_B is shown [3].

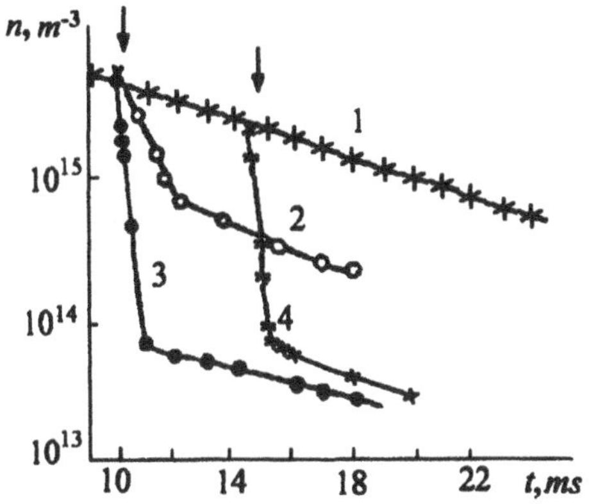

Figure 7.9. Plasma decay in a small cylinder in *He* [3], a=1cm, L=80cm, L_B=5cm, p=13Pa, B=0.18T. 1 -without short circuiting; 2 -φ_c=0; 3,4 - φ_c=15V. Arrows indicate the moment of short circuiting.

At the initial stage the decay time was $\tau \approx \tau_{i\perp}$ ('the long' device). The slow decay at the later stage corresponds to the ambipolar decay of plasma in the main tube volume, since the particles come to the small cylinder along the magnetic field.

Emission of electrons can also cause numerous interesting effects. For example, the diffusion of the complicated profile with two different scales across the magnetic field, considered in Section 7.2, can be significantly accelerated due to the arrival of electrons from the emitter to the peripheral regions of the tube. Emission can accelerate diffusion in the plasma volume, even in the tube with the dielectric side walls. The explanation is connected with the 2D motion of emitted electrons near the side walls, where the effective high conductive layer is created. In this layer electrons arrive from the emitter and diffuse to the side wall together with the ions, Fig. 7.10*a*, thus providing an ambipolar condition on the side walls. Such mixed diffusion has been observed [17] in the tube where sufficient voltage has been applied between the two end electrodes, so that the longitudinal current exceeds the diffusive electron current. One can see from Fig. 7.10*b* that the decay time in the dielectric tube is of the order of the corresponding value in the tube with the conductive walls, provided the emission is sufficiently large.

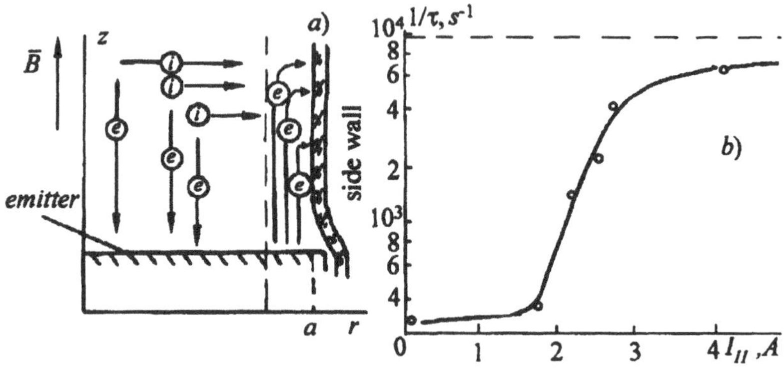

Figure 7.10. Diffusion in the presence of an emitting electrode; *a*) currents through the plasma; *b*) plasma decay time in a "short" tube as a function of the emitter heating current [17]; a=0.6 cm, L=30cm, B=0.25T, p=13Pa. Dashed line corresponds to the short circuiting regime.

7.4 ELECTROSTATIC PROBE IN MAGNETIC FIELD

In a strong magnetic field the motion of charged particles in the vicinity of a small electrode (probe) can be described by the transport equations even when their mean free paths exceed the probe dimensions. We shall show in this Section that the criterion of applicability of the electron description in terms of diffusion and mobility is

$$\rho_{ce} \ll a, \tag{7.42}$$

where ρ_{ce} is the electron gyroradius, and a is the probe size perpendicular to the magnetic field. The reason is that in strongly magnetized plasma for a positively biased probe the plasma density is perturbed in a region with the scale $l_{||}=a\omega_{ce}/\nu_{eN}=ax_e$ along the magnetic field. Hence, the condition $\lambda_{eN} \ll l_{||}$ is equivalent to the inequality (7.42). For magnetized ions the fluid description is valid if their gyroradius is smaller than the probe size. However, in the opposite situation part of the results obtained below remains valid for the positively biased probe. Therefore, the fluid approach in the magnetized plasma can be used even in plasma with relatively rare collisions.

We shall consider for simplicity the case of thin (with respect to the probe size) collisionless sheath. The probe is modelled by the ellipsoid of revolution with the major semiaxis b parallel to the magnetic field and minor semiaxes a perpendicular to $\vec{B}$. The major semiaxis is assumed to be smaller than ax_e which is the typical situation. The basic transport equations are

$$\vec{\nabla}\cdot(\hat{D}_e\vec{\nabla}n-\hat{b}_e n\vec{\nabla}\varphi)=0\ ;$$
$$\vec{\nabla}\cdot(\hat{D}_i\vec{\nabla}n+\hat{b}_i n\vec{\nabla}\varphi)=0\quad . \tag{7.43}$$

The boundary conditions at infinity, as in the absence of the magnetic field, are

$$n(\vec{r}\to\infty)=n_0\quad ;\ \varphi(\vec{r}\to\infty)=0\,.$$

The boundary condition at the probe surface $n=0$, $\varphi=\varphi_p$ are to be transformed to be applicable to the transport equations (7.43) by means of the expressions derived in Chapter 3.

Electron saturation current. If a sufficiently large positive potential is applied to the probe, then the electrons are gathered by the probe while ions are reflected. It is possible, therefore, to seek the solution of Eqs.(7.43) in a form which corresponds to zero ion flux in the volume and Boltzmann distribution for ions

$$e\varphi=-T_i \ln n/n_0. \tag{7.44}$$

Substituting Eq. (7.44) into Eq. (7.43), we obtain [9]

$$\frac{D_{e\perp}}{r}\frac{\partial}{\partial r}r\frac{\partial n}{\partial r}+D_{e\parallel}\frac{\partial^2 n}{\partial z^2}=0\,. \tag{7.45}$$

The boundary condition for potential is not needed, and the density should be zero on the probe surface.

The anisotropic equation (7.45) by the simple transformation of longitudinal coordinate $z'=z/x_e$ is reduced to the Laplace equation, and thus the case is equivalent to the problem of a probe in the absence of a magnetic field considered in Section 4.2. The ellipsoidal probe is transformed into a disk with the longitudinal semiaxis b/x_e, and the solution is sought in the elliptic coordinates

$$\frac{x^2+y^2}{a^2(1+\xi)}+\frac{z^2}{b^2(1+\xi x_e^2)}=1.$$

The plasma density is

$$n(\bar{r})\equiv n_0(1-n^{(e)})=n_0\left[1-\frac{arctg\sqrt{\dfrac{1-\gamma_e^2}{\gamma_e^2+\xi}}}{arctg\sqrt{\dfrac{1-\gamma_e^2}{\gamma_e^2}}}\right], \tag{7.46}$$

where $\gamma_e=b/ax_e$. The density is perturbed in the ellipsoid of revolution with major semiaxis ax_e and minor semiaxes a, its longitudinal scale is much larger than the corresponding probe size if

$$b \ll ax_e. \tag{7.47}$$

The electron saturation or Bohm current is

$$I_e^{(B)}=\frac{4\pi en_0 a x_e(1+T_i/T_e)D_{e\perp}\sqrt{1-\gamma_e^2}}{arctg\sqrt{(1-\gamma_e^2)/\gamma_e^2}}. \tag{7.48}$$

If the condition Eq. (7.47) is satisfied (which is the typical situation), we have

$$I_e^{(B)}=8en_0 a x_e(1+T_i/T_e)D_{e\perp}. \tag{7.49}$$

If $b \ll ax_e$ the saturation current can be expressed through the electric capacity C of a disk, since the basic equations are reduced to the Laplace equation. In the general case of different perpendicular semiaxis $a_1 \geq a_2$ the saturation current is

$$I_e^{(B)}=Cen_0 x_e(1+T_i/T_e)D_{e\perp}/\varepsilon_0; \tag{7.50}$$

$$C=\frac{4\pi\varepsilon_0 a_1}{K(\sqrt{1-a_2^2/a_1^2})};\ K(\zeta)=\int_0^{\pi/2}\frac{d\varphi}{\sqrt{1-\zeta^2\sin^2\varphi}}. \tag{7.51}$$

Here K is the elliptic integral of the first type. For $a_1=a_2$ the capacity $C=8\varepsilon_0 a$, and Eq. (7.50) coincides with Eq. (7.49).

The expression for the electron saturation current Eq. (7.49) has been checked experimentally in [18] - [20]. Dependence on the magnetic field predicted by Eq. (7.50) has been observed in these experiments. In Fig. 7.11 the electron saturation current obtained in the decaying plasma [20] is plotted.

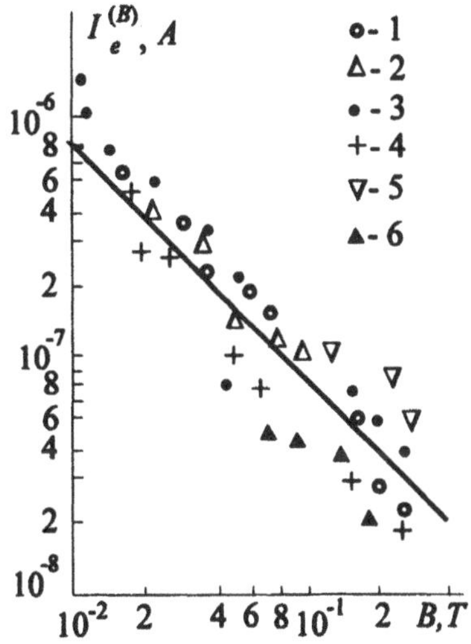

Figure 7.11. Electron saturation current versus magnetic field at $n=8\times10^{14}\mathrm{m}^{-3}$. Solid curve is calculated from Eq.(7.49) with the classical value of $D_{e\perp}$. Symbols 1-4 are experiments in *Ar* in a tube with the dielectric walls, 5 - *He*; 6 - *He*, the tube with the conducting walls.

A good agreement with theory is reported in spite of the fact that in some experiments plasma was turbulent.

The stationary profile Eq. (7.46) is established at a time scale of the order of electron transverse diffusion time $\tau_{e\perp}=a^2/8D_{e\perp}$. Therefore, when the probe is just inserted into the plasma, at $t<\tau_{e\perp}$, we have quite a different situation. Indeed, at small times the electron current is gathered along magnetic field from the distance $[(1+T_i/T_e)D_{e\parallel}t]^{1/2}$. For $t<\tau_{e\perp}$ this distance is smaller than the longitudinal scale of the depletion region ax_e which corresponds to the stationary case. Hence, at $t<\tau_{e\perp}$ the electron current to the probe $I_e(t)$ exceeds the Bohm current and decreases with time approximately as t^{-2} towards the value $I_e^{(B)}$. For $t<<\tau_{e\perp}$ the electron saturation current to the probe can be estimated as

$$\frac{I_e(t)}{I_e^{(B)}} \approx 1 + \frac{a x_e}{\sqrt{4\pi(1+T_i/T_e)D_{e\parallel}t}}. \tag{7.52}$$

In Fig. 7.12 the time history of electron current to the positively biased probe is displayed during the change of the probe polarity [21]. The dependence close to that described by Eq. (7.52) has been observed in these experiments.

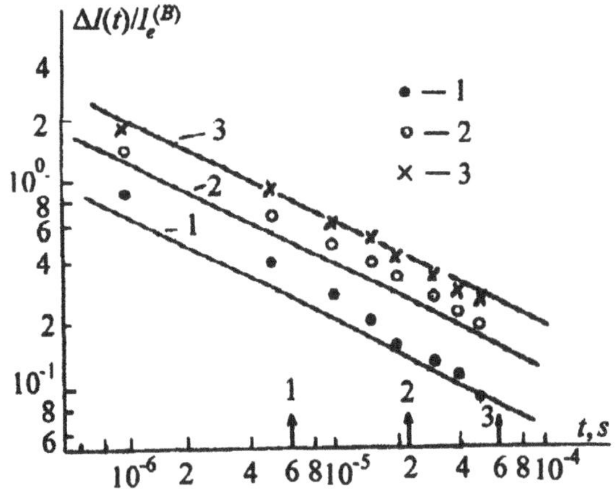

Figure 7.12. Temporal dependence of the value $\Delta I = I_e(t) - I_e^{(B)}$ in helium when the probe potential was changed by 1eV. Curves 1 - B=0.08T, 2 - B=0.15T, 3 - B=0.25T. Symbols are experiments, solid lines calculations Eq. (7.52). Arrows indicate the values $\tau_{e\perp}$.

Ion saturation current. Similarly for high negative potential of the probe the plasma potential corresponds to the Boltzmann distribution of electrons:

$$e\varphi = T_e \ln n/n_0. \tag{7.53}$$

The fluid description is valid if

$$\rho_{ci} \ll a. \tag{7.54}$$

In the case

$$\gamma_i = b/[(1+x_i^2)^{1/2} a] > 1 \tag{7.55}$$

the density perturbation is given by

$$n(r) \equiv n_0(1-n^{(i)}) = n_0\left\{1-\frac{\ln\left[\dfrac{1+\sqrt{(\gamma_i^2-1)/(\gamma_i^2+\xi)}}{1-\sqrt{(\gamma_i^2-1)/(\gamma_i^2+\xi)}}\right]}{\ln\left[\dfrac{1+\sqrt{(\gamma_i^2-1)/\gamma_i^2}}{1-\sqrt{(\gamma_i^2-1)/\gamma_i^2}}\right]}\right\}, \quad (7.56)$$

where the elliptic coordinates are defined according to

$$\frac{x^2+y^2}{a^2(1+\xi)}+\frac{z^2}{b^2+\xi a^2}=1. \quad (7.57)$$

The plasma density is perturbed in the ellipsoid of revolution with semiaxis b along the magnetic field and semiaxes $b/(1+x_i^2)^{1/2}$ perpendicular to the magnetic field. The ion saturation current is [22]

$$I_i^{(B)} = \frac{8\pi e a n_0(1+T_e/T_i)D_{i\parallel}\sqrt{\gamma_i^2-1}}{\sqrt{1+x_i^2}\,\ln\left[\dfrac{\gamma_i+\sqrt{1-\gamma_i^{-2}}}{\gamma_i-\sqrt{1-\gamma_i^{-2}}}\right]}. \quad (7.58)$$

If $\gamma_i<1$, the expression for the saturation current coincides with Eq. (7.48) where subscript e is to be replaced by i. The density is then perturbed inside the ellipsoid with the scales $a(1+x_i^2)^{1/2}$, a along and across $\vec{B}$.

Transitional part of the I-V characteristics. To obtain the current for intermediate probe potentials it is important to note the following. For large positive or large negative probe potentials, electrons and ions are gathered from the different regions in plasma, provided the inequalities Eq. (7.47) and (7.55) are satisfied. Electrons come to the probe from the electron ellipsoid with the scales ax_e, a along and across $\vec{B}$ respectively, while ions come from the ion ellipsoid with the scales b along $\vec{B}$ and $b/(1+x_i^2)^{1/2}$ across $\vec{B}$, Fig. 7.13.

The character of the diffusion in these regions has to be the same also for the intermediate values of φ_p [23]. Indeed, inside the electron ellipsoid the Boltzmann distribution for ions is to remain valid; otherwise there are large ion diffusive fluxes across the magnetic field. Analogously, in the ion ellipsoid the potential profile corresponds to the Boltzmann distribution for electrons. The density profile in the ellipsoids is thus given by Eq. (7.45) in the electron ellipsoid, and a similar equation with the ion diffusion coefficients in the ion ellipsoid. For intermediate probe potentials the solution of Eq. (7.45) ought to differ from Eq. (7.46) only by factor α:

$$n(\vec{r}) \equiv n_0(1-\alpha n^{(e)}). \tag{7.59}$$

Since for large positive probe potentials the solution Eq. (7.59) tends to Eq. (7.46), we have $\alpha=I_e/I_e^{(B)}$. Similarly, in the ion ellipsoid

$$n(\vec{r}) \equiv n_0(1-\beta n^{(i)}), \tag{7.60}$$

where $\beta=I_e/I_e^{(B)}$.

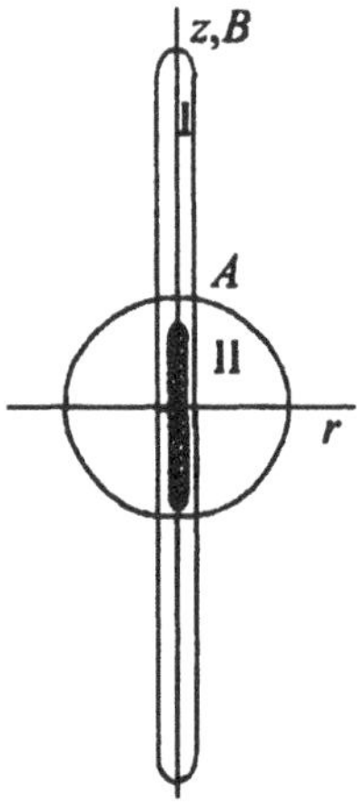

Figure 7.13. Schematic diagram of the regions from which electron and ion currents are gathered to the probe for the case of Eq. (7.55); I - electron ellipsoid, II-ion ellipsoid, the probe is cross-hatched.

In the overlap region which belongs to both ellipsoids, Eqs.(7.59), (7.60) are not valid, and it is very difficult to find an analytical solution here. At the edge of the overlap region the plasma density is $n=n_0(1-\alpha)$ on the z axis, Eq. (7.59), and $n=n_0(1-\beta)$ in the x,y plane, Eq. (7.60). Inside the overlap region the density decreases to zero with a large gradient. Nevertheless, it is possible to construct the I-V characteristic without detailed knowledge of the plasma profile in this zone [23].

Let us move along the z axis from infinity where the potential is zero towards the probe surface. Inside the electron ellipsoid the plasma density is depleted, so the potential, Eq. (7.53), is positive here and increases along z, Fig. 7.14. However, at small z, in the probe vicinity, electrons are trapped by the electric field, since their thermal current strongly exceeds the electron saturation current, Eq. (7.48). Hence, the potential is nonmonotonic along z, and at some point A, near the boundary of the overlap region, a maximum exists. This effect is known as potential overlap, and was first discussed in [24] for the case of fully ionized plasma (see below).

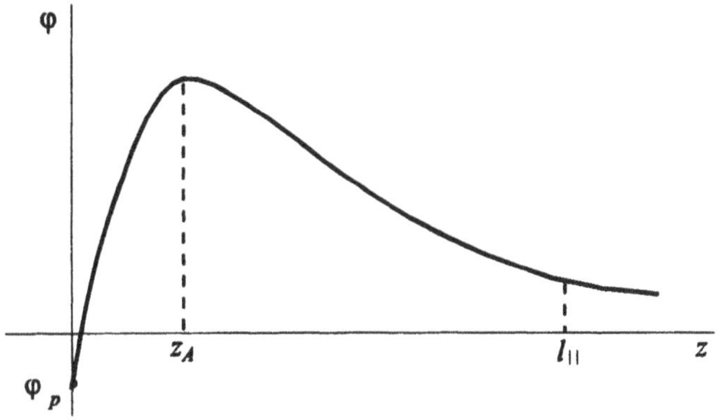

Figure 7.14. Potential profile along $\vec{B}$ for the transitional part of I-V characteristic.

At smaller z the potential distribution is the Boltzmann one for electrons, and for the collisionless sheath the flux of electrons to the probe can be expressed, similarly to Eq. (7.17), through the potential difference between a given point in the plasma and probe potential:

$$\varphi(z)-\varphi_p=-\frac{T_e}{e}\ln\frac{\sqrt{2\pi}\Gamma_{e\parallel}(z=b,x=y=0)}{n\sqrt{T_e/m_e}}. \tag{7.61}$$

The electron flux to the probe remains constant in the relatively small (with respect to the longitudinal scale of the electron ellipsoid αx_e) overlap region, and, hence, can be obtained from the solution in the electron ellipsoid, Eq. (7.59):

$$\Gamma_{e\parallel}(z=b,x=y=0)=\frac{\alpha I_e^{(B)}}{4\pi e a^2}. \tag{7.62}$$

Now combining Eqs.(7.53), (7.59), (7.61), (7.62), we find

$$e\varphi_p=-T_e\ln\frac{\sqrt{\pi/2}en_0a^2\sqrt{T_e/m_e}(1-I_e/I_e^{(B)})}{I_e}-T_i\ln(1-I_e/I_e^{(B)}) \tag{7.63}$$

Since the electron saturation current is much smaller than the electron thermal current, as follows from Eq. (7.63), the saturation is practically reached at negative probe potentials, Fig. 7.15. Further increase of the potential does not change the electron current significantly and results only in a decrease of the potential drop in the sheath and in reduction of the density at the edge of the overlap region, Eq. (7.59).

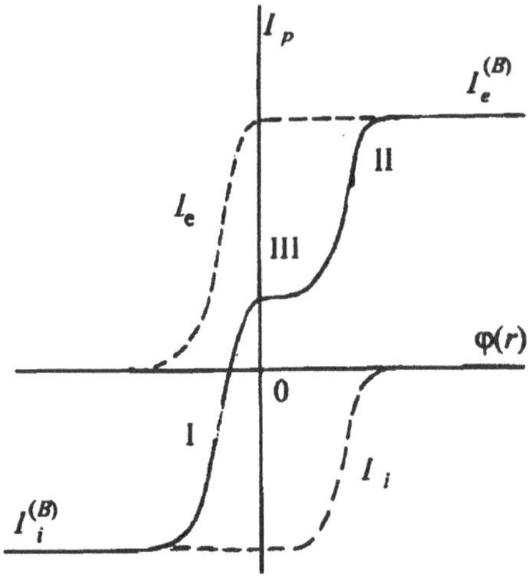

Figure 7.15. Qualitative *I-V* characteristic with $I_e^{(B)} \sim I_i^{(B)}$.

At some positive probe potential, which can be calculated from Eq. (7.63), the electron saturation current Eq. (7.49) becomes equal to the electron thermal current to the probe $n_0(1\text{-}\alpha)(2\pi)^{-1/2}(T_e/m_e)^{1/2}$. For larger potentials the overlap region disappears, electrons are accelerated towards the probe, and the potential profile becomes monotone everywhere.

For the case when inequality Eq. (7.55) is satisfied, moving towards the probe in the *xy* plane, we obtain analogously that the potential profile has the maximum at the edge of the overlap region, and that connection between the ion current and the probe potential is

$$e\varphi_p = T_i \ln \frac{4\pi e n_0 ab\tilde{\rho}_{ci}\nu_{iN}(1 - I_i / I_i^{(B)})}{I_i} + T_e \ln(1 - I_i / I_i^{(B)})\,. \quad (7.64)$$

The saturation of the ion current is practically reached at positive probe potentials. Therefore, in this case a plateau on the I-V characteristics exists, where two saturation currents flow simultaneously to the probe, Fig. 7.15. This effect is more pronounced for stretched probes with large size *b* along magnetic field, when $I_e^{(B)} \sim I_i^{(B)}$. Unfortunately, such probes are seldom used in spite of the fact that they contain rich information about the plasma parameters. Characteristics of this kind were observed in [25] for a specially shaped probe.

More typical is the case when the inequality inverse to Eq. (7.55) is satisfied. In this situation the ion ellipsoid is situated inside the electron ellipsoid, and the Eq. (7.64) for the ion current is not valid. Yet Eq. (7.63) remains valid, and since the ion saturation current in this case is small with respect to the electron saturation current, it is possible to use almost all the characteristic. In other words, parts II and III of the characteristic, Fig. 7.15, are hardly recognized, while the main part I is described by the Eq. (7.63) with $I_i = I_i^{(B)}$.

It is worthwhile to note that the transitional part of the characteristic Eq. (7.63) is almost a linear function of the potential in the interval $0.2 < I_e/I_e^{(B)} < 0.8$. Its slope with the accuracy of 5% can be approximated by [26], [27]

$$\frac{dI_e}{d\varphi_p} = \frac{eI_e^{(B)}}{T_e(1+\sqrt{1+T_i/T_e})^2}\,. \tag{7.65}$$

This expression in combination with the expressions for the saturation currents can be used for determining the temperatures of the charged particles. Such linear slope on I-V characteristics was observed in many experiments; the example is shown in Fig. 7.16.

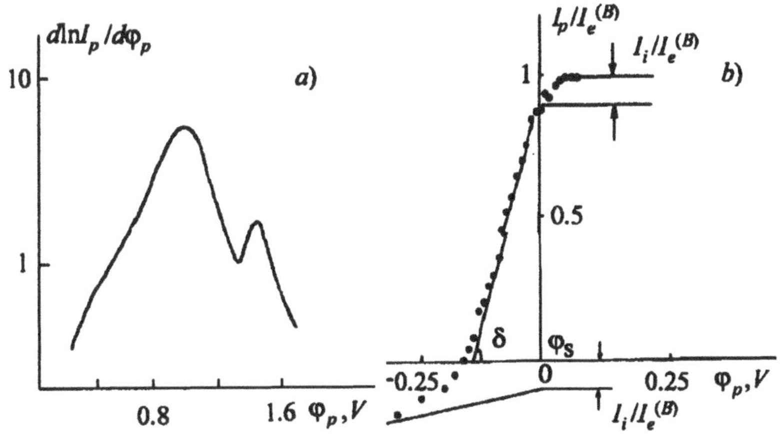

Figure 7.16. Peculiarities at probe characteristics in the magnetized plasma. *a)* $d\ln I_p/d\varphi_p$ versus φ_p in *Ar* [25]; *b)* *I-V* characteristic of a probe with the collisionless sheaths in the helium decaying plasma [26].

Collisional sheaths. In the case when the sheaths along and across the magnetic field are collisional, using the results of Section 3.6, it is possible to obtain

$$e\varphi_p = -T_e \ln \frac{4\pi n_0 (1 - I_e / I_e^{(B)}) b_{e\parallel} (3T_e a^2 b^2 e^2 I_i^{(B)} / b_{i\parallel})^{1/3}}{I_e} - T_i \ln(1 - I_e / I_e^{(B)}). \qquad (7.66)$$

The ion current, provided Eq. (7.55) is satisfied, has the form

$$e\varphi_p = T_i \ln \frac{4\pi n_0 (1 - I_i / I_i^{(B)}) b_{i\perp} (3T_i a^2 b^2 e^2 I_e^{(B)} / b_{e\perp})^{1/3}}{I_i} + T_i \ln(1 - I_i / I_i^{(B)}). \qquad (7.67)$$

Probe in fully ionized plasma. In fully ionized plasma the electron current to a small probe with perpendicular size a less than the ion gyroradius can be obtained in the same way as for weakly ionized plasma. The electron part of I-V characteristic has been calculated numerically in [24], and analytically in [28]-[29]. Since at small scales with respect to the ion gyroradius the Boltzmann distribution for ions is established, the process of gathering of electrons is described by an equation similar to Eq. (7.45):

$$\frac{1}{r}\frac{\partial}{\partial r}\left(r \frac{n}{n_0}\frac{\partial n}{\partial r}\right) + \frac{\omega_{ce}^2}{0.51 \nu_{ei}^2(n_0)} \frac{\partial}{\partial z}\left(\frac{n_0}{n}\frac{\partial n}{\partial z}\right) = 0\,, \qquad (7.68)$$

where ν_{ei} is the electron-ion collision frequency defined by Eq. (2.24). The Eq. (7.68) can easily be obtained from the electron momentum balance equation. In [28]-[29] it was demonstrated that the saturation current, which corresponds to the solution of the nonlinear Eq. (7.68), only differs from the saturation current Eqs.(7.48) - (7.51) by the numerical coefficient k=0.5 Moreover, using the same arguments as for weakly ionized plasma, it is easy to show that the whole characteristic is described by Eq. (7.63). Ideas and expressions similar to these obtained above were also discussed in [30]. In Fig. 7.17 the result of comparison with the numerical simulation performed for T_e=T_i=T [24] is presented.

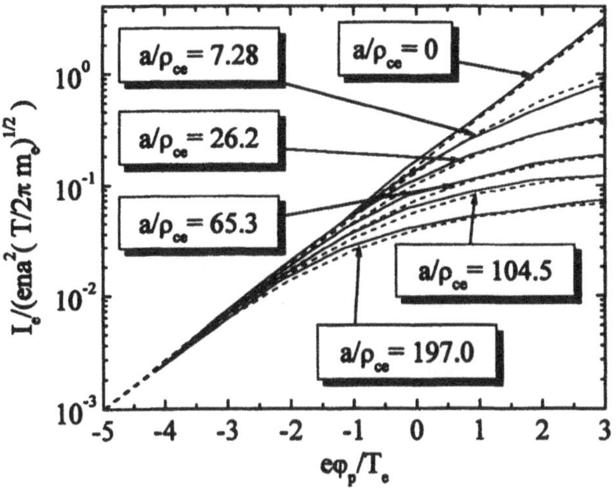

Figure 7.17. Comparison of the numerical simulations by Sanmartin [24] (dashed line) with the analytical solution given by presented model Eq. (7.63) (solid line).

Comparison with the experiments with small probes [31] also show reasonable agreement with the analytical model. The opposite limiting case $a \gg \rho_{ci}$ has been considered in [31] - [34].

REFERENCES

1. A. P. Zhilinskii, V. A. Rozhanskii, L. D. Tsendin, *Sov. J. Plasma Phys.* **4**: 317-321 (1978).
2. V. E. Golant, *Sov. Phys. Uspekhi* **5**: 161-197 (1963).
3. A. P. Zhilinsky, L. D. Tsendin, *Sov.Phys. Uspekhi* **23**: 331-355 (1980).
4. A. P. Zhilinskii, B. V. Kuteev, *Sov. Phys. Techn. Phys.* **20**: 163-171 (1975); 20: 1316-1321 (1975).
5. K. H. Geissler, *Phys. Rev.* **171**: 179-180 (1968).
6. A. P. Zhilinskii, A. S. Smirnov, *Sov. Phys. Techn. Phys.* **22**: 1208-1211 (1977).
7. A. Simon, *Phys. Rev.* **98**: 317-318; 100: 1557-1559 (1955).
8. R. K. Porteous, D. B. Graves, *IEEE Trans. Plasma Sci.* **PS-19**: 204-213 (1991).
9. D. Bohm, *The Characteristics of Electrical Discharges in Magnetic Field,* Editors: A.Guthrie, R. Wakerling , (New York: McGraw-Hill, 1949):346. Chapts. 1, 2, 9.

10. A. V. Zharinov, *Atomnaya energia*, 7: 215-219 (1959); 10: 368-369 (1961) (in Russian).
11. M. A. Vlasov, *Sov. Phys. JETP* **24**: 475-482 (1966).
12. F. Schwirzke, *Phys. Fluids* **9**: 2244-2251 (1966); 10: 183-188 (1967).
13. V. E. Golant, A. P. Zhilinski, *Sov. Phys. Tech. Phys.* 7: 84-85 (1962); 7: 970-977 (1963).
14. I. A. Vasil'eva, V. L. Granovski, A. F. Chernovolenko, *Radio Eng. Electron. Phys. (USSR)* **5**: 1508-1515 (1960).
15. M. A. Lieberman, A. J. Lichtenberg, *Principles of Plasma Discharges and Materials Processing* (New York: Wiley, 1994): 572.
16. A. P. Zhilinskii, B. V. Kuteev, *Sov. Phys. Techn. Phys.* **21**: 708-710 (1976).
17. A. P. Zhilinskii, B. V. Kuteev, *Sov. Phys. Techn. Phys.* **23**: 358-361 (1978).
18. T. Dote, H. Amemiya, *Japan. J. Appl. Phys.* **3**: 789-796 (1964).
19. F. G. Baksht, G. A. Duzhev, B. I. Tsirkel et al, *Sov. Phys. Techn. Phys.* **22**: 1313-1318 (1977).
20. A. P. Zhilinskii, B. V. Kuteev, N. V. Sakharov, A. S. Smirnov, *Sov. J. Plasma Phys.* **3**: 568-573 (1977).
21. V. M. Voronov, A. P. Zhilinskii, B. V. Kuteev, *Sov. J. Plasma Phys.* **3**: 573-577 (1977).
22. I. M. Cohen, *Phys. Fluids* **6**: 1492-1499 (1963).
23. V. A. Rozhansky, L. D. Tsendin, *Sov. Phys. Techn. Phys.* **23**: 932-936 (1978).
24. J. Sanmartin, *Phys. Fluids* **13**: 103-116 (1970).
25. M. Sato, *Phys. Fluids* **15**: 2427-2433 (1972); **17**: 1903-1904 (1974).
26. B. V. Kuteev, V. A. Rozhansky, *Sov. Tech. Phys. Lett.* **4**: 49-50 (1978).
27. B. V. Kuteev, V. A. Rozhansky, L. D. Tsendin, *Beitrage aus der Plasmaphysik* **19**: 123-126 (1979).
28. V. A. Rozhansky, A. A. Ushakov, *Contrib. Plasma Phys.* **38**: **38S**: 19-24 (1998).
29. V. A. Rozhansky, A. A. Ushakov, *Technical Physics Letters* **24**: 869--872 (1998).
30. P. C. Stangeby, J. Phys. D: Appl. Phys. **25**: 1007-1030 (1982).
31. J. P. Gunn, C. Boucher, B. L. Stansfield, C. S. MacLatchy, *Contrib. Plasma Phys.* **36**: 45-52 (1996).
32. V. Rozhansky, A. Ushakov, S. Voskoboynikov, *Contrib. Plasma Phys.* **36**: 391-395 (1996).
33. V. Rozhansky, A. Ushakov, S. Voskoboynikov, *Plasma Phys. Reports* **24**: 777-788 (1998).
34. V. Rozhansky, A. Ushakov, S. Voskoboynikov, *Nuclear Fusion* **39**: (1999).

Chapter 8

Thermodiffusion in Magnetic Field

In the course of local heating of plasma particles, the plasma density becomes redistributed by the gradient of the pressure and by the thermal force. The effect plays an important role in, for example, the process of ionosphere heating by intense radio waves. The expulsion of plasma from the wave absorption region results in a change of the reflection point, in rearrangement of the energy input geometry, and in a number of other complicated nonlinear effects. In the laboratory these phenomena are also very important in numerous applications of microwave discharges. In these discharges the thermodiffusion processes greatly influence the plasma parameters, density and temperature profiles, etc.

To analyze the problem it is necessary, in principle, to solve both the particle and heat balance equations simultaneously, since the density is redistributed during the heating of the particles. For small perturbations the problem can be split into separate parts (see Chapter 12). For pronounced perturbation these processes are coupled, and it is necessary to develop numerical codes. However, especially in a magnetic field, even the density evolution in the presence of temperature gradients is rather a complicated process. A great number of interesting and complicated problems arise in the gas discharge physics in the *dc*, RF, and in the microwave fields. In these situations the processes of energy input, of its subsequent transport, of the plasma generation, and of the transport of the charged particles are in a complex self-consistent manner intertangled with the distribution of the electromagnetic fields. The analysis of these extremely interesting and practically important topics goes far beyond the scope of this book [1-3]. Therefore, in this Chapter we shall restrict ourselves to considering the process of thermodiffusion for a given temperature profile and/or for a given heat source. The energy balance equation for a given density profile is analyzed in Chapter 12. The most pronounced effects are caused by nonuniform electron temperature profiles, so we shall assume for simplicity T_i=const. As in the previous Sections, we shall also consider the case of strong magnetic field, so that $b_{i\perp} >> b_{e\perp}$.

The transport of the charged particles in a magnetic field in the absence of particle sources and sinks and in the presence of both density and temperature inhomogeneity is described by the system of equations (Eq. (2.41))

$$\frac{\partial n}{\partial t} = -\vec{\nabla} \cdot \vec{\Gamma}_e = \vec{\nabla} \cdot (\hat{D}_e \nabla n - \hat{b}_e n \nabla \varphi + \hat{D}_e^{(T)} n \nabla \ln T_e) \ ;$$
$$\frac{\partial n}{\partial t} = -\vec{\nabla} \cdot \vec{\Gamma}_i = \vec{\nabla} \cdot (\hat{D}_i \nabla n + \hat{b}_i n \nabla \varphi). \qquad (8.1)$$

General expressions for the mobility, diffusion and thermodiffusion tensors are presented in Section 2.7.

8.1 ONE-DIMENSIONAL THERMODIFFUSION

For the pure plasma and $\vec{B}$=0, in the 1D case, as demonstrated in Section 4.1, the electric field can be excluded from the system Eq. (8.1). Accordingly, this system can be reduced to a single ambipolar equation for the density. In a slab geometry, for example, in full analogy to the case T_e=const, Section 4.1, we have

$$\frac{\partial n}{\partial t} - \frac{\partial}{\partial z}(D_a \frac{\partial n}{\partial z}) - \frac{\partial}{\partial z}(D_a^{(T)} n \frac{\partial \ln T_e}{\partial z}) = 0 , \qquad (8.2)$$

where D_a is the ambipolar diffusion coefficient Eq. (4.8), and the ambipolar thermodiffusion coefficient is given by

$$D_a^{(T)} = \frac{D_e^{(T)} b_i}{b_e + b_i} . \qquad (8.3)$$

The corresponding electric field is

$$\frac{\partial \varphi}{\partial z} = \frac{D_e - D_i}{b_e - b_i} \frac{\partial \ln n}{\partial z} + \frac{D_e^{(T)}}{b_e + b_i} \frac{\partial \ln T_e}{\partial z} - \frac{(b_e^{(0)} + b_i^{(0)}) n_0}{(b_e + b_i) n} E_0 . \qquad (8.4)$$

where E_0, n_0, $b_e^{(0)}$, $b_i^{(0)}$ are the field strength, plasma density and particle mobilities far from the perturbation region.

If at t=0 a local 1D electron temperature perturbation is created, the density starts to redistribute according to Eq. (8.1). In 1D case the length of the region in which plasma density is disturbed increases with time as $(D_{a\|}t)^{1/2}$. The magnitude and even the sign of this density perturbation depends on the value of $D_a^{(T)}$.

In the approximation of elementary theory, when the diffusion and thermodiffusion coefficients for electrons coincide (see Section 2.2) for positive temperature disturbance plasma is pushed out from the heated region. The stationary solution of Eq. (8.2) in this case corresponds to the constant total pressure

$$n(T_e+T_i)=\text{const.} \tag{8.5}$$

This can easily be understood from the momentum balance equations. Indeed, the elementary approximation is equivalent to the absence of thermal force, and, therefore, the electric field, Eq. (8.4), balances the electron pressure gradient. In the ion momentum balance equation this electric force is added to the ion pressure gradient, resulting for the stationary case in the condition Eq. (8.5). Since the same electric force balances both the ion and electron pressure gradients, and the ion temperature is assumed to be uniform, the steady-state potential perturbation corresponds to the Boltzmann distribution for ions.

In general, due to the thermal force we have $D_e \neq D_e^{(T)}$. If the electron collision frequency decreases with velocity, the thermal force is added to the $n\vec{\nabla}T_e$ term in the electron pressure gradient, and $D_e^{(T)}>D_e$, see Sections 2.2, 2.7. For example, in the important case of partially ionized plasma where for electrons Coulomb collisions dominate $\nu_{ee}>>\nu_{eN}$, the thermodiffusion coefficient is $D_e^{(T)}=1.7D_e$. Using the Einstein relation, we obtain for the stationary density profile

$$n(T_e+T_i)^{1.7}=\text{const.} \tag{8.6}$$

The magnitude of a density perturbation which arises in nonuniformly heated plasma is thus considerably larger than in the approximation of elementary theory, Eq. (8.7). In contrast, if the electron collision frequency increases with velocity, $D_e^{(T)}<D_e$, so that the density perturbation is smaller than that given by Eq. (8.5). In principle, the thermodiffusion coefficient can even become negative, and then the sign of the density perturbation coincides with the sign of the temperature perturbation.

For weakly ionized unmagnetized plasma, in the multidimensional case the initial system, Eq. (8.1), is also reduced to an ambipolar equation similar to Eq. (8.2). This reduction can be performed by multiplying the equations (8.1) by b_e, b_i correspondingly, and taking their difference. However, the particle fluxes are not equal to each other, as in 1D situation. The reason is connected with the fact that in the general case the potential which equalizes the particle fluxes does not exist. Therefore, the evolution of inhomogeneously heated plasma, even in the absence of a magnetic field, is accompanied by vortex currents. The example of such a situation is presented in Section 12.2.

The 1D diffusion perpendicular to the magnetic field is described as in the case $\vec{B}=0$. In the slab geometry, for example, we have

$$\frac{\partial n}{\partial t}-\frac{\partial}{\partial y}(D_{a\perp}\frac{\partial n}{\partial y})-\frac{\partial}{\partial y}(D_{a\perp}^{(T)}n\frac{\partial \ln T_e}{\partial y})=0, \tag{8.7}$$

where the ambipolar diffusion coefficient $D_{a\perp}$ is given by Eq. (5.7), and the ambipolar thermodiffusion coefficient is

$$D_{a\perp}^{(T)}=\frac{D_{e\perp}^{(T)}b_{i\perp}}{b_{e\perp}+b_{i\perp}}\approx D_{e\perp}^{(T)}. \tag{8.8}$$

The length of the region in which plasma density is disturbed increases with time as $(D_{a\perp}t)^{1/2}$. In weakly ionized magnetized plasma the approximation of elementary theory is fulfilled, Section 2.4, $D_{e\perp}^{(T)}=D_{e\perp}$, and the stationary solution of Eq. (8.7) is given by Eq. (8.5). In partially ionized plasma where $\nu_{ee}>>\nu_{eN}$, the thermodiffusion coefficient is negative: $D_{e\perp}^{(T)}=-1/2D_{e\perp}$, see Section 2.7. Hence, in contrast to the case $\vec{B}=0$, plasma is gathered to the heated region forming the stationary profile

$$n(T_e+T_i)^{-1/2}=\text{const}. \tag{8.9}$$

In the general case, when the temperature depends only on the coordinate ζ which forms an angle β with the magnetic field, it is convenient to introduce a new coordinate system (x, η, ζ), as in Section 5.3, see Fig. 5.3. The new system is obtained by rotation of the old one over the x axis. In the absence of a net current through a plasma inhomogeneity, excluding the potential analogously to the case of the diffusion, we find

$$\frac{\partial n}{\partial t}-\frac{\partial}{\partial \zeta}[D(\mu^2)\frac{\partial n}{\partial \zeta}]-\frac{\partial}{\partial \zeta}[D^{(T)}(\mu^2)n\frac{\partial \ln T_e}{\partial \zeta}]=0, \tag{8.10}$$

where $D(\mu^2)$ is defined according to Eq. (5.32), and

$$D^{(T)}(\mu^2) = \frac{[D_{e\parallel}^{(T)}\mu^2 + D_{e\perp}^{(T)}(1-\mu^2)][b_{i\parallel}\mu^2 + b_{i\perp}(1-\mu^2)]}{(b_{e\parallel} + b_{i\parallel})\mu^2 + (b_{i\perp}\mu^2 + b_{e\perp}(1-\mu^2)}. \quad (8.11)$$

The evolution in the presence of the net current is analyzed in Chapter 11.

8.2 MULTIDIMENSIONAL THERMODIFFUSION. SMALL TEMPERATURE PERTURBATION

One-dimensional ambipolar thermodiffusion occurs only under rather restrictive conditions on the dimensions ($\lambda_{\|,\perp}$) of the temperature perturbation. The ambipolar thermodiffusion along $\vec{B}$, similar to the case of ordinary diffusion (Section 6.3), occurs if the severe condition $\lambda_\perp/\lambda_\| >> (b_{i\perp}/b_{i\|})^{1/2}$ is satisfied. The perpendicular ambipolar thermodiffusion takes place when inequality $\lambda_\perp/\lambda_\| << (b_{e\perp}/b_{e\|})^{1/2}$ is valid. However, the length scales of the temperature perturbation due to a point source are equal to the energy relaxation lengths Eq. (2.20): $\lambda_\| = \lambda_e/\delta^{1/2}$, $\lambda_\perp = \rho_{ce}/\delta^{1/2}$; for further details see Chapter 12. Therefore, the thermodiffusion in this case is essentially multidimensional and is accompanied by the vortex currents.

In this Section we consider small temperature perturbations, nevertheless, in the general situation the energy balance and the particle continuity equations remain coupled. However, in many situations the problems of determining T_e and the density profile can be treated separately. We shall, for simplicity, consider this case. The perturbation of T_e generated by a point source would become

$$T_e(\vec{r},t) - T_e^{(0)} = AT_e^{(0)} \frac{\exp(-z^2/\lambda_\|^2 - r^2/\lambda_\perp^2)^{1/2}}{(z^2/\lambda_\|^2 + r^2/\lambda_\perp^2)^{1/2}}, \quad (8.12)$$

i.e., it would fall off exponentially with length scales $\lambda_{\|,\perp}$ along and across magnetic field. Since we are interested in the density distribution at times longer than the energy relaxation time, the themodiffusion problem is formulated in the following way. At t=0 the temperature profile is in the steady state, e.g., is given by Eq. (8.12) with A<<1. The initial and boundary conditions for the density and the potential are

$$n(\vec{r},t)\big|_{t=0} = n_0; \quad n(\vec{r}\to\infty,t) = n_0; \quad \varphi(\vec{r}\to\infty,t) = 0. \quad (8.13)$$

We seek the solution of the linearized Eq. (8.1) in the form of a Fourier integral

$$n(\vec{r},t) = n_0 + \frac{1}{(2\pi)^3}\int(\delta n)_{\vec{k}}\exp(i\vec{k}\vec{r})d\vec{k} \ ,$$

$$\varphi(\vec{r},t) = \frac{1}{(2\pi)^3}\int(\delta\varphi)_{\vec{k}}\exp(i\vec{k}\vec{r})d\vec{k} \ .$$

We find [4]

$$\begin{aligned} \delta n_{\vec{k}} &= \frac{n_0\vec{k}\hat{D}_e^{(T)}\vec{k}}{\vec{k}\hat{D}_e\vec{k} + \vec{k}\hat{b}_e\vec{k}\vec{k}\hat{D}_i\vec{k} / \vec{k}\hat{b}_i\vec{k}}\left\{\exp[-D(\mu^2)k^2t]-1\right\}\frac{(\delta T_e)_{\vec{k}}}{T_e^{(0)}}, \\ \varphi_{\vec{k}} &= \frac{\vec{k}(\hat{D}_e - \hat{D}_i)\vec{k}(\delta n)_{\vec{k}} / n_0 + \vec{k}\hat{D}_e^{(T)}\vec{k}(\delta T_e)_{\vec{k}} / T_e^{(0)}}{\vec{k}(\hat{b}_e + \hat{b}_i)\vec{k}}, \end{aligned} \tag{8.14}$$

where $D(\mu^2)$ is defined according to Eq. (5.32). The solution takes its simplest form in a weakly ionized plasma, in which case the Einstein relations hold, and in the elementary theory approximation where $\hat{D}_e^{(T)} = \hat{D}_e$. The coefficient in front of the exponential function in this case is $n_0A/(1+T_i/T_e^{(0)})$, and for the density and the potential we find the sum of a steady-state profile in the region of the temperature perturbation and a region of density perturbation which is increases with time:

$$\begin{aligned} n(\vec{r},t) &= n_0\left[1 - \frac{T_e(\vec{r}) - T_e^{(0)}}{T_e^{(0)}\left(1 + T_i / T_e^{(0)}\right)}\right] + \frac{n_d(\vec{r},t)}{\left(1 + T_i / T_e^{(0)}\right)}, \\ \varphi(\vec{r},t) &= \frac{T_i}{e}\frac{T_e(\vec{r}) - T_e^{(0)}}{T_e^{(0)}\left(1 + T_i / T_e^{(0)}\right)} + \frac{\varphi_d(\vec{r},t)}{\left(1 + T_i / T_e^{(0)}\right)} \ . \end{aligned} \tag{8.15}$$

Here $n_d(\vec{r},t)$ and $\varphi_d(\vec{r},t)$ are the familiar solutions of the problem of diffusion of a small perturbation with an initial distribution $\delta n(\vec{r},0) / n_0 = \delta T_e / T_e^{(0)}$, at a constant temperature which was analyzed in Sections 6.1, 6.3.

For time

$$t \gg \lambda_\perp^2/(1+T_e^{(0)}/T_i)D_{i\perp}, \tag{8.16}$$

the main density perturbation is a superposition of the densities in the electron ellipsoid $n^{(e)}$ and ion ellipsoid $n^{(i)}$, Fig. 8.1.

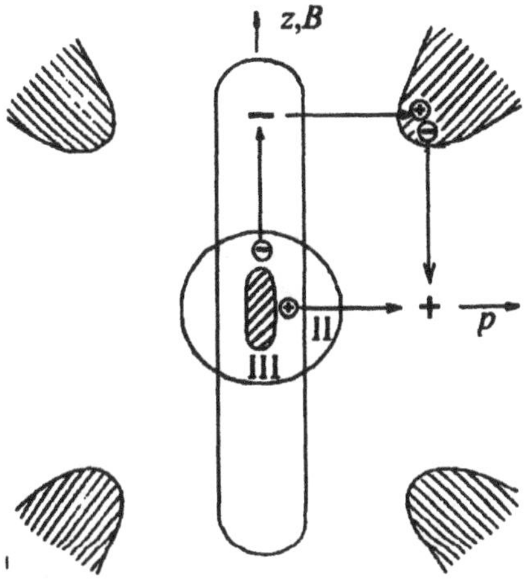

Figure 8.1. Schematic diagram of the regions of density perturbation. I is the electron ellipsoid; II is the ion ellipsoid; III is a region with dimensions $\lambda_{\|,\perp}$ in which the temperature is perturbed. The hatched are depletion regions.

The solution takes its most compact form for a Gaussian profile of temperature perturbation

$$T_e(\vec{r},t) = T_e^{(0)}[1 + A\exp(-z^2/\lambda_\|^2 - r^2/\lambda_\perp^2)]. \tag{8.17}$$

In this case in the neighboring zone ($|z|>\Lambda_{e\|}$, $r>\Lambda_{i\perp}$) we have

$$n_d(\vec{r},t) = n^{(e)} + n^{(i)}\ ;\ \ \varphi_d(\vec{r},t) = -T_i n^{(e)}/e + T_e^{(0)} n^{(i)}/e\ ; \tag{8.18}$$

$$n^{(\alpha)} = n_0 \frac{\int (T_e(\vec{r}) - T_e^{(0)})/T_e^{(0)} \exp(-\frac{z^2}{\Lambda_{\alpha\|}^2} - \frac{r^2}{\Lambda_{\alpha\perp}^2})d\vec{r}}{\pi^{3/2}\Lambda_{\alpha\|}\Lambda_{\alpha\perp}^2}, \tag{8.19}$$

where

$$\Lambda_{e\parallel,\perp}^2 = \lambda_{e\parallel,\perp}^2 + 4(1+T_i/T_e^{(0)})D_{e\parallel,\perp}t;\ \Lambda_{i\parallel,\perp}^2 = \lambda_{i\parallel,\perp}^2 + 4(1+T_e^{(0)}/T_i)D_{e\parallel,\perp}t\,.$$

The asymptotic expression for the density perturbation at $z \gg \Lambda_{e\parallel}$, $r \gg \Lambda_{i\perp}$ contains depletion regions:

$$\delta n(\vec{r},t) = \frac{A(1+T_e^{(0)}/T_i)D_{i\perp}t}{2\pi^{1/2}\mu_0^4(z^2+r^2)^{5/2}} \times \frac{12\cos^2\tilde{\alpha} - 36\cos^2\tilde{\alpha}\sin^2\tilde{\alpha}/\mu_0^2 + (9/2)\sin^4\tilde{\alpha}/\mu_0^4}{(\cos^2\tilde{\alpha} + \sin^2\tilde{\alpha}/\mu_0^2)^{9/2}}, \tag{8.20}$$

where $\mu_0=(b_{i\perp}/b_{e\parallel})^{1/2}$, $\tilde{\alpha}$ is the angle between radius vector and magnetic field. The spatial dependence of the density perturbation is the same as that of the Green's function for the case of diffusion obtained in Chapter 5.

In the limit $t\to\infty$ we are left with only the first terms in Eq. (8.15). In the case considered (in the absence of a thermal force) the steady-state density profile corresponds to a density decrease in the heated region, so that the gradient of the total pressure vanishes, Eq. (8.5). The potential perturbation Eq. (8.15), as in 1D case, corresponds to the Boltzmann distribution for the ions.

The physical picture of multidimensional thermodiffusion is reminiscent of particle diffusion. Because of thermodiffusion, the electrons are displaced from the region of temperature perturbation along the magnetic field, forming potential minima. Through diffusion they form an electron ellipsoid with a density $n^{(e)}$ (I in Fig. 8.1). The electric field extracts the ions from the depletion region across the magnetic field to the electron ellipsoid to provide quasineutrality. The electrons leave depletion regions along $\vec{B}$, while ions from the heated region are extracted by the electric field across $\vec{B}$ (the first term in the second equation (8.15)), forming a potential minimum. They then diffuse, forming an ion ellipsoid with density $n^{(i)}$ (II in Fig. 8.1). The density profile, as in the case of diffusion, is thus controlled by the mechanism of short circuiting through the plasma.

The nature of the solution does not change qualitatively when we switch to a partially ionized plasma and/or take the thermal force into account (the distinction between $\hat{D}_e^{(T)}$ and $\hat{D}_e$), but several new features do appear. The Fourier integrals, Eq. (8.14), were evaluated in [4] similarly to the evaluation of the integrals in Section 6.1. The solution under condition Eq. (8.16) is

$$\frac{n(\vec{r},t)}{n_0} = 1 - \alpha \frac{A\lambda_{\parallel}\lambda_{\perp}^2}{\Lambda_{i\parallel}\Lambda_{i\perp}^2} \exp(-y^2) + \Psi(t) - \Psi(0);$$

$$\Psi(t) = \frac{A\lambda_{\parallel}\lambda_{\perp}^2}{\tilde{\Lambda}_{e\parallel}\tilde{\Lambda}_{e\perp}^2} \{\beta \exp(-x^2) + (\alpha - \beta)[\frac{r^2 \exp(-x^2)}{2\tilde{\Lambda}_{e\perp}^2 x^2} \qquad , \quad (8.21)$$

$$+ (\frac{z^2}{\tilde{\Lambda}_{e\parallel}^2 x^2} - \frac{r^2}{2\tilde{\Lambda}_{e\perp}^2 x^2})(\frac{\exp(-x^2)(x^2+1)}{x^2} - \frac{\sqrt{\pi}}{2} erf(x)/x^3]\},$$

where

$$\alpha = D_{e\parallel}^{(T)} / \tilde{D}_{e\parallel} \; ; \beta = D_{e\perp}^{(T)} / \tilde{D}_{e\perp} \; ; \tilde{\Lambda}_{e\parallel,\perp} = (\lambda_{\parallel\perp}^2 + 4\tilde{D}_{e\parallel} t)^{1/2} \; ;$$

$$x^2 = z^2 / \tilde{\Lambda}_{e\parallel}^2 + \rho^2 / \tilde{\Lambda}_{e\perp}^2 \; ; \; y^2 = z^2 / \Lambda_{i\parallel}^2 + r^2 / \Lambda_{i\perp}^2 \; .$$

At the origin as $x \to 0$ we have

$$\Psi(t) = \frac{A\lambda_{\parallel}\lambda_{\perp}^2}{\tilde{\Lambda}_{e\parallel}\tilde{\Lambda}_{e\perp}^2} \frac{\alpha + 2\beta}{3} . \qquad (8.22)$$

In the important case $\nu_{ei} >> \nu_{eN}$, for example, we have $\alpha=1.7$, $\beta=-0.5$, and in the limit $t \to \infty$ the maximal depth of the depletion in the heated region is

$$(n^{(max)} - n_0)/n_0 = -\Psi(0) = -0.23(T_e^{(max)} - T_e^{(0)})/T_e^{(0)}. \qquad (8.23)$$

The density profile Eq. (8.21) differs qualitatively in certain ways from Eq. (8.15). In the approximation of elementary theory the density perturbation is negative in the heated region and positive in the electron and ion ellipsoids. In general, the profile is more complex. At later stages it is determined by $\Psi(0)$. In the heated region the density perturbation is again negative: $(n^{(max)} - n_0)/n_0 = A(\alpha + 2\beta)/3 < 0$. With $\tilde{\Lambda}_{e\parallel} >> |z| >> \lambda_{\parallel}$, $\tilde{\Lambda}_{e\perp} >> r >> \lambda_{\perp}$ the perturbation has a power-law asymptotic behavior

$$\frac{n - n_0}{n_0} = \frac{A(\alpha - \beta)}{x^3(t=0)x^2(t)} \left(\frac{z^2}{\lambda_{\parallel}^2} - \frac{r^2}{2\lambda_{\perp}^2} \right) . \qquad (8.24)$$

The steady state perturbation in the case $\alpha > \beta$ is thus positive for $|z| >> \lambda_{\parallel}$, $r=0$ and negative for $z=0$, $r >> \lambda_{\perp}$. The change of sign occurs at angles of the order of μ_0 to the magnetic field. In general, profile Eq. (8.21) is nonmonotonic having several extrema.

8.3 EVOLUTION OF A PRONOUNCED PERTURBATION

Thermodiffusion for a pronounced temperature perturbation occurs in qualitatively the same way as in the preceding case. The perturbation of the plasma density which is pushed from the heated region is of the order of $\delta n \sim n_0$. In contrast, the density perturbation in the depletion regions away from the temperature perturbation, from which the particles which maintain quasineutrality arrive, is shallow, see Section 6.1. Accordingly, the short circuiting mechanism is effective, as in the linear case.

A thermodiffusion spreading of this sort has been observed in laboratory experiments [5]-[8] where the heating of ionosphere by electromagnetic waves has been modelled. In [5]-[7] the plasma was produced in a tube with an insulating end wall. The initial density profile of the decaying plasma was given by

$$n_0(r,z) = n_0^{(\max)} \cos\frac{\pi z}{2L} J_0(2.4r / a), \tag{8.25}$$

and afterwards decreased with time. The length of the tube was $2L$=1.5m, and its radius was a=0.4m. The plasma decay time $\tau_{i\|}=L^2/[\pi^2(1+T_e/T_i)D_{i\|}]$ was long in comparison with the thermodiffusion time. At t=0 and z=0, r=0, an approximately point source of electron heating was turned on. Over a time of the order of few tens of microseconds, a steady state temperature profile was established. This profile can be approximated by the expression (8.17) with $\lambda_\|$=20cm, $\lambda_\perp$=2cm and A=2.1. The ions had the room temperature while the unperturbed electron temperature was $T_e^{(0)}$=0.4eV. The transport coefficients for the ions were determined by ion-neutral collisions, and those for electrons, by electron-ion collisions. According to the results of Sections 2.2 and 2.7, when the Coulomb collisions dominate, and $T_e >> T_i$, we have

$$D_{e\|} = \frac{2T_e}{m_e \nu_{ei}}; \; D_{e\perp} = \frac{T_e \nu_{ei}}{m_e \omega_{ce}^2}; \; D_{e\|}^{(T)} = 1.7 D_{e\|}; \; D_{e\perp}^{(T)} = -\frac{1}{2} D_{e\perp} . \tag{8.26}$$

Under the experimental conditions at $T_e^{(0)}$=0.4eV we would have $D_{e\|}=2\cdot 10^6 cm^2/s$, $D_{e\perp}=10^4 cm^2/s$, $D_{i\|}=1.7\cdot 10^4 cm^2/s$, $D_{i\perp}=0.6\cdot 10^4 cm^2/s$.

These experimental results are representative for a rather wide range of conditions in which $(1+T_e/T_i)D_{i\perp} >> D_{e\perp}$ and

$$\lambda_\perp^2/(1+T_e/T_i)D_{i\perp} >> \lambda_\|^2/D_{e\|}. \tag{8.27}$$

The last inequality has to be satisfied in the case of a point heating source because the dimensions of the heated region are determined by the energy relaxation length and, therefore, $\lambda_{\|}^2/\lambda_{\perp}^2 \sim D_{e\|}/D_{e\perp}$. Under these conditions the initial Eqs.(8.1) can be simplified substantially. It follows from the inequality (8.27) and from the condition $\vec{\nabla}\cdot\vec{\Gamma}_i = \vec{\nabla}\cdot\vec{\Gamma}_e$ that the transverse electric field is much weaker than $T_{e\perp}/(e\lambda_{\perp})$.

Hence, to determine the main density perturbation, we can neglect the transverse electric field. In other words, in the equation for the electrons we can set the transverse electric field equal to zero: $\varphi=\varphi(z,t)$. Therefore, the longitudinal electric field corresponds to the field in the unperturbed plasma at $r \gg (\lambda_{\perp}^2+4D_{e\perp}t)^{1/2}$

$$\frac{\partial\varphi}{\partial z} = T_e^{(0)} \frac{\partial \ln n_0(z)}{\partial z}. \tag{8.28}$$

Substituting Eq. (8.28) into the first of the Eqs. (8.1), we find one equation which is formulated in terms of the density:

$$\begin{aligned} \frac{\partial \tilde{n}}{\partial t} &= \frac{\partial}{\partial z}\left(\tilde{T}_e^{5/2}\frac{\partial \ln \tilde{n}}{\partial z} - \tilde{T}_e^{3/2}\frac{\partial \ln \tilde{n}_0}{\partial z} + 1.7\tilde{T}_e^{3/2}\frac{\partial \ln \tilde{T}_e}{\partial z}\right) \\ &+ \frac{C}{r}\frac{\partial}{\partial r}\left[r\left(\frac{\tilde{n}}{\tilde{T}_e^{1/2}}\frac{\partial \tilde{n}}{\partial r} - \frac{1}{2}\frac{\tilde{n}}{\tilde{T}_e^{3/2}}\frac{\partial \tilde{T}_e}{\partial r}\right)\right], \end{aligned} \tag{8.29}$$

where

$$\tilde{n} = n/n_0^{(\max)},\ \tilde{T}_e = T_e/T_e^{(0)},\ \tau = tD_{e\|}(T_e^{(0)}, n_0^{(\max)})/\lambda_{\|}^2,$$
$$C = (\lambda_{\|}^2/\lambda_{\perp}^2)D_{e\perp}(T_e^{(0)}, n_0^{(\max)})/D_{e\|}(T_e^{(0)}, n_0^{(\max)}).$$

Here longitudinal coordinate z is expressed in units of $\lambda_{\|}$, and r is expressed in units of $\lambda_{\perp}$.

Equation (8.29) has been solved numerically in [4] for the parameters of experiments [5] - [7], for $n_0^{(\max)}=3.5\cdot 10^{18}\,\mathrm{m}^{-3}$. Figures 8.2 and 8.3 show the evolution of the density profile.

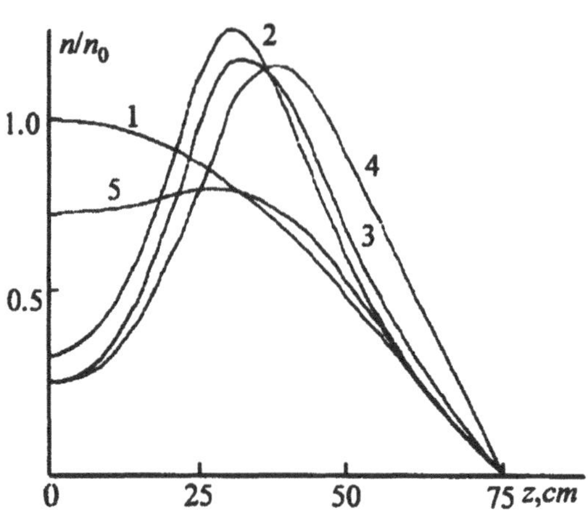

Figure 8.2. Longitudinal density profile at r=0 at various moments. 1 - t=0; 2 - 70μs; 3 - 250μs; 4 - limiting profile Eq. (8.28), 5 - 330μs after the heating source has been turned off.

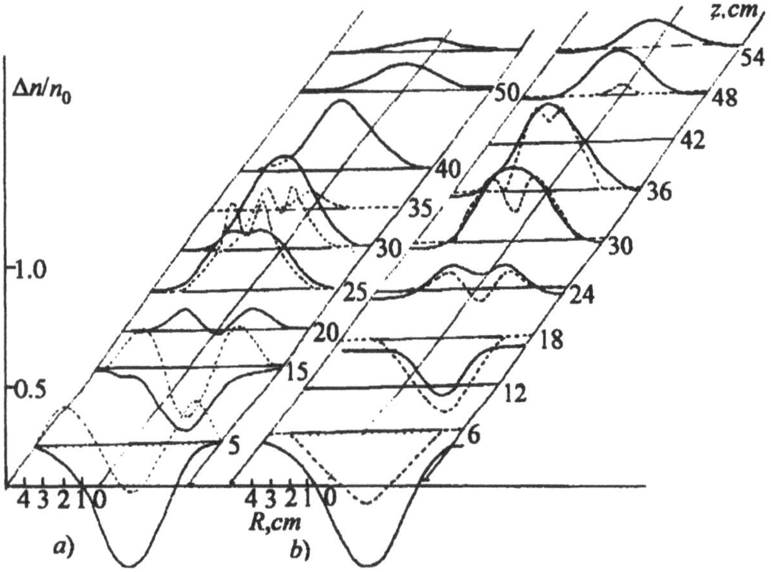

Figure 8.3. Transverse density profiles. a) t=70μs; b) t=250μs. Dashed lines are the experimental profiles for $n_0^{(max)}$=$2.0 \cdot 10^{18}$m^{-3}.

The length scales of the regions of positive and negative perturbations at times larger than the energy relaxation time are close to the results observed experimentally.

The fact that the calculated density is perturbed over a distance along $\vec{B}$ slightly greater than in the experiments is a consequence of an experimental temperature profile steeper than the Gaussian. The density maximum depth is greater than in the experiment due to the large uncertainty of experimental determination of the quantity $n_0^{(max)}$ (it was in a range 2-3.5·10^{18}m^{-3}). Under our simplifications, the density perturbation in the ion ellipsoid and in the depleted region from which the particles maintaining quasineutrality arrive, are negligibly small. The latter, nevertheless were observed in [5] - [7] (the relative density perturbation was less than 1%).

At the late stage a steady-state density profile corresponding to the vanishing of the left side of Eq. (8.29) is established in the region of the temperature perturbation. Since in Eq. (8.29) C≈0.5, and since the diffcoefficients $D_{e\parallel}$ and $D_{e\perp}$ depend on the density in different ways, the role of transverse diffusion decreases as the density decreases. Therefore, the steady state density profile roughly corresponds to $\Gamma_{e\parallel}$=0. From this we find (in analogy with Eq. (8.6))

$$n(T_e+T_i)^{1.7}=const\cdot\cos(\pi z/2L). \qquad (8.30)$$

This limiting profile is shown in Fig. 8.2.

Fig. 8.4 shows the time history of longitudinal electron fluxes. Similar fluxes have been observed in the experiment.

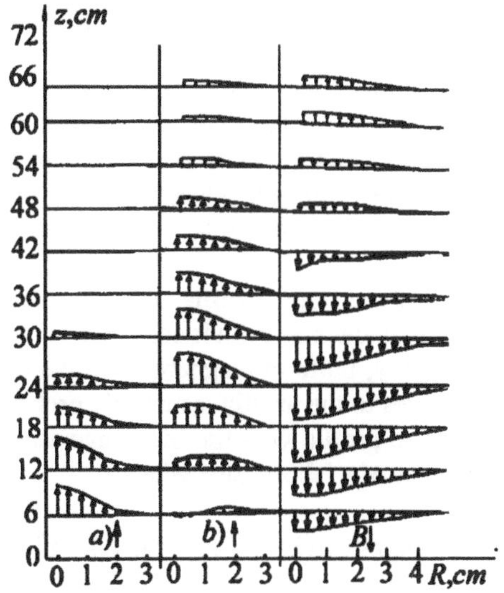

Figure 8.4. Longitudinal electron fluxes; the scale Γ is shown by the arrows in the bottom. *a)* t=0, Γ=5Γ_0; *b)* t=70μs, Γ=0.5Γ_0; c) t=330μs, after the heating source has been turned off, with Γ=0.1Γ_0. Here $\Gamma_0=D_{e\parallel}(T_e^{(0)},n_0^{(max)})n_0^{(max)}/\lambda_{\parallel}$.

At the time t_0=250μs, the heating source was turned off, the direction of the particle fluxes were observed to change, and an influx into the depletion region was observed. Figures 8.2 and 8.4c show the result of the simulation for $t>t_0$.

A common feature of the profiles which have been found, both in numerical simulations for the nonlinear case and in the linear case discussed in the preceding Section, is the presence of two density maxima in the transverse direction. This result is a consequence of the thermal force, i.e., the difference between the particle diffusion and thermodiffusion coefficients, $D\neq D^{(T)}$.

Numerical simulation of a coupled system of density and energy balance equations was performed recently in [9]. Results does not differ significantly from those discussed in this Section.

8.4 THERMODIFFUSION OF A MAGNETIZED PLASMA IN A CONDUCTING VESSEL

The problem of magnetized nonisothermal plasma transport in a conducting vessel is of considerable interest in connection with the numerous applications of RF discharges in the magnetic fields in plasma surface processing. The full problem of gas discharge modelling is rather complicated, so we shall consider here only the process of charged particle transport and their loss from the plasma volume for a given heat power input. This problem can be solved analytically [10]. At first we shall find the steady state density profile for a given electron temperature radial profile. The maximum temperature value is obtained in this approach as a solution of the eigenvalue problem. Then we shall obtain the electron temperature profile for a given profile of the power input.

Plasma transport for fixed $T_e(r)$ profile. We consider plasma with the given $T_e(r)$ profile, in a conducting cylindrical vessel of length L and radius a immersed in a stationary, uniform magnetic field parallel to the cylinder axis. We restrict ourselves to the axially-symmetric problem and to the case of monotone temperature profile with the maximum value of $T_e(r)$ at r=0. We shall analyze the case of a 'short' tube when

$$\tau_{e||}=L^2/(\pi^2 D_{e||})<<\tau_{i\perp}(1+T_e/T_i)=(1+T_e/T_i)a^2/(2.4^2 D_{i\perp}). \qquad (8.31)$$

Due to the large longitudinal electron thermal conductivity, the electron temperature should be independent of z. As in Section 7.2, in a strong magnetic field we neglect the electron transverse diffusion and thermodiffusion coefficients with respect to the transverse ion diffusion coefficient. Under this assumption the steady state analog of Eq. (8.1) has the form

$$\frac{\partial}{\partial z}\left(D_{e\parallel}\frac{\partial n}{\partial z}-nb_{e\parallel}\frac{\partial \varphi}{\partial z}\right)+\nu^{(ion)}(T_e)n=0\ ;$$

$$\frac{\partial}{\partial z}\left(D_{i\parallel}\frac{\partial n}{\partial z}+nb_{i\parallel}\frac{\partial \varphi}{\partial z}\right)+\frac{1}{r}\frac{\partial}{\partial r}r\left(D_{i\perp}\frac{\partial n}{\partial r}+nb_{i\perp}\frac{\partial \varphi}{\partial r}\right)+\nu^{(ion)}(T_e)n=0\ . \tag{8.32}$$

It is interesting to note, that in spite of the fact that the electron thermodiffusion coefficients are explicitly absent in Eq. (8.32), the effect of the temperature gradient remains important, and even dominates. It enters the final result through the radial electric field which is determined by the radial electron temperature profile.

For the ionization frequency $\nu^{(ion)}$ we choose the simple expression corresponding to the Maxwellian electrons in *Ar*:

$$\nu^{(ion)}(T_e)=N\sigma_{i0}[T_e(r)/2\pi m_e]^{1/2}\exp[-\varepsilon/T_e(r)], \tag{8.33}$$

where N is the density of the neutral particles, while σ_{i0} and ε are the adjustment constants (ε=16.3 eV, $\sigma_{i0}=3.1\cdot 10^{-20}\mathrm{m}^2$). This choice makes possible a comparison with the results of the Monte-Carlo numerical simulation [11]. Since the electrons are trapped along $\vec{B}$, they should have a Boltzmann distribution. The potential profile in plasma with respect to the conducting end wall, for the case of the collisionless sheath adjacent to it, (see Section 7.2, Eq. (7.17)) is

$$\varphi(r,z,t)=\frac{T_e(r)}{e}\ln\frac{n(r,z)\sqrt{T_e(r)/m_e}}{\sqrt{2\pi}\Gamma_{e\parallel}(r,z=0,L)}\ . \tag{8.34}$$

Here $\Gamma_{e\parallel}(r,z=0)=\Gamma_{e\parallel}(r,z=L)$ is the electron flux to the end walls. As we have neglected the transverse electron transport, the flux to the end walls is determined only by the ionization frequency:

$$2\Gamma_{e\parallel}(r,z=0,L)=\nu^{(ion)}(T_e)\int_0^L n(r,z)dz\ . \tag{8.35}$$

Substituting Eqs.(8.35), (8.33) into Eq. (8.34), we find

$$\varphi(r,z)=\varepsilon/e+\alpha(r,z)T_e(r)/e, \tag{8.36}$$

where

$$\alpha(r,z)=\ln\frac{n(r,z)}{2LN\sigma_{i0}\int_0^L n(r,z)dz}. \tag{8.37}$$

The Eqs.(8.36) and (8.37) express the potential in plasma in terms of the density and electron temperature profiles. If the density profile can be factorized, $n=f_1(z)f_2(r)$, the value of α becomes r-independent. Indeed, if, for example, $f_1(z)=\sin(\pi z/L)$, we have

$$\alpha=\ln\frac{\pi\sin(\pi z/L)}{4LN\sigma_{i0}}. \tag{8.38}$$

The value of α can be either negative or positive. We restrict ourselves to the case of positive α.

With the potential given by Eq. (8.36) one can now solve the ion transport equation - the second one in Eq. (8.32). From this equation the physical reason for the distinction between the cases of uniform and non-uniform $T_e(r)$ profiles is clearly seen. If $T_e(r)=const$, the radial electric field is determined, according to Eqs.(8.36), (8.37), only by the slow logarithmic dependence of $\alpha(r)$. The radial ion transport is thus caused by the mobility in this small electric field and by the radial ion diffusion. In contrast, in the opposite case, since usually $T_e>>T_i$, the radial potential profile is determined by the electron temperature profile. Moreover, it is possible to neglect the ion diffusion terms, both longitudinal and perpendicular, in Eq. (8.32). In the case of nonuniform temperature, substituting the potential Eq. (8.38) into Eq. (8.32), we find

$$\frac{b_{i\|}T_e}{e}\frac{\partial^2 n}{\partial z^2}+\frac{1}{r}\frac{\partial}{\partial r}r\left(n\frac{b_{i\perp}}{e}\frac{\partial\alpha T_e}{\partial r}\right)+\nu^{(ion)}(T_e)n=0. \tag{8.39}$$

Neglecting the weak logarithmic dependence of $\alpha(r,z)$, it is possible to seek the solution in the form

$$n=n^{(max)}(z=L/2,r=0)f_1(z)f_2(r). \tag{8.40}$$

Substituting into Eq. (8.39) yields

$$\frac{b_{i\parallel}}{e}\frac{\partial^2 f_1}{\partial z^2} = C_1\bar{\alpha} f_1 \ ;$$
$$\frac{b_{i\perp}}{e}\frac{1}{r}\frac{\partial}{\partial r}\left(rf_2\frac{\partial T_e}{\partial r}\right) = -\frac{\nu^{(ion)}(T_e)}{\bar{\alpha}} f_2 - C_1\bar{\alpha} f_2 \ , \tag{8.41}$$

where C_1 is the integration constant, and $\bar{\alpha}$ is the averaged over the volume value of α. From the boundary condition $f_1(z{=}0,L){=}0$ it follows

$$f_1 = \sin(\pi z / L) \ ; \qquad C_1 = -\frac{b_{i\parallel}(\pi / L)^{1/2}}{\bar{\alpha} e} .$$

After substituting into Eq. (8.41) we obtain

$$\bar{\alpha}\frac{b_{i\perp}}{e}\frac{1}{r}\frac{\partial}{\partial r}\left(rf_2\frac{\partial T_e}{\partial r}\right) = [-\nu^{(ion)}(T_e) + \tau_{a\parallel}^{-1}] f_2 \ , \tag{8.42}$$

where

$$\tau_{a\parallel}^{-1} = (T_e/e) b_{i\parallel}(\pi/L)^2 = D_{a\parallel}(\pi/L)^2 \tag{8.43}$$

is the inverse time of the longitudinal ambipolar diffusion. The solution of Eq. (8.42) for an arbitrary $T_e(r)$ profile is given by

$$f_2 = \exp\left[-\frac{\int_0^r \beta(r') + j(r')}{\frac{b_{i\perp}\bar{\alpha}}{e}\frac{\partial T_e}{\partial r'}} dr' \right] , \tag{8.44}$$

where $\beta(r) = \nu^{(ion)}(T_e) - \tau_{a\parallel}^{-1}$ and

$$j(r) = \frac{\bar{\alpha} b_{i\perp}}{e}\frac{1}{r}\frac{\partial}{\partial r}\left(r\frac{\partial T_e}{\partial r}\right) .$$

The value of $\bar{\alpha}$ may be replaced with logarithmic accuracy by

$$\alpha(0) = \ln\frac{\pi}{4LN\sigma_{i0}} . \tag{8.45}$$

Plasma density is finite in the vessel center, where $\partial T_e / \partial r = 0$, so β(0)+j(0)=0. As

$$j(0) = 2\frac{\bar{\alpha} b_{i\perp}}{e}\frac{\partial^2 T_e}{\partial r^2}\Big|_{r=0},$$

we obtain the equation for the central electron temperature $T_e^{(\max)}$:

$$\nu_i(T_e^{(\max)}) - \frac{T_e^{(\max)}\pi^2}{\mu_{iN}\nu_{iN}L^2} + 2\frac{\bar{\alpha} b_{i\perp}}{e}\frac{\partial^2 T_e}{\partial r^2}\Big|_{r=0} = 0, \tag{8.46}$$

where the ionization rate $\nu^{(ion)}$ is given by Eq. (8.33), μ_{iN} is the reduced mass, and ν_{iN} is the ion-neutral collision frequency.

Eq. (19) describes the one-dimensional balance between the generation and longitudinal diffusion of the electrons. For an arbitrary $T_e(r)$ profile the value of $T_e^{(\max)}$ is larger than for T_e=*const*, if $\bar{\alpha}>0$ (such a situation was considered in [11]), because $\partial^2 T_e / \partial r^2|_{r=0} < 0$. It means that the ion generation at r=0 is now balanced by both their diffusion along z and their radial outward convection in the electric field Eq. (8.36). (Note that for a long enough vessel $\bar{\alpha}$ is negative. Such a complicated situation is beyond consideration now.)

It should be particularly emphasized that all the main results which are derived in this Section are based on the assumption of the Maxwellian electron distribution function. But the electron losses on the vessel walls and inelastic electron-neutral collisions (e.g. excitation, ionization) result in considerable depletion of the electron distribution tail and reduce the values of $\Gamma_{e\parallel}$ and $\nu^{(ion)}$ with respect to Eqs. (8.33), (8.34). These expressions are valid only in plasma where the electron-electron collisions are frequent enough to restore the Maxwellian distribution. In the opposite case, the whole problem is essentially kinetic, and the radial potential profile is determined by the form of the tail of electron distribution function, see [10].

Energy balance. The profile of $T_e(r)$ is determined by the electron energy balance equation, averaged over z for the given profile of power input $P(r)$:

$$P(r) = \int_0^L p n dz = \int_0^L \nu_i n(\varepsilon + \frac{3}{2}T_e)dz - \int_0^L e\Gamma_{e\parallel}\frac{\partial\varphi}{\partial z}dz + 4\Gamma_{e\parallel}(z=0,L)T_e, \tag{8.47}$$

where $\rho(r,z)$ is the power input per electron. Here the first term in the right-hand side represents the energy losses for the ionization and subsequent heating of the new electrons. The second term is the energy which is transferred from the electrons to the ions by means of the longitudinal electric field, Eq. (8.36). The last term is the heat flux to the end walls which is calculated according to Eq. (2.13) with the Maxwellian distribution function.

Employing Eq. (8.36), the second term can be rewritten in the form

$$-\int_0^L e\Gamma_{e\parallel}\frac{\partial\varphi}{\partial z}dz = e\int_0^L \varphi\frac{\partial\Gamma_{e\parallel}}{\partial z}dz = \int_0^L(\varepsilon+\overline{\alpha}T_e)\frac{\partial\Gamma_{e\parallel}}{\partial z}dz$$
$$= 2(\varepsilon+\overline{\alpha}T_e)\Gamma_{e\parallel}(z=0,L). \tag{8.48}$$

Here we have replaced α by $\overline{\alpha}$ because of its slow logarithmic variation. From Eqs. (8.35), (8.47) and (8.48) one finally obtains

$$\rho(r) = \nu^{(ion)}[2\varepsilon+(\overline{\alpha}+\frac{7}{2}T_e)]. \tag{8.49}$$

Now, taking into account the expression for $\nu^{(ion)}$, Eq. (8.33), one can obtain the expression for $\partial^2 T_e/\partial r^2|_{r=0}$ from Eq. (8.49) as a function of $\partial^2\rho/\partial r^2|_{r=0}$:

$$\frac{\partial^2\rho/\partial r^2|_{r=0}}{\rho|_{r=0}} = \frac{\partial^2 T_e/\partial r^2|_{r=0}}{T_e^{(\max)}}A, \tag{8.50}$$

where

$$A = \frac{\varepsilon}{T_e^{(\max)}}+\frac{1}{2}+\frac{\overline{\alpha}+\frac{7}{2}}{(2\varepsilon/T_e^{(\max)})+\overline{\alpha}+\frac{7}{2}}. \tag{8.51}$$

At last, substituting Eqs.(8.50), (8.51) into Eq. (8.45), we obtain the equation for $T_e^{(\max)}$:

$$\frac{\exp(\varepsilon / T_e^{(\max)})}{T_e^{(\max)}} = \frac{2\sqrt{2}\mu_{iN} \nu_{iN} N\sigma_{i0} L^2}{m_e^{1/2} \pi^{5/2} \left[\frac{2\bar{\alpha} b_{i\perp} L^2}{\pi^2 b_{i\parallel}} \frac{\partial^2 \rho / \partial r^2 |_{r=0}}{\rho|_{r=0} A} + 1 \right]} . \quad (8.52)$$

The expressions obtained determine the plasma parameters. Indeed, the maximum temperature in the vessel is given by Eq. (8.52), the temperature profile for the arbitrary power deposition is determined by Eq. (8.49), the density profile is described by Eqs.(8.40), (8.44), and the potential profile is given by Eqs. (8.36) and (8.37). These analytic expressions were compared in [10] with the results of the Monte-Carlo simulation performed in [11] and good agreement was reported. The corrections due to the finite ratio of λ_{iN}/L are calculated in [10]. The problems of the electron energy balance and formation of the T_e profile is discussed in more detail in Chapter 12.

Applying a voltage between the end and the side walls may lead to various interesting effects analogous to those discussed in Section 7.3. Thus the possibility of local control of the plasma parameters arises.

REFERENCES

1. J. W. Coburn, *Plasma Etching and Reactive Ion Etching* (N.Y.: Americ. Vac. Soc., 1970): 133.
2. B. Chapman, *Glow Discharge Processes* (N.Y.: Wiley, 1980): 406.
3. *Plasma Etching: an Introduction*, Ed. by D. M. Manos, D. L. Flamm (N.Y.: Academic Press, 1989): 476.
4. S. P. Voskoboynikov, I. Yu. Gurvich, V. A. Rozhansky, *Sov. J. Plasma Phys*. 7: 479-484 (1989).
5. S. V. Egorov, A. V. Kostrov, A. V. Tronin, *Sov. Phys. JETP Lett.* **47**: 102 -106 (1988).
6. G. Yu. Golubatnikov, S. V. Egorov, A. V. Kostrov et al., *Sov. J. Plasma Phys.* **14**: 285-287 (1988).
7. G. Yu. Golubatnikov, S. V. Egorov, A. V. Kostrov et al., *Sov. Phys. JETP* **69**: 1134-1138 (1989).
8. J. M. Urrutia, R. L. Stenzel, *Phys. Rev. Lett.* **67**: 1867-1870 (1991).
9. L. E Kurina, *Plasma Phys. Rep.* **24**: 937-941 (1999).
10. L. L. Beilinson, V. A. Rozhansky, L. D. Tsendin, *Phys. Rev. E* **50**: 3033-3040 (1994).
11. R. K. Porteous, D. B. Graves, *IEEE Trans. Plasma Sci.* **PS-19**: 204-213 (1991).

Chapter 9

Influence of Net Current Through Plasma Inhomogeneity on Its Evolution in the Absence of Magnetic Field

9.1 PLASMAS WITHOUT IONIZATION AND RECOMBINATION PROCESSES. GENERAL CONSIDERATION

As it was mentioned above in Section 4.1, the net current through a plasma radically changes the evolution scenario of large scale inhomogeneities. In contrast to the standard diffusion process, when an initial inhomogeneity is leveled over the whole space, the nonlinearity of the problem usually leads to steepening of some parts of the partial density profiles. It was noticed as early as more than century ago, with respect to electrolytes [1] and [2], that this phenomenon has a profound hydrodynamic analogy in the widely known phenomenon of overturn of smooth profiles and shock formation [3].

In this Section we shall consider this situation for the simple 1D problem in boundless collision dominated multispecies plasma with scalar constant mobility and diffusion coefficients, in the absence of all the plasmachemical processes of generation and mutual transformation of the charged particles.

As in the absence of the net current, when the inhomogeneity scale L satisfies the condition $L \gg \max[(E_0\varepsilon_0/en); r_d]$, during the Maxwellian time Eq. (2.123) the quasineutral state is established, and the evolution process, according to Eqs.(2.7), (2.41), is described by the following set of equations:

$$\frac{\partial n_\alpha}{\partial t} - \frac{\partial}{\partial x}(D_\alpha \frac{\partial n_\alpha}{\partial x} - Z_\alpha n_\alpha b_\alpha E) = 0;$$
$$\frac{\partial n_e}{\partial t} - \frac{\partial}{\partial x}(D_e \frac{\partial n_e}{\partial x} + b_e n_e E) = 0; \qquad (9.1)$$
$$n_e = \sum_{\alpha}^{p-1} Z_\alpha n_\alpha ,$$

where the subscript α corresponds to the ion species. We shall consider, for simplicity, only the case when the charge numbers $Z_\alpha = \pm 1$. The boundary conditions are

$$n_\alpha(x \to \pm\infty) = n_\alpha^{(0)};$$
$$n_e(x \to \pm\infty) = n_0; \qquad (9.2)$$
$$E(x \to \pm\infty) = E_0 \neq 0.$$

The latter condition corresponds to the net current through the plasma inhomogeneity:

$$j = eE_0[b_e n_0 + \sum_{\alpha=1}^{p-1} b_\alpha n_\alpha^{(0)}] = eE[b_e n_e + \sum_{\alpha=1}^{p-1} b_\alpha n_\alpha] +$$
$$+ e[D_e \frac{\partial n_e}{\partial x} - \sum_{\alpha=1}^{p-1} D_\alpha Z_\alpha \frac{\partial n_\alpha}{\partial x}]. \qquad (9.3)$$

Equation (9.3) permits the exclusion of the electric field from the system Eq. (9.1) and the formulation the equation system for the plasma evolution:

$$\frac{\partial n_\alpha}{\partial t} + \frac{\partial}{\partial x} E_0 \sum_{\beta=1}^{p-1} n_\beta^{(0)} (b_e + b_\beta) \frac{Z_\alpha b_\alpha n_\alpha}{\sum_{\beta=1}^{p-1} (b_e + b_\beta) n_\beta}$$
$$+ \frac{\partial}{\partial x} \left[-\frac{\partial}{\partial x} D_\alpha \frac{\partial n_\alpha}{\partial x} - Z_\alpha b_\alpha n_\alpha \frac{\sum_{\beta=1}^{p-1} (D_e - D_\beta) Z_\beta \frac{\partial n_\beta}{\partial x}}{\sum_{\beta=1}^{p-1} (b_e + b_\beta) n_\beta} \right] = 0. \qquad (9.4)$$

If the plasma current depends on time, but does not changes its sign, we can reduce the system Eqs. (9.1) - (9.3) to the form Eq. (9.4) by replacing the time by $q = \int_0^t j(t')dt'$, i.e. by the total charge which has flown through plasma [1], [2]. For sufficiently smooth inhomogeneities, with a characteristic scale in the current direction

$$L >> T/(eE), \tag{9.5}$$

the diffusion terms in Eq. (9.4) can be neglected. Such an approach is referred to as the drift approximation. For the constant b_e, b_α in the drift approximation the combination

$$\psi(x) = \sum_{\alpha=1}^{p-1} n_\alpha \frac{b_e + b_\alpha}{b_e Z_\alpha b_\alpha} \approx \sum_{\alpha=1}^{p-1} \frac{n_\alpha}{Z_\alpha b_\alpha}, \tag{9.6}$$

which is determined by the initial plasma profile, is conserved in time. In the pure plasma this quantity is proportional to the profile $n_e(x)$, and its conservation is a trivial result of the fact that in the drift approximation the diffusion is neglected. In multispecies plasmas, on the other hand, the conservation of $\psi(x)$ implies that an arbitrary plasma density disturbance is split into one immobile and several moving modes.

Small perturbations. For the small ($\delta n_\alpha << n_\alpha$) perturbations, linearizing Eqs.(9.4) and neglecting b_β with respect to b_e (which means that the plasma conductivity is determined by electrons), for the Fourier components exp(-$i\omega t+ikx$) we obtain

$$(-\omega + Z_\alpha b_\alpha E_0 k) n_{\alpha k} = n_\alpha^{(0)} Z_\alpha b_\alpha E_0 k \sum_{\beta=1}^{p-1} n_{\beta k} / \sum_{\beta=1}^{p-1} n_\beta^{(0)}. \tag{9.7}$$

As in the case of plasma diffusion (Section 4.2), this linear system has a nontrivial solution only if its determinant equals zero. The resulting algebraic dispersion equation determines (p - 1) branches which connect ω and k for different evolution modes. One of them, which corresponds to $\psi(x)$ conservation, $\omega(k)=0$, describes the fixed perturbation. The remaining (p-2) branches describe different modes of propagating signals which are moving with different velocities and are slowly damped due to diffusion. In other words, an arbitrary localized disturbance is split into (p-2) separate signals. Each of them propagates independently from the others. The densities of all the plasma components and the electric field are perturbed in each of these modes.

If, for example, plasma consists of electrons and two species of positive ions, the dispersion equation for the propagating mode is

$$\omega/k=V=E_0(b_1 n_2^{(0)}+b_2 n_1^{(0)})/(n_1^{(0)}+n_2^{(0)}). \tag{9.8}$$

The velocity V is usually referred to as the ambipolar drift velocity, and the proportionality coefficient between V and E_0 is known as the ambipolar mobility [4]. The perturbation of the partial densities (α=1,2) has the form

$$\delta n_\alpha(x,t)=\frac{1}{2\pi}\int_{-\infty}^{\infty} n_{\alpha k}\exp(ikx)dk,$$

where

$$\begin{aligned} n_{1k}(t) &= \frac{b_2 n_1^{(0)}}{b_1 n_2^{(0)}+b_2 n_1^{(0)}}\{[n_{2k}(0)++n_{1k}(0)b_2/b_1] \\ &+[-n_{2k}(0)+n_{1k}(0)n_2^{(0)}/n_1^{(0)}]\exp(-ikVt)\}. \end{aligned} \tag{9.9}$$

The expression for δn_2 is obtained analogously by changing the subscripts. If the initial species density perturbations are proportional to each other, $\delta n_1(x,t=0)/n_1^{(0)}=\delta n_2(x,t=0)/n_2^{(0)}$, the second (time dependent) term in the curly brackets in Eq. (9.9) vanishes, and $\delta n_\alpha(x,t)=\delta n_\alpha(x,t=0)$. This case corresponds to the stationary profiles and the $\psi(x)$ conservation. If the initial profiles satisfy $\delta n_1(x,t=0)/\delta n_2(x,t=0)=-b_1/b_2$, the immobile perturbation is absent, and only the propagating signal remains.

Consider the situation when in pure plasma, consisting of electrons and ambient positive ions of species 2, a small quantity of positive ions of another species 1 is injected. Substituting $n_1^{(0)}=0$ into Eq. (9.8) we obtain that the ambipolar drift velocity equals b_1E_0. This result is obvious since the field perturbation influence on the motion of the injected particles is negligible. In semiconductor physics measurement of the ambipolar drift velocity is often used to find the mobility of the minority carriers [5] and [6]. From the expression for n_{2k}, in the same way as for Eq. (9.9), we obtain

$$n_{2k}(t)=n_{1k}(0)[1-\exp(-ikb_1E_0t)]b_2/b_1. \tag{9.10}$$

We see that in the background plasma a depletion region arises, which moves together with the injected test particles 1 (Fig. 9.1).

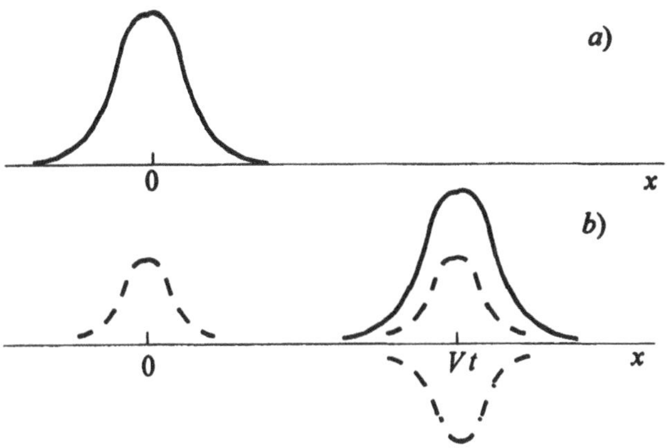

Figure 9.1. Disturbances of the partial densities after injection of the small Gaussian admixture of the test ions 1 into the ambient current carrying plasma of the species 2. The broken line is the total plasma density n_e; the dotted line corresponds to n_2; the full line to n_1. *a*) initial disturbance; *b*) disturbance at the moment $t=5a/(b_1E_0)$ after injection. Mobilities are: $b_1=2b_2<<b_e$.

The depletion depth differs by the factor b_2/b_1 from the initial density disturbance. This means that the moving electron density perturbation is positive for $b_2<b_1$ (Fig. 9.1) and negative in the opposite case $b_2>b_1$. At the injection place immobile density perturbation is located. Its magnitude is also b_2/b_1 times smaller or larger or than the initial perturbation. The immobile density perturbation consists of ambient ions, and its evolution process is controlled by the ambipolar diffusion, Eq. (4.7). In the special case when $b_2=b_1$, the depletion of the ambient ion density in the propagating mode fully compensates the peak of the test ions, and the total electron density n_e remains here unperturbed.

From the solution obtained above we can conclude that the fact that in the pure current carrying plasma the density evolution is described by the same ambipolar diffusion equation (4.7) as in the currentless plasma, results from the exact compensation of several rather complicated disturbances of ambient and injected particle densities. In plasmas of more complex composition, in addition to the fixed density disturbance, several modes of propagating weakly damping signals exist. In each of these modes the densities of all the plasma components are perturbed. In the subsequent Sections we proceed to analyze the nonlinear properties of these signals.

9.2 THREE-COMPONENT ELECTROPOSITIVE PLASMA WITH IMMOBILE IONS OF ONE SPECIES

We shall start the nonlinear analysis with the simplest case $b_2=0$ when the mobility of one species of positive ions in the plasma is negligible. The well-known situation of this kind corresponds to an n-type semiconductor, where positively charged fixed donor ions with spatially uniform density $n_2^{(0)}$ are present. Some situations in gaseous and dusty plasmas can also be described by this approximation. Assuming $n_2=n_2^{(0)}=const$, in the drift approximation from Eq. (9.4) it follows

$$\partial n_1 / \partial t + V(n_1)\partial n_1 / \partial x = 0, \tag{9.11}$$

where the nonlinear ambipolar drift velocity $V(n_1)$, Eq. (9.8), is related to the flux of mobile ions $\Gamma_1=b_1 n_1 E$ according to

$$\begin{aligned} V(n_1) &= \frac{d\Gamma_1}{dn_1} = \frac{jb_1 n_2^{(0)}}{eb_e(n_1 + n_2^{(0)})^2}; \\ \Gamma_1 &= \frac{jb_1 n_1}{eb_e(n_1 + n_2^{(0)})}. \end{aligned} \tag{9.12}$$

Equation (9.11) corresponds to propagation of a simple nonlinear wave [3]. Its solution is

$$n_1(x,t)=n_{10}(x-V(n_1)t), \tag{9.13}$$

where n_{10} is the initial density profile of the mobile ions. It means that each point of the initial profile moves with constant velocity $V(n_1)$. In other words, the plasma density is conserved along the characteristics - the straight lines on the (x,t) plane with tangent equal to $V(n_1)$.

As an example, we shall consider evolution of the Gaussian initial profile

$$n_{10}(x)=An_2^{(0)}\exp(-x^2/a^2). \tag{9.14}$$

From Eq. (9.12) it follows that the ambipolar drift velocity decreases with density, i.e. the plasma profile parts with high density move slower, and the low density parts move faster. As a result, the front side of the profile Eq. (9.14) becomes steeper, while the back becomes more gradual, see Fig. 9.2.

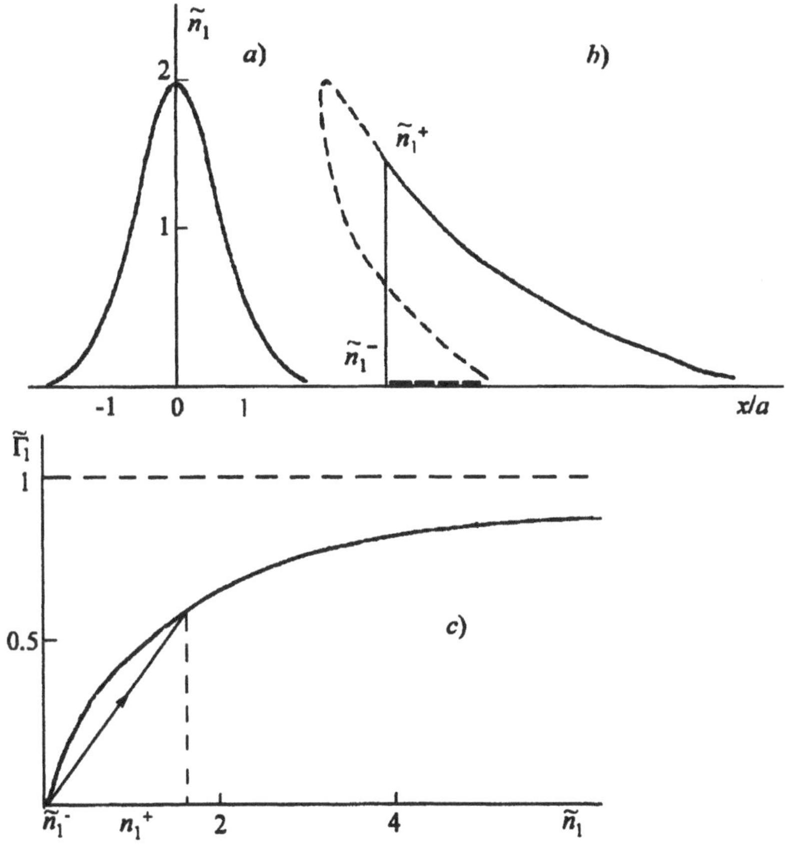

Figure 9.2. Evolution of the nonlinear Gaussian initial density profile in the three-component plasma with one species of fixed ions. The dimensionless density $\tilde{n}_1 = n_1/n_2^{(0)}$; time $\tilde{t} = tbE_0/a$; amplitude of the initial disturbance in Eq. (9.14) is A=2. *a*) initial profle; *b*) profile at $\tilde{t}$ =15 after shock formation; *c*) the flux of ions 1 in the drift approximation $\tilde{\Gamma} = \Gamma_1/(n_2^{(0)} bE_0)$; the arrow corresponds to shock which satisfies the evolution condition.

In the drift approximation a multivalued density profile is formed. This phenomenon is the same as the overturn of a finite amplitude wave in hydrodynamics. On the (x, t) plane the moment of the profile overturn corresponds to crossing of the characteristics. In reality, of course, no multivalued profile can exist. During the profile evolution steep profile sections are formed, where the drift approximation becomes invalid. Here the diffusion processes dominate. They compensate the profile overturn and determine the plasma profile in these diffusive shocks. A resulting plasma profile consists of the flat parts, which are described in the framework of the drift approximation, and of steep diffusive shocks. It is easy to obtain an estimate of the characteristic scale of the diffusive shocks l_T by equating of the field-driven and diffusive fluxes. It results in

$$l_T \sim T/(eE) << a. \tag{9.15}$$

According to [3], [7], [8] the shock arises at the moment

$$t_c = \min[dV(n_{10}(x))/dx]^{-1} \tag{9.16}$$

when the characteristics on the (x, t) plane begin to cross. A single-valued density profile which consists of sections of the drift solution separated by the diffusive shock, its position and the density values n_1^+ and n_1 on the right and left sides of the shock can be determined readily. It can be done by means of the drift solution in combination with the area rule [3], [7], [8] (see the Fig. 9.2b). Both the complete Eq. (9.4), and Eq. (9.11) of the drift approximation represent the differential forms of the particle conservation law. This means that $\int n_1 dx$, taken over the drift multivalued density profile $n_1(x,t)$, remains constant during the evolution process and equal to the integral over the real solution. The shock position in the drift profile is determined by the condition that the areas which are cut off from the three-valued drift profile are equal.

The shock velocity W can be determined from the ion flux conservation in the coordinate frame attached to the shock. This flux equals $\Gamma_1 - Wn_1$. Hence,

$$W = \frac{\Gamma_1(n_1^+) - \Gamma_1(n_1^-)}{n_1^+ - n_1^-}. \tag{9.17}$$

It is easy to see that in our problem only shocks which satisfy $n_1^+ > n_1^-$ can arise. This relation is often referred to as the evolution condition [3], [7] and [8].

The evolution of the profile Eq. (9.14) proceeds as follows. The profile overturn occurs at the moment t_c which satisfies Eq. (9.16). The density difference within the shock first increases, and then starts to decrease. The expression Eq. (9.17) for the shock velocity converts into Eq. (9.12) for the small signal velocity. At this almost linear evolution stage, the density perturbation δn_1 decreases and its length L increases with time with velocity $(dV/dn_1)\,\delta n_1$. The density profile at this stage tends to a triangular shape with the diffusive shock at the back side for $n_1 > n_1^{(0)}$, and at the front side - for $n_1 < n_1^{(0)}$ [3].

In the case of a more complex initial profile such phenomena as the appearance of several shocks, and their coalescence, etc. are possible. In the subsequent Sections we shall denote the drift sections of the density profiles by curly brackets, and the shocks by bold ones. So the profile with the shock which satisfies the condition (9.2) will be denoted as $\{n_1^{(0)}, n_1^-\}$, $[n_1^-, n_1^+]$, $\{n_1^+, n_1^{(0)}\}$.

Structure of the diffusive shock. Accounting for $b_e >> b_1; b_2=0$, from Eq. (9,4) we find [9]

$$\frac{\partial n_1}{\partial t}+\frac{\partial \Gamma_1}{\partial x}-\frac{\partial}{\partial x}[D(n_1)\frac{\partial n_1}{\partial x}]=0;$$
$$D(n_1)=D_1[n_1(1+T_e/T_i)+n_2^{(0)}]/(n_1+n_2^{(0)}). \tag{9.18}$$

Neglecting the slow temporal variation of W, in the reference frame which moves with the velocity W we obtain

$$-W\frac{dn_1}{dX}+\frac{d\Gamma_1}{dX}-\frac{d}{dX}[D(n_1)\frac{dn_1}{dX}]=0;$$
$$X=x-Wt. \tag{9.19}$$

This ordinary differential equation has a solution which tends asymptotically to $n_1^{\pm}$ at $X\to\pm\infty$, if W satisfies Eq. (9.17), and the evolution condition $n_1^+ > n_1^-$ is fulfilled. Integration of Eq. (9.19) yields

$$\frac{dn_1}{dX}=\frac{j}{b_e T_i}\frac{(n_1^+-n_1)(n_1-n_1^-)n_2^{(0)}}{[(1+T_e/T_i)n_1+n_2^{(0)}](n_1^++n_2^{(0)})(n_1+n_2^{(0)})}. \tag{9.20}$$

For a strong shock ($(n_1^+-n_1^-)\geq n_2^{(0)}$) the characteristic spatial scale equals $\sqrt{l_T^+ l_T^-}$ ($l_T^{\pm}$ are the values of l_T at the right and left sides of the shock). The potential difference over it is of the order of the T/e. Of course, the shock concept is useful and effective, if the shock width scale l_T or l_0 (see below, Eq. (9.24)) is negligible with respect to the global scale of a problem L.

In the case of weak shock $\Delta n_1 = n_1^+ - n_1^- << n_2^{(0)}$ the shock width is proportional to $(\Delta n_1)^{-1}$. From Eq. (9.20) for $T_e=T_i$ we have

$$n_1(X)=\frac{n_1^++n_1^-}{2}+\frac{\Delta n_1}{2}th\frac{X\Delta n_1 n_2^{(0)}}{2l_T(n_1^++n_2^{(0)})(2n_1^++n_2^{(0)})}. \tag{9.21}$$

It is to be noted that in a gas discharge plasma the electron temperature $T_e >> T_N$ is maintained by the electric field itself. In this case the shock width l_T coincides with the energy relaxation length, Eq. (2.20). In the absence of collisions between electrons this length is equal to the relaxation length of the electron distribution function, Eq. (2.59). Therefore, the analysis of strong shock, Eq. (9.20), cannot be performed in the framework of standard fluid equations, and demands rigorous kinetic treatment. From the other side, the form of the gradual global plasma profile, the shock position and limiting density values on the shock sides, etc., can be found from the drift approximation, even without account of the diffusion. The same situation arises in semiconductors with hot carriers. Both these cases are deeply analogous to standard gas dynamics, in which the concept of shocks turned out to be extremely effective and enlightening, but the strong shock structure can be found only as a result of extremely complicated kinetic calculations [7], [8].

If the plasma density is sufficiently low, the Debye radius exceeds l_T. In this case the quasineutrality condition is violated within the shock, and its width is determined by this process. The structure of such shocks was discussed in [10] (see also [9]). Neglecting diffusion in Eq. (9.19), and accounting for the Poisson equation, we have

$$-W\frac{dn_1}{dX}+\frac{d\Gamma_1}{dX}=0;$$
$$\frac{dE}{dX}=e(n_1+n_2^{(0)}-n_e)/\varepsilon_0. \tag{9.22}$$

The shock velocity W is determined by the drift solution according to Eq. (9.17) and does not depend on its structure. Substituting the expression for W into the first of the Eqs.(9.22), after integrating we can calculate $n_1(x)$. Substituting it into the second equation and accounting for the current conservation $n_e \approx j/(eb_eE)$, we obtain

$$\frac{dE}{dX}=-\frac{n_2^{(0)}(E-E^+)(E^--E)}{\varepsilon_0 E(E-E^+E^-n_2^{(0)}b_e e/j)}. \tag{9.23}$$

The denominator in this expression is positive and is equal to $\varepsilon_0(E^-)^2 n_1^+ E^+ eb_e/j$ and to $\varepsilon_0(E^+)^2 n_1^- E^- eb_e/j$ at $X\to\pm\infty$. Therefore, the solution of Eq. (9.23) which tends to E^+ and to E^- at $X\to\pm\infty$ exists only if $E^+ > E$. Since from Eq. (9.22) it follows

$$n_1=(b_1 n_1 E^- - W n_1^-)/(b_1 E - W),$$

the evolution condition for density has the same form, as for the quasineutral shock: $n_1^+ > n_1^-$.

The shock in this case represents the strong-field region with a large space charge density separating two quasineutral plasmas. In other words, it constitutes the double layer in collisional plasma. The shock width is of the order of

$$l_0 = \varepsilon_0 E / (en_0) = r_d^2 / l_T >> r_d \quad \text{for} \quad \Delta n_1 \sim n_2^{(0)}. \tag{9.24}$$

The potential difference within the shock is of the order of El_0. This value considerably exceeds the potential scale T/e, which is typical for the quasineutral shock.

Evolution of a discontinuity in initial conditions. The evolution problem of a discontinuity in the initial profile presents another example of a situation when the shock concept turns out to be effective. Let the initial density profile of the mobile ions 1 have a discontinuity with densities n_1^+ and n_1^- at the right and left sides, correspondingly. If a discontinuity satisfies the evolution condition, $n_1^+ > n_1^-$, it propagates as the diffusive shock with the velocity Eq. (9.17). In the opposite case the characteristics of Eq. (9.11) on the (x, t) plane are bunch-like and diverging [3], [7], [8]. This means that such a discontinuity vanishes at the first moment of evolution, and a smooth density profile arises.

It is convenient to perform the evolutionality analysis using the flux plot $\Gamma_1(n_1)$ in the drift approximation, Eq. (9.12); see Fig. 9.2c. In our problem this graph plays a role which is fully equivalent to the widely known Hugoniot adiabate in the traditional gas dynamics. The directed segment $[n_1^- \to n_1^+]$ on this plot corresponds to the shock. According to the Eq. (9.17), its tangent equals the shock velocity. The arrow means that from the evolution condition, if the $\Gamma_1(n_1)$ graph is convex upward, the density in the shock increases from left to right. According to Eq. (9.12), in the shock reference frame the small signal velocities on both sides of the shock are directed into it, in full analogy with traditional gas dynamics [7]. The adjacent plasma is not aware of the shock's existence.

From a practical point of view the most interesting are the fixed shocks which can be observed in stationary conditions. Such a situation is possible only in systems, for which the corresponding $\Gamma_1(n_1)$ graph has extrema. If $\Gamma_1(n_1)$ has a maximum, for example, the density in the immobile shock increases in the current direction. In other words, immobile shocks are possible in the case when ambipolar drift reversal exists. The analog of such a fixed shock in gas dynamics is the shock which arises in the critical cross-section of the Lavale nozzle.

The evolution condition can be interpreted as follows. In our example the $\Gamma_1(n_1)$ graph is convex upward. This means that the drift flux exceeds the total one, which in the shock reference frame equals (Γ_1-Wn_1). If, for example, the shock structure is diffusion dominated, this difference is compensated by the diffusive flux, which is negative. This condition implies that only shocks with $n_1^+ > n_1^-$ are possible.

Multidimensional case. In a multidimensional problem, instead of Es.(9.11), (9.12), in the drift approximation we have

$$\frac{\partial n_1}{\partial t} + \vec{\nabla} \cdot \frac{b_1 n_1 \vec{j}}{eb_e(n_1 + n_2^{(0)})} = 0, \tag{9.25}$$

where the current density $\vec{j} \approx -eb_e(n_1 + n_2^{(0)})\vec{\nabla}\varphi$ satisfies

$$\vec{\nabla} \cdot \vec{j} = 0. \tag{9.26}$$

Introducing the coordinates ζ, ξ which label the current lines, and λ - along them, from Eqs.(9.25), (9.26) we obtain

$$\frac{\partial n_1}{\partial t} + \frac{b_1 j(\zeta, \xi, \lambda) n_2^{(0)}}{eb_e(n_1 + n_2^{(0)})^2} \frac{\partial n_1}{\partial \lambda} = 0. \tag{9.27}$$

This form of equation is convenient for numerical calculations. The potential for a given plasma density profile is determined by Eq. (9.26), and Eq. (9.25) describes the density transport with the density dependent velocity. Profile overturn occurs along the current lines which vary with time. The shock velocity is controlled by the particle conservation in current tube.

The problem is greatly simplified in symmetrical problems, when the direction of the current lines is known beforehand. For spherical symmetry, for example, we have

$$\frac{\partial n_1}{\partial t} + \frac{b_1 n_2^{(0)} I}{eb_e(n_1 + n_2^{(0)})^2} \frac{\partial n_1}{\partial \Omega} = 0, \tag{9.28}$$

where $\Omega = 4\pi r^3/3$ is a volume, and I is the net radial current. From the particle conservation it follows that the area rule in the coordinates n_1, Ω is applicable.

9.3 MULTISPECIES PLASMA WITH CONSTANT MOBILITIES

In the general 1D case, when all the ion mobilities are finite, it is convenient to introduce dimensionless densities, which are referred to the immobile perturbation Eq. (9.6),

$$\tilde{n}_\alpha = n_\alpha (b_\alpha + b_e)/(b_\alpha b_e \psi(x)),$$

and spatial coordinate

$$s = \int_0^x b_e^2 \psi(x')dx' / \sum_{\beta=1}^{p-1} n_\beta^{(0)} (b_\beta + b_e). \tag{9.29}$$

Accounting for the $\psi(x)$ conservation in time, from Eqs.(9.4) we find

$$\begin{aligned} &\frac{\partial \tilde{n}_\alpha}{\partial t} + b_\alpha E_0 \frac{\partial}{\partial s} \frac{\tilde{n}_\alpha}{\sum_{\beta=1}^{p-1} \tilde{n}_\beta q_\beta} = 0; \\ &\sum_{\alpha=1}^{p-1} \tilde{n}_\alpha = 1; \quad q_\alpha = b_\alpha / b_e; \quad \alpha = 1, \ldots, p-1. \end{aligned} \tag{9.30}$$

If p=3, only one type of propagating signal exists, and Eqs.(9.30) reduce to the $\psi(x)$ conservation Eq. (9.6), and to the equation for the propagating mode which coincides with Eq. (9.11):

$$\begin{aligned} &\frac{\partial \tilde{n}_1}{\partial t} + \tilde{V}(\tilde{n}_1) \frac{\partial \tilde{n}_1}{\partial s} = 0; \\ &\tilde{V}(\tilde{n}_1) = b_1 E_0 q_2 / [\tilde{n}_1 (q_1 - q_2) + q_2]^2; \\ &\tilde{n}_2 = 1 - \tilde{n}_1. \end{aligned} \tag{9.31}$$

Its solution represents a simple nonlinear wave Eq. (9.13). Since in the new variables $\tilde{n}_\alpha$ and s the equation (9.30) corresponds to the conservation of $\int_{-\infty}^{\infty} \tilde{n}_\alpha ds$, the shock position can be found from the area rule in these variables. The shock structure can be obtained in the same manner, as in Eqs.(9.20) and (9.23) for the case b_2=0.

As an example we shall consider the evolution of a localized density perturbation. If the initial profiles are similar, $n_{10}(x)/n_1^{(0)}=n_{20}(x)/n_2^{(0)}$; $\tilde{n}_{\alpha 0} = n_\alpha^{(0)}$, then propagating disturbance is absent. This follows from Eq. (9.29), and from the fact of $\psi(x)$ conservation in the drift approximation.

In the opposite situation, when $\psi(x)=\psi_0$ is unperturbed, the perturbations of the partial densities satisfy

$$\frac{n_1(x,t)-n_1^{(0)}}{n_2(x,t)-n_2^{(0)}}=\frac{n_{10}(x)-n_1^{(0)}}{n_{20}(x)-n_2^{(0)}}=-\frac{b_1(b_2+b_e)}{b_2(b_1+b_e)}, \tag{9.32}$$

and values of $\tilde{n}_\alpha(x)$ are proportional to $n_\alpha(x)$. Such a perturbation is completely wiped out from its initial position, according to Eq. (9.31), and only the propagating profile with subsequently arising diffusive shocks exists.

An arbitrary initial profile is separated into the propagating and immobile parts

$$n_\alpha(x,t)=\bar{n}_\alpha(x)+\bar{\bar{n}}_\alpha(x,t)-n_\alpha^{(0)}.$$

The fixed part is

$$\bar{n}_\alpha(x)=b_\alpha\tilde{n}_\alpha^{(0)}\psi(x). \tag{9.33}$$

Evolution of the propagating profile

$$\bar{\bar{n}}_\alpha(x,t)=b_\alpha\psi(x)[\tilde{n}_\alpha(x,t)-\tilde{n}_\alpha^{(0)}]+n_\alpha^{(0)}, \tag{9.34}$$

is described by Eq. (9.31); the values of $\bar{\bar{n}}_\alpha$ are moving along s with the velocity $\tilde{V}(\tilde{n}_1)$. From Eq. (9.31) it follows that if $b_1>b_2$ the shock is formed at the rear side of a $\bar{\bar{n}}_\alpha(x,t)$ profile. In the opposite case the profile overturn occurs at its front.

Consider the case when only the initial density of ions 1 is perturbed: $n_{20}(x)=n_2^{(0)}$; $n_{10}(x)>n_1^{(0)}$. From Eq. (9.33) we have

$$(\bar{n}_2(x)-n_2^{(0)})/(n_{10}(x)-n_1^{(0)})=(n_1^{(0)}/n_2^{(0)}+b_1/b_2)^{-1}. \tag{9.35}$$

If $(n_1^{(0)}/n_2^{(0)}+b_1/b_2)\ll 1$, the perturbation of the ambient ion density n_2, which arises in the injection region, significantly exceeds the initial perturbation of the test ion density. For $\bar{\bar{n}}_\alpha(x,t)$ we find

$$\bar{\bar{n}}_1(x,t) = n_1^{(0)}[1 + b_1 n_2^{(0)} / b_2 n_1^{(0)}][\tilde{n}_1(x,t) - \tilde{n}_1^{(0)}] + n_1^{(0)} > n_1^{(0)};$$
$$\bar{\bar{n}}_2(x,t) = n_2^{(0)}[1 + b_2 n_1^{(0)} / b_1 n_2^{(0)}][1 - \tilde{n}_1(x,t) - n_2^{(0)}] + n_2^{(0)} < n_2^{(0)}. \tag{9.36}$$

If only the ambient ions of species 2 are present at infinity, then far from the region of the initial perturbation

$$\bar{n}_1 = 0; \quad \bar{\bar{n}}_2 = n_2^{(0)}[1 - b_2 \bar{\bar{n}}_1 / (b_1 n_2^{(0)})] > 0. \tag{9.37}$$

Therefore, $\bar{\bar{n}}_1(x,t)$ cannot exceed $b_1 n_2^{(0)}/b_2$. If the initial disturbance is rather strong, and maximum value of $n_{10}(x)$ exceeds $\bar{\bar{n}}_\alpha(x,t)$, then, for a long time $t \leq t_1 = n_{10}L/(b_1 n_2^{(0)}/b_2)$, where L is the scale of the initial disturbance, their profile remains practically unchanged. Their outflux in the field direction is controlled by the formation of depletion regions in the ambient ion density. In the perturbed plasma region which extends along the field, the ambient ions are almost absent, and the density of the test ones equals $b_1 n_2^{(0)}/b_2$. At $t >> t_1$ the plasma density peak which remains at the place of the initial disturbance contains only the ambient ions, and its evolution is determined by ambipolar diffusion.

In the general case the evolution process of immobile perturbation is determined by the equation for $\psi(x,t)$ taking into account the diffusion terms:

$$\frac{\partial \psi}{\partial t} = D_2(1 + T_e / T_i) \frac{\partial^2}{\partial x^2}[(1 + \tilde{n}_1(D_1 / D_2 - 1))\psi]. \tag{9.38}$$

In the vicinity of the injection point the profile of $\tilde{n}_1(x,t)$ does not exceed unity and remains gradual, since the shocks, which are liable to arise during the evolution, propagate out of this region. Thus, it is possible to extract the parenthesis $(1+\tilde{n}_1(x,t)(D_1/D_2-1))$ from the derivative sign in Eq. (9.38). This results in the diffusion equation with a slowly time-dependent effective diffusion coefficient. The n_1, n_2 profiles, which are necessary for its calculation, are connected with $\psi(x,t)$ using Eq. (9.33).

In the special case $b_1=b_2=b$ a shock does not arise, and the electron density profile $n_e(x)$ is immobile [11]. We have

$$\psi = n_e/b; \; s = \int_0^x (b_e n_e(x')) / (b n_0) dx'.$$

Eq. (9.31) becomes linear:

$$\frac{\partial \tilde{n}_1}{\partial t} + b_e E_0 \frac{\partial \tilde{n}_1}{\partial s} = 0. \tag{9.39}$$

The $\tilde{n}_1(x,t)$ profile is moving with constant velocity along s. In traditional variables the solution of Eq. (9.39) is

$$n_1(x,t) = n_e(x) n_1(x_0)/n_e(x_0), \tag{9.40}$$

where the coordinates x, x_0 are connected by

$$\int_0^x n_e(x')dx' = bE_0 n_0 t + \int_0^{x_0} n_e(x')dx'. \tag{9.41}$$

The evolution of density profiles n_1, n_2 in a strongly nonlinear case for the Gaussian electron density profile, Eq. (9.14), is shown in Fig. 9.3.

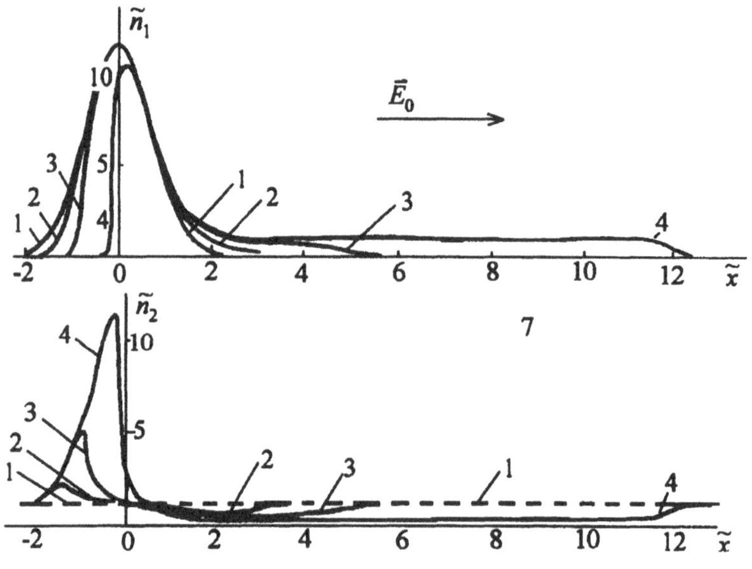

Figure 9.3. Evolution of the density profiles $\tilde{n}_1 = n_1/n_0; \tilde{n}_2 = n_2/n_0; b_1 = b_2 = b$ for the Gaussian initial disturbance Eq. (9.14) for $A = 20/\sqrt{\pi}; \tilde{t} = tE_0 b/a$. The numbers at the curves correspond to the values of the dimensionless time [11].

For a long time the majority of the test ions stay fixed, since in the vicinity of the injection point the field $E<<E_0$. The injected ions emerge from the front side of their initial profile, at the place where $n_1 \cong n_0$, with the drift velocity in the unperturbed field $V \cong bE_0$. The ambient plasma here is strongly depleted; its density is of the order of n_0/A. The ambient ions gradually fill up the initial density profile, starting from negative values of x. As a result, at $n_1>>n_2$ a profile of test ion density is spreading out in the external electric field considerably faster than due to the diffusion alone.

More detailed numerical calculations with account of the diffusion were performed in [12]. Analysis of the case $b_1<<b_2$ will be presented in Chapter 11, since the equation for the 1D evolution across the magnetic field coincides with Eq. (9.31).

9.4 PLASMAS WITH VARIABLE MOBILITIES

The absence of the ambipolar drift velocity in the pure plasma is a consequence of the fact that the combination $\psi(x)$, Eq. (9.6), is conserved in the drift approximation. It follows from the assumption that the particle mobilities are constant and spatially uniform. In such a degenerate situation the partial electron and ion drift fluxes are proportional to $n_e b_e E$, $n_i b_i E$. Since the approximation of quasineutrality usually holds with a high accuracy, both of these fluxes are proportional to the current density. In other words, the current conservation implies that density evolution in pure plasma vanishes in the drift approximation. It is determined only by the violation of this approximation, i.e. by the diffusion. Numerous factors - such as the Coulomb collisions of the charged particles, their heating by an electric field, the motion and nonuniformity of the neutral gas - result in the fact that the drift partial fluxes are no longer proportional. Therefore, in the general case, even in the pure current carrying plasma (p=2), the density perturbation corresponds in the drift approximation to the propagating mode. The diffusion determines only its weak damping. In multispecies plasmas all the (p-1) modes (for a linear perturbation all (p-1) roots of the dispersion equation (9.8)) correspond to the propagating signals. In such a situation immobile plasma disturbance is absent, and ambipolar drift velocity and ambipolar mobility exist even in the pure plasma.

This phenomenon was first investigated in the physics of electrolytes [1] and [2]. For a fully dissociated binary electrolyte the ion transport is described by an equation which practically coincides with Eq. (9.11):

$$\frac{\partial n}{\partial t}+\frac{j}{e}\frac{\partial}{\partial x}\left(\frac{b_1}{b_1+b_2}\right)=\frac{\partial n}{\partial t}+\frac{j}{e}\frac{dF}{dn}\frac{\partial n}{\partial x}=0. \qquad (9.42)$$

The ratio $F(n)=b_1/(b_1+b_2)$ is called the cation transference number. In dense solutions of strong electrolytes it is density dependent due to the solution nonideality. The concepts of profile overturn and shock formation were formulated by Weber for this case.

It is to be noted that in the ideal partially ionized pure plasma the mobilities are density dependent due to Coulomb collisions. Nevertheless, the dependencies of b_e and b_i, according to Eqs.(2.89), are similar, and ambipolar drift in this case is absent.

The phenomena which result from the $F(n)$ dependence in strong electrolytes were observed in [13]. In these experiments the evolution of a sharp boundary between two water solutions of the same substance with different concentrations was explored. According to Section 9.2, three different evolution scenaria are possible. If the function $F(n)$ is density independent, the ambipolar drift is absent, and the boundary between the solutions should remain immobile. Its spreading is determined by the ambipolar diffusion which was neglected in Eq. (9.42). The width of this transition region grows over time, as $t^{1/2}$, according to the standard diffusion law. If the dependence $F(n)$ is close to the linear one, directed motion of the boundary arises, which changes its sign with the current reversal. Its width grows with time due to diffusion. In the strongly nonlinear case, when $d^2F/dn^2\neq 0$, it is possible that such a sharp boundary comprises a diffusive shock. For $d^2F/dn^2<0$, for example, it corresponds to $n^+>n^-$. According to Eq. (9.17), the shock moves with the velocity

$$W=j[F(n^+)-F(n^-)]/[e(n^+-n^-)] \tag{9.43}$$

without diffusive spreading. The current reversal results in the fact that the same boundary does not satisfy the evolution condition. Its fast spreading is proportional to t.

In Fig. 9.4 the results of the measuring of the gradient of the refraction index (which is approximately proportional to the solution concentration) are presented. The arrows correspond to the current direction. It is definitely seen that the evolution pattern closely corresponds to the three cases described. For KCl solutions the transference number is practically independent of concentration. In the case of $CaCl_2$ the dependence $F(n)$ is almost linear. The ratio $F''(n^+-n^-)/(2F')$ which characterizes the nonlinearity of the $F(n)$ function equals 0.1. For CdI this value is considerable - of the order of 0.4.

In the plasma of a *dc* gas discharge the mobilities practically always depend on plasma density as well. Comparing the expressions for the shock width Eqs.(9.5), (9.15) with the formation criterion of the local distribution function Eq. (2.110) (using the expression for the mean electron energy Eq. (2.107) instead of the electron temperature), we can see that in the drift approximation Eqs.(2.105) and (2.109) are applicable.

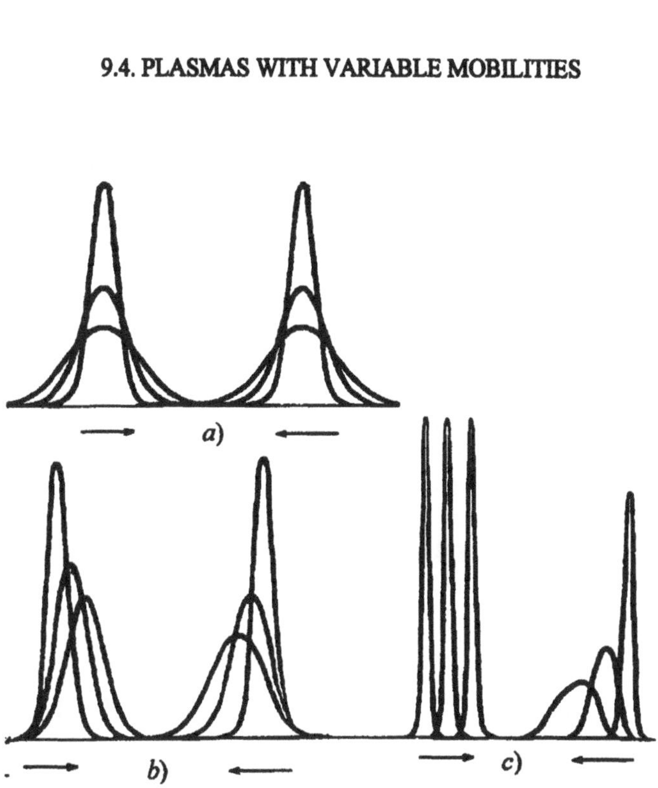

Figure 9.4. Evolution of the dk/dx (k is the refraction coefficient) profile, which is related to the concentration gradient, when the current flows through the initially sharp boundary between two solutions. The right and left families of curves refer to opposite current directions (arrows). The successive profiles were recorded at intervals of 2 hours. The solution densities n^+ and n^- are: *a*) 0.2*N*:0.5*N*, *KCl*; *b*) 0.1*N*:0.2*N*, $CaCl_2$; *c*) 0.1*N*:0.2*N*, *CdI*.

Since the (electron) current conservation in this case implies $j_e=eEb_e(E)n$=const, it follows that the effective dependence of electron mobility on plasma density arises. Neglecting, for simplicity, the dependence of ion mobility on the field strength, we obtain that the ambipolar drift velocity in this case equals ([14], [15])

$$V=d\Gamma_i/dn=b_iE\beta/(1+\beta); \quad \beta=d\ln b_e/d\ln E. \tag{9.44}$$

Usually $\alpha\leq 0$ in Eq. (2.109), and the electron mobility decreases with electric field and increases with plasma density. The ambipolar drift velocity is of the order of the ion drift velocity and is directed to the anode.

Similar phenomena take place in intrinsic semiconductors in rather strong fields, when Joule heating of the charge carriers becomes significant, and the mobilities become field dependent [6].

In gas discharges and in semiconductors with hot carriers, when the dependence Eq. (9.44) takes place, at low plasma densities the distribution functions are strongly non-Maxwellian. It is to be noted that in the absence of the Maxwellization of the electron distribution due to the collisions between electrons, at $T_e >> T_N$ (in semiconductors with hot carriers T_N corresponds to the lattice temperature) the expression for the electron distribution relaxation length Eq. (2.59) coincides with Eq. (9.15) for the width of the large amplitude diffusive shock. It means that its structure cannot be analyzed in the framework of the fluid approach, and rather complicated kinetic description is necessary. This situation is fully analogous to the situation widely known in traditional gas dynamics. The shock structure in an ordinary gas up to now can be investigated only numerically (with the only exception of the weak shock), but the global properties of the gas flow depend only on the positions and amplitudes of shocks, which can be found without analysing their structure [7].

9.5 STATIONARY DRIFT PROFILES WITH SHOCKS IN CURRENT CARRYING PLASMAS

When analyzing the stationary problems the plasmachemical processes cannot be neglected. It follows from the preceding Sections that in a current carrying plasma both the plasma density profiles with characteristic scales $b_\alpha E\tau$, where τ is an average particle lifetime, and $T_\alpha/(eE)$, are possible. In the large-scale (drift) profiles the diffusion can be neglected, and the segments with small scale $T_\alpha/(eE)$ can be interpreted as diffusive shocks. The position and even existence of such profiles crucially depends on the direction of the ambipolar drift velocity. We shall consider first pure plasma with only ionization and recombination of the charged particles.

Smooth drift profiles in the Faraday dark space. The investigation of such an object was performed in [16] and [17] for the case of the Faraday dark space in N_2. The experiment was performed at medium pressure 2.67kPa in glow discharge. In order to control the magnitude and sign of the ambipolar drift velocity, a gas flow with constant uniform velocity U was applied. In contrast to the preceding Sections, in this case the laboratory reference frame which is connected with the discharge tube and its electrodes is fixed, but not the neutral gas frame. If the characteristic scale exceeds l_T, l_0, Eqs. (9.15) and (9.24), in the drift approximation (compare to Eq. (4.18)) the 1D stationary problem can be formulated as

$$d\Gamma_i/dx = d(b_i nE + nU)/dx = (V(n)+U)dn/dx = n\nu^{(ion)}(E(n)) - S(n). \quad (9.45)$$

In this equation x is directed towards the cathode; the field strength $E(n)$ can be expressed in terms of the plasma density by the equation of the current conservation $j=eb_e(E)nE$. The $S(n)$ term corresponds to the effective recombination rate both in the plasma volume and at the tube walls, Section 4.1. The ambipolar drift velocity with respect to the gas flow $V(n)$ is given by Eq. (9.44). Since the electron mobility usually decreases with electric field and hence increases with plasma density, the ambipolar drift velocity is of the order of the ion drift velocity and is directed to an anode. In the large fields the ion drift flux $\Gamma_i(n)$ in the laboratory reference frame always strongly exceeds nU. Consequently, the right (high field) branch of the graph $\Gamma_i(n)$ decreases with the density, as in the absence of gas flow (Fig. 9.5). On the other hand, at a low field (i.e., at high plasma density) the flux nU dominates in the expression for Γ_i. Hence, in the case of the cathode directed gas flow the dependence $\Gamma_i(n)$ becomes nonmonotonic (see Fig. 9.5).

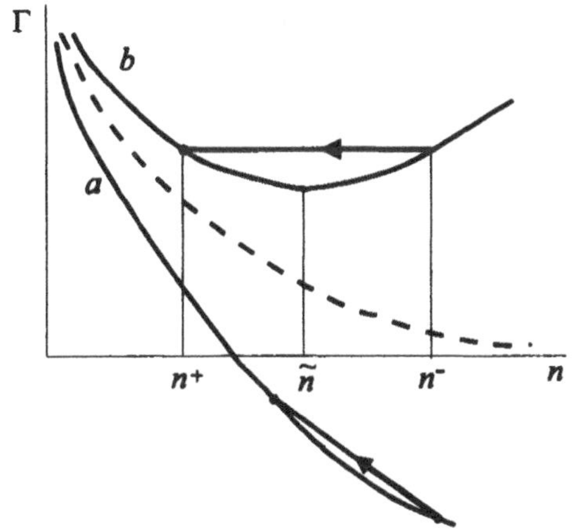

Figure 9.5. Drift flux of the plasma in a discharge with a gas flow: *a*) flow from cathode to anode ($U<0$); *b*) flow towards the cathode. Broken curve corresponds to $U = 0$. Arrows represent the possible shocks which are compatible with the evolution condition.

The boundary condition at the cathode side can be formulated as follows. As it was mentioned in Section 3.7, when the fluid approximation is valid, a cathode region is unstable. This instability leads to the formation of the kinetic regime which is accompanied by the formation of fast electrons in the cathode sheath. These fast electrons produce considerable nonlocal ionization in the plasma region adjacent to sheath. Since the electric field in plasma is low, due to this ionization source a plasma density peak arises at the plasma-sheath boundary. The field strength here is low not only with respect to the strong field in the sheath, but even in comparison with the moderate field in the positive column E_c, which corresponds to zero of the r.h.s. of the Eq. (9.45). The approximate boundary condition $n(x=0)=\infty$; $E(x=0)=0$ can thus be imposed at the cathode side of this weak field region which is known as the Faraday dark space. The plasma density and the field strength vary in it from these values to n_c, E_c.

If the gas flow velocity is anode directed or equals zero, the function $\Gamma_i(n)$ decreases with n. The drift solution of Eq. (9.45) at $U \leq 0$ is given by

$$x = \int_n^{\infty} \frac{(V(n') + U)dn'}{R(n') - n'\nu^{(ion)}(n')}, \tag{9.46}$$

and corresponds to smooth transition $\{\infty, n_c\}$ with the drift spatial scale

$$l_E \sim (V+U)n/R \tag{9.47}$$

which increases with U (lines 2 and 3 in the Fig. 9.6).

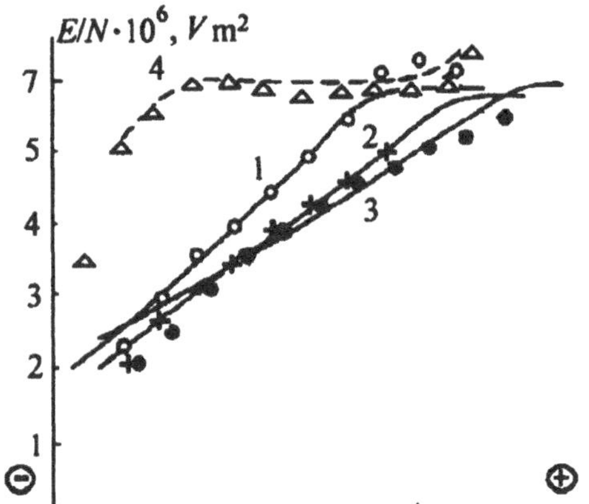

Figure 9.6. The field strength profiles in the Faraday dark space for the discharge with the gas flow in N_2 [17]. The bold lines are calculated according to Eq. (9.45). The curve and experimental points 1 correspond to the current density j=60A/m^2, gas flow velocity U=73m/s; 2 to j=40A/cm^2, U=73m/s; 3 to j=40A/m^2, U=164m/s; 4 to j=40A/m^2, U=164m/s. For the cases 1-3 the gas flow is anode directed; for the case 4 cathode directed.

These drift profiles correspond to propagation of the ambipolar drift signals generated by perturbation at the cathode; their length is determined by the damping of the ambipoalr drift. The distinction between lines 1 and 2 is attributed to the decrease of the drift scale l_E with the current rise due to a more intense volume recombination rate. Agreement between the calculation according to Eq. (9.46), and the experiment, seems quite satisfactory. If the neutral gas velocity U is cathode directed, the minimum at the $\Gamma_i(n)$ dependence arises at a certain density value $n=\tilde{n}$ (Fig. 9.5). This critical density $\tilde{n}$, which corresponds to the ambipolar drift reversal, decreases as U increases. If $\tilde{n}<n_c$, the plasma profile in the Faraday dark space is given by the large-scale drift profile Eq. (9.46), as in the case of anode directed flow. This profile corresponds to the cathode directed U, which does not exceed the critical value U_c (for the experimental conditions of [17] U_c was about 100m/s). For the cathode directed gas flow at $U>U_c$, the ambipolar drift velocity in the laboratory reference frame at $n\sim n_c$ becomes cathode directed. It means that at large cathode directed values of U, the signal velocity is also cathode directed, and the Faraday dark space with the large drift scale is practically absent (line 4 in Fig. 9.6). In other words, in such a situation plasma 'is not aware' of the disturbance which is produced by the cathode.

Plasma profiles with fixed shocks which arise due to the ambipolar drift reversal in *dc* discharge. As it was mentioned in Section 9.2, such plasma profiles are possible in any system with reversal of the ambipolar drift velocity. If in the *dc* discharge in N_2 described above the gas flow is directed towards the cathode, the $\Gamma_i(E)$ dependence is nonmonotonic, Fig. 9.7.

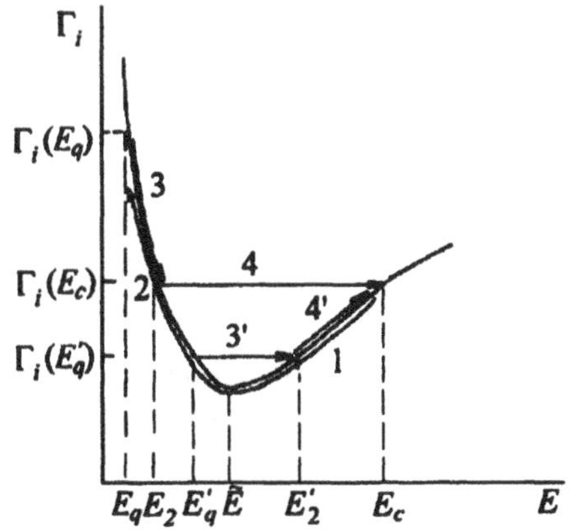

Figure 9.7. Dependence $\Gamma_i(E)$ for the gas flow anode-cathode.

The right branch of the graph is caused by the dependence $b_e(E)$, and the left one by the gas flow. At a certain field strength $\widetilde{E}(\widetilde{n})$ ambipolar drift reversal occurs. For $E>\widetilde{E}$ the drift profiles correspond to the displacement of perturbations towards the anode, and in the opposite case towards the cathode. At $E=\widetilde{E}$ the characteristic scale of the drift profile l_E tends to zero. According to the evolution condition, the fixed shocks in which the plasma density decreases (and the field strength increases) along the current direction are possible in this situation [18]. In this paper it was proposed to use some additional external ionization source in order to bring these fixed shocks into existence.

Consider the situation when the infinitely long, homogeneous positive *dc* column with $E=E_c>\widetilde{E}$, $n=n_c<\widetilde{n}$, which is described by Eq. (9.45), is subject to action of some external ionizer - say, an electron beam. We restrict ourselves, for simplicity, to the situation when the external ionization source $q(x)$ is Π-shaped. The source rate $q(x)$ is spatially uniform: $q(x)=q=const$ at $x_0<x<x_1$, and equals zero outside this interval. The length of the ionization region $l_q=(x_1-x_0)$ is presumed to be large with respect to the characteristic drift scale l_E, Eq. (9.47). It follows that in the main part of the region, where the external ionizer is active, the plasma density is uniform. The field strength here $E_q<\widetilde{E}$ corresponds to the ionization-recombination balance

$$n\nu^{(ion)}(E_q(n_q))-R(n_q)+q=0.$$

If both E_q, $E_c>\widetilde{E}$ (or both E_q, $E_c<\widetilde{E}$), a smooth drift solution $\{n_c, n_q\}$, $\{n_q, n_c\}$ without shocks occurs. In any case, when we approach the external ionizer zone from the anode side at $E_c>\widetilde{E}$, $n_c<\widetilde{n}$, the field fall, and the density rise are described here by the drift solution $\{n_c, n\}$ (arrow 1 in Fig. 9.7; segment 1 in Fig. 9.8):

$$x(n)=\int_{n_c}^{n}\frac{(V(n')+U)dn'}{n'\nu^{(ion)}(n')-R(n')}>0. \tag{9.48}$$

This corresponds to the plasma outflux from the ionizer zone with the ambipolar drift velocity, and to gradual transition to the column equilibrium density n_c due to the recombination.

But such a profile exists only at $E>\widetilde{E}$. If $E_q>\widetilde{E}$, it is impossible to achieve in such a way a smooth transition to the new equilibrium state n_q. On the other side, in this case at the anode part of the ionizer zone, the ambipolar drift is anode directed.

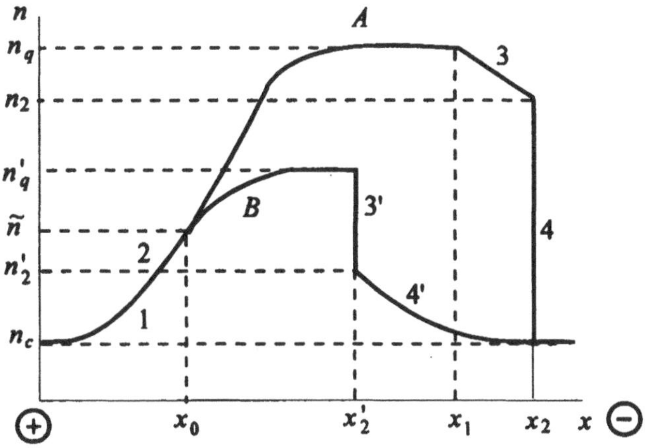

Figure 9.8. The plasma density profiles for the intensive (A) and weak (B) external ionizers.

Hence, in this region a drift solution with the density increasing towards the cathode, $\{n, n_q\}$, becomes possible (2 in Figs. 9.7 and 9.8):

$$x = \int_{n}^{n_q} \frac{(V(n') + U)dn'}{R(n') - n'\nu^{(ion)}(n') - q}. \tag{9.49}$$

It describes the smooth density rise up to the density value $n=n_q$, which satisfies the local ionization-recombination balance in the presence of the external ionizer. Its characteristic scale is of the order of $(V+U)\sqrt{qR}\,/\,n$. In full accordance with the gas dynamic problem of the Lavale nozzle, the boundary between the solutions Eqs.(9.48) and (9.49) coincides with the anode boundary of the ionizer profile x_0. This results in a smooth drift profile $\{n_c, n_q\}$. If the ionization rate $q(x)$ is x-dependent, the coordinate x_0, where the ambipolar drift reversal occurs, can be found from

$$q(x_0) + n\nu^{(ion)}(\tilde{n}) - R(\tilde{n}) = 0. \tag{9.50}$$

The diffusive shock in this problem arises on the cathode side of the ionizer zone. The corresponding scenario depends crucially on the ionizer intensity. In the case of a strong ionizer, when $\Gamma_f(E_q) > \Gamma_f(E_c)$, the plasma remains homogenous, $E = E_q$, up to $x=x_1$ - to the cathode edge of the ionizer zone. At $x>x_1$ the drift profile $\{n_q, n_2\}$ is situated where the plasma outflows from the ionization zone, and recombines (3 in Figs. 9.7, 9.8). At the point $x=x_2$, where $n=n_2$; $E=E_2$, which satisfies $\Gamma_i(E_2)=\Gamma_i(E_0)$, a diffusive shock occurs (4 in Figs. 9.7 and 9.8). The density profile (*A* in Fig. 9.8) corresponds to $\{n_c, \tilde{n}\}$, $\{\tilde{n}, n_q\}$, $\{n_q, n_2\}$, $[n_2, n_c]$.

If the ionizer is relatively weak, $\Gamma_i(E')<\Gamma_i(E_c)$, Fig. 9.8, the shock occurs inside the ionizator zone. The plasma from the anode side of the ionizer flows with velocity $(V+U)<0$ towards the cathode. Its density increases (4′ in Figs. 9.6 and 9.7) according to

$$x_1 - x = \int_n^{n_c} \frac{(V(n') + U)dn'}{q + n'\nu^{(ion)}(n') - R(n')}. \tag{9.51}$$

up to the shock (3′ in Figs. 9.7, 9.8). The field strength E_2' at its right side is determined by the condition $\Gamma_i(E_q')=\Gamma_i(E_2')$. Its position x_2' can be found by the substitution of $n_2'(E_2')$ instead of the lower integration limit in Eq. (9.51) The resulting density profile $\{n_c, \tilde{n}\}$, $\{\tilde{n}, n_q\}$, $[n_q, n_2']$, $\{n_2', n_c\}$ is sketched in Fig. 9.8 (line *B*).

Experimental photographs [19] of nitrogen *dc* discharge with cathode directed gas flow are presented in Fig. 9.9.

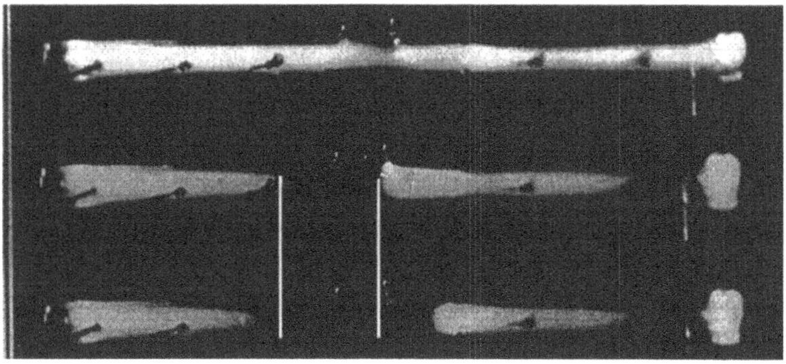

Figure 9.9. The positive column of the *dc* discharge in N_2 with the cathode directed gas flow with U=40m/s under the influence of the beam of fast electrons with energy 100KeV. The region of the beam injection is restricted by the white vertical lines in the tube middle. The discharge parameters are: p=2KPa; j=10A/m^2; a = 2cm. The beam current in the case *a*) was absent, *b*) 0.25μA/cm^2; *c*) 2μA/cm^2.

The external ionization was produced by a localized electron beam of variable intensity. Its x-dependence was close to Π-shaped. The luminosity, as well as the ionization rate, exponentially depends on the field strength. Hence, the dark regions in the photographs correspond to high plasma density. In full agreement with the foregoing analysis, the density profile at $x \sim x_0$, increases in the transition region from n_c to n_q in the cathode direction (the left parts of Figs. 9.9*b,c*). In Figs. 9.7 and 9.8 it corresponds to the smooth transition through the density $\tilde{n}$ which is analogous to the critical cross-section of the Lavale nozzle. The sharp boundary between the bright and the dark regions was observed at the cathode side of the profile. For the weak beam (Fig. 9.9*b*) this boundary emerged inside the ionizator zone, at $x_2' < x_1$. In the opposite case of the intense beam this shock was situated rather far away from the ionizator (Fig. 9.9*c*).

9.6 STATIONARY DRIFT PROFILES WITH SHOCKS IN SEMICONDUCTOR PLASMA. INJECTION AND ACCUMULATION OF HOT CARRIERS

Another simple situation in which stationary shocks were observed was met in semiconductor plasmas. The charged particles there can be generated in two different ways. The first way, which consists in the direct excitation of an electron from the valence band to the conductivity band, results in the formation of an electron-hole pair. The second source is supplied by impurities. This source results in the formation of a population of free electrons in the conductivity band, and of fixed positively charged donor centers, or in the generation of holes in the valence band, and of fixed negative acceptors. The fraction of positively charged donors (and of negatively charged acceptors) is determined by the temperature (see, for example, [20]). The situation, when the donors (acceptors) are fully ionized, coincides with that considered in Section 9.2. We shall restrict ourselves to this case, and consider the behavior of electron-hole plasma in *n*-type sample with uniform density of fully ionized donors N_D, and the situation, when $b_e >> b_h$. It was shown above in Section 9.2 that at constant mobilities the ambipolar drift velocity is cathode directed, and only moving shocks are possible in this case (see Fig. 9.1c). On the other hand, at high field strength the mobilities become field-dependent, according to Eq. (2.109). Consider the case when only the electron mobility is field-dependent:

$$b_e = b_e^{(0)}, \qquad \text{at } E < E_h ; \tag{9.52}$$

$$b_e = b_e^{(0)} \sqrt{E_h / E}, \qquad \text{at } E > E_h .$$

The field E_h corresponds to the situation when the mean electron energy is of the order of the lattice temperature, Eq. (2.108). At $E > E_h$ the anode directed ambipolar drift velocity exists even in the intrinsic sample with $n=p$. It follows that in an n-type sample with field dependent mobility ambipolar drift reversal and stationary shocks are possible. At low current $j<j_h=2eN_D b_e^{(0)} E_h$, the holes flux Γ_p in the drift approximation is given by

$$\Gamma_p = \frac{j b_p p}{e b_e^{(0)} (p+N_D)}. \tag{9.53}$$

This flux increases monotonously with p (line 1 in Fig. 9.10).

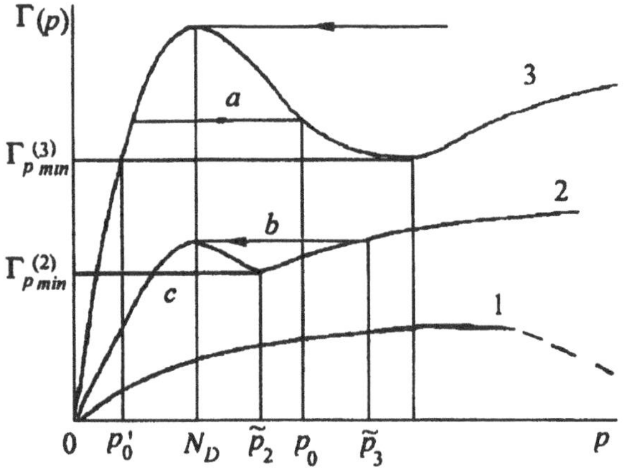

Figure 9.10. The flux of holes in the drift approximation in a n-type semiconductor with fully ionized donors. The curve 1 corresponds to the absence of electron heating; 2 to currents $j_h<j<2j_h$; 3 to high currents $j>2j_h$.

When the current is high, at low plasma densities the electrons are hot, and from Eq. (9.52) it follows that $\Gamma_p(p)$ is nonmonotonous:

$$\Gamma_p = \frac{b_p j^2}{(e b_e^{(0)})^2} \frac{p}{(p+N_D)^2}. \tag{9.54}$$

It is maximum at $p=N_D$, and decreases at higher values of p up to $p=\tilde{p}=j/(e b_e^{(0)} E_h)-N_D>N_D$. At $p>\tilde{p}$ the electrons are cold, and

$\Gamma_p(p)$ satisfies the Eq. (9.53) (see lines 2 and 3 in Fig. 9.10). From the evolution condition it follows that fixed shocks with increasing (towards the cathode) plasma density are possible 'under the hump', and shocks with density drop - 'above the pit' (*a* and *b* in Fig. 9.10). The maximum value of $\Gamma_p(p)$, according to Eq. (9.54), is proportional to j^2, and at moderate current, $j_h<j<2j_h$, is smaller than $\Gamma_p(\infty)$, Eq. (9.53). At $j>2j_h$ the maximum value of $\Gamma_p(p)$ exceeds $\Gamma_p(\infty)$ (line 3 in Fig. 9.10).

The profiles with shocks were observed first in [21] and [22], and interpreted in [23] and [24]. In [22] and [23] the double injection of holes into semi-infinite sample of *n*-type *InSb* was studied. As an anode, a strongly doped sample of *p*-type was used. The situation corresponded to the boundary condition $p(x=0)=\infty$ at the anode surface [25]. This process in the drift approximation is described by

$$\frac{d\Gamma_p(p)}{dx}=-\frac{p-p^{(0)}}{\tau}=-R(p,p^{(0)}), \tag{9.55}$$

where τ is the lifetime of the excess carriers; $p^{(0)}<<N_D$. When $j<j_h$, the ambipolar drift is cathode directed. From Eq. (9.55) the smooth reduction in density, according to $\{\infty, p^{(0)}\}$, down to the equilibrium state $p^{(0)}$ occurs with a scale of the order of the drift length $l_E=b_pE\tau$. However, the situation changes as the current increases. At $j>j_h$ such a continuous drift solution becomes impossible. Close to the anode, where plasma density is high, the carriers are cold, and due to the recombination both Γ_p and p decrease from the anode up to the point where the heating starts (curve 1 in Fig. 9.10). Actually, the hole flux should decrease further, as a result of recombination, whereas according to Eq. (9.54) the plasma density should rise here. As a result, a stationary diffusive shock 'above the pit' emerges. The solution at $j_h<j<2j_h$ is of the form $\{\infty, p'\}$, $[p', N_D]$, $\{N_D, p^{(0)}\}$, Fig. 9.10. The density p' is determined according to $\Gamma_p(p')=\Gamma_p(N_D)$. At high current $j>2j_h$ the shock is merged into the anode diffusive region. The drift profile $\{N_D, p^{(0)}\}$ corresponds to the low injection level $p<N_D$, and to high field $E>E_h$; curve 3 in Fig. 9.10.

Shocks 'under the hump' were observed in the experiment with the anomalous accumulation of holes in *n*-type *InSb* [21]. Under these conditions the cathode of practically semi-infinite sample was made of strongly doped n^+-*InSb*, and as a source of minority carriers (holes) impact ionization was used. Since the ionization rate increases very rapidly with the electric field, we assume simply that at some critical value $E_B>E_h$ of the electric field strength the ionization is 'switched on', and the field strength cannot exceed E_B. The plasma density adjusts itself to maintain $E=E_B$. In the experimental conditions of [21] and [22] $E_h\cong 20$V/cm; $E_B\cong 240$V/cm. Far from the contacts the plasma density profile tends to

$$p=p^{(0)}=j/(eEb_e(E))-N_D\approx 0 \quad \text{at } j<j_B=eN_D b_e^{(0)}\sqrt{E_h E_B}\ ,$$

$$p=p^{(0)}=j/(eE_B b_e)-N_D \qquad \text{at } j>j_B\ . \tag{9.56}$$

In order to obtain the boundary condition at the cathode, let us notice that the junction n-n^+, which separates two quasineutral regions, corresponds to the double layer (Section 3.4). The space charge at the n^+ side of the layer is positive. It means that the cathode corresponds to a high potential barrier for holes, and their density and Γ_p in the n^+ region are negligible. Since in the bulk the flux of holes is $\Gamma_p(p^{(0)})$, the recombination dominated accumulation region in the vicinity of the cathode arises, where the plasma density is high, and Γ_p decreases. In the absence of the electron heating the width of this region is small - of the order of the diffusion length $L_D=(4D\tau)^{1/2}$. The corresponding short scale solution, in which $p(x)$ is high and due to the recombination the hole flux disappears, is presented schematically by the horizontal lines *c* in Fig. 9.11.

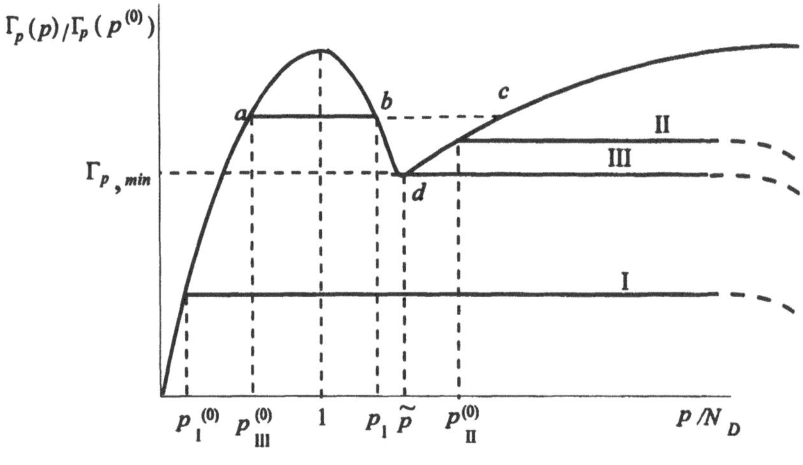

Figure 9.11. Dependence of the holes drift flux on hole density in *n-InSb* for $j=2.5j_h$ [24]..

In the presence of electron heating, at $j>j_h$, a similar plasma profile occurs, if the equilibrium density of the holes $p^{(0)}$ satisfies $\Gamma_p(p_I^{(0)})<\Gamma_{p,min}$ (line *I* in Fig. 9.11). The situation changes drastically at large values of $p^{(0)}$. If $N_D>p_{III}^{(0)}>p_c^{(0)}$, where $\Gamma_p(p_c^{(0)})=\Gamma_{p,min}$, the horizontal line $\Gamma_p(p_{III}^{(0)})=\Gamma_p$ intersects the graph of the drift flux of the holes at the point $p=p_1<\tilde{p}$ (Fig. 9.11, curve III, segment *ab*). It corresponds to the diffusive shock $[p_{III}^{(0)}, p_1]$. The shock 'above the pit' (Fig. 9.11, dashed line *bc*) is impossible due to the evolution condition. Since the total hole flux is conserved in the shock, the diffusive flux (the difference between the total flux and the drift one) is to be positive. But it is incompatible with an increase of $p(x)$ towards the cathode. Therefore, at $p=p_1$ the long scale drift profile $\{p_1, \tilde{p}\}$ begins (the segment *bd* of the curve III, Fig. 9.11). This phenomenon was named in [24] 'the anomalous accumulation'. The plasma density increases quickly at $p>\tilde{p}$, where the flux $\Gamma_{p,min}$ vanishes due to recombination. If $1<p(0)<\tilde{p}$, the profile consists of the drift segment $\{p^{(0)},\tilde{p})$ and the cathode-adjacent diffusive region III. In the case $p_{II}^{(0)}>\tilde{p}$ the drift segment disappears (II in Fig. 9.11).

In the experiments [21], due to $E_h << E_B$, the uniform plasma density satisfied $p^{(0)}<<N_D$. At $\tilde{p}>p^{(0)}>N_D$ the drift profile starts immediately from uniform plasma, and the shock disappears. If $\tilde{p}<p^{(0)}$, the drift segment is absent, and the standard accumulation profile with short spatial scale remains.

In the experiment [21] the voltage U between a pair of probes as a function of current density was measured. The first probe was situated at the accumulating cathode, and the second at a distance l from it in a long sample of *n-InSb* at 77^{0}K. It was found that that the conductivity in rather long region of the order of $l_B=b_p\tau E\sim 1$mm near the cathode was considerably higher than elsewhere in the sample. The results are presented in Fig. 9.12a.

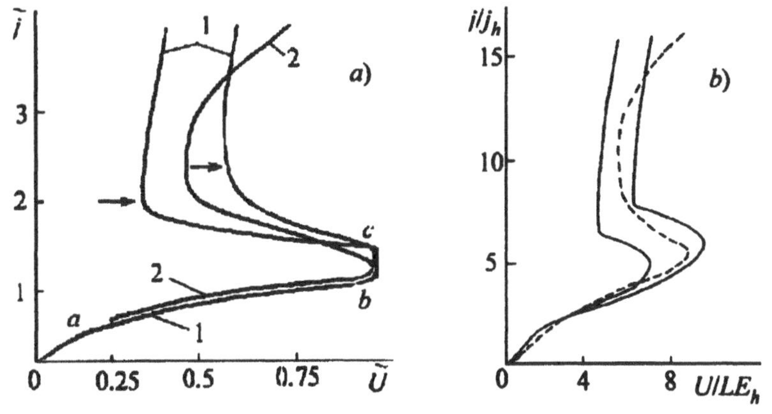

Figure 9.12. *a*) The calculated (1) and experimental (2) current-voltage characteristics between cathode and probe at distance l=0.44mm in *n-InSb* at 77K in the presence of the impact ionization and anomalous accumulation. *b*. Current-voltage characteristic of an *n-InSb* specimen of finite length. Solid lines are calculated for L/L_h=1 for the left-hand line and L/L_h=1.5 for th right-hand line. Broken curve is experiment [22]; E_h=15V/cm; L/L_h=1.3.

Since the diffusion length l_D~0.1mm was small with respect to l, the contribution of a small region of high conductivity to U in the case of the standard accumulation is negligible. The branches Oa and ab correspond to ordinary Ohm's law for $n=N_D$; $p^{(0)}=0$ with account of the electron mobility dependence Eq. (9.52). The vertical branch bc corresponds to the onset of the impact ionization at $E=E_B$. As a result of the anomalous accumulation, a long (of the order of l_E) region of high conductivity arises at the cathode. It first becomes broader, as the current increases, and occupies all the gap between the probes. The voltage over l falls steeply with current.

Instead of impact ionization, in a sample of finite length the minority carriers can be created also by injection. Such a situation was realized in the experiments [22]. The electron-hole plasma in *n-InSb* at 77^{0}K was studied with the hole-injecting contact as an anode, and the accumulating contact as a cathode. Depending on the current value and the relation between the sample length and characteristic scale $L_h=b_pE_h\tau$, diverse scenaria are possible (Fig. 9.13).

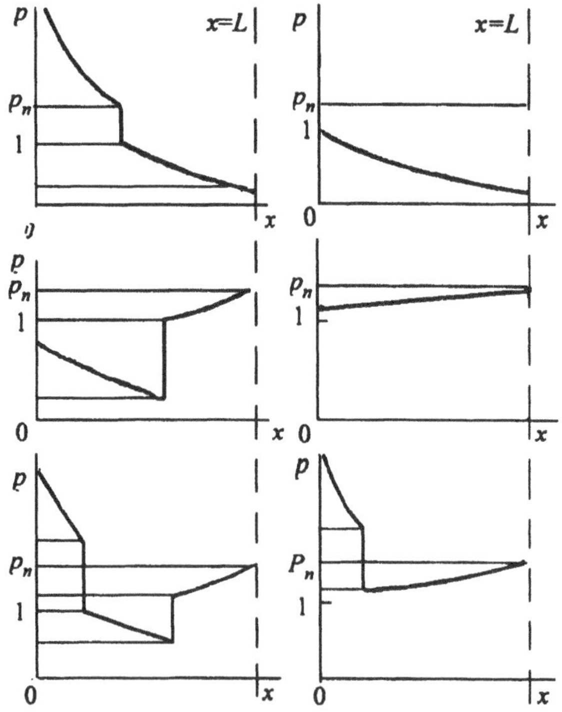

Figure 9.13. Plasma density profiles in a specimen of *n-InSb* of finite length. The density of holes is in units of N_D.

The plasma profiles without shocks, with one shock 'under the hump', or 'over the pit', and with both of them can be created. The evolution condition for the shocks, and requirement that the sum of the lengths of the drift segments should coincide with the sample length, allows us to construct unambiguously these drift profiles with shocks [26]. Due to the fact that, as a result of the anomalous accumulation, a long region of high conductivity plasma is formed, the I-V characteristics, in accordance with the experiment, are *S*-shaped. The comparison is presented in Fig. 9.12b. The solution of the full equation system with account of the diffusion for different situations is presented in [27] and [28]. These results can also be interpreted in terms of drift profiles with shocks [9].

9.7 DIFFUSIVE SHOCKS IN SYSTEMS WITH FAST PLASMACHEMICAL REACTIONS. SEMICONDUCTORS WITH CARRIERS CAPTURED BY TRAPS.

The general problem of the inhomogeneity evolution in current carrying, chemically active plasma is very complicated. The situation is greatly simplified if the characteristic plasmachemical time scales are strongly different, so it is possible to group them into fast and slow ones. The influence of the slow reactions can be accounted for in the framework of the drift approximation, as with the ionization and recombination in the two preceding Sections. The fast (with respect to the evolution rate) processes result in the establishment of equilibrium plasma composition. If, in the course of evolution, the scale of inhomogeneity becomes steeper than the scales of these processes, a delay between the real plasma composition and equilibrium arises. In this situation an analogy with traditional gas dynamics turns out to be useful. It is well known that in gas dynamics the impact of chemical reactions on the shock structure can often be described in terms of a relaxation zone [3], [7], [8]. This corresponds to the division of the shock into a narrow region, which is determined by the viscosity and thermal conductivity of the 'frozen' chemical composition, and the zone in which slow relaxation of chemical processes occurs. These considerations can be effectively applied to the diffusive shocks in current carrying gaseous or semiconductor plasma.

Real semiconductors frequently contain impurities that produce deep levels in the forbidden band. They act as carrier traps and, in general, the situation is rather complicated. However, if the trapping times are very different, and the current is high enough, the plasma profile consists of drift segments and of sharp shocks. For its analysis the gas dynamic analogue can be used again. Suppose, for example, that the specimen contains not only shallow fully ionized donors, but also N_t traps. The hole trapping time and the interband recombination time will be assumed to be long in comparison with all the characteristic times of the problem, and the electron trapping time τ_e will be assumed to be small. In the drift approximation we then have the following equation for the holes:

$$\frac{\partial p}{\partial t}+\frac{\partial \Gamma_p}{\partial x}=\frac{\partial p}{\partial t}++\frac{\partial}{\partial x}(b_p pE)=0, \tag{9.57}$$

and for the electrons:

$$\frac{\partial n_e}{\partial x}+\frac{\partial \Gamma_e}{\partial x}=\frac{\partial n_e}{\partial t}-\frac{\partial}{\partial x}(b_e n_e E)=[(n_e+n_1)f-n_e]/\tau_e. \tag{9.58}$$

The value of f corresponds to the degree of filling of the traps, and n_1 is the equilibrium density of free electrons when the Fermi level coincides with the level of the trap [20].

When the density profile is sufficiently smooth ($L>>b_e E\tau_e$), the electrons are in equilibrium with the traps, so according to Eq. (9.58)

$$f=\tilde{f}=\frac{n_e}{n_e+n_1}. \tag{9.59}$$

Substituting Eq. (9.59) into the quasistationarity condition

$$n_e=p+N_t(f_0-f); \qquad f_0=1+N_D/N_t \tag{9.60}$$

we obtain the relation between n_e and p:

$$p=n_e-N_t(f_0-n_e/(n_e+n_1)). \tag{9.61}$$

Eliminating the electric field from the system Eqs.(9.58) - (9.61), we have equation which coincides with Eq. (9.11):

$$\frac{\partial p}{\partial t}+V(p)\frac{\partial p}{\partial x}=0\ ;$$
$$V(p)=\frac{jb_e b_p(n_e-p\,dn_e/dp)}{e(b_e n_e+b_p p)}=\frac{d\Gamma_p}{dp}. \tag{9.62}$$

According to Eq. (9.61),

$$d\Gamma_p / dp > 0;\ d^2\Gamma_p / dp^2 < 0.$$

The corresponding graph of the flux of the holes in the drift approximation is presented schematically in Fig. 9.14. Evolution and overturn of the density profile occur, in this case, essentially as described in the Section 9.2. A shock which moves with velocity W is formed. Its structure is, however, more complicated.

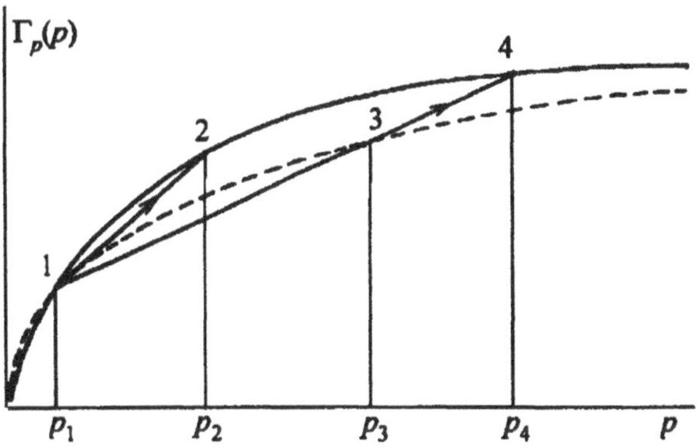

Figure 9.14. The drift flux of holes in a n-semiconductor containing traps. The broken line is a drift flux at constant level of filling of traps which corresponds to the point 1.

The degree of filling of the traps undergoes a change in a shock. The characteristic time of this process is τ_e. When $l_e=W\tau_e>>\max\{l_T, l_0\}$, Eqs. (9.15) and (9.24), the problem is analogous to that of shock wave propagation in a medium with chemical relaxation, see for example [3], [7] and [8].

For a moderate density gradient, the quantities within the shock vary relatively slowly on a scale of the order of l_e. This type of shock is illustrated by the segment 1→2 in Fig. 9.14. According to the evolution condition, the density must increase in the shock.

Let us now transform to $X=(x-Wt)$ in Eqs. (9.57) and (9.58). Neglecting the diffusion, we obtain

$$\frac{\partial}{\partial X}(-Wp + b_p pE) = 0, \tag{9.63}$$

$$\frac{\partial}{\partial X}(-Wn_e - b_e n_e E) = [(n_e + n_1)f - n_e]\tau_e. \tag{9.64}$$

Eliminating the electric field from Eq. (9.64), and then integrating Eq. (9.63), we obtain the relation between p and n_e in the shock:

$$-Wp+\frac{b_p pj}{e(b_e n_e+b_p p)}=\frac{\Gamma_p(p^-)p^+-\Gamma_p(p^+)p^-}{p^+-p^-}\,;\qquad W=\frac{\Gamma_p(p^+)-\Gamma_p(p^-)}{p^+-p^-}\,,\tag{9.65}$$

where p^+, p^- are the hole densities on the right and left boundaries of the shock. Since, according to Eq. (9.63), $d\Gamma_p/dX=d\Gamma_e/dX=Wdp/dX$, we have instead of Eq. (9.64)

$$W\left(1-\frac{dn_e}{dp}\right)\frac{dp}{dX}=[(n_e+n_1)f-n_e]<0.\tag{9.66}$$

The values of n_e and p are connected by Eq. (9.65). When $dn_e/dp>1$, Eq. (9.66) defines a monotone density profile in a shock with scale of the order of l_e. The trap filling function in it is given by Eq. (9.60).

If the density gradient in the shock is high enough, the quantity $(1-dn_e/dp)=N_t df/dp$, obtained by solving Eq. (9.66), changes its sign. This means that a smooth profile with a scale l_e cannot exist. In fact, the resulting profile consists of the diffusive shock 1→3 with the short scale $\max\{l_T, l_0\}$, and of the segment 3→4 of the solution of the Eq. (9.66) with the long scale l_e (Fig. 9.14). This region is fully analogous to the relaxation zone in an ordinary shock wave.

There is a large number of practically important situations that can be described in terms of a model in which, in addition to shallow donors and traps, there is an acceptor level (compensator) [4], [20]. The dependence of the drift flux on density may then be also nonmonotonic, and both mobile and stationary shocks are possible. The situation is, however, complicated due to the fact that, even for relatively low currents, when $l_D{\sim}l_E{\sim}l_T$, and beginning with which it is only meaningful to introduce the subdivision into the diffusive shocks and drift profiles, recombination instability occurs [29] and [30].

The development of instability must, of course, modify the nature of the plasma profile. However, a sharp subdivision of a specimen into two regions with different plasma densities must remain. Experiments to observe the recombination instability [30] have, in fact, shown the presence of well-defined separation of the specimen into two regions with high and low electric fields. Large-amplitude oscillations were observed in the high-field region in accordance with the instability criterion [29]. Therefore, it is not surprising that the position of the boundary between these regions did not coincide precisely with the position of the shock predicted by the stationary theory.

9.8 SHOCKS IN REACTIVE GASEOUS PLASMAS

In the simplest case of plasma with two species of positive ions in stationary uniform neutral gas, accounting for $b_e >> b_1, b_2$, instead of Eq. (9.4) we have

$$\frac{\partial n_1}{\partial t}+\frac{\partial}{\partial x}[\frac{b_1 j}{eb_e}(\frac{n_1}{n_1+n_2})]=I_1-S_1;$$
$$\frac{\partial n_2}{\partial t}+\frac{\partial}{\partial x}[\frac{b_2 j}{eb_e}(\frac{n_2}{n_1+n_2})]=I_2-S_2, \quad (9.67)$$

where the functions I_1, I_2, S_1, S_2 are determined by the plasmachemical kinetics and depend on n_1, n_2. Let the characteristic lifetime of the particles of species 1 far exceed the corresponding value for the second species.

Ionization and recombination processes in plasma maintained by external source. Consider the situation when the evolution of initial disturbance of the ions 1 occurs at $I_1=S_1=0$; $I_2=const$; $S_2 \sim n_2$ [31]. In this case we have in Eq. (9.67) $(I_2-S_2)=(n_2^{(0)}-n_2)/\tau$. If $\tau \to 0$, we have $n_2=n_2^{(0)}$, and from Eq. (9.67) it follows the equation which coincides with Eq. (9.11):

$$\frac{\partial n_1}{\partial t}+b_1 E_0 n_2^{(0)}\frac{\partial}{\partial x}(\frac{n_1}{n_1+n_2^{(0)}})=0. \quad (9.68)$$

The corresponding plasma profile is presented in Fig. 9.2. It undergoes the overturn, and a shock moving with the velocity W Eq. (9.17) emerges. The correction $\delta n_2=n_2-n_2^{(0)}$ everywhere outside the shock vicinity is small; its relative value is of the order of $b_2E\tau/L$. It can easily be found from Eqs. (9.67):

$$\delta n_2=\frac{b_2}{b_1}\int_0^t\frac{\partial n_1}{\partial t'}\exp(\frac{t'-t}{\tau})dt'. \quad (9.69)$$

Integrating this expression by parts, we obtain at $t>>\tau$

$$\delta n_2=-\frac{b_2\tau}{b_1}\frac{\partial n_1}{\partial t}=\frac{b_2\tau E_0(n_2^{(0)})^2}{(n_1+n_2^{(0)})^2}\frac{\partial n_1}{\partial x}. \quad (9.70)$$

This small correction diverges at the shock. In order to calculate the ambient density perturbation in the shock vicinity, we rewrite Eqs. (9.67) in a reference frame which moves with the shock velocity W. Assuming for simplicity $b_1=b_2=b$, we have

$$W = \frac{bE_0(n_2^{(0)})^2}{(n_1^+ + n_2^{(0)})(n_1^- + n_2^{(0)})}. \tag{9.71}$$

Introducing $X=x-Wt$ and net density $n_e=n_1+n_2$, for $l_r=W\tau>>\max\{l_T, l_0\}$, Eqs.(9.15), (9.24), from Eqs. (9.67) we obtain

$$\begin{aligned} \frac{W}{bE_0}\frac{dn_1}{dX} &= d\left(\frac{n_1 n_2^{(0)}}{n_e}\right)/dX; \\ \frac{W}{bE_0}\frac{dn_e}{dX} &= (n_e - n_1 - n_2^{(0)})/\tau. \end{aligned} \tag{9.72}$$

Integrating the first of the Eqs. (9.72), we have

$$n_1(W - bE_0 n_2^{(0)}/n_e) = n_1^-(W - bE_0 n_2^{(0)}/n_e^-) = -Wn_1^+ n_1^-/n_2^{(0)} = K. \tag{9.73}$$

Substituting Eq. (9.73) into the second of Eqs. (9.72), we find

$$\frac{dn_e}{dX} = -\frac{(n_e - n_e^-)(n_e - n_e^+)}{\tau(bE_0 n_2^{(0)} - Wn_e)}. \tag{9.74}$$

The denominator in Eq. (9.74) always remains positive. It means that for $n_1^+ > n_1^-$ the value of n_e increases with X from $n_e^- = n_1^- + n_2^{(0)}$ up to $n_e^+ = n_1^+ + n_2^{(0)}$. The characteristic scale of this recombination shock does not depend on n_1^+, n_1^- and equals l_r. In other words, since the velocity of the short-wave signals with a wavelength shorter than l_r (which equals simply bE) at the right side of the shock exceeds W, the shock width is equal to the damping length of these signals. A similar relation between the width of the relaxation zone and the damping length of the short-wave signals also holds in traditional gas dynamics [3], [7] and [8].

An important exception from this simple rule is the case of the strong shock $n_1^+ >> n_1^-$. Such a situation arises, for example, in course of the plasma profile evolution at $n_1(x \to \pm\infty) = 0$, when K=0, $W=bE_0 n_2^{(0)}/n_1^+$ in

Eqs.(9.71) and (9.73). At $X \to +\infty$ the densities n_e, n_1 in this case vary very steeply. According to Eqs. (9.73), (9.74), on the right side of the shock

$$\begin{aligned} n_e &= n_e^+ - n_2^{(0)} \exp[-n_1^+ X/(n_1^- bE_0 \tau)]; \\ n_1 &= n_1^+ n_1^- n_e^+ / \{n_1^+ n_1^- + (n_2^{(0)})^2 \exp[-n_1^+ X/(n_1^- bE_0 \tau)]\}. \end{aligned} \tag{9.75}$$

At $n_1^- \to 0$ the diffusion or the quasineutrality violation starts to play a role in this region. The profiles $n_1(X)$, $n_2(X)$ here are the combination of two segments. The first (smooth) one corresponds to

$$n_1 = 0; \quad n_2 = n_e = n_2^{(0)}[1 + B\exp(X/W\tau)]. \tag{9.76}$$

Its characteristic scale is l_r. The value of n_2 increases in this region from $n_2^{(0)}$ up to $n_2^{(0)} + n_1^+$. At the right boundary of this region there is a diffusive shock. Since its width is small, n_e is continuous in it.

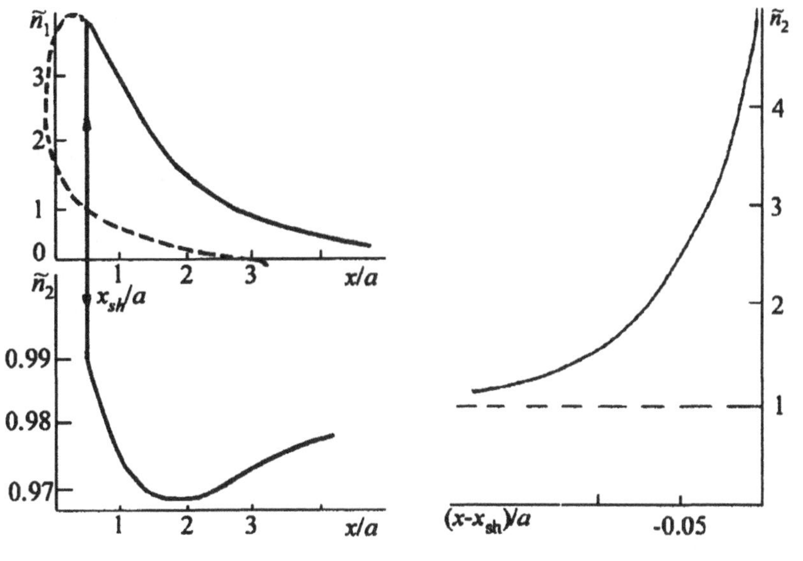

Figure 9.15. The profiles of $\tilde{n}_{1,2} = n_{1,2}/n_2^{(0)}$ for the evolution of the Gaussian initial profile of the test ions 1 and fast ionization-recombination processes in the ambient plasma. The parameters are: A=4; b_1=b_2=b; t=6a/bE_0; τ=0.2a/bE_0 *a*). Solution outside the shock. The arrows denote direction of the density variation in the shock. *b*). Variation of n_2 in the relaxation zone at the left side of the shock.

The value of n_2 falls in this shock to $n_2^{(0)}$; and n_1 rises up to n_1^+. Hence, in Eq. (9.76) the value of B is

$$B = (n_1^+ / n_2^{(0)})\exp(-X_{sh} / W\tau),$$

where X_{sh} is the position of the diffusive shock (Fig. 9.15).

Redistribution of metal ions in the polar ionosphere. The traditional explanation of the formation of sporadic layers in the inhomogeneous ionosphere plasma consists in the so-called wind shear model [32]. In the reference frame which moves with the neutral gas across the magnetic field, due to the Lorenz transformation, an electric field arises. The inhomogeneous Hall fluxes in these crossed fields result in piling up of the plasma density. In the lower part of the polar ionosphere narrow layers of high plasma density with sharp gradients are often observed. In order to explain these layers in terms of the wind shear theory it is necessary to assume that the neutral wind changes its direction on a very short spatial scale. From the other side, the lifetime of these layers is very long, reaching several hours. It contradicts the fact that the characteristic ionization and recombination times of the main ionospheric ions (NO^+, O_2^+) is of the order of 10-20s. Therefore, these sporadic layers probably consist of the metal ions of meteor origin which are characterized by very slow recombination times. Furthermore, in the polar ionosphere a considerable current parallel to the Earth's magnetic field flows. This current in the multispecies ionospheric plasma can cause the formation of recombination shocks.

The situation is, in principle, similar to the preceding example. One of the main distinctions consists in the fact that the ionization-recombination processes in real plasma are rather diverse. In particular, the removal of the ambient ions is controlled by dissociative recombination. In this case the second of the Eqs.(9.67) is to be rewritten as

$$\frac{\partial n_2}{\partial t} + \frac{b_2 j}{eb_e}\frac{\partial}{\partial x}\frac{n_2}{n_1 + n_2} = \alpha[(n_2^{(0)})^2 - n_2(n_1 + n_2)], \qquad (9.77)$$

where α is the coefficient of the dissociative recombination. The r.h.s. of the first of the Eqs.(9.67) for the metal ions can be considered as zero. Since the recombination of the ambient ions is fast, outside the shock the value of n_2 is determined by balancing the photoionization and recombination:

$$n_1(n_1+n_2)=(n_2^{(0)})^2. \qquad (9.78)$$

The dependence of the metal ion flux $\Gamma_1(n_1)$ on their density in the drift approximation coincides, in principle, with that described in the previous example for the case of linear recombination. The shock arises at the rear side of a density hump (if the current outflows from the ionosphere, the shock is formed at the lower side of an inhomogeneity). The shock structure and the density profile are close to presented in Fig. 9.15. For typical ionospheric parameters the values of l_n, l_T are of the order of several hundred meters and L is of the order of several kilometers. These numbers are within the experimentally observed range.

Stationary and nonstationary profiles in plasma with constant source. Consider the situation when a stationary localized external source of ions of species 1 is present in the current-carrying plasma. It is described by the term $q_1(x)$ in the r.h.s of the first of the Eqs.(9.67). It turns out that the stationary drift solution of this problem in the absence of sinks of injected particles 1 is possible only in the case of a relatively weak source:

$$\int_{-\infty}^{\infty} q_1(x)dx < \Gamma_1^{(\max)} = jb_1 / eb_e, \tag{9.79}$$

where $\Gamma_1^{(max)}$ is the maximum drift flux which is reached at $n_1 \gg n_2^{(0)}$. If the Eq. (9.79) holds, we have

$$\Gamma_1(n_1) = \int_{-\infty}^{x} q_1(x')dx'. \tag{9.80}$$

Expressing n_1 in terms of Γ_1, we obtain

$$n_1(x) = \frac{n_2 \int_{-\infty}^{x} q_1(x')dx'}{\Gamma_1^{(\max)} - \int_{-\infty}^{x} q_1(x')dx'}, \tag{9.81}$$

where, due to the fast ionization-recombination processes of the ambient ions, $n_2 = n_2^{(0)}$ in the case of linear recombination, and is determined by Eq. (9.78) for the dissociative recombination. Eq. (9.81) determines the smooth stationary profile. At $x \to -\infty$ we have $n_1 \to 0$ (the left region in Fig. 9.16a), and at positive x the density n_1 is uniform (the right region in Fig. 9.16a). The solution thus corresponds to the outflux of the generated particles 1 to infinity. In the case of an intense source, when the condition opposite to Eq. (9.79) holds, the ambipolar drift is unable to withdraw the generated particles 1, and the stationary solution Eq. (9.81) fails.

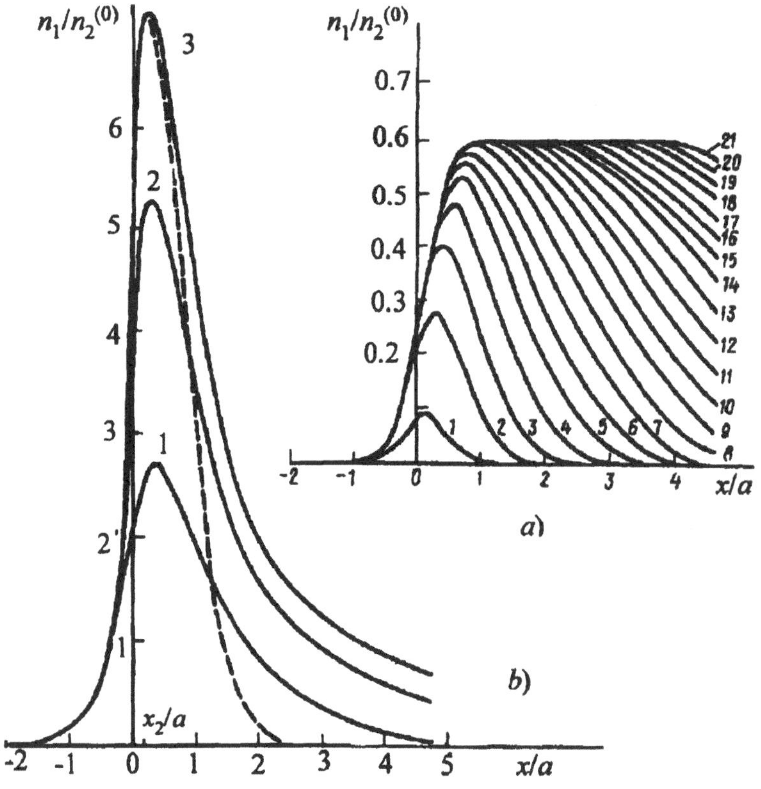

Figure 9.16. Evolution of the density profile of the injected ions in the presence of the Gaussian stationary source $q_1(x.t>0)=(Ajb_1/eb_e a)\exp(-x^2/a^2)$ and fast ionization-recombination processes in the ambient plasma. Parameter $l_T/a=[D_1/(ab_1E)]<<1$ characterizes the diffusion role. *a*). Weak source, A=0.25; $l_T/a=10^{-2}$; time in units $a/(b_1E_0)$ is equal to 1/3; 4/3; 7/3 etc. *b*). Strong source, A=1; $l_T/a=10^{-3}$; time in units of $a/(b_1E_0)$ is equal to: 1 - t=3; 2 - t=6; 3 - t=8. The broken line corresponds to the dependence $n_1(x,t)=q_1(x)t$ for t=8.

At $x=x_2$ which satisfies

$$\int_{-\infty}^{x_2} q_1(x')dx' = \Gamma_1^{(\max)}, \tag{9.82}$$

the denominator in Eq. (9.81) equals zero. In this point $dn_1/dx=(d\Gamma_1/dx)/(d\Gamma_1/dn_1)=q_1/(d\Gamma_1/dn_1)\to\infty$. Hence, for a relatively long time the profile $n_1(x)$ consists of three smooth portions separated by a sharp density drop.

The first smooth part corresponds to the interval $(-\infty, x=x_2)$. The density profile here is practically stationary and coincides with the one given by Eq. (9.81) at $t>>a/(b_1E_0)$, where a is the source scale.

The second smooth profile portion, which is situated at $x>x_2$, is nonstationary. At $t<<a^2/(4D_1)$ this part corresponds to the density rise caused by the source. The density here increases linearly with time: $n_1=q_1t$. This dependence corresponds to the situation (dashed line in Fig. 9.16b) when all ions of the species 1 created by the source remain in the place of their generation.

The third smooth segment corresponds to the outflow of a small fraction of these ions with the maximum ion flux $\Gamma_1=\Gamma_1^{(max)}$ at the right boundary of the source region ($x=a$ in Fig. 9.16b). At this boundary $x\sim a$ the profile moves with low velocity of the order of $b_1E_0/(q_1t)$.

It is possible to treat the profile at the left boundary of the source (Fig. 9.16b) as a shock. At $t<<(b_1E_0)^{2/3}(n_2^{(0)})^{4/3}q_1^{-4/3}D_1^{-1/3}$ the diffusion in it is negligible, and the transition occurs in the vicinity of the point $x=x_2$. The shock width is characterized by $l_1=b_1E_0(n_2^{(0)})^2/(q_1^2t)<<a$. Later on the shock width is determined by the diffusion and its scale is $l_2=b_1E_0D_1(n_2^{(0)})^2/q_1^{2/3}<<a$.

Fig. (9.16) presents the numerical calculation of outflowing plasma profiles for the Gaussian localized source in different moments. The fast dissociative recombination of the ambient ions which results in the dependence $n_2(n_1)$ according to Eq. (9.78) was supposed. Shock is observed at the left boundary of the source.

Formation of the stationary drift profiles in gas discharge electronegative plasmas. It was stated in Section 9.2 that for the formation of stationary shocks the ambipolar drift reversal is necessary. As an example when such a situation arises in plasmas of a *dc* gas discharge due to the plasmachemical processes, we shall consider below the 1D problem of a *dc* discharge in an electronegative gas. The corresponding equation system coincides with Eqs. (4.46), where the ionization and attachment rates $\nu^{(ion)}(E_z)$, $K_{att}(E_z)$ depend on the longitudinal electric field strength:

$$-\frac{1}{r}\frac{d}{dr}[r(D_p\frac{dp}{dr}-b_pE_zp)]=I_p-S_p$$
$$=\nu^{(ion)}(E_z)n_e-\alpha_i np-\alpha_e n_e p+q(x),$$
$$-\frac{1}{r}\frac{d}{dr}[r(D_n\frac{dn}{dr}+b_nE_zn)]=I_n-S_n \qquad (9.83)$$
$$=K_{att}(E_z)N_n n_e-K_{dt}n-\alpha_i np;$$
$$p=n+n_e.$$

In these equations K_{dt} is the detachment rate, and α_e and α_i are the electron-ion and ion-ion volume recombination coefficients. In the drift approximation, neglecting diffusion and ion mobility in Eqs. (9.83), we have $j=eb_en_eE_z$. If we consider the case when the attachment-detachment equilibrium is characterized by the fast time K_{dt}^{-1} with respect to long ionization and recombination times, it is possible to construct drift long-scale profiles, assuming attachment-detachment equilibrium, on the same lines as in the previous Sections:

$$n/n_e=K_{att}(E_z)N_n/K_{dt}\ ;$$
$$\Gamma_p=jb_p(1+K_{att}(E_z)N_n/K_{dt})/(eb_e). \qquad (9.84)$$

Numerous situations exist when the dependence $K_{att}(E_z)$ is non-monotonous. In air, for example, at low field strength the three-body attachment dominates which decreases with field as E_z^{-1}. At higher fields the dissociative attachment prevails, and $K_{att}(E_z)\sim\exp(-const/E_z)$ [33]. The dependence $K_{att}(E_z)$ has a minimum at $\tilde{E}_z/N_n\approx 3\cdot 10^{-20}\,Vm^2$. The field strength of this order of magnitude is characteristic for *dc* discharges at pressures $p\sim 10^4 Pa$. According to Eq. (9.84), the dependence of the drift flux of the positive ions $\Gamma_p(p)$ is qualitatively similar to the one presented in Fig. 9.7. Hence, in the electronegative gas under the influence of a local external ionizer, the strongly inhomogeneous plasma profiles emerge. They are similar to those presented in Fig. 9.8 which exist in the absence of gas flow due to the dependence $b_e(E_z)$. The large scale smooth drift segments are determined by the combination of Eq. (9.84) with the equations

$$(d\Gamma_p/dE)(dE/dx)=I_p-S_p\ ;$$
$$p(E)=\Gamma_p(E)/b_pE;\quad n(E)=p(E)-j/(eEb_e). \qquad (9.85)$$

It is possible that the dependences $p(E)$, $n(E)$, Eq. (9.85), are non-monotonic. In this situation, in contrast to the case of Section 9.6 (Fig. 9.7), the segments $x<x_0$; $x_2'<x<x_1$ of the drift profiles $p(x)$, $n(x)$ can have minima.

The shock structure in this case is far more complicated. The scenario depends on the relative value of the characteristic scales $l_{ad}=b_nE/K_{dt}$ on which, due to the fast plasmachemical attachment-detachment processes, the equilibrium plasma composition is established, and l_T, l_0, Eqs. (9.15) and (9.24). To analyze the shock structure at $l_T<<l_0$, for example, we assume $\Gamma_p=\Gamma_0=const$. It is possible to express the flux Γ_n from the Poisson equation. After substituting it into the second of the Eqs. (9.83), we have

$$\frac{d^2E_z^2}{dX^2}=\frac{2jK_{att}(E_z)N_n}{\varepsilon_0 b_e b_n E_z}-\frac{2eK_{dt}}{\varepsilon_0 b_n}\left(\frac{\Gamma_0}{b_pE_z}-\frac{j}{eb_eE_z}\right)+\frac{2K_{dt}}{b_n}\frac{dE_z}{dX}. \tag{9.86}$$

This equation describes the motion of a nonlinear oscillator with negative friction (the last term in the r.h.s. of Eq. (9.86)). The 'potential' $W(E_z^2)$ is presented in Fig. 9.17.

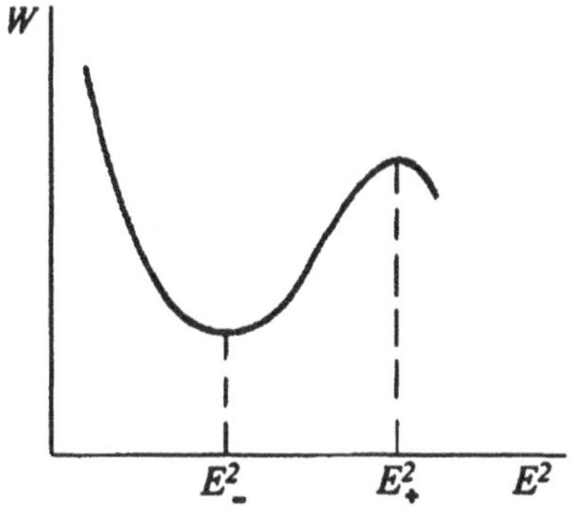

Figure 9.17. Effective 'potential' $W(E^2)$ for Eq. (9.86)

The values of E_z^+, E_z^- on the cathode and anode sides of the shock correspond to the maximum and minimum of the 'potential' $W(E_z^2)$. A similar equation was investigated in Section 3 of the book [34] for the case of the shock structure in cold plasma across a magnetic field. If the 'friction' dominates in Eq. (9.86), the term with the second derivative in it can be omitted. At $n\sim p\sim n_e$ for the strong shock $E_z^+-E_z^-\sim E_z^+$ it corresponds to $l_{ad}<<l_0$. The resulting equation describes monotone transition from E_z^- to E_z^+ with the scale l_0.

In the case $l_{ad}>>l_0$ the 'friction' in Eq. (9.86) is small, and the shock structure becomes oscillatory. On the left (anode) side of the shock the density oscillations are small and harmonic. Their length is of the order of $\sqrt{l_0 l_{ad}}$. Towards the right shore the amplitude of the oscillations grows, and their form tends to be soliton-like (Fig. 9.18). The distance between the first and the second (from E_z^+) solitons is of the order of

$$\sqrt{l_0 l_{ad}} \ln(l_0 / l_{ad}).$$

The whole shock width is of the order of l_{ad}.

At small values of the ratio n/n_e the plasma at the right side of the shock is stable. The plasma at the left side of the shock is unstable with respect to the attachment instability [35], and a turbulent state can develop in this region. If $l_{ad} >> l_T >> l_0$, the results are, in principle, the same with the replacement of l_0 by l_T..

If at least one of the particle mobilities (for example, of the electrons) is field dependent, the field variation in the shock is monotonous with a characteristic scale equal to $\max\{l_{ad}, l_T, l_0\}$. In the presence of the gas flow at $l_{ad} >> l_T$, l_0, a situation is possible, when the shock consists of the relaxation zone with a width $\sim l_{ad}$, and of a narrow (with the width of the order of $\max\{l_T, l_0\}$) diffusive shock.

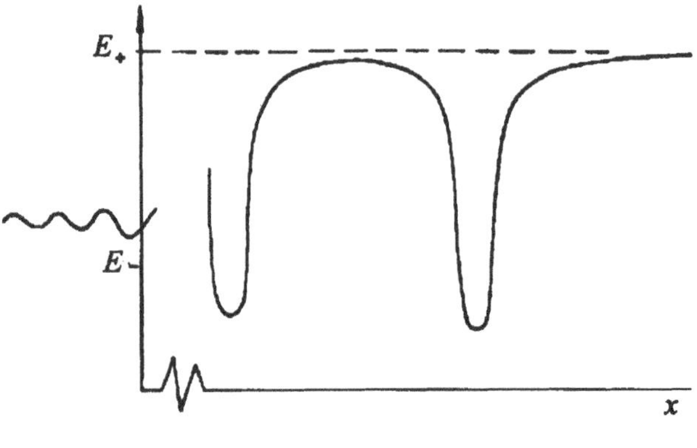

Figure 9.18. Oscillatory field profile in the shock in an electronegative gas.

Formation of periodic structures. It was formulated first in [36] that in multispecies current-carrying plasmas with ambipolar drift reversal, the formation of periodic structures is possible. Such structures consist of drift segments separated by diffusive shocks. Let us consider the case when in the drift approximation the dependence of the flux on the density is S-shaped (see Fig. 9.19). Such a situation is possible, if the equilibrium electric field E_c which satisfies the condition $I_p(E_c)=S_p(E_c)$ corresponds to the falling part of the dependence $\Gamma_p(E)$. Furthermore, from the continuity equation for the positive ions

$$\frac{\partial}{\partial t}\left(\frac{\Gamma_p(E)}{b_p E}\right)+\frac{\partial \Gamma_p(E)}{\partial x}=I_p(E)-S_p(E)$$

it follows that this state is unstable.

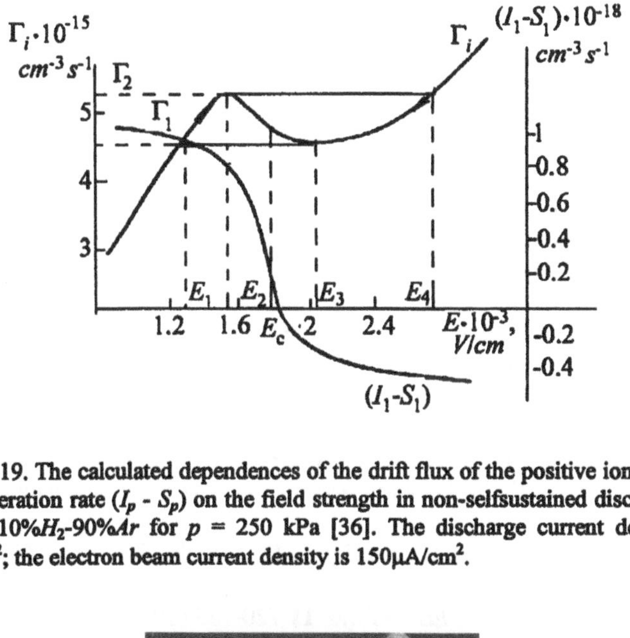

Figure 9.19. The calculated dependences of the drift flux of the positive ions Γ_p and their generation rate (I_p - S_p) on the field strength in non-selfsustained discharge in mixture 10%H_2-90%Ar for p = 250 kPa [36]. The discharge current density is 0.9A/cm^2; the electron beam current density is 150μA/cm^2.

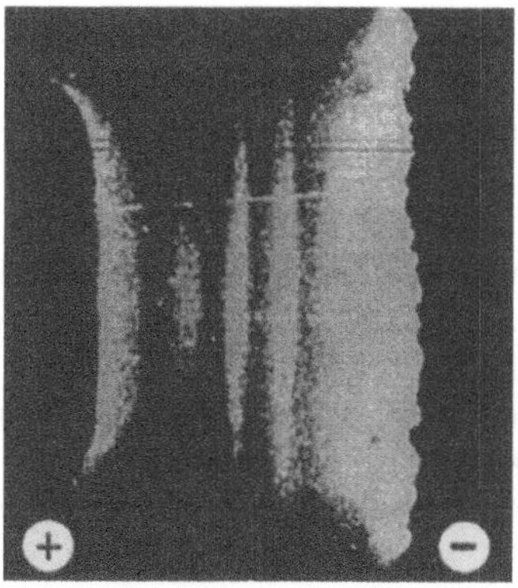

Fig. 9.20. Photograph of the non-selfsustained discharge in mixture 10% D_2-90% Ar for p=250 kPa [36].

The development of this instability can result in the formation of a limit cycle which corresponds to a profile consisting of shocks and drift segments: $[E_2; E_4]$, $\{E_4; E_3\}$, $[E_3;E_1]$, $\{E_1; E_2\}$. In [36] it was shown that such dependences $\Gamma_p(E)$, $I_p(E)$ arise in the non-selfsustained discharges in mixtures H_2-Ar, D_2-Ar due to the fast attachment-detachment processes which involve the negative ions H^-, D^-. Under these conditions the distinct periodic structures of discharge were observed in the experiment (Fig. 9.19). The period of the structures was of the order of the drift scale $l_E = b_p E \tau_p$.

REFERENCES

1. F. Kolrausch, *Ann. Phys. Chem.* **62**: 209-231 (1897).
2. H. Weber, *Sitz. Akad. Wiss. Berlin* **44**: 936-945 (1897).
3. G. B. Whitham, *Linear and Nonlinear Waves* (N. Y.: Wiley, 1974): 636.
4. S. G. Bonch-Bruevich, S. Kalashnikov, *Physics of Semiconductors* (Moscow: Nauka, 1990): 1965 (in Russian).
5. J. Haynes, W. Shockley, *Phys. Rev.* **81**: 835-843 (1951).
6. A. C. Prior, Proc. Phys. Soc. **76** (4904) 465-480 (1960).
7. L. D. Landau, E. M. Lifshits, *Fluid Mechanics* (Oxford: Pergamon, 1987): 539.
8. Ja. Zeldovich, Yu. P. Raiser, *Physics of Shock Waves and High-Temperature Hydrodinamic Phenomena* (New York: Academic, 1967): 916.
9. A. P. Dmitriev, V. A. Rozhansky, L. D. Tsendin, *Sov. Phys. Uspekhi* **28**: 467-483 (1985).
10. I. Visikailo, *Sov. J. Plasma Phys.* **11** 720-723 (1985)
11. V. A. Rozhansky, L. D. Tsendin, *Geomagnetism and Aeronomia* **24** 755-759 (1984).
12. M. Scholer, H. Haerendel, *Planet. Space Sci.* **19**: 915-927 (1971).
13. A. G. Longsworth, *J. Amer. Chem. Soc.* **65**: 1755-1765 (1943).
14. D. W. Ross, *Phys. Rev.* **146**: 178-185 (1966).
15. L. D. Tsendin, Sov., *Phys. Techn. Phys.* **14** 1013-1018 (1969).
16. Yu. S. Akishev, F. I. Visikailo, A. P. Napartovich et al., *Sov. Phys. - High Temp.* **18**: 216-221 (1980).
17. A. V. Bondarenko, F. I. Visikailo, Cohan V. I. et al., *Teplofizika Vysokih Temperatur* **21**: 388-389 (1983) (in Russian).
18. F. I. Visikailo, L. D. Tsendin, *Sov. J. Plasma Phys.* **12**: 696-699 (1986).
19. V. N. Babichev, F. I. Visikailo, S. A. Golubev, *Sov. Phys. Techn. Phys. Lett.* **12**: 409-411 (1986).
20. J. S. Blakemore, *Semiconductor Statistics* (Oxford: Pergamon, 1962): 381.

21. S. Tosima, *J. Phys. Soc. Japan* **22**: 1025 - 1031 (1967).
22. S. Tosima, K. Ando, *J. Phys. Soc. Japan* **23**: 812 - 819 (1967).
23. A. P. Dmitriev, A. E. Stephanovich, L. D. Tsendin, *Sov. Physics - Semiconductors* **9**: 894-901 (1975).
24. A. E. Stephanovich, L. D. Tsendin, *Sov. Physics - Semiconductors* **10**: 406-409 (1976).
25. M. Lampert, P. Mark, *Current Injection In Solids* (N. Y.: Academic, 1970): 351.
26. A. P. Dmitriev, A. E. Stephanovich, L. D. Tsendin, *Phys. Stat. Solidi* **46a**: 45 - 53 (1978).
27. A. A. Akopjan, Z. S. Gribnikov, *Sov. Physics - Semiconductors* **9**: 981-985 (1975).
28. A. A. Akopjan, Z. S. Gribnikov, *Solid State Slectron.* **19**: 41-45 (1976).
29. O. V. Konstantinov, V. I. Perel', *Sov. Phys. - Solid State* **6**: 2691-2698 (1965).
30. I. V. Karpova, S. G. Kalashnikov, O. V. Konstantinov, V. I. Perel', G. V. Tsarenkov, *Phys. Status Solidi*, **33**: 863-869 (1969).
31. V. A. Rozhansky, L. D. Tsendin, *Geomagnetism and Aeronomia* **24**: 491-493 (1984).
32. B. N. Gershman, Dynamika ionosphernoy plasmy (Moscow: Nauka: 1974) 256 (in Russian).
33. N. L. Alexandrov, F. I. Visikailo, R. S. Islamov, et al., *Sov. Phys. - High Temp.* **19**: 485-450 (1981).
34. R. A. Sagdeev, in *Rewiews of Plasma Physics*, vol.**4**, Editor: M.A. Leotovich (New York: Consultants Bureau, 1966): 23-91.
35. A. P. Napartovich, A. N. Starostin, in *Plasma Chemistry*, vol.**6**. Editor: B. M. Smirnov (Moscow: Atomizdat, 1979) 153-208 (in Russian).
36. A. V. Demjanov, I. V. Kochetov, A. P. Napartovich et al., *Sov. Phys. Techn. Phys. Lett.* **12**: 351-352 (1986).

Chapter 10

Inhomogeneities in Currentless Multispecies Plasmas with Great Distinction in Partial Temperatures

In numerous applications extremely nonisotermal plasmas arise. In gas discharges at low and moderate pressures, for example, the characteristic values of T_e/T_i is of the order of 10^2. We shall demonstrate in this Chapter that this factor, in analogy with a *dc* current, in multispecies inhomogeneous plasmas can result in the propagation of weakly damping perturbations: signals. Since in the majority of practically important discharges rather complex gas mixtures are used, the situation when several ion species are present in the plasma is typical. Very important case corresponds to the discharges in the electronegative gases. They are widely used in the technology and surface processing. It is to be noted that the ordinary air is also considerably electronegative. Therefore, the concept of overturn of smooth plasma density profiles, formation of strongly inhomogeneous shock-like plasma structures and the flux conservation in these shocks turns out to be also very effective in this important case of currentless strongly nonisothemal plasmas. As an example, we shall chose the gas discharge plasmas with $T_e >> T_i$. In the first Section of this Chapter we shall consider the dynamical problem in the absence of plasmachemical reactions. The stationary positive column of a *dc* discharge in an electronegative gas is studied in the second Section. The third Section treats the capacitively coupled RF discharges in electronegative gases. In order to simplify the mathematics, we shall restrict ourselves to the plasmas with only two species of ions, both positive, or one positive and one negative.

10.1 MULTISTAGE EVOLUTION AND SHOCK FORMATION IN PROCESS OF DIFFUSION OF MULTISPECIES PLASMA WITH HOT ELECTRONS

According to Section 4.2, we have

$$\frac{\partial n}{\partial t}=-\frac{\partial}{\partial x}\Gamma_n=\frac{\partial}{\partial x}\left(D_n\frac{\partial n}{\partial x}\pm b_n nE\right),$$
$$\frac{\partial p}{\partial t}=-\frac{\partial}{\partial x}\Gamma_p=\frac{\partial}{\partial x}\left(D_p\frac{\partial p}{\partial x}-b_p pE\right),$$
$$\frac{\partial n_e}{\partial t}=-\frac{\partial}{\partial x}\Gamma_e=\frac{\partial}{\partial x}\left(D_e\frac{\partial n_e}{\partial x}\pm b_e n_e E\right),$$
$$n_e=p\mp n. \tag{10.1}$$

The upper sign corresponds to the case when ions of species n are negative, and the lower sign to the situation in which they are positive.

From the absence of current we have

$$E=\frac{D_p p'\mp D_n n'-D_e n_e'}{b_p p+b_n n+b_e n_e}, \tag{10.2}$$

where prime corresponds to spatial derivative. If the electron density and its gradient are not too small, the electrons are described by a Boltzmann distribution:

$$E=-T_e/e(\ln n_e)'. \tag{10.3}$$

In the following Sections we shall restrict ourselves mainly to this case. The ion density in the denominator of Eq. (10.2) can be neglected in a plasma which is not too electronegative:

$$b_e n_e >> b_n n,\ b_p p. \tag{10.4}$$

Since the typical ratio of mobilities b_e/b_n is of the order of 300, Eq. (10.4) implies that the ratio n/n_e<<300. The ion diffusion can be disregarded in Eqs. (10.1), if the electron density gradient is not very small:

$$b_n nE\sim\frac{T_e b_e n}{en_e}\frac{\partial n_e}{\partial x}>>D_n\frac{\partial n}{\partial x}=\frac{b_n T_i}{e}\frac{\partial n}{\partial x}.$$

If relative slopes of n, n_e profiles were comparable, this condition would be far less restrictive than Eq. (10.4) and would be fulfilled up to $n/n_e << D_e/D_n \sim 30\,000$. On the other hand, the situations often arise when the electron density profile in the plasma bulk becomes very flat, and the ion diffusion becomes important even at not quite high electronegativity (see below this Section and Section 10.2).

After neglecting the ion diffusion and substituting the field Eq. (10.3) into Eq. (10.1), the system of equations which controls the evolution process is reduced to

$$\pm\frac{\partial n}{\partial t}+\frac{\partial}{\partial x}\left(\frac{d_n n}{n_e}\frac{\partial n_e}{\partial x}\right)=0, \tag{10.5}$$

$$\frac{\partial p}{\partial t}-\frac{\partial}{\partial x}\left(\frac{d_p p}{n_e}\frac{\partial n_e}{\partial x}\right)=0, \tag{10.6}$$

$$p\mp n=n_e, \tag{10.7}$$

where the partial ambipolar diffusion coefficients (the partial ion diffusion coefficients with the electron temperature) are $d_{n,p}=b_{n,p}T_e/e$. The specific property of the system Eqs. (10.5-10.7) consists in the fact that the ion fluxes are determined only by the electron density gradient.

Let us start the analysis of the system Eqs. (10.5) - (10.7) by studying propagation of the small signals in an unbounded homogenous plasma ($n^{(0)}, n_e^{(0)}, p^{(0)}$=*const*(*x*)). Ion and electron density variations δn_α are taken proportional to exp($-i\omega t+ikx$). Since the system Eqs. (10.5-10.7) contains two temporal derivatives, there are two perturbation modes $\omega_{1,2}(k)$. The diffusive decaying mode, in which the perturbations satisfy

$$\delta n/(d_n n^{(0)})=\mp\delta p/(d_p p^{(0)})=\mp\delta n_e/(d_n n^{(0)}+d_p p^{(0)}),$$

evolves with frequency

$$\omega_1=-D_{eff}ik^2, \qquad D_{eff}(n^{(0)}/n_e^{(0)})=(d_n n^{(0)}+d_p p^{(0)})/n_e^{(0)}. \tag{10.8}$$

This frequency corresponds to monotone decay of a density perturbation with the effective diffusion coefficient D_{eff}. The evolution scenario in this case is close to the situation in the pure two component plasma with the diffusion coefficient D_{eff} instead of D_a, Eqs. (4.7) - (4.10). In the limit of the pure plasma, $n\to 0$, the coefficient D_{eff} tends to $D_a\equiv d_p$, i.e. to the ordinary ambipolar diffusion coefficient, Eq. (4.10). The presence of negative ions strongly enhances the value of D_{eff}. Since $n>0$, it always holds $D_{eff}>D_a$. At large electronegativity, $p/n_e>>1$, $D_{eff}>>D_a$. This inequality results in the considerable peculiarities in the evolution of electronegative plasma profiles as will be discussed in Section 10.2. It is easy to guess that the disturbances, in which the electron density does not change, $\delta n = \pm\delta p; \delta n_e = 0$, do not evolve in time (their slow dissipation is described by the ion diffusion with low ion temperature T_i neglected here).

The situation is different if we account for inhomogeneity of a background plasma: $(1/n_e^{(0)})(\partial n_e^{(0)}/\partial x) = 1/L \neq 0$. In the limit $|kL|>>1$ the diffusive decaying mode with ω_1, Eq. (10.8), remains unchanged. On the other hand, the slowly decaying mode with $\omega=\omega_2$ converts into the propagating mode. Perturbation of the electron density remains small (of the order of $(kL)^{-1}$) with respect to perturbations of the ion densities. These signals propagate now with the velocity

$$\omega_2 = u_{eff}k, \quad u_{eff}(n^{(0)}/n_e^{(0)}) = \pm\frac{d_n d_p n_e^{(0)}}{L(d_n n^{(0)} + d_p p^{(0)})}. \tag{10.9}$$

For small density of negative ions $n^{(0)}/n_e^{(0)} << 1$, u_{eff} coincides with the drift velocity of negative ions $u_{eff}=-b_nE$ in the field Eq. (10.3). If $n^{(0)}/n_e^{(0)} >> 1$, u_{eff} is significantly smaller than the ion drift velocity. The reason consists in the coupling of Eqs. (10.5) - (10.7), due to the influence of the self-consistent electric field on $n(x)$ profile during evolution.

Subtracting Eq. (10.5) from Eq. (10.6), we obtain

$$\frac{\partial n_e}{\partial t} - \frac{\partial}{\partial x}\left(D_{eff}\frac{\partial n_e}{\partial x}\right) = 0, \tag{10.10}$$

where effective diffusion coefficient $D_{eff} = (d_n n + d_p p)/n_e$; for small perturbations D_{eff} is given by Eq. (10.8). Eq. (10.10) has the form of a simple diffusion equation for n_e, but the effective diffusion coefficient is determined by the ion mobility, Eq. (10.3), and depends on the ratio n/n_e.

In our limit $|kL|>>1$, frequency of perturbations which correspond to the second (propagating) mode satisfy $\omega_2<<\omega_1$. From Eq. (10.10) for perturbations of the partial densities we find

$$\omega_2\delta n_e = D_{eff}k^2\delta n_e + \left(\frac{\partial D_{eff}}{\partial n}\delta n + \frac{\partial D_{eff}}{\partial p}\delta p\right)\frac{ikn_e}{L}.$$

This means that the perturbation of the electron density in these signals is small (of the order of $(kL)^{-1}$) with respect to $\delta p \cong -\delta n$. Since $\omega_2 << D_{eff}k^2$, electron flux $\Gamma_e = -D_{eff}\ \partial n_e / \partial x$ remains practically uniform in these perturbations: $\Gamma_e = \Gamma_e^{(0)}$. After substituting $\partial n_e^{(0)} / \partial x$ from the expression for Γ_e, Eq. (10.5) can be rewritten as

$$\frac{\partial n}{\partial t} - \frac{\partial}{\partial x}\Gamma_n = 0, \Gamma_n(n / n_e) = \pm d_n n \Gamma_e / (d_n n + d_p p). \qquad (10.11)$$

Linearisation of Eq. (10.11) accounting for Γ_e conservation results in the expression for u_{eff} Eq. (10.9). Equation (10.11) is very convenient for analysis of a plasma profile evolution, since it is also correct for nonlinear signals. The plot of $\Gamma_e(n/n_e)$ is shown schematically in Fig. 10.1.

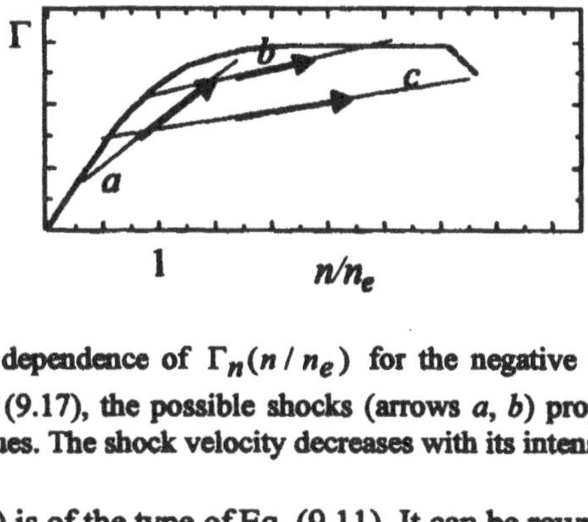

Figure 10.1. The dependence of $\Gamma_n(n / n_e)$ for the negative ions n and $\Gamma_e < 0$. According to Eq. (9.17), the possible shocks (arrows a, b) propagate towards the maximal $n_e(x)$ values. The shock velocity decreases with its intensity.

Equation (10.11) is of the type of Eq. (9.11). It can be rewritten as

$$\frac{\partial n}{\partial t} + u_{eff}(n)\frac{\partial n}{\partial x} = 0. \qquad (10.12)$$

Each part of $n(x)$ profile moves with its own velocity $u_{eff}(n)$, and profile overturn and shock formation are possible. The flux $\Gamma_e(n/n_e)$, Eq. (10.11), tends to constant at $n/n_e \to \infty$. The small signal propagation velocity $u_{eff} = \partial\Gamma_n / \partial n$ depends on the ion density n; it increases with the rise of n. Shock velocity, which equals to slope of tangent of $\Gamma_n(n/n_e)$ curve, decreases with its amplitude and tends to zero for intense shocks.

We shall consider here several numerical examples of different evolution scenaria in inhomogeneous non-isothermal plasmas in the absence of particle sources and sinks. Even in such a simple formulation the problem turns out nontrivial. The evolution process of an arbitrary plasma density perturbation can be subdivided into several stages with different temporal scales. Measuring the spatial coordinate in units of the characteristic length scale L of the initial electron density perturbation and the time in units of ambipolar diffusion time of the negative ions $T_i L^2/D_n T_e$ and denoting $D=D_p/D_n$, we find:

$$\frac{\partial p}{\partial t}=\frac{\partial}{\partial x}\left(Dp\frac{\partial}{\partial x}\ln(n_e)\right);\ \frac{\partial n}{\partial t}=\mp\frac{\partial}{\partial x}\left(n\frac{\partial}{\partial x}\ln(n_e)\right);$$
$$\frac{\partial n_e}{\partial t}=\frac{\partial}{\partial x}\left((Dp+n)\frac{\partial}{\partial x}\ln(n_e)\right);n_e=p\mp n. \tag{10.13}$$

We begin with considering the case $D<<1$, and assume the second (light) ion species to be negative.

As was demonstrated in Section 4.2 for two positive components, the evolution of an arbitrary perturbation takes place in two stages. The first (fast) stage corresponds to the ambipolar time of the light (negative) ions, $t\sim 1$. It is determined by their motion in the field Eq. (10.3), which corresponds to the Boltzmann law for electrons. During this stage the density profile of the heavy (positive) ions remains practically fixed. The negative ions are piled up by the field Eq. (10.3) towards the maxima of $p(x)$ according to Eq. (10.6). The diffusive flux of the negative ions and the temporal variation of $p(x)$ can be neglected, and the Eq. (10.5) can be rewritten as

$$\frac{\partial n_e}{\partial t}=\frac{\partial}{\partial x}d_n\frac{(\pm(p_0(x)-n_e)}{n_e}\frac{\partial n_e}{\partial x}.$$

It has the form of diffusion equation for electrons with the density-dependent effective diffusion coefficient. This means that electron density in the end of this first (fast) stage, which is characterized by the diffusive time scale of the light ions with the electron temperature, becomes uniform. Electron density tends to its spatially averaged value $<n_e>$. Evolution of ion profiles during the second (slow) stage is determined by the system

$$\frac{\partial p}{\partial t}=\frac{\partial}{\partial x}\left(D_p\frac{\partial p}{\partial x}-b_pEp\right),$$
$$\frac{\partial n}{\partial t}=\frac{\partial}{\partial x}\left(D_n\frac{\partial n}{\partial x}-b_nEn\right),$$
$$p=n+<n_e>.$$

Since electron density during this stage is uniform, the absence of the net current results in the electric field which corresponds to the ion ambipolarity $\Gamma_p = \Gamma_n$:

$$E = \frac{\partial p}{\partial x} \frac{D_p - D_n}{b_p p + b_n n}.$$

This stage corresponds to the ion-ion diffusion with the effective density-dependent diffusion coefficient

$$\frac{\partial p}{\partial t} = \frac{\partial}{\partial x}\left[\frac{D_p b_n n + D_n b_p p}{b_n n + b_p p} \frac{\partial p}{\partial x}\right]. \tag{10.13a}$$

The second stage is characterized by the slow ion-ion ambipolar time $T_e/(DT_i)$ (L^2/D_p in the ordinary time scale); it corresponds to the diffusion of less mobile ions which is enhanced by the electric field.

In spite of the fact that during this second (slow) stage electron Boltzmann law, Eq. (10.3), remains valid, electron profile is practically flat, and the electric field corresponds now to the ion-ion diffusion with the low ion temperature. Electron Boltzmann law, Eq. (10.3), can be used only to find the remaining small inhomogeneity of the electron density, which follows adiabatically the evolution of $n(x,t)$, $p(x,t)$. In our case D<<1 the more mobile negative ions approach equilibrium at the end of the first stage, so that their Boltzmann distribution is established.

The character of evolution becomes substantially different if profile segments exist where the spatially averaged electron density $\langle n_e \rangle$ exceeds initial positive ion density $p_0(x)$ [1] and [2]. In this case the third, intermediate, stage of evolution can be singled out. In the course of the first (fast) stage the electric field extracts negative ions out of these segments. Consequently, after a time $1/D >> t >> 1$ (at the end of the first stage) the resulting $n_e(x)$ profile consists of flat parts (where $n \neq 0$ holds and the slow ion-ion ambipolar diffusion takes place) and the parts, in which negative ions are absent while the electron profile remains inhomogeneous, $n_e(x) = p_0(x)$ (traces 1, 1' in Fig. 10.2). At the boundary between these segments a kink in the electron density profile arises, where the derivative dn_e/dx has a discontinuity (point A in Fig. 10.2).

The next, intermediate stage of evolution of such a profile is characterized by ambipolar time of electrons and positive ions, $1/D$. In region II (Fig. 10.2), where the negative ions are located, $dn_e/dx \approx 0$ holds. Here $p(x,t)$ and $n(x,t)$ profiles remain practically unchanged during the first and the second stages; they coincide with the initial profile $p_0(x)$. The region I (to the right of the point A in Fig. 10.2) during this second stage of the evolution is occupied by the pure electron-ion plasma where $n\approx 0$. In this region evolution is described by the traditional ambipolar diffusion equation (4.7).

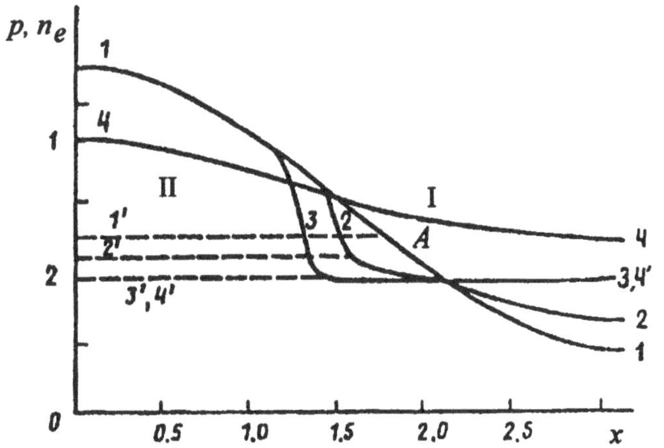

Figure 10.2 Density profiles of the heavy positive ions (1, 2, 3, 4) and electrons (1′, 2′, 3′, 4′) for D=0.1, T_i/T_e=0.01 at the dimensionles time moments t=0, 12.8, 20.5, 300 in the electronegative plasma. The shock thickness was connected with numerical diffusion.

The characteristic time scale of this process is of the order of $1/D$ (i.e. L^2/d_{ep}). Let us now consider the processes at the boundary between regions I and II. Positive ion flux Γ_p on the left side of this boundary, where $\partial n_e/\partial x \approx 0$ holds, is negligible (during this stage it is determined by the slowest process of ion-ion ambipolar diffusion). On the other hand, $\Gamma_p = -D[(p/n_e)(\partial n_e/\partial x)] = -D\partial p/\partial x$ on the right side of this transition region. This discontinuity in the value of the flux results in a jump forming both in the positive and negative ion densities. This jump moves towards larger $p_0(x)$ values (to the left in Fig. 10.2; traces 2, 2` and 3, 3`). In the limit T_i=0 its thickness equals zero because it is determined by the ion-ion ambipolar diffusion.

Electron density cannot be discontinuous in this approximation, since its derivative determines the ion fluxes (a jump in n_e would yield infinite ion fluxes). This means that jumps in densities of positive and negative ions are equal. Similarly to Section 9.2, this jump can be treated as shock.

It is more suitable in this case to consider the conditions for the flux of the positive ions Γ_p. Shock velocity W, in accordance with Eq. (9.17), is determined from the condition of Γ_p conservation in the reference system moving with the shock, which is situated at $x=x_{sh}(t)$:

$$W=\frac{dx_{sh}}{dt}=\frac{D(\partial p/\partial x)^{+}}{p^{-}-p^{+}}, \tag{10.14}$$

where superscripts plus and minus correspond, as in Chapter 9, to density values on the right and left sides of the shock.

The positive ion profile in the region II, Fig. 10.2, remains unchanged at this stage: $p=p_0(x)$ for $x<x_{sh}$. The profile of the electron-ion plasma for $x>x_{sh}$ satisfies the ambipolar diffusion equation

$$\frac{\partial p}{\partial t}=D\frac{\partial^2 p}{\partial x^2}, \tag{10.15}$$

with the boundary condition

$$p(x=x_{sh}-0)=n_e(x=x_{sh}-0). \tag{10.16}$$

Additional condition at the shock can be found from conservation of the total number of negative ions. If the problem is symmetrical with respect to $x=0$, we have

$$\frac{d}{dt}\int_0^{x_{sh}} n\,dx=(p_0(x_{sh})-p^{+})\frac{dx_{sh}}{dt}-x_{sh}\frac{dp^{+}}{dt}=0 \tag{10.17}$$

In Fig. 10.2 result of the numerical solution of Eqs. (10.14) - (10.17) is presented. Initial conditions correspond to the end of the fast stage when the profile $p(x)$ is still smooth and coincides with $p_0(x)$, but the kink of the electron profile has already been formed.

Shock moves towards large values of $p_0(x)$. Negative ions during this stage are gathered towards the plasma center. The kink in the electron density $n_e(x,t)$ which is responsible for shock motion, according to Eq. (10.14), decreases with time. As a result of this intermediate stage, after a long interval with respect to the ambipolar diffusion time of the electrons and positive ions (which is T_i/T_e times shorter than their own diffusion time), the $n_e(x,t)$ profile flattens out, and shock stops (traces 3 and 3′ in Fig. 10.2).

In order to obtain an equation for shock structure we must consider ion's own diffusion. Since ion fluxes in the reference frame which moves with the shock velocity W are conserved, we obtain

$$\Gamma_n = \frac{n}{n_e}\frac{dn_e}{dx} - \frac{T_i}{T_e}\frac{dn}{dx} - Wn = 0, \quad \Gamma_p = -\frac{Dp}{n_e}\frac{dn_e}{dx}$$
$$-\frac{DT_i}{T_e}\frac{dp}{dx} - Wp = D\left(\frac{dn_e}{dx}\right)^+ (1+\frac{T_i}{T_e}) - Wn_e^+. \tag{10.18}$$

Since on the right side of the shock n_e=p, and on its left side $\partial n_e / \partial x \approx 0$, expression for the shock velocity, Eq. (10.14), with accuracy up to T_i/T_e<<1, follows immediately from the second of the Eqs. (10.18). Substituting expression for dn_e/dx from the first of Eqs. (10.18) into the second one, accounting for the fact that the n_e variation in the shock is negligible, and D<<1, T_i/T_e<<1, we obtain

$$\frac{dn}{dx} = \frac{WT_e}{DT_i}\frac{n(n^- - n)}{p+n}. \tag{10.19}$$

As n^+=0, this equation coincides with Eq. (9.20), and describes a steep decrease of the ion densities with the spatial scale of the order of LT_i/T_e<<L.

The last stage corresponds to smearing out of the remaining irregularity of the ion density profile, according to Eq. (10.13a), with the plasma density-dependent nonlinear effective diffusion coefficient $D_{pn}=T_i b_p b_n (p+n)/(b_p p+b_n n)$. This process results in the uniform density profile (the trace 4′).

In the case of plasma with two species of positive ions of different mobilities the situation is substantially similar to that described above. The main distinction is that the highly mobile positive ions do not pile up under the influence of the field Eq. (10.3) at the points of n_e maxima, but rapidly leave these regions. After a time of the order of unity (ambipolar time of electrons and the mobile ions) n_e profile can be developed with kink, and shock can arise in the ion densities. Figure 10.3 shows an example of such profiles with shock that has developed at the end of the intermediate stage as a result of smooth inhomogeneity evolution in such a plasma.

These profiles represent initial conditions for the ion-ion ambipolar diffusion, which results in the final leveling of all three densities. Shocks described above can also develop in the less artificial case of comparable species mobilities. In this case, however, the time for the electron profile to flatten out is comparable with the time scale on which the $p(x, t)$ profile evolves, so that kink in n_e and shock are able to form only if rather special initial and boundary conditions are imposed. In [2] it was shown that shock formation is possible in three cases: i. when the ion mobilities are strongly different (already treated); ii. for D~1 provided the initial profiles satisfy

$|dn_0/dx|>|\partial n_e / \partial x|$; and iii. for the case when n_e, for some reason, is small in the vicinity of the point where shock is formed. The latter situation corresponds, in particular, to restricted plasmas, when zero boundary conditions at the absorbing boundaries are imposed. This case will be studied in detail in the next Sections.

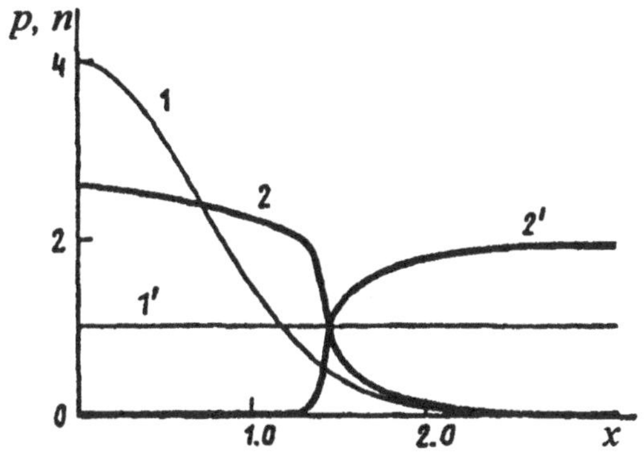

Figure 10.3. Evolution of plasma inhomogeniety in electropositive plasma with the ratio of mobilities D=0.1. The initial density profiles of less mobile ions (trace 1), and more mobile ions (trace 1′) are: 1) $p(x, 0)=4\exp(-x^2)$; 1′) $n(x, 0)=1$. The traces 2, 2′ correspond to dimensionless time t=5. Shock thickness is connected with numerical diffusion.

In Figs. 10.4 and 10.5 examples of the second scenario are presented. The boundary conditions of zero density derivatives $\partial n_e / \partial x = 0$ were imposed at x=0, π in Fig. 10.4.

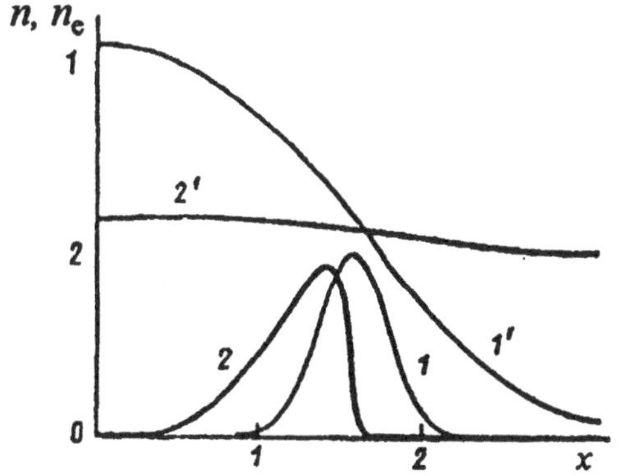

Figure 10.4. Density profiles of the negative ions (traces 1, 2) and of electrons (traces 1′, 2′) at D=1. The curves 1, 1′ correspond to the initial conditions $n(x, 0)=2\exp[-11(x-\pi/2)^2]$; $n_e(x, 0)=2(1.1-\cos x)$; the curves 2, 2′ - to t=5.

This corresponds to symmetrical spatially periodic problem. An initial profile of the negative ion density in Fig. 10.3 was chosen (trace 1), which was considerably steeper than the initial n_e profile (trace 1′). Negative ions are sucked into the region of maximum electron density. A shock-like profile steepening develops at the trailing edge of the ion density profiles (trace 2).

In Fig. 10.5 the density evolution in a plasma with two species of positive ions of comparable mobility is presented. The shock in this case is formed due to the fact that the more mobile ions were absent at small x; the self-consistent field additionally sweeps them out of this region.

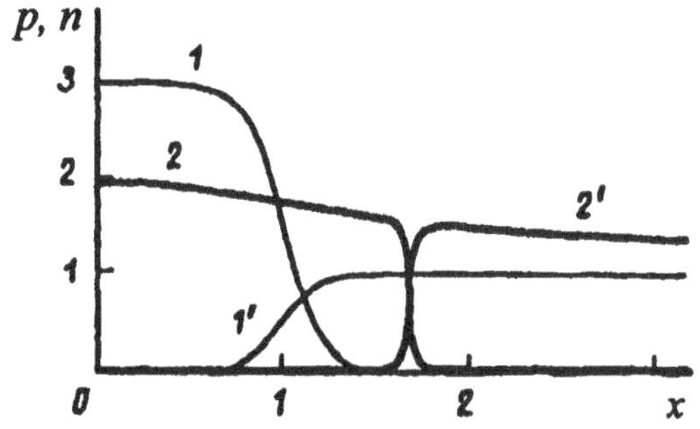

Figure 10.5. Density profiles for two species of the positive ions with comparable mobilities (p are the less mobile, traces 1,2, and n are more mobile, traces 1′, 2′; D=0.5). The initial density profiles were: 1) $p(x, 0)$=1.5(1-tanh5(x-1)); 1′) $n(x, 0)$=0.5(1+tanh5(x-1)). The curves 2, 2′ correspond to the time t=2.1.

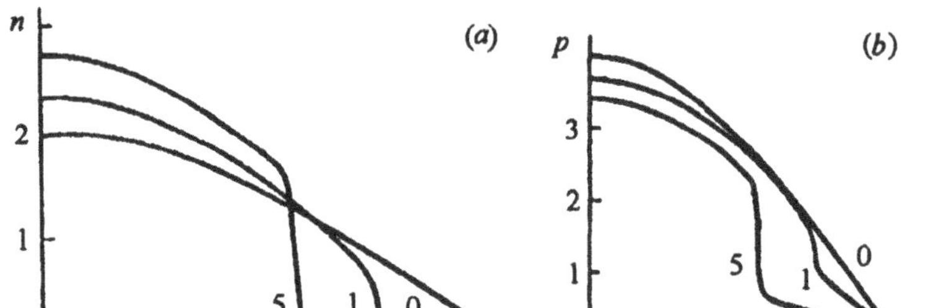

Figure 10.6. Density profiles of the negative (a) and of the positive (b) ions in course of the diffusion in three-species plasma in presence of an absorbing boundary at x=π. Ratio of mobilities is D=0.5. The initial density profiles were: negative ions $n(x, 0)= 2\cos(x/2)$; $p(x, 0)= 4\cos(x/2)$. The numbers at the profiles correspond to values of the dimensionless time t

A situation which corresponds to the third scenario arises in the vicinity of an absorbing wall in an electronegative plasma. In Fig. 10.6 an example of such a shock is presented.
It emerges in the vicinity of the wall and moves towards the plasma center.

In all our examples density of the mobile ions was zero on one of the shock sides. Such profiles result from the fact that, according to Eqs. (9.16), (10.9) and (10.12), the duration of shock formation is given by

$$t_{sh}=\min\left(\frac{\partial u_{eff}}{\partial n}\frac{dn^0}{dx}\right)^{-1}. \tag{10.20}$$

Quite sharp minimum of

$$\left|\frac{\partial u_{eff}}{\partial n}\right|=\frac{2d_p^2 d_n(d_p+d_n)n_e\Gamma_e}{(d_p p+d_n n)^3} \tag{10.21}$$

is situated in the vicinity of n=0. This means that, if the initial profiles are not steep enough, shocks are formed preferably in the regions where initial values of n are small. Even if such regions were absent initially, they could arise in the course of evolution [2]; see also Figs. (10.2) - (10.6). In principle shocks without such regions can exist. However, such shocks are created only in quite special initial conditions, when the initial density profiles are extremely nonuniform.

Fast plasmachemical reactions can lead to shock broadening and to more complicated shock structure. Some examples are considered in [1], [2].

In [3] (see also Section 15.1 of the book [4]) an approach was proposed based on the assumption that, simultaneously with the Boltzmann's relation Eq. (10.3) for electrons, negative ions density also satisfies the Boltzmann's relation $e\varphi=T_i\ln n$. Since these relations demand $\Gamma_e=\Gamma_n=0$, in order to obtain the nonzero fluxes Γ_e, Γ_n, the Boltzmann's distributions for electrons and negative ions were replaced by equivalent condition

$$T_i\vec{\nabla}\ln n=T_e\vec{\nabla}\ln n_e. \tag{10.22}$$

In such a way expressions for effective partial diffusion coefficients , i.e. the density-dependent proportionality coefficients $D_\alpha^{(eff)}$ were derived in the linear relations $\Gamma_\alpha=-D_\alpha^{(eff)}\vec{\nabla}n_\alpha$. Since the field Eq. (10.2) is determined by gradients of all the partial densities, in general case such a procedure, when every partial flux of particles α is proportional only to gradient of their own density n_α, is impossible. In reality this approximation, as it follows from the above-mentioned examples, has rather restricted applicability. First of all, in the 2D or 3D problems Eq. (10.22) is applicable only in special plasma configurations, where very restricting condition of collinearity of the partial densities gradients holds. In the general case the field Eq. (10.2) results in the direction of the flux Γ_α which does not coincide with the direction of the partial density gradient. The fast evolution stage, when the total flux Γ_n of the negative ions can be directed even against their diffusive flux, i. e. the coefficient $D_n^{(eff)}$ becomes negative, also cannot be analyzed in this framework. Electric field given by Eq. (10.13a) at the second and third evolution stages for the case, when the positive ions are less mobile, corresponds to the Boltzmann's law for the negative ions, see Fig. 10.2. But for the case of similar problem with less mobile negative ions their Boltzmann's distribution is never reached. Situation, when in part of the plasma volume the Boltzmann's law for the negative ions holds, is described in the end of next Section; see Eq. (10.47), and Figs. 10.9, 10.10.

In [5] (see also[8]) expressions for $D_\alpha^{(eff)}$ were derived using assumption that the partial density profiles are similar. As we have seen, in the typical situations the regions exist, in which an initial inhomogeniety in plasma composition tends even to increase with time. This means that, even if the initial partial profiles were similar, this similarity is inevitably lost in the course of evolution process. Accordingly, in the general case such an approach is misleading.

10.2 POSITIVE COLUMN IN ELECTRONEGATIVE PLASMAS

Since the positive column is uniform in the current direction, the plasma density profile in it corresponds to the currentless case. Situation in the extremely electronegative plasma, when the self-consistent electric field does not satisfy Boltzmann law for electrons and is determined by the ion temperature, was analyzed in Section 4.2. It corresponds to negligible contribution of electrons to plasma conductivity and to weak electric field of the order of $T_i/(ea)$. In this case electric field has no influence on the electron motion, which is characterized by an extremely fast free electron diffusion. The electron density in such a regime is negligible, while ion motion practically corresponds to the ion-ion ambipolar diffusion.

We shall consider in this Section the more realistic case of a positive column which contains a substantial fraction of electrons. However, the overall electronegativity is assumed high enough, so that in the central plasma region density of the negative ions significantly exceeds n_e (see below Eqs. (10.38)).

For simplicity we restrict ourselves to the plane 1D problem with cold absorbing walls which are situated at $x=\pm L$. As in Section 4.2, the reduced longitudinal electric field (E_z/N) maintains the plasma and represents an eigenvalue of the problem. The ionization rate by the electron impact $\nu^{(ion)}$, as well as the attachment and recombination coefficients, depend on the form of the electron distribution as field- and composition independent constants. In reality, of course, the attachment, recombination, and especially the ionization rates depend on the form of electron distribution function, which is determined by (E_z/N). The detachment rate depends on the density of the exited neutral molecules. We shall neglect this fact and the relatively weak dependencies of attachment and recombination rates of (E_z/N), and consider that on the field (E_z/N) depends only $\nu^{(ion)}(E_z/N)$. Contrary to the case of pure plasma, when this dependence of (E_z/N) on the external parameters of a discharge (pressure, current, vessel size, etc.) is given by simple expressions Eqs. (4.19a,b), in the case considered the dependencies of (E_z/N) on these parameters turns out to be rather complicated. The problem is described by the system, which is close to Eq.(4.46):

$$
\begin{aligned}
&-\frac{d}{dx}\left(D_p\frac{dp}{dx}-b_pE_xp\right)=\nu^{(ion)}n_e-K^{(rec)}np;\\
&-\frac{d}{dx}\left(D_n\frac{dn}{dx}+b_pE_xn\right)=K_{att}n_e-K_{dt}n-K^{(rec)}np; \qquad (10.23)\\
&p=n+n_e.
\end{aligned}
$$

From the absence of net current in the x direction it follows that the field E_x is given by Eq. (10.3). In the following analysis we shall consider the ion mobilities as comparable.

Since the total number of ions in electronegative gases usually considerably exceeds the number of electrons, the contribution of ion-ion volume recombination in the overall plasma balance usually dominates over the recombination rate between positive ions and electrons. Accordingly, we have accounted only for the first process in Eqs. (10.23). As we are interested in the case when the electron fraction is considerable, the electric field which hinders the electron fast removal is reduced to electron Boltzmann distribution, Eq. (10.3). Since $T_i<<T_e$, strong electric field at the column periphery, where the electron density is low and the logarithm in Eq. (10.3) becomes large, sucks the negative ions into the plasma, and their own diffusion cannot oppose this field-driven flux. Negative ions in this regime

are trapped in the column's interior. Consequently, in the wall-adjacent peripheral plasma negative ions which appear due to the attachment reaction are driven quickly by this strong field. Since their density is small here, the detachment and recombination of negative ions are strongly suppressed at the plasma periphery, and the inward directed flux of negative ions is formed. Accordingly, their removal is determined by electron detachment and by ion-ion recombination in the central plasma region. In contrast, the outward directed flux of positive ions is formed in the central plasma and recombines at the walls.

If we neglect the sheath thickness and the ion mean free path with respect to the characteristic scales of the problem, the boundary conditions for the system Eqs. (10.23) are reduced to

$$n=p=\Gamma_n=0 \quad \text{at } x=\pm L; \qquad \frac{\partial n}{\partial x}=\frac{\partial p}{\partial x}=0 \quad \text{at } x=0. \tag{10.24}$$

If the ratio λ_{in}/L of the ion mean free path to the plasma scale L is not too small, the Bohm criterion, Eqs. (3.18, 3.26), gives more accurate results than Eq. (3.20); see also Section 3.9.

The system of second-order nonlinear Eqs. (10.23) with the field Eq. (10.3) and boundary conditions Eq. (10.24) seems quite simple, and was investigated numerically in numerous publications [3-20]. In the papers [5] - [8] the assumption of a constant electron/ion density ratio was made. As we have seen in Sections 4.2, 10.1, such an approach is reasonable only at extremely high electronegativity. In more realistic situations this density ratio is extremely nonuniform, and a more adequate approximation, as it was demonstrated in previous Section, consists in division of a column into two regions with different ion composition, separated by a sharp boundary [1]. The extreme diversity of the arising scenaria is greatly surprising. Depending on the values of the plasmachemical rates K_{att}, K_{dt}, $K^{(rec)}$ and on external discharge parameters, such as the vessel size, pressure and current, etc., strongly inhomogeneous plasma profiles arise. It is sometimes difficult to find out how their characteristics and the eigenvalue $\nu^{(ion)}(E_z/N)$ depend on these parameters. We shall present below several examples and a crude qualitative classification of the different possible situations.

Detachment dominated column In this case

$$K_{dt} >> K^{(rec)} p. \tag{10.25}$$

The removal both of electrons and positive ions is controlled by the diffusion towards the tube wall, and negative ions are trapped by the space charge field and disappear only due to the detachment in the central plasma.

Flux of negative ions is inward directed and equals zero both at the wall and at the tube center. Since the outflux of the negative ions is absent in this regime, the densities, spatially averaged over the column cross-section, satisfy

$$K_{att}\langle n_e\rangle = K_{dt}\langle n\rangle. \tag{10.26}$$

This regime with electric field which obeys electron Boltzmann relation, Eq. (10.3), arises, if the plasmachemical reaction rates satisfy inequality

$$K_{dt} > K_{dt}^{c} = 2K_{att}b_p/b_e, \tag{10.27}$$

which is inverse to Eq. (4.48). In this case two plasma regions with strikingly different properties exist (compare to Figs. 10.2-10.6). In the peripheral region the field, Eq. (10.3), is strong, and the density of negative ions is low because they are driven into the central plasma by the field. In this region the detachment is small, and the attachment process dominates. Flux of negative ions increases inward from zero at the wall. Neglecting the ion diffusion, detachment and recombination terms, and summing up Eqs. (10.23), we obtain a linear equation for the electron density in the peripheral plasma:

$$\frac{d^2 n_e}{dx^2} = -[\nu^{(ion)}/d_p + K_{att}/d_n]n_e = -\kappa^2 n_e, \tag{10.28}$$

where the effective ambipolar diffusion coefficients d_p, d_n are defined as in Eqs. (10.5, 10.6). The analogous region of the electron-ion plasma, where the negative ions are removed by the field Eq. (10.3), and their density is low, was also analyzed in the preceding Section (see Fig. 10.2, the right branches of the curves 1, 2, 3). The solution of Eq. (10.28), which satisfies zero boundary condition at the vessel walls $x = \pm L$, is

$$n_e \sim \sin \kappa(L - |x|). \tag{10.29}$$

From Eq. (10.29) it follows that at the distance

$$l = (\pi/2)\kappa^{-1/2} = (\pi/2)(\nu^{(ion)}/d_p + K_{att}/d_n)^{-1/2} \tag{10.30}$$

from the wall the electron density reaches maximum. Electric field Eq. (10.3) decreases towards this point. Nevertheless, the flux of negative ions Γ_n, which in this region of comparatively high field coincides with the field-driven flux $\Gamma_n = -b_n n E_x$, equals zero at the tube wall, and increases towards the plasma center. Since the negative ion density $n(x)$ is proportional to $\Gamma_n/(dn_e/dx)$, in this approximation it becomes singular at $L-x=l$. On the other hand, according to Eq. (10.26), the spatially averaged value of $n(x)$ can greatly exceed the electron density. Hence the central region is occupied by the ion-ion plasma with high density of the negative ions. In other words, in the vicinity of the point $L-x=l$ the transition occurs between the peripheral electron-ion plasma, where the flux Γ_n is generated, and the central ion-ion plasma, where the electron density is relatively low, and the removal of the negative ions takes place. Similarly to the preceding Section, the transition region between these two plasmas at not too low a pressure can be treated as a diffusive shock, in which n_e and the fluxes Γ_n, Γ_p are continuous.

The reason for such a sharp transition can be understood better if we consider the behaviour of these fluxes with x. Since there is no ion sources at the tube wall, and the density of the negative ions equals zero here, the flux Γ_n, as we have seen, increases from the plasma boundary towards its center due to the attachment. On the other hand, in the inner region the flux Γ_p increases outward from zero in the column center. If the ion diffusion is negligible both in the central and in the peripheral plasma, (it is not so only at comparatively low pressures, see below, Eq. (10.46)), in the central region with $n \approx p$

$$\Gamma_n/b_n = -\Gamma_p/b_p \tag{10.31}$$

approximately holds. In order to equalise the divergences of the fluxes Γ_n/b_n, Γ_p/b_p, the source terms in the right-hand sides of Eqs. (10.23) are to be equal. Since the right-hand sides of the balance equations for positive and negative ions, Eqs. (10.23), in the absence of the recombination are linear in densities, the expression for Γ_p in the field Eq. (10.3) is

$$\Gamma_p = -d_p \frac{p}{n_e}\frac{dn_e}{dx} = -d_p \frac{dp}{dx}.$$

Accordingly, in the central i-i region for positive ions we have the familiar ambipolar diffusion equation. The source term in this equation $\nu^{(ion)} n_e = \nu^{(ion)}(p/n_e)p$ is (p/n_e) times smaller, than in Eq. (4.18), since the major part of electrons disappears due to the attachment. On the other hand, the flux $|\Gamma_n|$ increases here with x. The main distinction with the ambipolar diffusion in the pure plasma consists in the fact that the solution of the ambipolar diffusion equation for $p(x)$ does not satisfy the zero boundary condition at the tube wall, but is to be matched with the solution Eq. (10.29). Depending on this matching, the diverse scenaria are possible. For example, the flat $p(x)$ profiles can arise even in the presence of plasma sources. The matching point $L-|x|=l$, Eq. (10.30), corresponds to relatively sharp change in $\Gamma_n(x)$ behaviour: the flux $\Gamma_n(x)$ has maximum in this point.

Fig. 10.7 presents the results of the numerical solution [12] of Eqs. (10.23) for a model gas in a cylindrical tube. Quite sharp boundary, in which variations of the densities $n(x)$, $p(x)$ are steep with respect to the neighbouring regions, is clearly seen. We can interpret it as shock. It separates the electron-ion peripheral plasma, where $n(x)$ is negligible and the flux $\Gamma_n(x)$ increases inward, and the ion-ion central plasma, where $\Gamma_n(x)$, $\Gamma_p(x)$ fall to zero in the discharge center. The position of this boundary, defined as the position of maximum of (dn/dx), (dp/dx), coincides with the maximum of the flux $\Gamma_n(x)$. This shock occurs at distance ~0.3a from the wall; it corresponds to the result of Eq. (10.37) (see below) $(\alpha)^{-1/2}a \approx 0.25a$. The value of $Z=7.8b_p/b_n$ in the calculations [12] is also close to the result of cylindrical analogy of Eq. (10.41), which gives the coefficient 5.8.

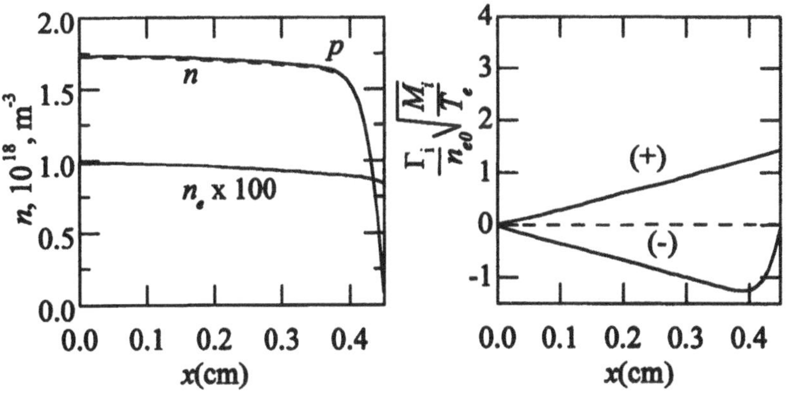

Figure 10.7 The variation of densities of electrons, of negative ions, and of the fluxes of the negative and of the positive ions in the calculations [19] for the recombination dominated plane column in Cl_2 at L=0.45cm, $n_{e0}=10^{10}$cm^{-3}, pressure 266Pa.

In order to find the plasma profile and to calculate the eigenvalue $\nu^{(ion)}(E_z/N)$, we note that in the inner region at $|x|<(L-l)$ Eq. (10.31) holds. This means that the partial densities in the inner plasma satisfy

$$(K_{att}n_e-D_{dt}n)/b_n=-\nu^{(ion)}n_e/b_p. \qquad (10.32)$$

Substituting Eq. (10.32) into the first of Eqs. (10.23), we obtain that the partial densities profiles in the central region are similar and are given by

$$\frac{d^2p}{dx^2}+\kappa_1 p=0,$$
$$\kappa_1^2=\frac{K_{dt}\nu^{(ion)}(E_z/N)}{d_p(K_{dt}+b_n\nu^{(ion)}(E_z/N)/b_p)}. \qquad (10.33)$$

The solution of Eq. (10.33) is

$$p(x)\sim n(x)\sim n_e(x)=n_{e0}\cos(\kappa_1 x). \qquad (10.34)$$

It is to be matched at $x=\pm(L-l)$ with the solution Eq. (10.29). The electron density $n_e(x)$ and the fluxes Γ_n, Γ_p are continuous at this point. These conditions are formulated in terms of the beforehand unknown values κ, κ_1, and determine the eigenvalue $\nu^{(ion)}(E_z/N)$. In order not to write cumbersome general expressions, we shall analyse different limiting cases.
The ingoing flux Γ_n, and the outgoing flux Γ_p at $x=\pm(L-l)$ satisfy Eq. (10.31):

$$2lK_{dt}n_e(x=L-l)/(\pi b_n)=\nu^{(ion)}(E_z/N)\int_0^{L-l}n_e(x)dx/b_p$$
$$=\frac{T_e}{e}\left(\frac{p}{n_e}\frac{dn_e}{dx}\right)\bigg|_{x=(L-l)}. \qquad (10.35)$$

If the attachment rate is large,

$$K_{att}>>d_n/L^2, \qquad (10.36)$$

the electrons in the peripheral region of the electron-ion plasma quickly generate the flux Γ_n. Accordingly, this region is relatively thin, see Eq. (10.30),

$$l\approx\pi L/2(K_{att}L^2/d_n)^{-1/2}<<L. \qquad (10.37)$$

Due to the integral balance of negative ions, Eq. (10.26), the ion densities in the central region satisfy

$$n\approx p\approx K_{att}n_e/K_{dt}>>n_e. \qquad (10.38)$$

The characteristic scale l, Eq. (10.37), can be interpreted as diffusive displacement of negative ions during the attachment time. It seems paradoxical that the criterion Eq. (10.36) has the form of the relation between the *electron* attachment probability and the *ion* diffusion lifetime. It follows from the fact, that due to quasineutrality the electron and ion currents Γ_e and $(\Gamma_p-\Gamma_n)$ (and their divergences) are to be equal. Thus, the condition $\mathrm{div}\Gamma_e=\mathrm{div}(\Gamma_p-\Gamma_n)$ is imposed both on the electron and ion characteristic times.

We can estimate from Eq. (10.35)

$$K_{att}n_{e1}l/b_n \sim \nu^{(ion)}n_{e0}(L-l)/b_p \sim \frac{T_e}{e}\left(\frac{p}{n_e}\frac{dn_e}{dx}\right)\Bigg|_{x=(L-l)} \sim \frac{T_e K_{att}(n_{e0}-n_{e1})}{eK_{dt}(L-l)}. \tag{10.39}$$

Here we denoted the densities in the electron-ion peripheral plasma by subscript 1, and the central values by subscript 0, and accounted for the fact that the electron density is continuous in the whole plasma volume. Using Eqs. (10.36), (10.37), we obtain that if

$$K_{dt}(L^2/(d_nK_{att}))^{1/2}>1, \tag{10.40}$$

the central plasma profile is convex, ($\kappa_1 L\approx\pi/2$), the electron density at the shock position is small, $n_{e0}/n_{e1}\sim K_{dt}(L^2/(d_nK_{att}))^{1/2}>1$. The ionization frequency is

$$\nu^{(ion)}=\frac{\pi^2 d_p K_{att}}{4L^2 K_{dt}}. \tag{10.41}$$

If at constant attachment, which satisfies Eq. (10.36), detachment decreases, then the width of the peripheral plasma l, Eq. (10.37), remains constant and small, the central plasma becomes more and more electronegative and its profile is flattened. If the inequality inverse to Eq. (10.40) holds, the central plasma profile becomes flat, and the ionization rate equals

$$\nu^{(ion)}=\frac{\pi b_p}{2b_n}(K_{att}d_n/L^2)^{1/2}. \tag{10.42}$$

If the condition inverse to Eq. (10.36) holds, the peripheral region of the pure (electron-ion) plasma occupies practically the whole column cross-section. In this situation the condition of discharge maintenance coincides with the condition for the pure plasma, Eq. (4.19),

$$\nu^{(ion)} = \frac{\pi^2 d_p}{L^2} \tag{10.43}$$

both in the cases of the detachment-dominated and of the recombination-dominated plasma core. It follows from the fact that the main part of the ionization, as well, as of the diffusion, takes place in the pure electron-ion plasma, and the processes in the relatively small ion-ion central core give negligible contribution in the overall balance of the positive ions.

Substituting the calculated expressions for $\nu^{(ion)}$ and using the shock condition Eq. (10.31) together with the overall balance of negative particles, Eq. (10.26), with the partial density profiles Eqs. (10.29) and (10.34) it is easy to find the coefficients in Eqs. (10.29), (10.34) and the ion density jump at the shock.

The width l_1 of the central ion-ion plasma core is

$$l_1 = 2LK_{att}\tau_{an}D / \pi .$$

The central density of the negative ions is

$$n_0 = n_{e0}(K_{dt}\tau_{an}D)^{-1} .$$

Different possible regimes are shown in a schematic map Fig. 10.8.

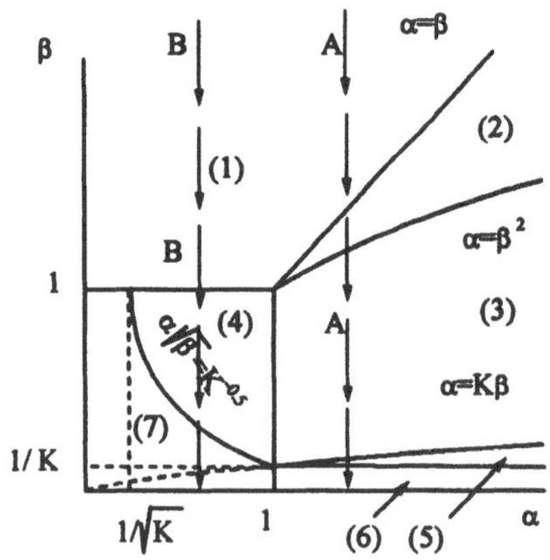

Figure 10.8. Map of the dimensionless parameters α, β, Eq. (10.43), which classifies the different column density profiles.

Below we shall normalize all the densities to the electron density in the discharge center; $N=n_e/n_{e0}$, the coordinate to half of the discharge gap L, and all the plasmachemical rates to inverse ambipolar lifetime of the negative ions

$$\tau_{an}=L^2T_i/(D_nT_e)=L^2/d_n;\quad K=T_e/T_i>>1;\quad D=D_p/D_n\sim 1;.$$

$$\alpha=K_{att}\tau_{an};\quad \beta=K_{dt}\tau_{an};\quad Z=\nu^{(ion)}\tau_{an};\quad \gamma=K^{(rec)}n_e^0\tau_{an}. \tag{10.44}$$

In these variables Eqs. system (10.23) takes the simple dimensionless form:

$$\begin{aligned} &D[-(n+N)'/K-(n+N)N'/N]'=ZN;\\ &[-n'/K+nN'/N]'=\alpha N-\beta n. \end{aligned} \tag{10.45}$$

For $\beta>\max[\alpha,1]$ (region 1 in Fig. 10.8) negative ion density is small in the whole plasma, and their influence on the plasma parameters is unimportant; $\nu^{(ion)}$ is given by Eq. (10.43). In the interval $\beta^2>\alpha>\beta$ (region 2) all the partial profiles are convex, and the electron density in the central plasma greatly exceeds its value in the peripheral pure plasma; $\nu^{(ion)}$ is given by Eq. (10.41). Since the ion densities undergo a rapid change (shock), their central densities far exceed the density of positive ions at the column periphery. For $\alpha>\beta^2,1$ (region 3) the ionization rate is given by Eq. (10.42), and all the density profiles become flat. In the region 4 (α, β <1) the ionization frequency is given by Eq. (10.43), but the narrow peak with $n\approx p>>n_e$ is formed in the plasma center.

The ion diffusion plays an important role primarily in the shock region. It can be considered as thin only if its width, which is of the order of

$$L_f\sim L(K\beta)^{-1/2} \tag{10.46}$$

(of the negative ion displacement during its lifetime) is small both with respect to the central and peripheral regions. With the increase of the ion diffusion the shock width increases primarily towards the plasma center, where the field, Eq. (10.3), is weak. If L_j exceeds the central ion-ion plasma core, the removal of the trapped negative ions (detachment) occurs slowly with respect to their diffusive displacement, and their density in all the central ion-ion region obeys ion Boltzmann's law Eq. (10.22).

For $\alpha>1$ the peripheral electron-ion plasma is thin, $l\sim L/\sqrt{\alpha}$, Eq. (10.37), and the more restrictive criterion for the presented above scenario to be valid demands the shock width, Eq. (10.45), to be small with respect to this length. In the opposite case ($\alpha>\beta K$, region 5 in Fig. 10.8) the main part of the flux Γ_n, which flows into the central plasma, is generated in this transition region which is controlled by the ion diffusion. It becomes impossible to treat it as a thin shock; in this situation $Z \sim \alpha / \sqrt{K\beta}$.

If this region occupies the whole cross-section of the central ion-ion plasma (at $\beta < K^{-1}$, $\alpha > 1$, or at $K\alpha^2\beta > 1$, $\alpha < 1$, regions 6,7), the whole central plasma is dominated by the ion diffusion. The electron density profile remains flat, and the ion profiles become convex. Summing Eqs. (10.45) we have (since in the central plasma $n >> N \approx 1$) that the ion profiles are parabolic: $n(x) \approx p(x) = ZK(1-x^2)/4$. Integration the first of Eqs. (10.45) over the discharge gap gives $Z = 6\alpha/(\beta K)$ in region 6; $Z = \pi^2/4$ in the region 7. Since both the detachment probability K_{dt} and the diffusive lifetime τ_{an} are, roughly speaking, proportional to the gas pressure p, the criterion for this regime

$$1 >> K\beta = (T_e/T_i)K_{dt}\tau_{an} \sim p^2 \tag{10.47}$$

is fulfilled at low pressure.

At such a low pressure the zero boundary condition Eq. (10.24) becomes less accurate, and is to be replaced by the Bohm criterion, Eqs. (3.18), (3.26). Since collisions are not too frequent, the situation is possible when the electronegative ion-ion plasma with the flat profile $n_e(x)$ and low space-charge field of the scale of T_i in the central plasma volume is separated from the peripheral electron-ion plasma by the collisionless double sheath (see Section 3.9, [18] and [19]). At high electronegativity the ion-ion plasma core spreads up to the wall-adjacent sheath. The strong electric field of the scale of T_e which prevents the electron escape to the vessel wall remains only in this sheath [18], [19].

In order to get a rough idea, how the variation of the plasma properties manifests itself in the density profiles, let us consider the hypothetical situation of variation of single parameter β at fixed α (along the arrows A-A, B-B in Fig. 10.8).

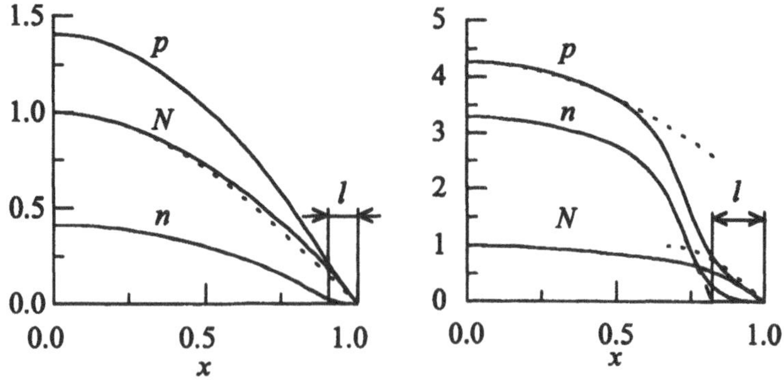

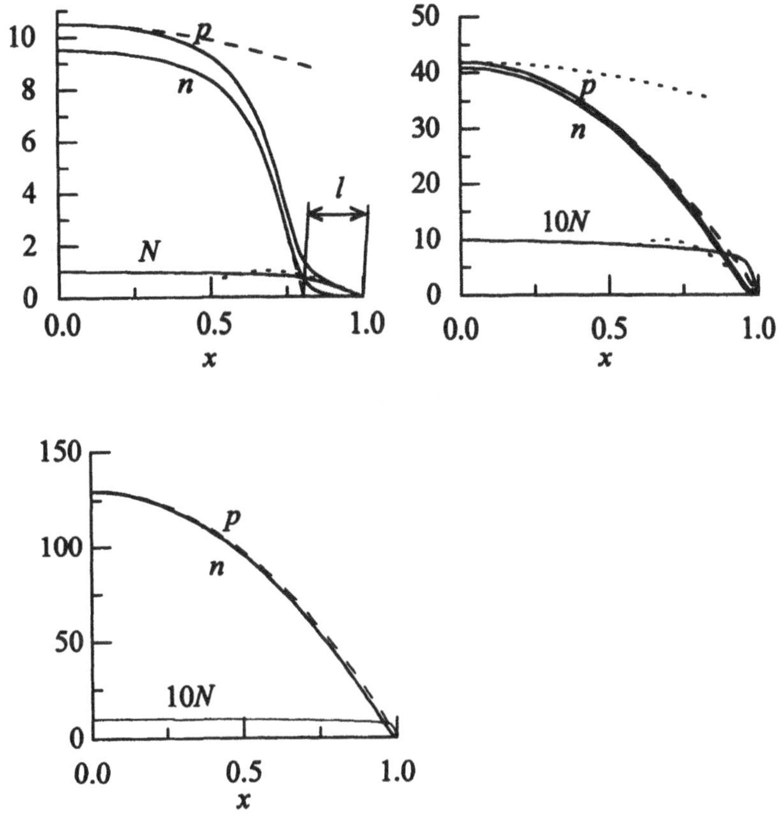

Figure 10.9. The partial density profiles in the *dc* positive column for α=20, D=1, β=50(*a*); 7(*b*); 2.5(*c*); 0.5(*d*); 0.05(*e*). The values of $K=T_e/T_i$ are 100 (*a, b, c*), and 10 (*d, e*). The dotted lines for $p(x)$ in *b, c, d, e* correspond to the profile $p(x)\sim\cos x\sqrt{Z\beta/(\alpha+Z)}$, Eq. (10.33), without ion diffusion. The dotted line in *a* corresponds to the ambipolar profile $N(x)$. The dotted electron density profiles in *b, c, d* represent the profiles Eq. (10.29). The dashed lines in *d, e* correspond to the diffusive profile $p(x)\sim(1-x^2)$.

The case *a* in Fig. 10.9 corresponds to a situation when negative ions, which comprise a small admixture to the pure electron-ion plasma, are piled up by the ambipolar field to the plasma center.

The value of Z=3.12 in this case (as well, as 2.59 in the Fig. 10.10*a*, and 2.5 in Fig. 10.10*b, c* is close to value $\pi^2/4$ for the pure plasma. In the cases Fig. 10.9*a,b, c, d* the extrapolated $n(x)$ profile tends to zero at small distances l=0.1(*a*), 0.18(*b*), 0.2(*c*) and 0.03(*d*) from the wall; the l value varies slowly with β. In the cases *b,c* this distance is of the order of the value $\pi/(2\sqrt{\alpha}) = 0.35$ which follows from Eq. (10.37). The smaller values of l in Figs.10.9*d,e* can be attributed to the ion diffusion influence. Fig. 10.9*b* corresponds to Eq. (10.40); all the profiles $n(x)$, $p(x)$, $N(x)$ are convex and close to similar in the central plasma; the calculated value of Z=4.3 is of the order of 7.1, which follows from the estimate Eq. (10.41). The central density profiles in Fig. 10.9*c*, which corresponds to the region 3, Fig. 10.8, are flat; Eq. (10.42) gives Z=7.02 instead of the calculated value 4.7. The scale of the transition regions in these cases (see also Fig. 10.10*b*) is considerably smaller than the scales of the central and peripheral plasmas, this can be interpreted as shock. The case *d* corresponds to the region 5; the electron profile remains flat, and the $n(x)$, $p(x)$ profiles become considerably convex due to the ion diffusion. The expression $Z = \sqrt{K\beta}$ gives 7 instead of the calculated 5.1. The shock thickness in this case is of the order of $\sqrt{K\beta} \sim 0.14$, Eq. (10.37). In the case Fig. 10.9*e* the peripheral electron-ion plasma is practically absent, and the ion profiles are parabolic; the estimate gives very high value $Z \sim 6\alpha/(K\beta)$=240 (the calculated value of Z is 43). In other words, the monotone decrease of β along A-A (Fig. 10.8) results in the monotone flattening of the electron density profile and to monotone decrease of the peripheral electron-ion plasma region. The ion profiles behave in a more complicated manner: they are peaked at large β, are flat at moderate β and become peaked again at small β.

Figs. 10.10*a*, *b*, *c* all correspond to strongly peaked $n(x)$ profiles. In the case *c* it is parabolic (determined by the ion diffusion) and in the case *b* it corresponds to Eq. (10.33) with shock.

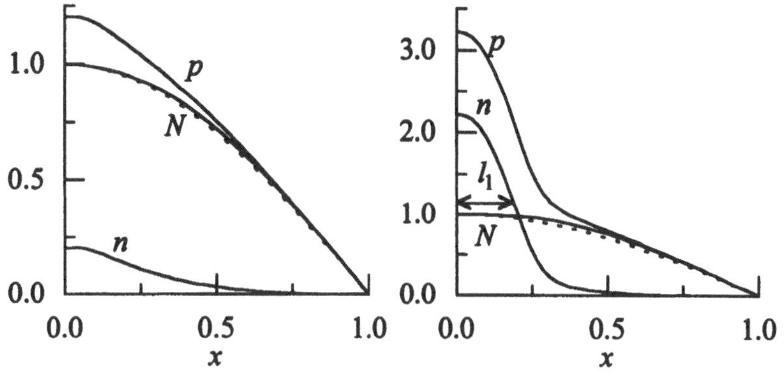

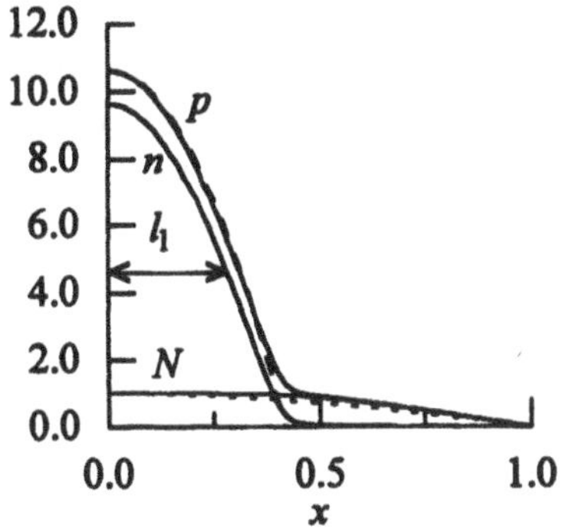

Figure 10.10 The partial density profiles for low detachment, α=0.3, D=1, β=3.0 (*a*); 0.3 (*b*); 0.03 (*c*). The dashed lines correspond to the diffusive parabolic profiles. The dotted line in Fig. *a* represents the ambipolar profile.

The length of the ion-ion plasma core l_1 for Fig 10.10*b* can be estimated as the coordinate of the point where Eq. (10.31) holds:

$$Z\int_0^{l_1} N(x)dx = l_1 Z = l_1\pi^2/4 = \int_{l_1}^{1}\alpha N(x)dx = 2\alpha/\pi. \qquad (10.48)$$

The resulting value l_1=0.12 roughly corresponds to value 0.2, Fig. 10.10*b*. Since the ion-ion plasma core in Fig. 10.10*c* is dominated by the ion diffusion, the profiles $n(x)$, $p(x)$ are parabolic here. These profiles can be hardly distinguished from the parabolic ones in Fig. 10.10*c*.

The shock-like behaviour can be clearly seen in the Figs. 10.9*b*, *c*, 10.10*b*. Shock position roughly corresponds to the point, where extrapolation of the electron profile, Eq. (10.29), tends to maximum and the electric field decreases fast.

Recombination dominated column. In this case, instead of Eqs. (10.45), we have

$$\begin{aligned}&[-n'/K + nN'/N]' = \alpha N - \gamma np,\\ &[-(n+N)'/K - (n+N)N'/N]' = ZN/D - \gamma np/D.\end{aligned} \qquad (10.49)$$

Instead of Eq. (10.46), the characteristic scale of the ion diffusion is given by

$$L_j = L(K\gamma n_0)^{-1/2}. \qquad (10.50)$$

If this scale is small, shock arises between the central ion-ion and peripheral electron-ion plasmas. Neglecting the recombination in the peripheral plasma, the summation of Eqs. (10.49), in the absence of the ion diffusion, results in the solution Eq. (10.29). Accordingly, the peripheral region is thin (of the order of $\alpha^{-1/2}$) at $\alpha \gg 1$, Eq. (10.36), and occupies almost all discharge gap in the opposite case. Instead of Eq. (10.33), in the central plasma holds

$$-N'' = (\alpha + Z/D)N - \gamma np/D. \tag{10.51}$$

We shall consider first the case $\alpha \gg 1$. Since in the central region the left-hand side of Eq. (10.51), representing the ion transport, is small (it can not exceed unity), neglecting it, and having in mind that in this region at high overall electronegativity $n \approx p$, from Eq. (10.31) it follows the local connection between the species densities

$$N = n^2 \frac{\gamma(D+1)}{\alpha D + Z}. \tag{10.52}$$

Substitution of Eq. (10.52) into Eq. (10.49) yields

$$n'' + \frac{Z - \alpha}{2(Z + \alpha D)} \gamma n^2 = 0. \tag{10.53}$$

This equation coincides with the equation of motion of nonlinear oscillator. The density profile of the central ion-ion plasma is determined by

$$x = \int_{n_0}^{n(x)} \frac{dn'}{\sqrt{\frac{(Z-\alpha)\gamma(n_0^3 - n'^3)}{3(Z + \alpha D)}}} \tag{10.54}$$

with the central density

$$n_0 = \sqrt{\frac{(Z + \alpha D)}{(1 + D)\gamma}}. \tag{10.55}$$

Equations for the ionization frequency Z, plasma density at the shock n_1 and central density $n_0 \approx p_0$ can be obtained as follows. If the central density profile is flat, since the electron density is continuous at the shock, the solution Eq. (10.29) is

$$N(x) = \sin \kappa x.$$

From Eq. (10.54) at the shock boundary we have

$$\delta = 1 - n_1 / n_0 = \frac{\gamma n_0 (Z - \alpha)}{4(Z + \alpha D)}. \tag{10.56}$$

The condition Eq. (10.31) at the shock results in

$$\begin{aligned} |\Gamma_n| &= \sqrt{\frac{\alpha D}{Z/\alpha + D}} = \Gamma_p / D = \left|\frac{n_0 N'}{DN}\right| \\ &= |2n'/D| = 2n_0 / D \sqrt{\frac{\gamma n_0 \delta (Z - \alpha)}{Z + \alpha D}}. \end{aligned} \tag{10.57}$$

Combining Eqs. (10.55) - (10.57), we find

$$\begin{aligned} &Z = \alpha + \sqrt{\alpha D^3 (D+1)}; \\ &n_0 = (\alpha / \gamma)^{1/2}; \\ &\delta = 1/4(\gamma D^3 / (D+1))^{1/2}. \end{aligned} \tag{10.58}$$

It follows from Eq. (10.58) that the flat profiles in the plasma center correspond to $\gamma \ll 1$. In the opposite case, in order to find these profiles it is possible to impose the zero boundary condition at $x=1-l$ in Eq. (10.54) for $n(x)$, and to use the net balance of the negative ions

$$\int_0^1 (\gamma n p - \alpha N) dx = 0 .$$

With accuracy up to numerical factor we obtain

$$Z \sim \alpha; \qquad n_0 \sim (\alpha / \gamma)^{1/2}. \tag{10.59}$$

For $\alpha \ll 1$ the width of the central ion-ion region is of the order of α. In this case $Z \approx 1$, $n_0 \sim \gamma^{-1/2}$.

Increase of the ion diffusion results in the shock expansion in both directions: to the tube wall and towards the column center. Since in the central plasma the electric field is weak, shock expands predominantly in the second direction. If its width, Eq. (10.50), exceeds the central ion-ion core, in this region the parabolic diffusive profile emerges.

10.3 CAPACITIVELY COUPLED RADIO FREQUENCY (RFC) DISCHARGE IN ELECTRONEGATIVE GAS

Microwave and radio frequency discharges are widely utilised in plasma technology for deposition and etching semiconductor coatings. Such discharges with negative ions are described by many parameters and, consequently, a great variety of situations are possible. Due to the practical importance of these discharges, they have been treated analytically [21] and numerically [22-27] in numerous publications. The qualitatively new feature of these discharges consists in the fact that dynamics of the RF sheath is more complicated, and the ion motion here is determined not by the Boltzmann's field, Eq. (10.3), but by the far more strong space charge field (see Section 3.8). We shall investigate here the most typical cases and qualitative effects, according to [21]. We shall consider the RFC discharge at moderate pressures when the electron distribution function is controlled by the local oscillatory electric field E, which determines the ionization frequency $\nu^{(ion)}(E/p)$. We restrict ourselves, for simplicity, to the case, when the mobilities of negative and positive ions are equal, $b_p=b_n=b_i$.

In the previous Section we have focused on the influence of negative ions on the plasma profiles. As we shall see below, their effect on the sheath properties can be even more pronounced. The sheath thickness in the RFC discharges (see Section 3.8) usually considerably exceeds the Debye radius, which determines its characteristic scale in the *dc* case. As well, as in the preceding Section, the choice of scenaria depends crucially on the electron attachment frequency to the neutral species. We assume that the discharge frequency ω satisfies the same conditions, as in Section 3.8. In this case the electron motion is determined by an instantaneous RF electric field, and ions follow the field $\langle E \rangle$ which is averaged over the fast RF oscillations.

The time interval τ_i which characterises the ion drift through the sheath (with a thickness L) can easily be estimated from the Poisson equation and is equal to the Maxwellian time

$$\tau_i^{-1} \sim \frac{b_i < E >}{L} \sim \frac{b_i < E >}{< E > /(\varepsilon_0 e n_e)} = \varepsilon_0 e b_i n_e. \qquad (10.60)$$

Since the plasma density profiles in electronegative plasmas are, as we have seen in the preceding Sections, extremely nonuniform, the value of n_e in the sheath can differ significantly from the central or average plasma density. As we shall see below, depending on the relation between the attachment and the partial Maxwellian times, the scenaria of the transport processes in the RF sheath are strikingly different.

It was shown in Section 3.8, that at every moment the RF sheath can be subdivided into the following two regions: first, the plasma phase region, where the quasi-neutrality condition $p \approx n+n_e$, is valid, and, second, the region of the ion volume charge, where $n_e=0$. According to the quasineutrality condition, in the plasma phase $n_e=p-n$. In the plasma phase the electron conductivity current dominates, and during the space charge phase the ion current is negligible, and the current is transported in the form of the displacement current. Ion densities in the RF sheath slightly vary in time. These perturbations δp, δn produce the perturbation in the electron density δn_e during the plasma phase,

$$\frac{\delta n_e}{n_e}=\frac{\delta p-\delta n}{n_e}=\frac{\delta(p-n)}{p}\frac{p}{n_e}. \tag{10.61}$$

In strongly electronegative case δn_e can be comparable to the electron density itself, even if the relative perturbations of the ion densities are small [24]. Below we shall see that in most cases these ion density perturbations can be ignored. Therefore we shall consider the quantity $n_e=n_e(x)=p(x)-n(x)$ as time independent during the plasma phase of the sheath.

The basic equations. The assumptions mentioned substantially simplify the basic equations, and allow to formulate them in the form close to Eqs. (3.92), (10.23). The resulting system is reduced to time-averaged ion equations and to the Poisson equation. The ion equations are

$$\begin{gathered}\frac{d\Gamma_p}{dx}=\frac{d}{dx}\left(Vp-Dp\left\langle\frac{dn_e}{n_e dx}\right\rangle\right)=\left\langle\nu^{(ion)}n_e\right\rangle-K^{(rec)}np,\\ \frac{d\Gamma_n}{dx}=\frac{d}{dx}\left(-Vn-Dn\left\langle\frac{dn_e}{n_e dx}\right\rangle\right)\\ =\left\langle K_{att}n_e\right\rangle-K^{(rec)}np-K_{dt}n,\end{gathered} \tag{10.62}$$

where $V=b_i<E>$ is the time-averaged ion velocity, $D_a=b_iT_e/e$ is the ambipolar diffusion coefficient. For the sake of simplicity, we consider moderately high pressures, when microwave diffusion (Section 3.9) can be neglected, and ignore the ion diffusion because $T_e>>T_i$. We shall consider the case of the recombination-dominated discharge, when far from electrodes a uniform positive column exists with the ion densities p^c, n^c. Accordingly, we shall neglect the detachment $K_{dt}<<K^{(rec)}p$. It corresponds

to large density of negative ions and is typical for highly electronegative gases like SF_6 and Cl_2 [24].

The boundary conditions for Eqs. (10.62) correspond to $\Gamma_p=\Gamma_n=0$ in the discharge center, due to symmetry, and to $\Gamma_n=0$ at the electrode surface. From the second boundary condition it follows that the integral of the right-hand side of the second Eq. (10.62) should be zero over the discharge period.

In the space-charge phase the displacement current transports the current in the sheath, and the arising field is shielded by the positive ion charge. If the current density depends on time as $j(t)=-j_0\sin\omega t$, the field in the sheath is determined by the Poisson equation:

$$E(x,t)=\frac{\varepsilon_0 j_0}{\omega}(\cos\omega t-\cos z(x)), \tag{10.63}$$

where $z(x)$ is the current phase corresponding to passage of a given point x by the sharp plasma-sheath boundary. At a certain moment electric field at the electrode should be small, in order to transport electrons to the electrode and to neutralize the current of positive ions (the current of negative ions at electrode is zero). Consequently, the phase $z=\pi$ corresponds to the electrode. Using z, and taking into account that the quasi-neutrality condition holds in the plasma phase, we can write the Poisson equation in the form

$$\sin z\frac{dz}{dx}=e\omega(p(x)-n(x))/j_0. \tag{10.63a}$$

The boundary conditions are: $x=L$ and $z=\pi$ at the electrode and $x=0$ and $z=0$ at the sheath boundary. Using the phase z, we can easily average Eq. (10.63) over time :

$$<E>(x)=\frac{4j_0}{\omega}(\sin z-z\cos z). \tag{10.64}$$

In strongly electronegative gases the ion densities p^c and n^c in the plasma column are much larger than the electron density n_e^c ($K_{att} >> K^{(rec)}n^c$). The plasma region and/or part of the sheath, in which $n >> n_e$, will be referred to as the ion-ion region (IIR), and the regions, in which $n\le n_e$, will be referred to as the electron-ion region (EIR). In the sheath negative ions drift from the electrode towards plasma. Accordingly, near the electrode practically always there should exist a region with a low negative ion density, an EIR. Below we shall see that the intermediate region, where $n \sim n_e$, is fairly narrow, and jumps in n and p arise there.

The structure of the electrode sheath. Depending on the discharge parameters, the sheath consists only of EIR or of both regions, IIR and EIR. This depends on the attachment frequency. The IIR arises in the sheath, if the time of ion motion through the EIR, which is of the order of $\tau_{i1} = (eb_i n_e^c / \varepsilon_0)^{-1}$, exceeds the time K_{att}^{-1} of the ion and electron densities relaxation up to the equilibrium densities p^c, n^c, n_e^c in the column. As in the preceding Sections, from the quasineutrality condition it follows that the *electron* attachment time K_{att}^{-1} controls the establishment of the *ion* profile. Below we shall see that the electron density in the sheath can vary significantly from the value of the order of n_e^c up to the value $\sim p^c$. As a result, the quantity τ_i also varies from $\tau_{i1} = (eb_i n_e^c / \varepsilon_0)^{-1}$ up to $\tau_{i2} = (eb_i p^c / \varepsilon_0)^{-1}$. Hence, the sheath structure, as in the preceding Section, depends on values of the products $K_{att}\tau_{i1}$ and $K_{att}\tau_{i2}$.

In the case of a low attachment frequency,

$$K_{att}\tau_{i1} \ll 1, \tag{10.65}$$

the negative ion density is low over the entire sheath, and the theory developed in Section 3.8 for the pure two-component plasma can be applied. The quantity $\nu^{(ion)}(\tilde{E})$ depends strongly on the electric field in the plasma phase $\tilde{E}$, which, due to the current conservation, is determined by the electron density here, $n_e(x)=p(x)-n(x)$. This implies that the field $\tilde{E}$ and the electron density vary slightly, that is, $\tilde{E} \approx \tilde{E}^c = j_0 / (eb_e n_e^c); n_e \approx n_e^c$. The quasineutrality condition yields $p \approx n_e \approx n_e^c$. The flux of the negative ions in the sheath is generated by attachment in its part, which is adjacent to electrode:

$$\begin{aligned} n(x) &= \int_x^L \langle K_{att} n_e \rangle dx / V(x) \\ &\sim K_{att} n_e^c L_{sh} / (b_i \langle E \rangle) \sim K_{att}\tau_{i1} n_e^c \ll n_e^c \end{aligned} \tag{10.66}$$

is low almost everywhere except for the close vicinity of the plasma-sheath boundary.

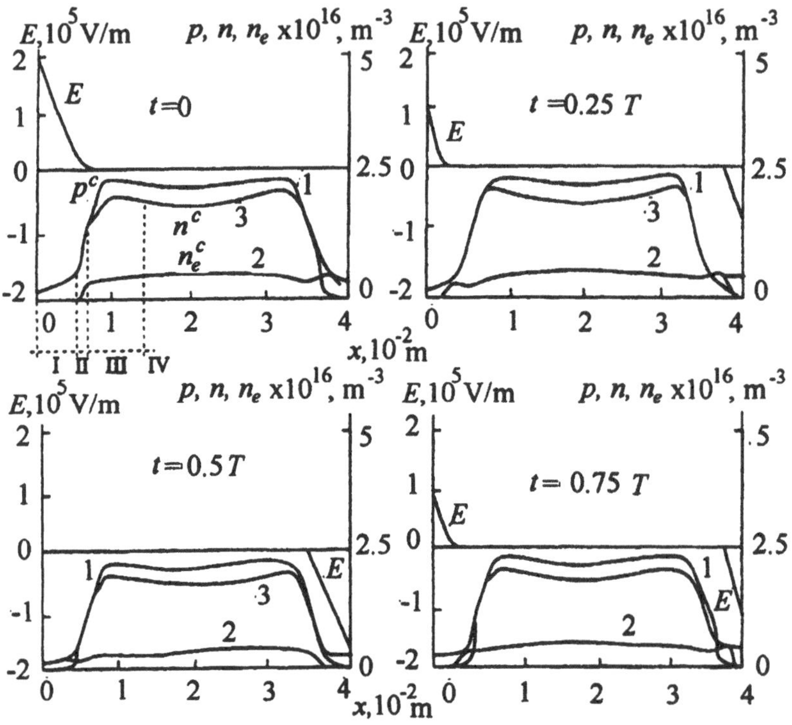

Figure 10.11. The spatial profiles of the electric field and charged particle densities (1 - p; 2 - n_e; 3 - n) at four moments during the *RF* period in the simulation [21]. The discharge parameters at frequency 10 MHz for the model electronegative gas based on helium corresponded to Eq. (10.39); $K_{att}\tau_{i1}$=0.08. In the simulations clearly seen are: *I* - the sheath region, which part of the *RF* period exists in the quasineutral plasma phase, and remaining part of it in the phase of positive ion space charge; *II* - the region of the shock-like transition from *EIR* to *IIR*; *III* - the nonuniform plasma region between the sheath *I-II* and the uniform column *IY*.

The negative ion density grows sharply at this boundary to the plasma. The sheath properties turn out to be similar to those for the discharge in electropositive gases (see the table). For example, the thickness of the sheath is $L_{sh} \sim 2b_e\tilde{E}^c/\omega$. This situation was studied in numerical calculations [23]. Figure 10.11 illustrates the results of this paper.
The parameter $K_{att}\tau_{i1}$ was equal to 0.08. The electron density in the sheath during the plasma phase is shown to be close to n_e^c. Correspondingly, for a 0.6cm thick sheath, an analytical estimate ($2b_e\tilde{E}^c/\omega$.) gives fairly good agreement, specifically, 0.45 cm.

If the electron attachment frequency is so high that the restricting inequality

$$K_{att}\tau_{t2} >> 1 \tag{10.67}$$

holds, the situation is also simple. In this case almost everywhere the plasmachemical processes dominate over the ion transport. The IIR with p, n, n_e close to p^c, n^c, n_e^c occupies practically the entire sheath. In a small region near electrode with a thickness of about $V(L)/K_{att} << L$, the EIR arises, in which negative ion density is low. As in the preceding Section (see Eq. (10.62)), its thickness is determined by displacement of negative ions during the electron attachment time. The quasineutrality condition in the plasma phase implies that the electron density in the EIR is of the order of $n_e \approx p^c$ (see Fig. 10.12).

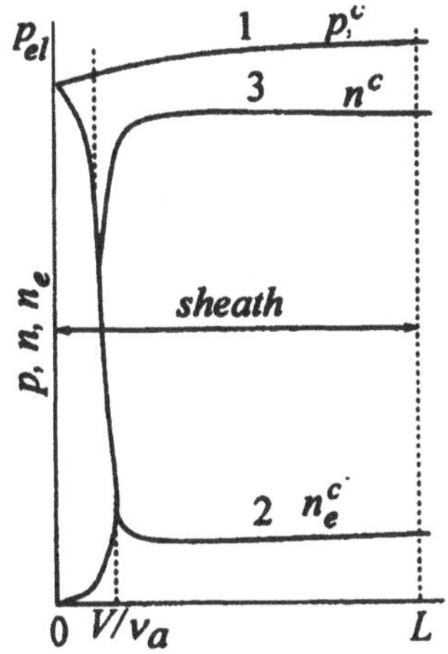

Figure 10.12. Sketch of the charged-particle density profiles in the strongly electronegative case $K_{att}\tau_{t2}>>1$.

The electron density decreases with the negative ion density growth towards the plasma, where the value of electron density is much smaller: $n_e \approx n_e^c << p^c$. In spite of the large ion charge in the EIR, an integral volume charge in this region during the space charge phase, when electrons are absent, $\int(p-n)dx$, is small (of the order of $p_c V(L)/K_{att} << n_e^C L_{sh}$) due to inequality (10.67). Hence, the field is only slightly shielded in the EIR. In the EIR, the positive ion density also changes insignificantly owing to the slow variation of both the ion flux and ion velocity. The main sheath parameters are listed in the table.

In the intermediate case,

$$\tau_{i1}^{-1} << K_{att} << \tau_{i2}^{-1}, \tag{10.68}$$

the ratio p^c / n_e^c can be very large (of the order of 100 - 1000 in the strongly electronegative gases, for example, in Cl_2 and SF_6). Hence, the condition (10.68) is typical for the discharges in such gases [22, 24 – 26]. In the cases described above, the EIR occupied either the entire sheath, or a small part of the sheath. Numerical simulations carried out in [24] for the intermediate case show an almost exact coincidence of the EIR with the sheath. The boundary between the EIR and IIR practically coincides with the plasma-sheath one. The positive ion density near the electrode p_{el} and in the sheath turns out to be such that the time of the ion drift through the sheath $\tau_i = (eb_i p^c / \varepsilon_0)^{-1}$ is comparable to K_{att}^{-1} and the condition $p_{el} \sim K_{att}\tau_{i2} p^c < p^c$ is satisfied. This effect stems from the strong coupling of the Poisson equation to the ion equations. We shall prove that the transition from the EIR to the IIR is really possible only near the plasma-sheath boundary. Using *reductio ad absurdum,* we propose that this boundary lies somewhere inside the sheath.

The left inequality (10.68) means that the EIR arises inside the sheath, i.e. negative ion density cannot be small everywhere in it. Indeed, in the opposite case, the sheath structure would be similar to the case of low attachment, Eq. (10.65): the electron density is about n_e^c, the ion drift time through the sheath is determined by the quantity τ_{i1}. This time is larger than the time K_{att}^{-1}. Accordingly, the density of the negative ions grows fast inside the sheath, and becomes $> n_e$. This contradicts the basic assumption.

On the other hand, the right inequality Eq. (10.68) shows that the positive ion density p_{el} near the electrode and in the main sheath region should be much less than p^c. In the EIR, we have $n_e \cong p >> n_e^c$ (see below), and the ionization is exponentially low. Consequently, the flux Γ_p here varies insignificantly. In the EIR with a thickness of $\sim V K_{att}^{-1}$ the average ion space charge can be estimated as

$$e \int_{(EIR)} (p - n)dx \sim ep_{pl}V / K_{att} \sim ep_{el} b_i E_{\max} / K_{att}, \tag{10.69}$$

where E_{max} is the maximum field value at the electrode. Since the electric field is screened by entire space charge in the sheath, and we have assumed that EIR is a part of the sheath,

$$e \int_{(EIR)} (p-n)dx < \varepsilon_0 E_{max}. \tag{10.70}$$

Comparing Eqs. (10.68) - (10.70), we obtain

$$p_{el} / p^c < eb_i p^c / (\varepsilon_0 K_{att}) = (K_{att} \tau_{i2})^{-1} \ll 1. \tag{10.71}$$

So we see that positive ion density at electrode is bounded: $n_e^c < p_{el} < p^c / (K_{att} \tau_{i2})$. In [21] it was shown that $p_{el} \approx p^c / (K_{att} \tau_{i2})$, and the sheath practically coincides with the EIR.

Now all characteristics of the sheath can be calculated. Summing Eqs. (10.62), we find

$$\frac{d}{dx}(\Gamma_p + \Gamma_n) = \left\langle \left(K_{att} + \nu^{(ion)}\right) n_e \right\rangle - 2K^{(rec)} pn, \tag{10.72}$$

where diffusion was neglected in the sheath region with respect to convection. From equation (10.72) we can find the exact value of p_{el} and the profiles of $p(x)$ and $n_e(x)$. Expressing the derivatives with respect to the coordinate through the derivatives with respect to z, Eq. (10.63a), and using the Poisson equation, we obtain

$$\frac{d}{dx} V(p-n) = \frac{j_0 \sin z}{e(p-n)\omega} \left[\left\langle \left(K_{att} + \nu^{(ion)}\right) n_e \right\rangle - 2K^{(rec)} pn \right]. \tag{10.73}$$

In the EIR the last terms in Eq. (10.73) are small, and the flux of positive ions to the electrode is determined by the integration of the first term in the square brackets:

$$\Gamma^* \equiv \Gamma_p(L) = p_{el} V(L) = \frac{j}{e\omega} \int_0^{\pi} \frac{\langle K_{att} n_e \rangle}{p-n} \sin z dz \sim \frac{K_{att} j_0}{e\omega}. \tag{10.74}$$

Using (10.64), we substitute the expression for the velocity at the electrode in formula (10.74), and find the quantity p_{el}:

$$p_{el} = K_{att} / (\varepsilon_0 e b_i). \tag{10.75}$$

The table lists the results of integration of Eqs. (10.73) and (10.74). In Fig. 10.13 the results of numerical simulations [24] are shown.

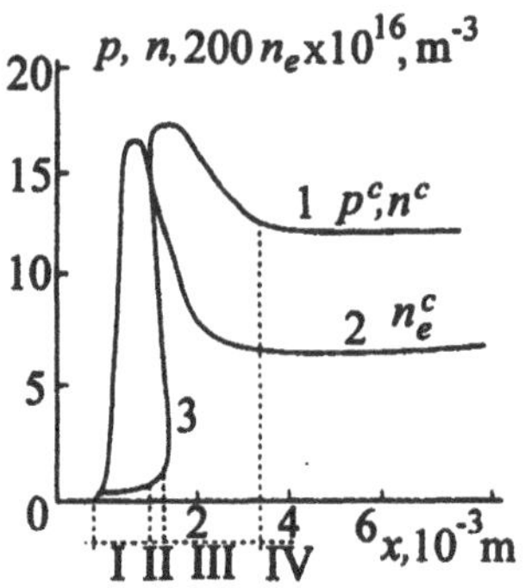

Figure 10.13. The profiles of the charged particle densities for the intermediate case Eq. (10.44) obtained in the simulations [22] for SF_6. The trace 1 corresponds to the positive ion density, and 2 to the time-averaged electron density. The discharge parameters (pressure 18Pa, frequency 13.6MHz, current density 2mA/cm^2) corresponded to $K_{att}\tau_{i1}$=9.0; $K_{att}\tau_{i2}$=0.03. As in Figs. 10.9 and 10.10, the distinctions between the sheath *I*, shock-like transition *II* from the *EIR* to the *IIR*, the nonuniform plasma region *III*, and the uniform recombination dominated column *IY* are clearly seen.

The parameter values are $K_{att}\tau_{i1} = 9$ and $K_{att}\tau_{i2} = 0.03$, which correspond to Eq. (10.68). The calculated from Eq. (10.74) value of p_{el} is 3.10^{15} m^{-3}, and the simulations [24] give 2.6.10^{15} m^{-3}.

In the case considered the sheath thickness is proportional to the current and is inversely proportional to the attachment frequency. This explains the small sheath thickness in strongly electronegative gases observed in numerical calculations [11], [24] and [27]. For low currents and large values of K_{att} we must take into account the finiteness of the Debye radius (as in [24]), because the sheath thickness can become comparable to it.

The ion and electron density profiles in the plasma. The negative ion flux induced by the electron attachment in the EIR, Eq. (10.74), should fall off to zero when passing to the column. Consequently, a nonuniform plasma region, in which the charge particle densities approach the equilibrium values in the column, should exist. When the attachment velocity is very high, Eq. (10.67), this transition occurs in the sheath near the electrode. In the cases Eqs. (10.65) and (10.68), the transition region can arise in the plasma or in the sheath near the plasma-sheath boundary, in which the ion convective velocity $V=b_i$<E> is low. In the sheath bulk the convection dominates over the diffusion. Approaching the plasma-sheath boundary convective velocity rapidly tends to zero, and the negative ion density grows very fast $n = \Gamma^*/V(x)$. At the sheath edge the diffusion terms (the second terms in the left-hand sides of Eqs. (10.62)) start to contribute. The transition region from the convective to the diffusive ion transport is rather complicated. One can estimate the importance of diffusion terms comparing at the point where $n \sim n^c$ the total velocity $V(x^*) = \Gamma^*/n^c$ with the diffusive one $D < d\ln n_e / dx >$. Depending on their ratio, two possible scenarios of transition to column appear. The diffusive flux exceeds the convective one, if [21]

$$D > \frac{\hat{\nu}^{(ion)} jV(L)}{en_e\omega}\left(K_{att}\tau_{i2}\right)^{5/3}, \tag{10.76}$$

where $\hat{\nu}^{(ion)} = d\ln \nu^{(ion)}(n_e)/d\ln n_e$. In this case, the flux Γ_n is determined by $(dn_e/dx)/n_e$. In Section 10.1 it was shown that in this case the ion density jumps arise near the plasma-sheath boundary. The ion density shocks are seen on the results of computations carried out in [24], [25] and [27]. In the region of a jump the ion fluxes are practically constant, and the ion densities increase steeply towards the plasma (the region II in Figs. 10.11, 10.13 corresponds to the location of the jumps) and reach values comparable to those in the column.

Now we can analyze the density profiles in plasma. After a jump the ion densities are much higher than the electron one. Consequently, the absolute values of the fluxes Γ_p, Γ_n practically coincide, Eq. (10.31). This results in the balance condition, which gives the relation between n and n_e:

$$\left\langle\left(K_{att} + \nu^{(ion)}\right)n_e\right\rangle = 2K^{(rec)}pn. \tag{10.77}$$

As the second condition we can take the equation for negative ions (10.62). At moderate pressures it can be simplified using strong dependence of the ionization frequency on n_e. Neglecting the small variation of n_e, the second of Eqs. (10.62) reads

$$\frac{d}{dx}\left(\frac{Dp}{n_e^c}\frac{dn_e}{dx}\right) = K_{att}n_e^c - K^{(rec)}p^2. \tag{10.78}$$

Eqs. (10.77) and (10.78), together with the value of the ion flux $\Gamma_p(L)$, Eq. (10.74), at the plasma-sheath boundary describes the nonuniform plasma properties. The negative ion flux should fall off toward the column. Hence, the right-hand side of Eq. (10.65) should be negative here.

Consequently, the ion density should be higher here, than that in the column (see Figs. 10.11, 10.13 and 10.14).

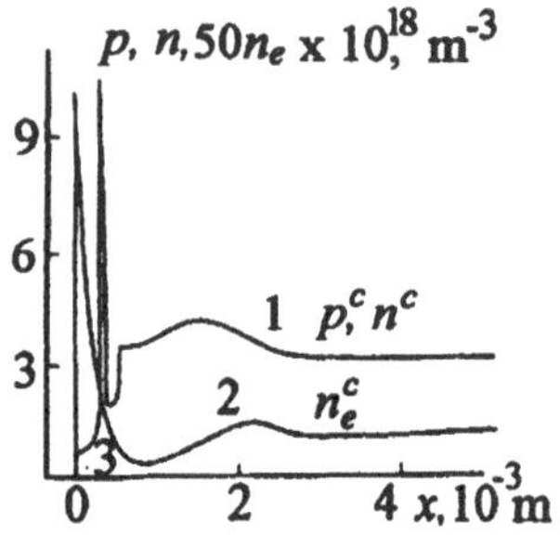

Figure 10.14. The formation of the ion density peaks in the simulation [24]. The conditions are the same, as in Fig. 10.12, with the exception of higher pressure (180Pa) and of higher current (100mA/cm^2).

This is in contrast to two-component pure plasma, Section 3.8, where the ion density decreases monotonically at transition from plasma to sheath. These bumps on ion density were observed experimentally in [30]; see Fig. 10.18. Eq. (10.78) implies that the thickness of the nonuniform ion-ion plasma region is

$$l \sim \left(D / \left(2K^{(rec)} p\hat{\nu}^{(ion)}\right)\right)^{1/2},$$

and the excess value of the ion density at the plasma side of the jump is

$$\Delta p \sim \left(\hat{\nu}^{(ion)} / \left(2DK^{(rec)} p\right)\right)^{1/2}.$$

For example, for the conditions of Fig. 10.13 these estimates give l=5mm and Δp=5.I 10^{16} m^{-3}; the numerical simulations [24] give l=2 mm and Δp=5.2.10^{16} m^{-3}.

For moderate pressures (when $\hat{\nu}^{(ion)} >> 1$) the electron density varies slightly in plasma, $(dn_e/dx)/n_e$ is small, the jumps arise at the plasma-sheath boundary, and the ion velocity in this region is low. In contrast, the flux Γ^*, Eq. (10.74), is sufficiently large (it is proportional to the attachment frequency and respectively to pressure), and the ion density grows steeply when passing from the sheath to the plasma. The case of low pressures is to be analysed separately, because of a large value of the diffusion coefficient, and the flux Γ^* is small due to the small value of K_{att}. Also, the nonlocality of the electron distribution becomes important. Roughly speaking, this corresponds to the case when $\nu^{(ion)}$ does not depend on n_e. Accordingly, the jump can arise in the plasma at a distance of the order of $(D/K_{att})^{1/2}$ from the plasma-sheath boundary. This distance is inversely proportional to the pressure and can be large at low pressures. At this distance the negative ion velocity falls off abruptly, as in the positive column of the *dc* discharge, see the previous Section. This situation was observed in numerical simulations carried out in [27].

The structure of the jump can be found by analogy with Section 10.1. Since this region is very narrow, we can ignore the variation of n_e in comparison with dn_e/dx. The growth in the ion densities is related to the decrease in the ion velocity V=$-D/n_e\, dn_e/dx$. Using Eq. (10.60) to express the change in the velocity dV/dx~$(\nu^{(ion)}+K_{att})$, we estimate the jump width:

$$\delta l \sim V\left(\frac{dV}{dx}\right)^{-1} \sim \frac{\Gamma^*}{n}\left(\nu^{(ion)}\right)^{-1} = \frac{K_{att}}{\nu^{(ion)}}\frac{Pel}{n} < \frac{Pel}{n}L \sim \frac{Pel}{p^c}L \qquad (10.79)$$

The jump width Eq.(10.79) is much smaller than the sheath thickness, because the ratio p_{el}/p_c is small. For example, under conditions corresponding to Fig. 10.13, the ratio is p_{el}/p_c~ 0.03, and the transition is practically jump-like. The width of this jump can be increased by ion diffusion; one can find the corresponding expressions in [1] and [2].

Another sheath-plasma transition scenario corresponds to the case, when the inequality, opposite to Eq. (10.76) holds. Under this condition the second terms in the left-hand sides of Eqs. (10.62) (the diffusion terms) can be neglected. Under the assumptions made the second Eq. (10.62) reads

$$-\frac{d\Gamma_n}{dx} = -\frac{d}{dx}(Vn) = K_{att}n_e^c - K^{(rec)}np, \qquad (10.80)$$

where velocity V=b_i<E> is given by Eq. (10.64). Approaching the plasma-sheath boundary the convection rapidly tends to zero. Consequently, the negative ion density grows very fast n= $\Gamma^*/V(x)$. Similarly to the previous case, the ion density in this region is considerably higher than the equilibrium one. The ion density greatly exceeds n^c until the negative ion flux decreases considerably, due to the ion-ion recombination. Since in the absence of diffusion the velocity V at the plasma-sheath boundary is zero, all the negative ion flux Γ^* should recombine in this small part of the sheath near the plasma-sheath boundary.

The growth rate of the ion density which is determined by the derivative of the velocity dV/dx, turns out in this region to be higher than the rate of ion losses $K^{(rec)}p$. This gives rise to the formation of narrow peaks in the ion densities. Analyzing the ion equation, we obtain the expressions for the density maximum and for the peak width $\widetilde{L}$. Because $dn/dx=0$ at this maximum, we have $(dV/dx)=-K^{(rec)}p_{max}$, the flux is of the order of $nV(x)\leq\Gamma^*$. This yields

$$p_{\max} \sim p^c\left(\left(\frac{3}{4}\right)^{1/2} \frac{1}{\pi} K_{att}\tau_{i2}\right)^{-1/2} . \tag{10.81}$$

Because the negative ion flux Γ^* should recombine in this peak, we have $K^{(rec)}np\widetilde{L} \sim \Gamma^*$, and

$$\widetilde{L} = \frac{j(K_{att}\tau_{i2})}{en_e^c\omega}. \tag{10.82}$$

The formation of the ion density peaks was found in the simulations [24]. Under the conditions of Fig. 10.14 the Eq. (10.81) gives $p_{max}=7p^c$, whereas the calculations carried out in [24] give $p_{max}=3.3p^c$. This difference is likely to be connected with the ion diffusion. The second reason stems from the ion density oscillations, which result in considerable perturbation of the electron density according to Eq. (10.61).

Table

	$K_{att}\tau_{i1}^{-1} \ll 1$	$\tau_{i1}^{-1} \ll K_{att} \ll \tau_{i2}^{-1}$	$K_{att}\tau_{i2}^{-1} \gg 1$
L	$\frac{2j_0}{e\omega n_e^c}$	$\frac{0.63 * b_i j_0}{\omega K_{att}\varepsilon_0}$	$\frac{2j_0}{e\omega n_e^c}$
p_{el}	n_e^c	$\sim \frac{K_{att}\varepsilon_0}{eb_i}$	p^c
U	$\frac{2j_0^2}{e\omega^2 n_e^c \varepsilon_0}$	$\frac{1.94 j_0^2}{e\varepsilon_0\omega^2 n_e^c K_{att}\tau_{i1}}$	$\frac{2j_0^2}{e\omega^2 n_e^c \varepsilon_0}$

The experimental data about the partial density profiles in the electronegative plasmas are rather scarce. The existing information confirms the concept of the strongly peaked profiles of the negative ions. The examples are presented in Figs. 10.15 - 10.18.

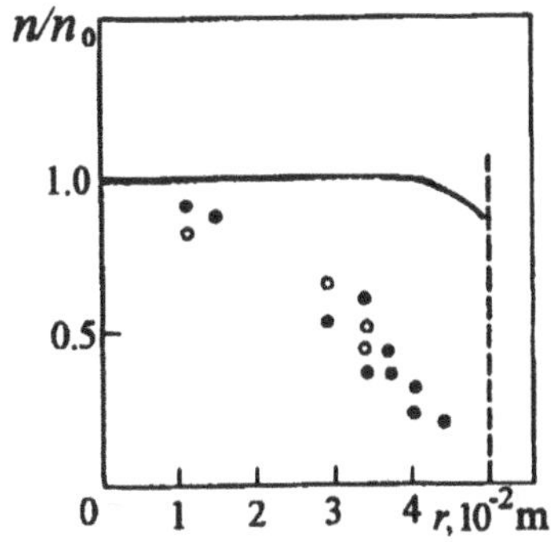

Figure10.15. The experimental radial density profiles of the electrons (full line), of the negative (full points) and of the positive (empty points) ions in the positive column of the *dc* discharge in O_2 at 5.3Pa, 4mA, in the tube with radius a=5cm [7].

Figure 10.15 demonstrates the flat electron density profile in positive column.

Figure 10.16 corresponds to the case of rather low gas pressure, when condition Eq. (10.47) holds.

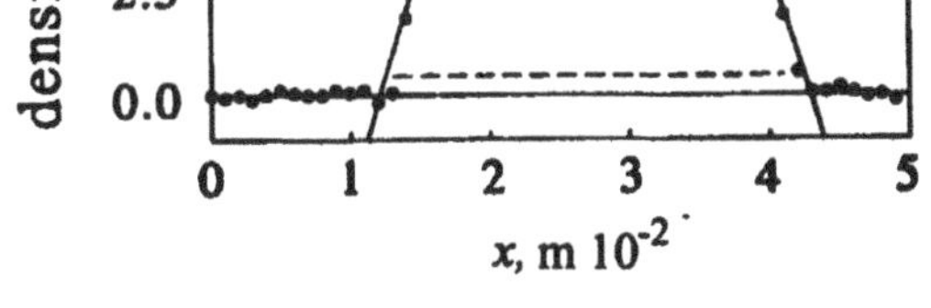

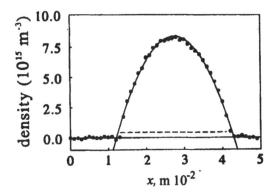

Figure 10.16. The experimental negative ion density profiles (the dots) in the *RF* discharge in O_2 at frequency 13.56MHz at pressure 1.33Pa and at 10W input power in cylindrical vessel of 17.5cm diameter and 5cm height [28]. At x=0 the driven electrode is situated. At it the current density is high, and the sheath is thick. The parabolic isolated from wall $n(x)$ profile is clearly seen. The flat electron density profile is shown by the dashed line.

The parabolic central plasma profile is determined by the ion diffusion. In the peripheral plasma and in the RF sheath, which were not resolved in this experiment, the negative ions are absent.

The plasma separation into peripheral electron-ion and the central electronegative plasma with convex in density profile can be seen in Fig. 10.17.

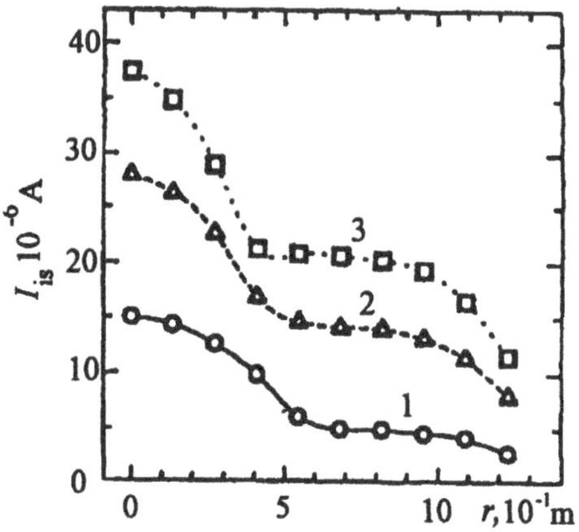

Figure10.17 The probe ion saturation current, which is proportional to the density of the positive ions, in discharge in Ar/O_2 mixture with density ratio 13:100 [29]. The cylindrical vessel of 46cm length and 13cm radius contained magnetically confined plasma of *dc* discharge with hot cathode at pressure 10mtorr. The discharge current is 10mA (1); 50mA (2); 90mA (3).

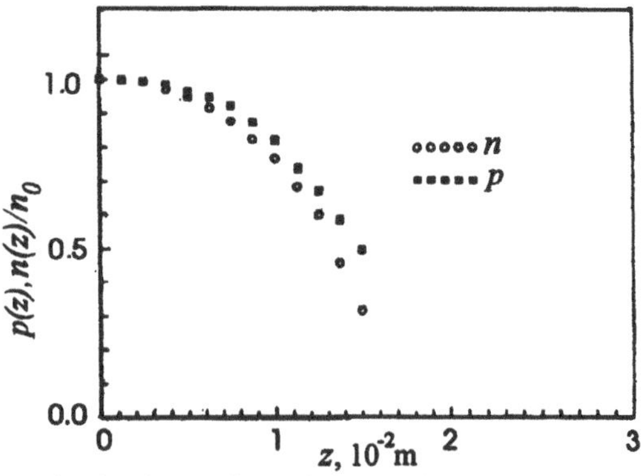

Figure 10.18*a*. The density profiles of the positive and negative ions in cylindrical RF discharge in O_2 [30]. The vessel dimensions are: height 6cm, radius 15cm; pressure is 6Pa. Almost parabolic profiles and the piling up of the negative ions to the discharge midplane are clearly seen.

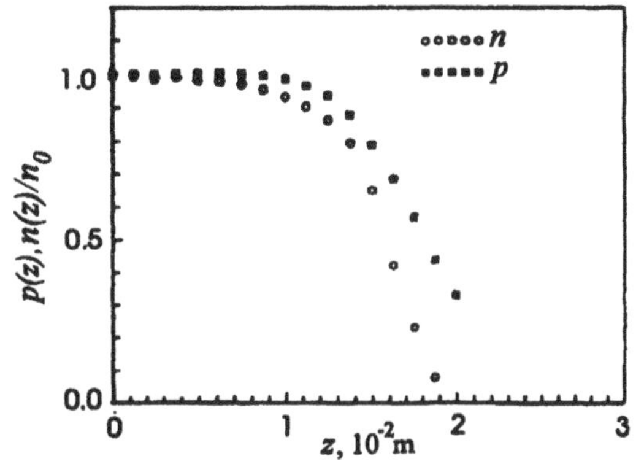

Figure 10.18*b*. The density profiles of the positive and negative ions in the same disharge as in Fig. 18*a*, at higher pressure 10Pa. The formation of the flat profiles sections in the central region is well pronounced.

REFERENCES

1. L. D. Tsendin, *Sov. Phys. Techn. Phys.* **30**: 1377-1379 (1985); **34** (1), 11-15, (1989).
2. I. D. Kaganovich, L. D. Tsendin, *Plasma Phys. Reports*, 19, 645 - 650 (1993).
3. J. B. Thompson, *Res. Notes, Proc. Roy. Soc.* **73**: 818-830 (1959); 262: 503-528 (1961).
4. H. Massey, Negative Ions (Cambridge: University Press) 1976 754.
5. R. Seeliger, *Ann Phys. Lpz.* **6**: 93-96 (1949).
6. R. Holm, *Z. Phys.* **75**: 171-190 (1932).
7. M. V. Konjukov, *Sov. Phys. JETP* **34**: 908-911; 1634-1635 (1958). (in Russian).
8. H. J. Oskam, *Philips Res. Repts.*, v.13: 335-400 (1958); G. A. Galechjan, in *Plasma Chemistry*, Editor: B. M. Smirnov, v.7 (Moscow: Atomizdat) 1980, 218-251. (in Russian).
9. H. Sabadil, *Beitr. Plasmaphys.*, **13**: 235-251 (1973).
10. P. D. Edgeley, A. von Engel, *Proc. Roy. Soc.* **A370**: 375-387 (1980).
11. R. N. Franklin, *Plasma Phenomena in Gas Discharges*, (Oxford, Clarendon:1976)
12. C. Ferreira, G. Gousset, M. Touseau, *J. Phys. D: Appl. Phys.* **21**:1403-1413 (1988).
13. C. Ferreira, G. Gousset, *J. Phys. D: Appl. Phys.* **24**: 775-778 (1991).

14. P. R. Daniels, R. N. Franklin, *J. Phys. D: Appl. Phys.* **22**:780-785 (1989).
15. P. R. Daniels, R. N. Franklin, J. Snell, *J. Phys. D: Appl. Phys.* **23**: 823-831 (1990); 26: 1638-1649 (1993).
16. R. N. Franklin, J. Snell, *J. Phys. D: Appl. Phys.* **25**: 453-457 (1992); **27**: 2102-2106 (1994).
17. A. J. Lichtenberg, V. Vahedi, M. A. Lieberman, T. Rognlien, *J. Appl. Phys.* **75**: 2339-2347 (1994).
18. I. G. Kouznetsov, A. J. Lichtenberg, *Plasma Sources Sci. Technol.* **5**: 662-676 (1996); V. I. Kolobov, D. J. Economou, *Appl. Phys. Lett.* **72**: 656-658 (1998).
19. A. J. Lichtenberg, I. G. Kouznetsov, T. D. Lee, M. A. Lieberman, I. D. Kaganovich, L. D. Tsendin, *Plasma Sources Sci. Technol.* **6**: 437-449 (1997).
20. L. L. Beilinson, I. D. Kaganovich, L. D. Tsendin, *Proc. Russian Conf. Low-Temperature Plasma, Petrozavodsk*, 1995, L. A. Luisova, Ed. (Petrozavodsk: University Press) 1995, v.3, 313-314.
21. I. D. Kaganovich, *Plasma Phys. Reports*, **21**, 410-417 (1995).
22. E. Gogolides, H. H. Sawin, *J. Appl. Phys.*, **72**: 3971-4002 (1992).
23. J. P. Boeuf, *Phys. Rev.* **A36**: 2782-2792 (1987).
24. V. A. Schweigert, *Plasma Phys. Repts.* **17**: 493-499 (1991).
25. Y. H. Oh, N. H. Choi, D.I. Choi, *J. Appl. Phys.* **67**: 3264-3268 (1990).
26. M. Meyyappan, T. R. Govindan, *IEEE Trans. Plasma Sci.* **19**: 122-128 (1992).
27. J. P. Passchier, W. J. Goedheer, *J. Appl. Phys.* **73**: 1073-1079 (1993).
28. D. Wender, W. W. Stoffels, E. Stoffels, G. M. W. Kroesen, F. J. De Hoog, *Phys. Rev.* **E51**: 2436-2444 (1995).
29. T. Kimura, K. Inagaki, K. Ohe, *J. Phys. D: Appl.Phys.* 31. 2295-2304 (1998).
30. Buddemeier, *Thesis*, Ruhr-Uni-Bochum 1997, 102.

Chapter 11

Evolution of Magnetized Plasma in the Presence of Net Current

In a magnetic field when a net current flows through the inhomogeneity, the character of the evolution is much more diverse and complicated than for $\vec{B} = 0$. First of all, in a magnetic field ambipolar mobility exists in the 1D problem even in the pure plasma. In the multidimensional case the perturbation of the pure plasma is split into several perturbations moving in different directions with different velocities. The spreading of each plasmoid in the linear case is determined by the combination of the diffusion process and the evolution in the self-consistent electric field generated by the net current. For the nonlinear evolution the overturn of the density profile, formation of the shocks and the nonlinear deformation are typical. In the presence of ions of different species the character of evolution becomes far more diverse and interesting.

The evolution of the magnetized plasma in the presence of the net current is a key element in numerous applications. The whole physics of the ionosphere is to large extent associated with this problem, since ionospheric plasma is known as a medium where strong electric fields are generated by neutral winds and/or by interaction with the magnetosphere. Ionospheric irregularities of different nature are responsible for the scattering of radio waves, and therefore, the investigations of their generation and evolution is a separate important task. In the laboratory such fundamental problems as plasma acceleration in various plasma guns, the energy transformation in MHD generators, etc., are connected with the net currents in magnetized plasma.

The study of the plasma evolution in each of these situations deserves a special book. Therefore, in this Chapter we focus on the most instructive general problems of plasma evolution in a magnetic field in the presence of a net current. Almost everywhere (with exception of Section 11.5) we shall consider, as for the diffusion problem, the case of the magnetized plasma $x_e x_i >> 1$. Since the homogeneous motion of the neutral gas across $\vec{B}$, due to the Lorentz transformation, is equivalent to the generation of electric field in the moving reference frame, we shall consider the evolution in the reference frame of the neutral gas. We restrict ourselves to the case of a plasma with constant uniform temperatures $T_{e,i}=const.$ The basic equations (in the absence of sources and sinks) are the same as in the absence of the net current, Eq. (6.1):

$$\frac{\partial n}{\partial t} - \vec{\nabla}\cdot(\hat{D}_e\vec{\nabla}n - \hat{b}_e n\vec{\nabla}\varphi) = 0 ;$$
$$\frac{\partial n}{\partial t} - \vec{\nabla}\cdot(\hat{D}_i\vec{\nabla}n + \hat{b}_i n\vec{\nabla}\varphi) = 0 \quad . \tag{11.1}$$

However, the boundary conditions differ from those of Chapter 6.

11.1 ONE-DIMENSIONAL EVOLUTION IN PARTIALLY IONIZED PLASMA

We shall consider here the 1D evolution of inhomogeneity of a uniform background plasma. Thus the boundary conditions for Eqs. (11.1) are

$$n(\vec{r} \to \infty) = n_0 \ ; \ \ \varphi(\vec{r} \to \infty) = -\vec{E}_0 \, . \tag{11.2}$$

We shall study the situation when the plasma density depends only on the coordinate ζ which forms an angle β with magnetic field. The latter is parallel to the z axis. It is convenient to transform the equations to the new coordinate system (x, η, ζ), which was introduced in Section 5.3, Fig. 5.3. The new system is obtained by rotating the old one over the x axis. For the arbitrary direction of external electric field E_0 in the 1D case the Eqs.(11.1) are reduced to

$$\frac{\partial n_\alpha}{\partial t} = \frac{\partial}{\partial \zeta}\left[D_{\alpha\zeta\zeta}\frac{\partial n_\alpha}{\partial \zeta} \mp b_{\alpha\zeta\zeta} n_\alpha \frac{\partial \varphi}{\partial \zeta} \mp b_{\alpha\zeta x} n_\alpha \frac{\partial \varphi}{\partial x} \right] ; \ \alpha = e, i \, , \tag{11.3}$$

where the components of the diffusion and mobility tensors are given by Eq. (5.29). With account of the boundary conditions at infinity it is possible to obtain the electric field:

$$\vec{E} = -\vec{\nabla}\varphi_d + \vec{E}_{0\eta} + \vec{E}_\zeta + \vec{E}_{0x}. \tag{11.4}$$

Here the diffusive potential φ_d is defined by Eq. (5.33). Components of the electric field perpendicular to the density gradient, E_η and E_x, are independent of ζ because of the potential (electrostatic) character of the electric field. The E_ζ component of electric field is

$$E_\zeta = -\frac{(b_{i\zeta\eta} + b_{e\zeta\eta})E_{0\eta} + (b_{i\zeta x} + b_{e\zeta x})E_{0x}}{(b_{i\zeta\zeta} + b_{e\zeta\zeta})} + \frac{n_0[(\hat{b}_e^{(0)} + \hat{b}_i^{(0)})\vec{E}_0)_\zeta}{n(b_{i\zeta\zeta} + b_{e\zeta\zeta})}. \tag{11.5}$$

The part of the ion flux caused by the field $\vec{E}_0$ (ion flux in the drift approximation) is thus

$$\Gamma_{i\zeta} = n(\hat{b}_i\vec{E})_\zeta = \Gamma_{i\zeta}^{(0)} - \frac{b_{e\zeta\zeta}\Gamma_{i\zeta}^{(0)} + b_{i\zeta\zeta}\Gamma_{e\zeta}^{(0)}}{b_{e\zeta\zeta} + b_{i\zeta\zeta}} + \frac{n[b_{e\zeta\zeta}(\hat{b}_i\vec{E}_0)_\zeta - b_{i\zeta\zeta}(\hat{b}_e\vec{E}_0)_\zeta]}{b_{e\zeta\zeta} + b_{i\zeta\zeta}},$$

where $\vec{\Gamma}_{e,i}^{(0)} = n_0\hat{b}_{e,i}\vec{E}_0$. The components $b_{\alpha\zeta\zeta}$ of the mobility tensor are defined by Eq. (5.29):

$$b_{\alpha\zeta\zeta} = b_{\alpha\|}\mu^2 + b_{\alpha\perp}(1-\mu^2); \quad \mu = \cos(\vec{\zeta},\vec{B}) .$$

The equation for the density evolution is obtained by excluding the potential from Eqs. (11.3). The resulting equation has the form (compare with Eq. (5.31)):

$$\frac{\partial n}{\partial t} - \frac{\partial}{\partial\zeta}D(\mu^2)\frac{\partial}{\partial\zeta} + V_\zeta(\mu^2)\frac{\partial n}{\partial\zeta} = 0, \tag{11.6}$$

where the ambipolar drift velocity is

$$\vec{V}(\mu^2) = \frac{d}{dn}\{n_0 \frac{[b_{i\|}\mu^2 + b_{i\perp}(1-\mu^2)]\hat{b}_e^{(0)}\vec{E}_0}{(b_{e\|} + b_{i\|})\mu^2 + (b_{i\perp} + b_{e\perp})(1-\mu^2)}$$
$$- \frac{[b_{e\|}\mu^2 + b_{e\perp}(1-\mu^2)]\hat{b}_i^{(0)}\vec{E}_0}{(b_{e\|} + b_{i\|})\mu^2 + (b_{i\perp} + b_{e\perp})(1-\mu^2)} \tag{11.7}$$
$$+ n\frac{[b_{e\|}\mu^2 + b_{e\perp}(1-\mu^2)]\hat{b}_i^{(0)}\vec{E}_0 - [b_{i\|}\mu^2 + b_{i\perp}(1-\mu^2)]\hat{b}_e^{(0)}\vec{E}_0}{(b_{e\|} + b_{i\|})\mu^2 + (b_{i\perp} + b_{e\perp})(1-\mu^2)}\},$$

and $D(\mu^2)$ is defined by Eq. (5.32). We see that, in contrast to the case of $\vec{B} = 0$, ambipolar drift velocity exists even in the pure plasma.

In weakly ionized plasma ambipolar drift velocity is density independent [1]

$$\vec{V}(\mu^2) = \frac{[b_{e\|}\mu^2 + b_{e\perp}(1-\mu^2)]\hat{b}_i^{(0)}\vec{E}_0 - [b_{i\|}\mu^2 + b_{i\perp}(1-\mu^2)]\hat{b}_e^{(0)}\vec{E}_0}{(b_{e\|} + b_{i\|})\mu^2 + (b_{i\perp} + b_{e\perp})(1-\mu^2)}. \tag{11.8}$$

In this case only the components of the electric field which are perpendicular to the ζ direction give nonzero contributions to $V_\zeta(\mu^2)$.

The general expression for the ambipolar drift velocity is rather complicated, so here we shall briefly discuss only some special situations. If the external electric field belongs to the plane (ζ, z), after substituting the expressions for $b_{\alpha\zeta\eta}$, Eq. (5.29), into Eq. (11.7), for the magnetized plasma we obtain

$$V_\zeta(\mu^2) = -\frac{b_{e\|}b_{i\perp}\mu(1-\mu^2)^{1/2}E_{0\eta}}{b_{e\|}\mu^2 + b_{i\perp}(1-\mu^2)}. \tag{11.9}$$

This velocity equals zero when $\vec{\zeta}\|\vec{B}, \vec{\zeta}\perp\vec{B}$, but is important in the intermediate cases. Its largest value of the order of $b_{e\|}\mu_0 E_{0\eta}$ is reached at $\mu\sim\mu_0$. At such angles a very strong electric field is generated in plasma:

$$E_\zeta = -\frac{b_{e\|}\mu(1-\mu^2)^{1/2}E_{0\eta}}{b_{e\|}\mu^2 + b_{i\perp}(1-\mu^2)}\frac{n-n_0}{n}. \tag{11.10}$$

Therefore, for $\mu\sim\mu_0$ the external electric field is strongly amplified in the vicinity of the inhomogeneity up to the values $E_\zeta\sim E_{0\eta}/\mu_0$ for $|n-n_0|\sim n_0$.

In the case $\vec{E}\perp[\vec{B}\times\vec{\zeta}]$

$$V_\zeta(\mu^2) = \frac{E_{0x}(1-\mu^2)^{1/2}}{B} - \frac{[b_{e\|}\mu^2 + b_{e\perp}(1-\mu^2)]}{b_{e\|}\mu^2 + b_{i\perp}(1-\mu^2)}(b_{e\perp}x_e - b_{i\perp}x_i)E_{0x}(1-\mu^2)^{1/2}. \tag{11.11}$$

For the magnetized ions x_i>>1 this expression for $V_\zeta(\mu^2)$ is simply the ζ projection of the $\vec{E}\times\vec{B}$ drift of the charged particles.

In partially ionized plasma the ambipolar drift velocity, according to Eq. (11.7), depends on the plasma density. As a result, in the general situation the profile overturn and shock formation take place even in pure plasma [2]. In contrast to the case of weakly ionized plasma, the component E_ζ now also contributes to the ambipolar drift velocity, because the first and the second terms on the r.h.s. of Eq. (11.7) do not vanish. In the important case when the density gradient is parallel to the magnetic field, the ambipolar drift velocity turns to zero in spite of the density dependence of the mobilities. This can be explained by the fact that the ratio of the longitudinal mobilities does not depend on the ionization degree. Indeed, according to Eq. (2.89),

$$\frac{b_{e\|}}{b_{i\|}} = \frac{\mu_{iN}\nu_{iN}}{c_{eN\|}^{(\ddot{u})} m_e \nu_{eN}}.$$

Therefore, from Eq. (11.7) we obtain $V_{\|}$=0, and the transport of the pure partially ionized plasma independently of the net current is described by the ambipolar diffusion equation (4.7). In another important case $\vec{\zeta}\perp\vec{B}$ the contribution from the component E_ζ of the electric field is finite, but rather modest. Indeed, for $\vec{E}\|\vec{\zeta}$ Eq. (11.7) yields

$$V_\zeta(n) = b_{i\perp}E_0 n_0 \frac{d(b_{e\perp}/b_{i\perp})}{dn}. \tag{11.12}$$

The maximum value of the ambipolar drift velocity of the order of $b_{e\perp}E_0$ is reached for $\nu_{eN}\sim\nu_{ei}$. For $\vec{E}\perp\vec{\zeta}$, unmagnetized ions, x_i<<1, and $\nu_{eN}\sim\nu_{ei}$, we have

$$V_\zeta(n) \sim \frac{E_0}{B}\frac{b_{e\perp}}{b_{i\perp}}. \tag{11.13}$$

Therefore, the dependence $V_\zeta(n)$ is most pronounced in weakly magnetized plasma with $x_e x_i \sim 1$, where $\nu_{eN} \sim \nu_{ei}$.

11.2 ONE-DIMENSIONAL EVOLUTION OF WEAKLY IONIZED PLASMA WITH TWO POSITIVE ION SPECIES ACROSS MAGNETIC FIELD

The fact that Eq. (11.6) in the pure weakly ionized plasma is linear, as in the case of the ambipolar diffusion (Chapter 4) is a result of complete compensation of the nonlinear effects. We shall illustrate this by considering the evolution of the ions of species 1 injected into the ambient plasma consisting of ions of the different species 2. The temperatures of all charged particles are assumed to be equal and uniform: $T_e=T_1=T_2=T$. We shall analyze two examples when the ambient electric field is parallel and perpendicular to the density gradient: $\vec{E} \| \vec{\zeta} \perp \vec{B}$, $\vec{E} \perp \vec{\zeta} \perp \vec{B}$. In strongly magnetized plasma we shall neglect the electron perpendicular (Pedersen) mobility and diffusion, $b_{e\perp}=(e/T)D_{e\perp}=0$, and consider the case $m_e\nu_{ei} \ll \mu_{iN}\nu_{iN}$ when the perpendicular ion mobility is density-independent.

Inhomogeneity infinite in Hall direction. Here we consider the 1D situation when both the density gradient and the ambient electric field are parallel to the y axis, $\vec{y} \perp \vec{B}$. The equation system for the injected (subscript 1) and the ambient (subscript 2) ions has the form

$$\frac{\partial n_{1,2}}{\partial t} + \frac{\partial}{\partial y}\left(b_{1,2\perp} E n_{1,2} - D_{1,2}\frac{\partial n_{1,2}}{\partial y}\right) = 0 \qquad (11.14)$$

with the boundary conditions

$$n_1(y \to \pm\infty) = 0; \; n_2(y \to \pm\infty) = n_2^{(0)}; \; E(y \to \pm\infty) = E_0 . \qquad (11.15)$$

Since we neglected the electron perpendicular transport, the electron density remains constant:

$$n_e = n_1 + n_2 = n_{10}(y) + n_2^{(0)} = n_0(y), \qquad (11.16)$$

where $n_{10}(y)$ is the initial perturbation of injected ion density. The electric field can be excluded from Eq. (11.4). From the condition of current conservation in Pedersen direction, $j_y = const$, we find

$$E=\frac{b_{2\perp}n_2^{(0)}E_0}{b_{1\perp}n_1+b_{2\perp}n_2}+\frac{T}{e}\frac{\partial\ln(b_{1\perp}n_1+b_{2\perp}n_2)}{\partial y}. \tag{11.17}$$

Substituting Eq. (11.17) into Eq. (11.14) yields

$$\begin{aligned}&\frac{\partial n_{1,2}}{\partial t}+b_{2\perp}n_2^{(0)}E_0\frac{\partial}{\partial y}\frac{b_{1,2\perp}n_{1,2}}{b_{1\perp}n_1+b_{2\perp}n_2}\\&=\frac{\partial}{\partial y}\{D_{1,2\perp}[\frac{\partial n_{1,2}}{\partial y}-n_{1,2}\frac{\partial\ln(b_{1\perp}n_1+b_{2\perp}n_2)}{\partial y}]\}.\end{aligned} \tag{11.18}$$

We shall analyze the solution of Eq. (11.18) in the drift approximation [3] when the diffusion terms are negligible. For the case of equal ion mobilities $b_{1\perp}=b_{2\perp}=b$, Eq.(11.18) coincides with Eq. (9.31), and the evolution is reduced to the case considered in Section 9.3, see Fig. 9.3. Therefore, we shall consider in more detail the case $b_{2\perp}>> b_{1\perp}$ which is important for ionospheric applications (see the subsequent Section). In the dimensionless variables, after neglecting the diffusion terms, Eq. (11.18) is reduced to

$$\frac{\partial\tilde{n}_1}{\partial\tilde{t}}+\frac{\partial}{\partial\tilde{y}}\frac{\tilde{n}_1}{\beta\tilde{n}_1+\tilde{n}_2}=0\;;\quad \tilde{n}_1+\tilde{n}_2=\tilde{n}\;, \tag{11.19}$$

where

$$\tilde{n}_{1,2}=n_{1,2}(y)/n_2^{(0)};\quad \beta=b_{1\perp}/b_{2\perp}<<1;\quad \tilde{n}=n_0(y)/n_2^{(0)};$$
$$\tilde{y}=y/a\;;\;\tilde{t}=tb_{1\perp}E_0/a.$$

Here a is the characteristic spatial scale of initial perturbation. Substituting $\tilde{n}_2$ from the second of Eqs. (11.19) into the first one, introducing the new variables

$$s=\int_0^{\tilde{y}}\tilde{n}d\tilde{y};\qquad q=(1-\beta)\tilde{n}_1/\tilde{n}$$

we find

$$\frac{\partial q}{\partial\tilde{t}}+V(q)\frac{\partial q}{\partial s}=0\;;\;V(q)=\frac{1}{(1-q^2)}. \tag{11.20}$$

This equation for the simple nonlinear wave is analogous to Eq. (9.11). It has the solution

$$q(s,\tilde{t}) = q_0[s - V(q)\tilde{t}]. \qquad (11.21)$$

In the initial variables for $\beta<<1$ we have

$$n_1(\tilde{y}) = \frac{n(\tilde{y})}{n(\tilde{y}_0)} n(\tilde{y}_0), \qquad (11.22)$$

where the coordinates $\tilde{y}$ and $\tilde{y}_0$ are linked in a given moment t by the relation

$$\int_0^{\tilde{y}} \tilde{n}(\tilde{y})d\tilde{y} = \int_0^{\tilde{y}_0} \tilde{n}(\tilde{y}_0)d\tilde{y}_0 + \frac{\tilde{t}}{[1-(1-\beta)\tilde{n}_1(y_0)/\tilde{n}(\tilde{y}_0)]^2}. \qquad (11.23)$$

The velocity $V(q)$ in Eq. (11.20) increases with q. Hence, the profile overturn and the shock formation occur on the front side of the density profile. The structure of the shock can be analyzed, as in the absence of a magnetic field, see Section 9.2. The difference, however, is connected with the structure of the diffusive terms. In contrast to Eq. (9.4), two terms on the r.h.s. of Eq. (11.18) have opposite signs. It means that the self-consistent electric field now reduces the diffusive fluxes of ions. Therefore, for $b_{1\perp}n_1>>b_{2\perp}n_2$ the shock width is significantly smaller than in the absence of a magnetic field, and for large values of n_1 the small perpendicular electron diffusion may be important. Since the diffusion flux always exceeds the electron flux $-D_{e\perp}\partial n_e/\partial y$, the shock scale is restricted from below by $Tb_{e\perp}/eEb_{i\perp}$.

In the linear case $n_1<<n_2$ the initial perturbation just moves with the velocity $b_{1\perp}E_0$. So let us analyze the character of the nonlinear evolution of the Gaussian profile

$$n_{10}(y)=An_2^{(0)}\exp(-y^2/a^2)$$

provided the inequality

$$1<<A<<b_{2\perp}/b_{1\perp}=\beta^{-1} \qquad (11.24)$$

is satisfied. At first, for $|\tilde{y}_0|<1$ the displacement $\tilde{y}-\tilde{y}_0$ is small. Expanding Eq. (11.23) in series in $\tilde{y}-\tilde{y}_0$, we obtain $\tilde{y}-\tilde{y}_0=\tilde{n}(\tilde{y}_0)\tilde{t}$.

Hence, at the beginning the velocity which corresponds to the displacement of the points of the q profile is rather high - of the order of $b_{1\perp} E_0 \tilde{n}(\tilde{y}_0)$. At the moment $\tilde{t}_1 = A^{-1}$ the displacement $\tilde{y} - \tilde{y}_0$ becomes of the order of unity. The corresponding variation of q is A^{-1}, while change in $\tilde{n}_1$ is also of the order of unity. The maximum density of the injected ions which can outflow in the direction of the electric field is limited by the density of the ambient ions, $n_1 \approx n_2^{(0)}$ because the electron density remains constant. Therefore, at the beginning, $t < t_1 = A^{-1} a/(b_{1\perp} E_0)$, the outflow velocity on the front side of the inhomogeneity increases up to the values $b_{1\perp} E_0 A$. Later this velocity decreases. Let us determine the moment t_2 when the density at $y=0$ becomes twice less than the initial value. For the Gaussian profile with account of the condition Eq. (11.24) we obtain from Eq. (11.23):

$$\tilde{y} + A\frac{\sqrt{\pi}}{2} erf(\tilde{y}) = \tilde{y}_0 + A\frac{\sqrt{\pi}}{2} erf(\tilde{y}_0) + [1 + A\exp(-\tilde{y}^2)]^2 \tilde{t}^2 . \tag{11.25}$$

Substituting $\tilde{y} = 0$ into Eq. (11.22) with account of $\tilde{n}_1(0, \tilde{t}_2) = A/2$, we have

$$(1 + A)A\exp(-\tilde{y}_0^2)/[1 + A\exp(-\tilde{y}_0^2)] = A/2 .$$

Hence $A\exp(-\tilde{y}_0^2) = 1$. Since $erf(-\tilde{y}_0) \approx -1$, Eq.(11.25) yields

$$t_2 \approx A\pi^{1/2} a/(8b_{1\perp} E_0). \tag{11.26}$$

The outflow velocity of injected ions at later times is thus much smaller than at the beginning, and is determined by the velocity $b_{1\perp} E_0$.

The reason for the modification of outflow velocity of the injected ions consists in the following. If inequality (11.24) is satisfied, the electric field, Eq. (11.17), is determined by the density of the ambient ions. At the beginning, when $n_2 \sim n_2^{(0)}$, the initial profile of injected ions starts to move with the velocity $b_{1\perp} E_0$. Simultaneously, a depletion of the ambient ions is created in the front side region, Fig. 11.1. The ambient ions from the depletion region fill the initial profile reducing the electric field here. Significant decrease of the electric field begins at $t \sim t_1$ when the variation of n_2 becomes of the order of $n_2^{(0)}$. At the same time the electric field in the front region, where the ambient ions are depleted, increases up to the values $\sim AE_0$. Accordingly, the ambient ion density in the depletion region is reduced to the values $n_2 \sim A^{-1} n_2^{(0)}$. The time scale of the shock formation is also t_1.

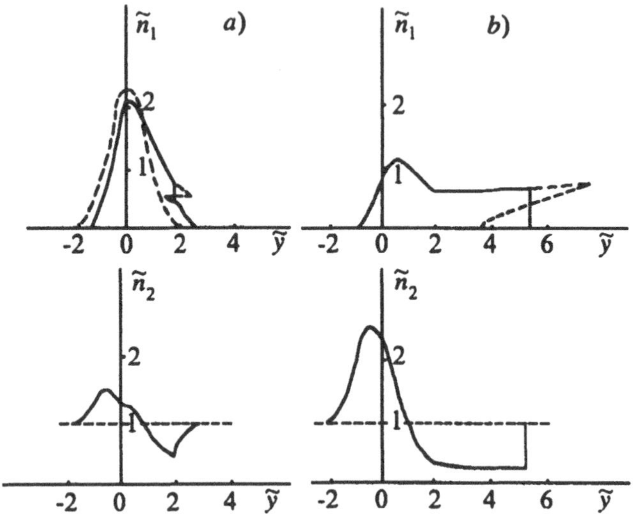

Figure 11.1. Density profiles of injected and ambient ions; $b_{1\perp}/b_{2\perp}=1/5$; $A=4/\pi^{1/2}$; *a)* broken line - initial profile, solid line - profiles at $t=0.5a/b_{1\perp}E_0$; *b)* profiles at $t=2a/b_{1\perp}E_0$ (solid line); dashed line - multivalued solution in the drift approximation.

Later, at $t>t_1$, the further increase of the ambient ion density inside the initial injection region occurs. The electric field here decreases accordingly. As a result, the time scale for injected ion outflow becomes independent of the ambient ion mobility. The corresponding time scale t_2 is determined by Eq. (11.26). The time history of the density profiles is shown in Fig. 11.1, Fig. 11.1*a* corresponds to the moment t_1. In Fig. 11.2 the evolution of injected ion density at $y=0$ is shown. It is compared with the corresponding evolution in the ambient electric field. It is clearly seen that the motion of most of the injected ions is strongly reduced with respect to the unperturbed motion (e.g., in the linear case) due to the self-consistent electric field.

For the very strong inhomogeneity

$$A\beta\gg 1 \qquad (11.27)$$

the solution can be analyzed in the same way. In this case

$$t_1=A\beta^2a/(b_{1\perp}E_0), \qquad t_2=A\pi^{1/2}a/(8b_{1\perp}E_0).$$

At first the outflow velocity on the front side increase with time, and the maximum velocity $b_{2\perp}E_0$ is reached at $t \sim t_1$ for $A \sim \beta^{-1}$. This maximum velocity decreases with the increase of A, and at $A\beta^2 > 1$ becomes of the order of $b_{1\perp}E_0$. Later, at $t > t_1$ the outflow velocity decreases from the maximum value up to the values of the order of $b_{1\perp}E_0$.

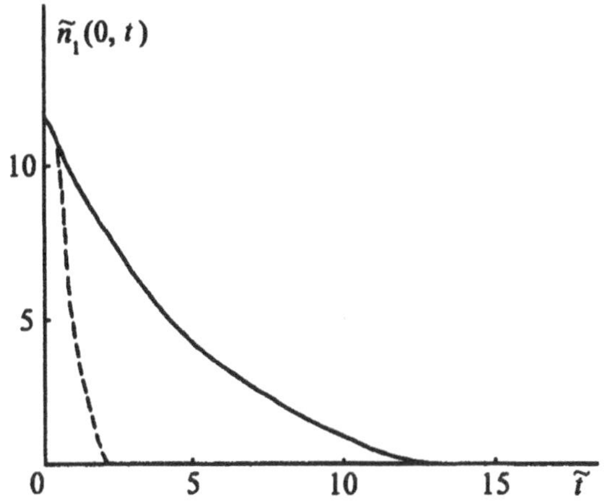

Figure 11.2. Temporal dependence of injected ions density at $y=0$ (solid line); $b_{1\perp}/b_{2\perp}=1/5$; $A=20/\pi^{1/2}$; dashed line corresponds to the motion in the unperturbed electric field E_0.

Finally, we can conclude that the strong perturbation of injected ions independently of the relation between the mobilities of injected and ambient ions moves much slower than the small perturbation. The outflux for most of the injected ions is of the order of $n_2^{(0)}b_{1\perp}E_0$ and hence is determined by the mobility of injected ions, the ambient density and the ambient electric field.

Since in the drift approximation Eq. (11.18) can be reduced to Eq. (9.31) by changing the variables, the analysis of this Section is applicable to the case of plasma without the magnetic field with different mobilities of injected and ambient ions.

Impact of ionization and recombination processes in the ambient plasma. The sources and sinks in the ambient plasma, similar to the case $\vec{B}=0$ which is analyzed in the preceding Chapter, leads to new qualitative effects. Here we consider the simplest processes in the ambient plasma in the τ approximation. In the drift approximation instead of Eqs. (11.18) we have

$$\frac{\partial n_1}{\partial t} + b_{2\perp} E_0 n_2^{(0)} \frac{\partial}{\partial y} \frac{b_{1\perp} n_1}{b_{1\perp} n_1 + b_{2\perp} n_2} = 0 ;$$
$$\frac{\partial n_2}{\partial t} + b_{2\perp} E_0 n_2^{(0)} \frac{\partial}{\partial y} \frac{b_{2\perp} n_{12}}{b_{1\perp} n_1 + b_{2\perp} n_2} = \frac{n_2^{(0)} - n_2}{\tau} . \tag{11.28}$$

For the case of equal mobilities $b_{1\perp}=b_{2\perp}=b$ equation system (11.28) coincides with Eqs. (9.67), so the corresponding analysis of Section 9.8 is applicable to Eq. (11.28). For the different mobilities and $\tau \to 0$ the ambient density in the first equation can be considered as a constant one. Hence, for injected ions [4]

$$\frac{\partial \tilde{n}_1}{\partial \tilde{t}} + \frac{1}{(\beta \tilde{n}_1 + 1)^2} \frac{\partial \tilde{n}_1}{\partial \tilde{y}} = 0 . \tag{11.29}$$

Analogously to the case of equal mobilities, the shock is formed on the rear side of the inhomogeneity. For $b_{2\perp}>b_{1\perp}$ and the perturbation magnitude which satisfies Eq. (11.24), the time scale of the shock formation, Eq. (9.16), is large - of the order of $t_c \sim ab_{2\perp}/(Ab_{1\perp}^2 E_0)$. The correction δn_2 can be obtained from the second Eq. (11.28), analogously to the case $\vec{B}=0$, Section 9.8. For $A\beta<<1$ at times $t>>\tau$ we find

$$\delta n_2 = b_{1\perp} E_0 \tau \frac{\partial n_1}{\partial y} .$$

The condition $\delta n_2 << n_2^{(0)}$ is fulfilled if

$$\tau << A^{-1} a/(b_{1\perp} E_0). \tag{11.30}$$

Hence, the motion of injected ions in the unperturbed electric field which is described by Eq. (11.29) is possible only at the beginning under rather severe restriction Eq. (11.30) and at times smaller than t_c.

For the very strong inhomogeneities $\beta A>>1$ shock at the rear side is created during $t \sim t_c \sim a/(b_{1\perp} E_0)$, and the condition $\delta n_2 << n_2^{(0)}$ corresponds to

$$\tau << a/(b_{2\perp} E_0). \tag{11.31}$$

Numerical modelling [5] demonstrates that for the large characteristic times of recombination τ, when the inequalities opposite to Eqs. (11.30), (11.31) are valid, shocks arise both on the front and rear sides of the inhomogeneity.

The shock at the rear side exists up to the values $\tau \sim a/(b_{1\perp}E_0)$. An example of modelling is shown in Fig. 11.3.

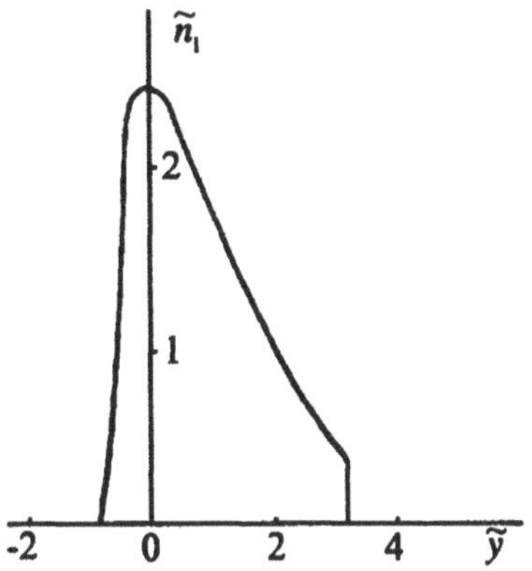

Figure 11.3 Numerical modelling of the evolution of the density profile of injected ions in the presence of ionization and recombination processes in the ambient plasma. Mobilities of the ions are different: $b_{1\perp}/b_{2\perp}=1/5$; $A=4$; $t=1.2a/b_{1\perp}E_0$. The quadratic recombination in Eq. (11.28) was chosen: $S=\alpha n_2(n_1+n_2)$, the coefficient $\alpha = b_{1\perp}E_0/(an_2^{(0)})$.

Inhomogeneity infinite in Pedersen direction. Consider now the 1D situation when the density gradient is parallel to the x axis, $\vec{x}\perp\vec{B}$, while the ambient electric field E_y is parallel to the y axis. In the absence of the ionization and recombination the equation system in the drift approximation has the form

$$\begin{aligned}
&\frac{\partial n_e}{\partial t} + b_{e\wedge}E_y\frac{\partial n_e}{\partial x} = 0;\\
&\frac{\partial n_1}{\partial t} + b_{1\wedge}E_y\frac{\partial n_1}{\partial x} + b_{1\perp}\frac{\partial n_1 E_x}{\partial x} = 0;\\
&\frac{\partial n_2}{\partial t} + b_{2\wedge}E_y\frac{\partial n_2}{\partial x} + b_{2\perp}\frac{\partial n_2 E_x}{\partial x} = 0.
\end{aligned} \tag{11.32}$$

Here we neglected the small Pedersen mobility of electrons: $b_{e\perp}=0$, while the other Hall and Pedersen components of the mobility tensors are defined by Eq. (2.91). The E_y component of electric field, as has been already mentioned in the previous Section, should be constant due to the potential character of the field. The inhomogeneous electric field E_x arises in this geometry due to the difference in the Hall mobilities of electron and ions. The electric field can be obtained from the condition of current conservation in the x direction:

$$E_x = E_y \frac{b_{e\wedge} n_e - b_{1\wedge} n_2 - (b_{e\wedge} - b_{2\wedge}) n_2^{(0)}}{b_{1\perp} n_1 + b_{2\perp} n_2}. \tag{11.33}$$

In the case of equal ion mobilities, $\hat{b}_1 = \hat{b}_2 = \hat{b}$, we have

$$E_x = E_y \frac{b_{e\wedge} - b_{\wedge}}{b_{\perp}} (1 - n_2^{(0)} / n_e). \tag{11.34}$$

For the unmagnetized ions (x_i<<1) the ratio of two components of electric field $E_x/E_y \sim b_{e\wedge}/b_{\perp}$>>1. In other words, the ambient electric field is strongly amplified in the presence of the inhomogeneity and changes its direction (from y-directed to x-directed). The perturbation of electron density, according to the first of Eqs. (11.32), in this case moves with the constant velocity $b_{e\wedge} E_y$. Furthermore, substituting Eq. (11.34) into Eqs. (11.32), in the reference frame moving with the electron velocity $b_{e\wedge} E_y$, we find

$$\begin{aligned} &\frac{\partial n_1}{\partial t} - (b_{e\wedge} - b_{\wedge}) E_y \frac{\partial}{\partial x} \frac{n_1 n_2^{(0)}}{n_e} = 0; \\ &n_e(x) = n_1 + n_2 = n_{e0}(x). \end{aligned} \tag{11.35}$$

This equation differs from Eq. (9.39) only by a constant factor. Therefore, in the laboratory reference frame the solution has the features discussed in Section 9.3. For A>1 the majority of the injected ions move with electrons with their velocity $b_{e\wedge} E_y$ due to the strong electric field E_x. The tail with the density $n_1 \approx n_2^{(0)}$ is left behind, its end moves with the velocity $b_{\wedge} E_y$. The character of the evolution is shown in Fig. 9.3 (in this case it corresponds to the negative velocity $b_{e\wedge} E_y$).

In the case of different ion mobilities $\hat{b}_1 \neq \hat{b}_2$ the substitution of Eq. (11.34) into Eq. (11.32) results in rather cumbersome expressions. The flux of injected ions is a nonlinear function of their density, and the shocks on the ion density profiles are formed.

11.3 ONE-DIMENSIONAL EVOLUTION OF LARGE-SCALE IONOSPHERIC IRREGULARITIES IN TWO-LAYER MODEL OF IONOSPHERE

We shall show here that the solutions obtained in the two previous Sections reproduce the most essential features of the 1D evolution of natural and artificial large-scale ionospheric irregularities. In the ionosphere the plasma density becomes considerable starting from heights of the order of 100km. It increases further up to heights of the order of 300km, and decreases afterwards. On the other hand, the neutral gas density decays vertically with spatial scale of the order of 10km. Therefore, starting from the heights ~ 80km electrons become magnetized by the Earth's magnetic field, and beginning with heights ~ 100-120km ions become magnetized too. Since the electron and the ion temperatures in the ionosphere have the same order of magnitude, the plasma is magnetized (the ion transverse mobility exceeds the electron one) starting from heights of the order of 100km. The transverse (Pedersen) conductivity has thus strong vertical dependence and reaches its maximum value somewhere at heights of the order of 120km [6]. Indeed, below 100km the ionospheric transverse conductivity is practically negligible due to the small plasma density, while above the layer with the maximal value of the transverse conductivity it is proportional to the ion-neutral collision frequency ν_{iN} and, hence, strongly decreases with height.

In this Section we consider the evolution of plasma inhomogeneities which consist of ions of the same or different species than the ionospheric ions. For simplicity the magnetic field is assumed to be vertical, $\vec{z}\|\vec{B}$, which models the polar or middle-latitude ionosphere. The finite size of the ionosphere can be taken into account by introducing two boundaries: z=-L below the height 100km, and the upper boundary $z=z_1$ above the inhomogeneity position, Fig. 11.4. The initial inhomogeneity is centered at z=0 and has the longitudinal scale $a_{\|}$. We chose $L \gg z_1 \gg a_{\|}$. In accordance with the steep decrease of the transversal conductivity above and below the conducting layer, the imposed boundary conditions correspond to the dielectric surfaces:

$$\Gamma_{ez}\Big|_{z=-L,z_1} = (\Gamma_{iz} + \Gamma_{Iz})\Big|_{z=-L,z_1} \approx \Gamma_{iz}\Big|_{z=-L,z_1}, \tag{11.36}$$

where the subscripts I, i denote accordingly the injected and the ambient ions.

It should be noted that in the process of evolution all additional perpendicular conductivity remains localized in the upper ionosphere and the magnetosphere. It is connected both with collisional mechanism, as well as with the polarization drift of ions in the Alfven wave which is launched from the ionosphere upwards to the magnetosphere.

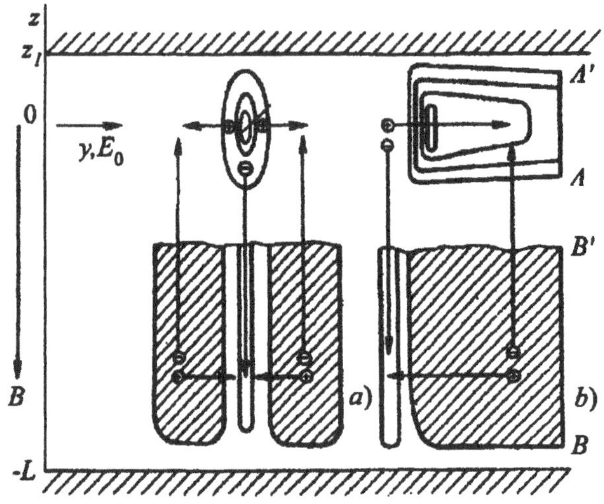

Figure 11.4. Scheme of the short-circuiting of the ionospheric currents. In the upper part the equidensities of injected ions are shown, the lower part corresponds to the ambient ionosphere. *a*) the case of pure diffusion; *b*) evolution in the strong ambient electric field; hatched are the depleted regions; *A-A'*, *B-B'* the diffusive shocks; arrows indicate the short-circuiting currents.

The conductivity associated with Alfven wave is caused by the temporal variation of the electric field in the ionosphere and the magnetosphere which cause the polarization drift of ions. As a result, the boundary condition at the surface $z = z_1$ has a more complicated form than Eq. (11.36) [7]. The corresponding effects were analyzed in [8] - [10]. However, when the integral conductivity of the ionosphere is sufficiently large, these effects have little influence on the character of the density inhomogeneity evolution.

In the reference frame of the neutral gas the initial equations has the form

$$\frac{\partial n_\alpha}{\partial t} - \vec{\nabla}\cdot(\hat{D}_\alpha \nabla n_\alpha \pm \hat{b}_\alpha n_\alpha \vec{\nabla} n_\alpha) = I_\alpha - S_\alpha\,; \qquad (11.37)$$

$$n_e = n_i + n_I;\; I_e = I_i + I_I;\; S_e = S_i + S_I;\; \alpha = i, I, e.$$

The homogeneous uniform motion of the neutral gas can be taken into account in the moving reference frame by supplement the additional electric field $[\vec{u}_N \times \vec{B}]$. The other boundary conditions for Eqs. (11.37) are

$$n_I(x,y \to \infty) = 0; \; n_i(x,y \to \infty) = n_0; \; \varphi(x,y \to \infty) = -\vec{E}_0\vec{r}. \tag{11.38}$$

Since in the ionosphere the condition $\mu_{iN}\nu_{iN} >> m_e\nu_{ei}$ is usually satisfied, the Einstein relation is valid for the parallel transport coefficients, see Eq. (2.89).

Two-layer model of the ionosphere. For the ionospheric inhomogeneities with large transverse scales $a_\perp$ (for criterion see below Section 11.9), the region, where the electric field and the plasma density are perturbed, is very stretched along the magnetic field. In the absence of ambient electric field $\vec{E}_0$ this longitudinal scale, as was shown in Chapter 6, is determined by the electron diffusion: $l_\| \sim [4(1+T_i/T_e)D_{e\|}t]^{1/2}$. During the short time of the order of the few tens of seconds (which is not sufficient for large-scale perturbation to spread considerably) this scale exceeds 100km, i.e., becomes comparable with the width of the ionosphere. In a strong electric field the propagation of the electric field perturbation along $\vec{B}$ occurs even faster, as it will be shown in the subsequent Sections. Even at the initial moment this scale in the homogeneous plasma is given by the estimate $l_\| \sim a_\perp(b_{e\|}/b_{i\perp})^{1/2}$. For $a_\perp \sim 1$km this scale far exceeds the vertical size of the ionosphere. On the other hand, the longitudinal spreading of the initial perturbation of injected ions density is controlled by the ion diffusion. Hence, the longitudinal scale $a_\|$ of the injected ions perturbation usually remains much smaller than L.

Since the electron signal can very quickly reach the ionospheric boundaries, the potential should correspond to the Boltzmann distribution for electrons

$$\varphi = \frac{T_e}{e}\ln\frac{n_e}{n_0} + \Psi(x,y). \tag{11.39}$$

On the other hand, since it takes a lot of time for ions to reach the ionospheric boundaries, the boundary condition Eq. (11.36) is reduced to $\Gamma_{e\|} \approx 0$. We shall restrict our consideration to the case when the strong perturbation of injected ions $n_I >> n_i$ is created above the zone of maximum conductivity (in the opposite case the injected ions just move in the unperturbed ambient electric field, spreading due to their unipolar diffusion). Substituting Eq. (11.39) into Eq. (11.37), we obtain

$$\frac{\partial n_e}{\partial t} + \vec{\nabla} \cdot (\hat{b}_e n_e \vec{\nabla} \Psi) = I_e - S_e;$$
$$\frac{\partial n_I}{\partial t} - \vec{\nabla} \cdot [(1 + \frac{T_e}{T_i}) \hat{D}_I \vec{\nabla} n_I + \hat{b}_I n_I \vec{\nabla} \Psi] = I_I - S_I; \qquad (11.40)$$
$$\frac{\partial n_i}{\partial t} - \vec{\nabla} \cdot [(1 + \frac{T_e}{T_i}) \hat{D}_I \vec{\nabla} n_i + \hat{b}_i n_i \vec{\nabla} \Psi] = I_i - S_i.$$

The second equation here describes the evolution of the injected ions in the region with the longitudinal scale $a_{||}$, and the third equation corresponds to the evolution of the ambient ions in the whole ionosphere. The perpendicular electron diffusion and mobility are neglected in the first equation because electrons are strongly magnetized.

Two-layer model of the ionosphere has been introduced in [12], see also [13], [3]. In this model the ambient ionosphere below the initial inhomogeneity was approximated by the plasma layer with large average transverse mobility and diffusion tensors $\hat{b}_2$, $\hat{D}_2$. The electron Hall mobility here is $b_{e\wedge}=1/B$. We shall refer to this region as to the layer 2, while we shall call the region which contains the initial inhomogeneity the layer 1. To pass to such description, we integrate Eqs. (11.40) over z neglecting the parallel ion fluxes to the boundaries z=L, z_1. Integration yields

$$\frac{\partial N}{\partial t} + \vec{\nabla}_\perp \cdot (\hat{b}_e N \vec{\nabla}_\perp \Psi) = \bar{I}_e - \bar{S}_e;$$
$$\frac{\partial N_1}{\partial t} - \vec{\nabla}_\perp \cdot [(1 + \frac{T_e}{T_i}) \hat{D}_1 \vec{\nabla}_\perp N_1 + \hat{b}_1 N_1 \vec{\nabla}_\perp \Psi] = I_1 - S_1; \quad . \qquad (11.41)$$
$$\frac{\partial N_2}{\partial t} - \vec{\nabla}_\perp \cdot [(1 + \frac{T_e}{T_i}) \hat{D}_2 \vec{\nabla}_\perp N_2 + \hat{b}_2 N_2 \vec{\nabla}_\perp \Psi] = I_2 - S_2.$$

where

$$N_1 = \int_{-\infty}^{\infty} n_I dz; \; N_2 = \int_{-\infty}^{\infty} n_i dz; \; \hat{b}_1 \equiv \hat{b}_I; \; \hat{D}_1 \equiv \hat{D}_I; \; N = N_1 + N_2;$$

$$I_{1,2} = \int_{-L}^{z_1} I_{I,i} dz; \; R_{1,2} = \int_{-L}^{z_1} R_{I,i} dz; \; \bar{I}_e = I_1 + I_2; \; \bar{R}_e = R_1 + R_2.$$

The average mobility and diffusion tensors in the layer 2 are defined as

$$\hat{b}_2 = \int_{-L}^{z_1} \hat{b}_i n_0(z)dz / N_2^{(0)}\ ; \quad \hat{D}_2 = \frac{T_i}{e}\hat{b}_2. \tag{11.42}$$

It is necessary to emphasize that this definition does not follow rigorously from the Eqs. (11.40). Instead, the value

$$\hat{\tilde{b}}_2 = \int_{-L}^{z_1} \hat{b}_i n_i(x,y,z)dz / N_2^{(0)}\ ; \quad \hat{\tilde{D}}_2 = \frac{T_i}{e}\hat{\tilde{b}}_2. \tag{11.43}$$

is obtained by the integration, which depends on the instantaneous density profile during the evolution and thus does not coincide with Eq. (11.42). Hence, in general, the two-layer model of the ionosphere can give only qualitative description of the inhomogeneity evolution. Only in the special cases of infinite in the Hall direction inhomogeneity in the absence of ionization and recombination or without the ambient electric field, this model can be rigorously justified, see below. In these cases Eq. (11.43) coincides with Eq. (11.42). The boundary conditions for Eqs. (11.41) can be derived from Eq. (11.38):

$$\begin{aligned} &N_1(x,y\to\infty)=0;\ N_2(x,y\to\infty)=N_2^{(0)}; \\ &\Psi(x,y\to\infty)=-\vec{E}_0\vec{r}. \end{aligned} \tag{11.44}$$

Diffusive spreading. In the absence of the ambient electric field from Eq. (11.41) for the case I_α=S_α=0 we have

$$N=N_1+N_2=N(x,y). \tag{11.45}$$

Summing up two ion Eqs. (11.41), employing the boundary conditions Eq. (11.44), in 1D case we find

$$\Psi = -\frac{T_e+T_i}{e}\ln\frac{D_1N_1+D_2N_2}{D_2N_2^{(0)}}. \tag{11.46}$$

Substitution of Eq. (11.46) into Eqs. (11.41) yields

$$\begin{aligned} \frac{\partial N_{1,2}}{\partial t} &= \vec{\nabla}\cdot\{(1+T_e/T_i)D_{1,2\perp}[\vec{\nabla}N_{1,2} \\ &- N_{1,2}\vec{\nabla}\ln(D_{1\perp}N_1+D_{2\perp}N_2)]\}. \end{aligned} \tag{11.47}$$

This equation only by a factor $(1+T_e/T_i)$ differs from Eq. (5.57) which was analyzed in Section 5.4 for the case of equal mobilities.

We shall show here that the profile of N_1 does not change significantly also in the case $D_{2\perp}>>D_{1\perp}$. We consider the infinite in x direction pinch, and we assume that for the Gaussian profile

$$N_{10}(y)=A(-y^2/a^2)N_2^{(0)} \tag{11.48}$$

Eq. (11.24) is satisfied. As in the case $D_{1\perp}=D_{2\perp}$, for $t>>a^2/4D_{1\perp}$ in the inner region $N_{1,2}=C_{1,2}N$. At the boundary of the inner and the outer regions, $|y|\sim a$, where the depletion of the ambient ions is maximal, $N_2\geq A^{-1}N_2^{(0)}>>\beta^{-1}N_2^{(0)}$ $(\beta=b_{1\perp}/b_{2\perp})$. Therefore, expanding Eq. (11.47) in series in $\beta N_1/N_2$, taking into account the fact that in the outer region $N_1\approx N_2^{(0)}$, we obtain the heat conduction equation with the nonlinear heat capacity:

$$\exp(M)\frac{\partial M}{\partial t}=(1+\frac{T_e}{T_i})D_{i\perp}\frac{\partial^2 M}{\partial y^2}; \quad M=\ln N_2. \tag{11.49}$$

When $C_1>>C_2$ the value $N_2\to 0$ at the boundary of the regions, $|y|\sim a$, so that in the outer region $\partial^2 M/\partial y^2\approx 0$. As a result, in the outer region

$$N_2(y)=N_2(0)\exp\left[\frac{\Gamma|y|}{(1+T_e/T_i)D_{i\perp}N_2^{(0)}}\right],$$

where Γ is the flux of the ambient ions 2 from the depletion region to the inner region. The boundary value of $N_2(0)$ can be found by matching the outer and the inner solutions: $N_2(0)=(C_2/C_1)\,N_2^{(0)}$. The condition of the particle conservation in the inner region yields

$$2\Gamma=(dC_2/dt)\pi^{1/2}/A.$$

Now we can estimate the length of the depletion region L_y, where N_2 becomes of the order of unity:

$$L_y\approx\frac{4AN_2^{(0)}D_{1\perp}}{a\sqrt{\pi}dC_2/dt}\ln\frac{C_1}{C_2}\approx\frac{\sqrt{\pi}}{2}\frac{C_2}{AN_2^{(0)}}.$$

It follows finally that with logarithmic accuracy

$$L_y\approx 2\{D_{1\perp}t\ln[C_1(t)/C_2(t)]\}^{1/2}. \tag{11.50}$$

Therefore, the outflow of injected ions from the central peak occurs analogously to the case $D_{1\perp}=D_{2\perp}$ provided the condition Eq. (11.24) is satisfied. When most of injected ions leave the central peak, the value of $\tilde{N}_2 = N_2 / N_2^{(0)}$ becomes of the order of unity. Then, expanding $\tilde{N}_2$ in series in $1-\tilde{N}_2$, we obtain for $\tilde{N}_2$ (and also for $\tilde{N}_1$) the diffusion equation with the effective diffusion coefficient $(1+T_e/T_i)D_{1\perp}$. It is the same equation as in the case $D_{1\perp}=D_{2\perp}$. The N_1 profile in the inner region is proportional to N. As well, as for $D_{1\perp}=D_{2\perp}$, the solution has the form of Eq. (5.55) where D is replaced by $(1+T_e/T_i)D_{1\perp}$.

However, in contrast to the situation analyzed in Section 5.4, the ions injected into the ionosphere and the ambient ions are separated in space. The injected ions spread across the magnetic field at the height $z=0$, and the return current of the ambient ions is distributed over the ionosphere below the initial inhomogeneity, Fig. 11.4*a*. The electrons are redistributed along the magnetic field line to provide quasineutrality, and their integrated over z density is conserved, Eq. (11.45). The character of the diffusion thus resembles the magnetized plasma diffusion in a dielectric vessel analyzed in Section 7.1. The difference is connected with the condition $a_{\parallel} \ll L$.

Let us show now that the expressions for the average mobility in the layer 2 given by Eqs. (11.42) and (11.43) coincide provided the condition Eq. (11.24) is satisfied. In other words, in the absence of the ionization and recombination processes, Eqs. (11.41) of the two-layer model of the ionosphere can be rigorously derived from Eqs. (11.37). Indeed, the profiles of $N_{1,2}$ for $D_{1\perp} \ll D_{2\perp}$ are independent of $D_{2\perp}$, and the characteristic time scale of their evolution (and of the evolution of the profiles $n_{1,2}$) is $t < a^2/D_{1\perp}$. Below the initial inhomogeneity we can assume $D_{i\perp} \gg D_{1\perp}$, while the contribution from the ionosphere above the inhomogeneity to the return current is negligible. Therefore, neglecting the temporal derivative in Eq. (11.37) for the ambient ions, with the accuracy up to $D_{1\perp}/D_{2\perp}$, we have

$$(1+T_e/T_i)D_{i\perp}\vec{\nabla}_\perp n_i + n_i b_{i\perp}\vec{\nabla}_\perp \Psi = 0 . \tag{11.51}$$

Comparing Eq. (11.51) with Eq. (11.46), for the condition Eq. (11.24), we find

$$\frac{n_i(y,z,t)}{n_0(z)} \approx \frac{\int_{-L}^{z_1} D_{i\perp}(z) n_i(y,z,t)dz}{\int_{-L}^{z_1} D_{i\perp}(z) n_0(z)dz},$$

and finally

$$\frac{n_i(y,z,t)}{n_0(z)} = \frac{N_2(y,t)}{N_2^{(0)}}. \tag{11.52}$$

Comparing Eq. (11.52) with Eq. (11.43), we can now conclude that the mobility tensor, Eq. (11.43), is time independent and coincides with Eq. (11.42). Eq. (11.52) gives the height distribution of the perturbed ambient density as a function of the height integrated density N_2 and unperturbed density profile.

If $A >> b_{2\perp}/b_{1\perp}$, the depth of the depletion regions in the ambient ionosphere becomes significant. Eq. (11.52) is not valid in this case, and to obtain the height distribution of the ambient density it is necessary to solve the equation for n_i. The n_i profiles at different heights are not similar, and the two-layer model gives only qualitative description of the process. In the inner region the distributions both of N_1 and of n_i are again the Boltzmann distributions: $n_i = C_i(z,t)N$. The outer region can be subdivided into two subregions. In the first of them the value of n_i is small, and the field-driven fluxes can be neglected in Eq. (11.40) for n_i. This equation describes the diffusion of ambient ions with large diffusion coefficient $(1+T_e/T_i)D_{i\perp}(z)$ and with the zero boundary condition at the boundary between the inner and outer regions. The corresponding scale is

$$L_y \sim [4(1+T_e/T_i)D_{i\perp}(z)t]^{1/2}. \tag{11.53}$$

This scale determines the outflux of the injected ions. Their density in the outer region is given by an estimate

$$N_1(y) = \int_{-L}^{z_0(y)} n_0(z)dz, \tag{11.54}$$

where z_0 is determined by the condition $L_y(z_0) \sim y$. In the second subregion the profile $n_i(z,y)/n_0(z)$ is z-independent, and the density $n_i(z,y)$ tends to $n_0(z)$. The length of the second subregion is smaller than L_y, Eq. (11.53).

The number of injected ions which outflow from the ambipolar peak can be estimated by integrating Eq. (11.54) over y. Here the main contribution gives the region of maximum transverse conductivity with the maximum diffusion coefficient $D_{i\perp}^{(m)}$. The outflow time is given by the estimate $t \sim (\pi\tilde{A}a)^2 / [16(1+T_e/T_i)D_{i\perp}^{(m)}]$ (see Section 5.4), where $\tilde{A}$ is the ratio of the value of N_1 to the integral of $n_0(z)$ over the high conductivity zone.

Afterwards the outflow velocity is controlled by the diffusion coefficient $D_{I\perp}$.

We can conclude that in all the cases considered the spreading of injected ion density is determined by the amplitude A (or $\tilde{A}$). In other words, the key parameter is the ratio of integrated along the magnetic field injected ion density to the integral of ambient ion density over the region below the initial perturbation. If $A<1$, the spreading of injected ions occurs with the diffusion coefficient $(1+T_e/T_i)D_{I\perp}$. In the opposite case the diffusion is decelerated considerably, since ambient density becomes strongly depleted. The longitudinal spreading of n_I and n_i profiles is determined by the parallel ambipolar diffusion. At later stages, when injected ions are already spread to a large volume, the density perturbation is localized in the lower ionosphere, and its evolution is ambipolar both along and across the magnetic field.

In many cases evolution of injected ions is of the main interest. As an important example we present here the expression for injected ion density profile for the intermediate nonlinearity, Eq. (11.24), at $t>>a^2/[4(1+T_e/T_i)D_{I\perp}]$, $D_{e\perp}=0$:

$$n_I(\rho,t)=\frac{A\exp\left[-\dfrac{z^2}{a_\parallel^2+4(1+T_e/T_i)D_{I\parallel}t}\right]}{4(1+T_e/T_i)D_{I\perp}t[a_\parallel^2+4(1+T_e/T_i)D_{I\parallel}t]^{1/2}}, \qquad (11.55)$$
$$\times\left\{A\exp(-\frac{\rho^2}{a^2})+\exp\left[-\frac{\rho^2}{4(1+T_e/T_i)D_{I\perp}t}\right]\right\}.$$

The ionization in the low ionosphere enhances amount of injected ions which can outflow with the unipolar diffusion coefficient. The recombination in the low ionosphere reduces the lifetime of the ambipolar peak which is formed in this region.

Evolution in the strong ambient electric field. Situations when the spreading of the strong inhomogeneities is determined by the diffusion alone are rather rare. More frequent are the cases, when for the inhomogeneities with the transverse scales $a_\perp>100$m the condition $eE_0a_\perp>T_{e,i}$ is satisfied. This condition coincides with the criterion of the validity of the drift approximation in the 1D problem. Moreover, the impact of the ambient electric field on the inhomogeneity deformation in the Hall direction becomes important at far smaller values of electric fields or $a_\perp$, due to the differences in the Hall and Pedersen mobilities.

We consider here inhomogeneity infinite in the Hall direction. As demonstrated in the subsequent Sections, such a plasma band is formed at the final stage of the evolution of a dense plasma cloud of arbitrary initial shape. From Eqs. (11.41) we obtain the equation system which coincides with Eq. (11.18), where the densities $n_{1,2}$ are replaced by the integral densities $N_{1,2}$, and the diffusion coefficients $D_{1,2\perp}$ by the coefficients $(1+T_e/T_i)D_{1,2\perp}$ ($I_\alpha=S_\alpha=0$):

$$\frac{\partial N_{1,2}}{\partial t}+b_{2\perp}E_0N_2^{(0)}\frac{\partial}{\partial y}\left(\frac{b_{1,2\perp}N_{1,2}}{b_{1\perp}N_1+b_{2\perp}N_2}\right)$$
$$-\frac{\partial}{\partial y}\left\{(1+\frac{T_e}{T_i})D_{1,2\perp}\left[\frac{\partial N_{1,2}}{\partial y}-N_{1,2}\frac{\partial}{\partial y}\ln(b_{1\perp}N_1+b_{2\perp}N_2)\right]\right\};$$
$$N=N_1+N_2=N_0(y). \tag{11.56}$$

This equation system was analyzed in detail in the previous Section. Therefore, as in the case of the diffusion, the character of the evolution depends critically on the parameter A in Eq. (11.48). If $A<<1$, the ambient electric field remains almost unperturbed, and initial inhomogeneity moves in this electric field and spreads due to the diffusion. For $A>1$ the nonlinear evolution of an inhomogeneity determines its motion and spreading across the magnetic field, see Fig. 11.1, the spatial positions of the perturbations are shown in Fig. 11.4*b*.

From the solution of the 1D problem of Section 11.2 it follows that the outflow velocity of injected ions in $\vec{E}_0$ direction is independent on $b_{2\perp}$ provided the condition Eq. (11.24) is satisfied. Hence, as in the case of the diffusion, the flux of the ambient ions towards the inner region remains approximately constant (with the accuracy $\sim b_{1\perp}/b_{2\perp}$). In the drift approximation from Eq. (11.40) we thus obtain

$$-b_{i\perp}n_i(y,z)\frac{\partial\Psi}{\partial y}=b_{i\perp}n_0(z)E_0. \tag{11.57}$$

Eq. (11.57) with account of Eq. (11.17) for the condition Eq. (11.24) yields

$$\frac{n_i(y,z)}{n_0(z)}=-\frac{E_0}{\partial\Psi/\partial y}=\frac{N_2(y)}{N_2^{(0)}},$$

and Eqs. (11.42) and (11.43) coincide. In other words, the two-layer model of the ionosphere is rigorously justified in this case. On the contrary, for $A>>b_{2\perp}/b_{1\perp}$ the value $\hat{\hat{b}}_2$ should be employed, and the two-layer model only quantatively describes the situation.

The pattern of the currents short-circuiting, which is shown in Fig. 11.4*b*, has been put forward in [14] where experiments with the injection of the barium clouds into the ionosphere were analyzed. In this paper the possibility of the depletion regions formation in the ambient ionosphere was also discussed. For understanding of the character of motion of the ionized plasma clouds it is very important to know the magnitude of the perturbation of the electric field which determines the cloud velocities. According to Eq. (11.17), its magnitude is given by

$$E = \frac{b_{2\perp} N_2^{(0)} E_0}{b_{1\perp} N_1 + b_{2\perp} N_2} = \frac{\Sigma_{2\perp}^{(0)} E_0}{\Sigma_{1\perp} + \Sigma_{2\perp}}. \tag{11.58}$$

Here $\Sigma_{1\perp}=eb_{1\perp}N_1$, $\Sigma_{2\perp}=eb_{2\perp}N_2$ are the integral Pedersen conductivities of the injected cloud and the ambient ionosphere accordingly. It follows from Eq. (11.58) that the electric field perturbation is small if $\Sigma_{1\perp}<<\Sigma_{2\perp}$. In [24] the simplified condition $\Sigma_{1\perp}<<\Sigma_{2\perp}^{(0)}=eb_{2\perp}N_2^{(0)}$ has been suggested. However, as shown in Section 11.1, in the absence of the ionization and recombination processes, the criterion for small electric field perturbation is much more severe: $N_1<<N_2^{(0)}$. The criterion $\Sigma_{1\perp}<<\Sigma_{2\perp}^{(0)}$ corresponds to small electric field perturbation only at the initial stages of the evolution, $t<t_1=aN_2^{(0)}/(b_{1\perp}E_0N_1)$, or in the presence of large ionization source which satisfies Eqs. (11.30) and (11.31). Generally, the correct criterion is $N_1<<N_2^{(0)}$, because the two conditions, $\Sigma_{1\perp}<<\Sigma_{2\perp}$. and $\Sigma_{1\perp}<<\Sigma_{2\perp}^{(0)}$, strongly differ from each other due to the strong depletion of the ambient ionosphere.

11.4 TWO-DIMENSIONAL MOTION AND DEFORMATION OF PLASMA CLOUDS

In the 2D case when the Hall fluxes are important the evolution pattern becomes far more complicated. We shall consider here the main physical mechanisms of the inhomogeneity evolution taking as an example the evolution of the plasma clouds in the two-layer model of the ionosphere. In other words, we shall analyze the motion and spreading of the cloud of injected ions, while ions of the ambient plasma will have the same or larger perpendicular mobility. In the absence of sources and sinks, in the drift approximation the 2D starting equations are obtained from Eqs. (11.37):

$$\frac{\partial N}{\partial t} - b_{e\wedge}\left(\frac{\partial \Psi}{\partial y}\frac{\partial N}{\partial x} - \frac{\partial \Psi}{\partial x}\frac{\partial N}{\partial y}\right) = 0; \tag{11.59}$$

$$\frac{\partial N_{1,2}}{\partial t} - b_{1,2\wedge}\left(\frac{\partial \Psi}{\partial y}\frac{\partial N_{1,2}}{\partial x} - \frac{\partial \Psi}{\partial x}\frac{\partial N_{1,2}}{\partial y}\right)$$
$$-\frac{\partial}{\partial x}\left(b_{1,2\perp}N_{1,2}\frac{\partial \Psi}{\partial x}\right) - \frac{\partial}{\partial y}\left(b_{1,2\perp}N_{1,2}\frac{\partial \Psi}{\partial y}\right) = 0; \tag{11.60}$$
$$N = N_1 + N_2 .$$

Equation for the potential can be obtained by subtracting Eqs. (11.60) from Eq. (11.59):

$$\vec{\nabla}_\perp \cdot [(b_{1\perp}N_1 + b_{2\perp}N_2)\vec{\nabla}_\perp \Psi] - \frac{\partial \Psi}{\partial y}\frac{\partial (b_{e\wedge}N - b_{1\wedge}N_1 - b_{2\wedge}N_2)}{\partial x}$$
$$+\frac{\partial \Psi}{\partial x}\frac{\partial (b_{e\wedge}N - b_{1\wedge}N_1 - b_{2\wedge}N_2)}{\partial y} = 0. \tag{11.61}$$

Pure plasma in the absence of Hall current. We start analysis of Eqs. (11.59) and (11.60) with the case of the pure plasma where injected and ambient ions are of the same nature, so that $b_{1\perp}=b_{2\perp}=b_\perp$ (the so-called one-layer model of the ionosphere). We also consider the case of the equal Hall mobilities of electrons and ions $b_{1\wedge}=b_{2\wedge}=b_{e\wedge}=b_\wedge$ which corresponds to the case of magnetized ions ($x_{1,2}>>1$). Hence, the Hall current is absent. Then Eq. (11.61) coincides with the equation for the electrostatic potential in the medium with the nonuniform dielectric permittivity:

$$\vec{\nabla}_\perp \cdot (N\vec{\nabla}_\perp \Psi) = 0 . \tag{11.62}$$

The potential can be easily obtained for the initial density profile of the form

$$N_1 = AN_2^{(0)} \quad for\ \rho = \sqrt{x^2 + y^2} \le a ;$$
$$N_1 = 0 \qquad for\ \rho > a. \tag{11.63}$$

It corresponds to the potential of the 2D dipole [15] ($\vec{E}_0 \| \vec{y}$)

$$\vec{E} = \frac{2\vec{E}_0 N_2^{(0)}}{N + N_2^{(0)}} \qquad for\ \rho \le a ;$$
$$\vec{E} = \frac{2\vec{E}_0 N_1^{(0)}}{N + N_2^{(0)}}\frac{a^2}{\rho^2}\frac{\vec{\rho}\vec{E}_0}{\rho E_0} + \vec{E}_0 \quad for\ \rho > a. \tag{11.64}$$

Electric field is homogeneous inside the circle of radius a. For the Gaussian initial condition

$$N_{10}=A\exp(-\rho^2/a^2)N_2^{(0)} \tag{11.65}$$

electric field has been simulated in [12]. The result only slightly deviates from Eq. (11.64), and in the far zone $a>>\rho$ the two fields coincide, since the dipole moments for the profiles Eq. (11.65) and Eq. (11.63) are equal.

Electron density evolution is given by Eq. (11.59). Electric field is determined by electron density profile N, Eq.(11.62). Therefore, Eq. (11.59) corresponds to the propagation of electron density along the characteristics. On the other hand, the characteristics of Eq. (11.59) are determined by the whole N profile and vary with time. The simulation of Eqs. (11.59), (11.62) has been performed in [12], [16], the results are displayed in Fig. 11.5.

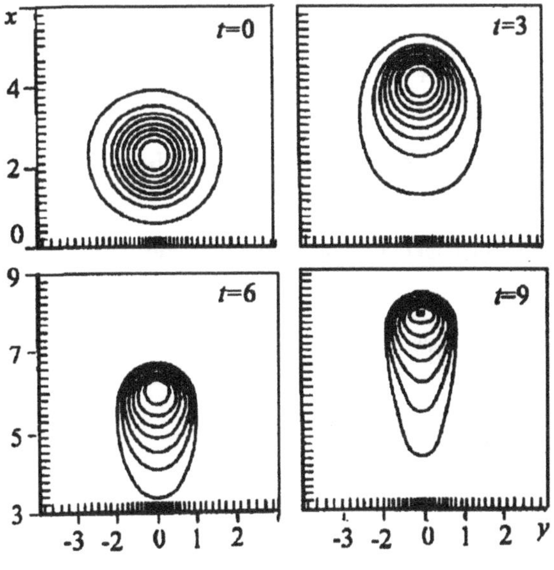

Figure. 11.5. Equidensities $\tilde{N}=N/N_2^{(0)}$ for the Gaussian initial profile, $A=4$, in the reference frame moving down with the velocity $\vec{V}_0=[\vec{E}\times\vec{B}]/B^2$ [16]. The distance is measured in the units of a, the time - in units of a/V_0.

It is clearly seen that the density maximum moves in the Hall direction slower than the ambient $\vec{E}_0\times\vec{B}$ drift. The lower density parts of the cloud move faster than the maximum. As a result, the rear side of the cloud (with respect to the $\vec{E}_0\times\vec{B}$ direction) becomes steeper, while the front side becomes more gradual. This can be understood from the fact that electric field of the type given by Eq. (11.64) has inverse density dependence, and, hence, the velocity in the Hall direction also decreases with N. At later stages the profile strongly elongates in the Hall direction. When the scale in the Hall direction d significantly exceeds the scale a in Pedersen direction, electric field can be evaluated from the condition of the current conservation in the y direction with account of the potential character of the field:

$$E_y = \frac{E_0 N_2^{(0)}}{N}\,;\ E_x = \int_0^y E_0 N_2^{(0)} \frac{\partial N^{-1}}{\partial x} dy\,. \tag{11.66}$$

In the vicinity of the x axis the transfer velocities of different parts of the N profile are determined by field E_y, because here $E_x \to 0$ and $\partial N / \partial y \to 0$. At the x axis we thus have the 1D equation

$$\frac{\partial N}{\partial t} + \frac{b_\wedge E_0 N_2^{(0)}}{N}\frac{\partial N}{\partial x} = 0\,; \qquad N(x,y) = N_{10}\left(x - \frac{b_\wedge E_0 N_2^{(0)}}{N}t, y = 0\right). \tag{11.67}$$

The example of such an evolution is shown in Fig. 11.6.

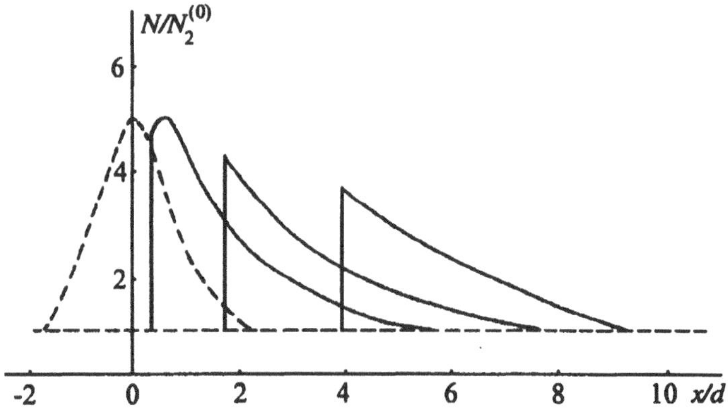

Figure 11.6. Density profile $N(x,y=0)$ for the initial perturbation $N_{10}=AN_2^{(0)}\exp(-x^2/d^2)$; $A=4$, $\tilde{t} = tb_\wedge E_0 / d$. The velocity of the density maximum for $d>>a$ is almost twice smaller than in Fig. 11.5, $d=a$, since the electric field Eq. (11.66) is almost twice smaller than the field Eq. (11.64). Dashed line corresponds to initial profile, solid lines represent profiles at different moments.

We see that shock is created at the rear side of the profile. Since the ion diffusion is absent in Eq. (11.59), the shock width should be extremely small and should be determined by the electron diffusion coefficient $D_{e\perp}$. This effect apparently was camouflaged by numerical diffusion in [12], [16]. Nevertheless, the shock width in Fig. 11.6 is significantly smaller than a. Moreover, shock in Hall direction is unstable with respect to the gradient-drift instability, see for example [12]. Its nonlinear stage was also simulated in [12], [16]. As a result of the instability development, the rear side of the cloud becomes split into separate striations. However, the average density profiles remain almost the same as the profiles presented in Fig. 11.5. The striations in the barium clouds in the ionosphere where observed in the experiments (see, for example, [17] and references therein).

The described evolution pattern corresponds to electron density evolution. The evolution of injected and ambient ions are far more complicated. Besides the drift in the Hall direction injected ions outflow into the y (Pedersen) direction, while the ambient ions are gathered towards the x axis. At the final stage $t>aA/b_{\perp}E_0$ the profile Eq. (11.67) consists mainly of ambient ions.

Plasma with two species of positive ions in the absence of Hall current. We consider here the case $b_{2\perp}>>b_{1\perp}$; $b_{1\wedge}=b_{2\wedge}=b_{e\wedge}=b_{\wedge}$. Let us analyze qualitatively the character of the evolution on the basis of 'quasionedimensional' approximation $d>>a$ when the component E_y is determined by Eq. (11.66). If $1<<A<<b_{2\perp}/b_{1\perp}$, then at first $E_y\approx E_0$, and the density profile of injected ions starts to move in the Hall direction without deformation, with the velocity $V_0=b_{\wedge}E_0$. The outflow in the Pedersen direction occurs simultaneously. During the characteristic time $t_1=a/(b_{1\perp}E_0A)$ the significant perturbation of the ambient plasma arises, and shock is formed at $y>0$ in the Pedersen direction, see Section 11.2. At $t\sim t_1$ the electric field is also perturbed significantly being of the order of E_0. The field increases at $a\geq y>0$ and decreases at $-a\leq y<0$. The Hall fluxes in such electric field cause the deformation in the x direction: the part of the profile which corresponds to $y>0$ moves faster than the region $y<0$. As a result, the profile elongates in the x direction, as shown in Fig. 11.7.

The cloud length in the x direction becomes of the order of its shift:

$$x_1=V_0t_1=ab_{\wedge}/(b_{1\perp}A). \tag{11.68}$$

Simultaneously the profile of injected ions N_1 is compressed in the y direction, owing to the drifts caused by the E_x component. Later, at $t>t_1$, the density of ambient ions N_2 increases further in the region $y>0$, so that the drift of this area in the Hall direction is practically terminated. As a result, the elongation of the cloud in the Hall direction continues.

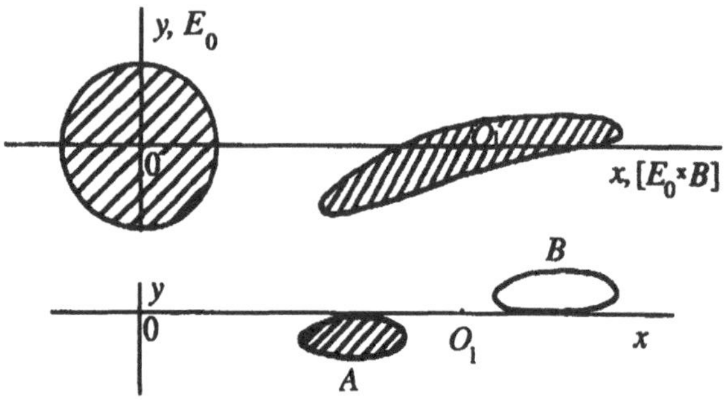

Figure 11.7. Drift and deformation of the plasma cloud in the absence of the Hall current; $1 \ll A \ll b_{2\perp}/b_{1\perp}$. In the upper part the injected ions are shown, in the lower part are the ambient ions; A corresponds to the region of increased ambient density, B to the depletion regions. The moment corresponds to $t=t_1=a/(b_{1\perp}E_0A)$, so that $OO_1=x_1$.

For $A \gg b_{2\perp}/b_{1\perp}$ the field E_y, Eq. (11.58), is small from the very beginning. If it is possible to consider the ionization and recombination rates as infinite ones, the ambient ion density remains unperturbed: $N_2=N_2^{(0)}$. Then, analogously to the previous case, near the x axis we have

$$\frac{\partial N_1}{\partial t}+\frac{b_\wedge E_0}{(1+b_{1\perp}N_1/b_2N_2^{(0)})^2}\frac{\partial N_1}{\partial x}=0. \tag{11.69}$$

The character of the deformation is similar to that presented in Figs. 11.5 and 11.6. The N_1 profile remains practically at rest, while injected ions outflow in the x direction and create a tail with the density of the order of

$$N_1^{(t)}=N_2^{(0)}b_{2\perp}/b_{1\perp}.$$

Shock is formed at the rear side ($x<0$), as in the case $b_{1\perp}=b_{2\perp}$, at a time scale $d/(b_\wedge E_0)$.

In the absence of the ionization and recombination processes the density of ambient ions in the tail is changed at a time scale

$$t_1=aN_2^{(0)}/(b_{1\perp}N_1^{(t)})=a/b_{2\perp}.$$

The electric field, Eq. (11.17), decreases in the tail, and the outflow velocity in the x direction is reduced. As a result, at $t>t_1$ the tail of the density $N_1^{(t)} \sim N_2^{(0)}$ is created, and its end moves with the unperturbed velocity $V_0 = E_0/B$.

Plasma with two species of positive ions with Hall current. In this case the electric field is given by general Eq. (11.61). We shall calculate the electric field for the step-like initial density perturbation Eq. (11.63). For such a perturbation both for $\rho > a$ and $\rho < a$ the potential Ψ satisfies the Laplace equation. In the inner region we shall seek the solution in the form of a homogeneous field inclined by a certain angle with respect to the ambient electric field E_0. In the outer region we seek the solution as a dipole electric field which decreases with distance as ρ^{-2}. Following [7] we introduce the complex variables $W=x+iy$ and the complex quantities

$$E = E_x + iE_y \quad ; \Sigma_{1,2} = \Sigma_{1,2\perp} - i\Sigma_{1,2\wedge} \ ,$$

where the Pedersen $\Sigma_{\alpha\perp}$, and the Hall $\Sigma_{\alpha\wedge}$, conductivities are defined by the expressions: $\Sigma_{\alpha\perp} = eb_{\alpha\perp}N_\alpha$, $\Sigma_{\alpha\wedge} = -e(b_{e\wedge} - b_{i\wedge})N_\alpha < 0$. The complex current is thus defined as

$$\vec{j} = (\Sigma_1 + \Sigma_2)\vec{E} \,. \tag{11.70}$$

The radial component of the current is

$$j_r = \frac{jW^* + j^*W}{2|W|} \tag{11.71}$$

(star denotes the complex conjugate value). The condition of radial current conservation gives the relation between the E_1 and E_2, the values inside ($\rho < a$) and outside the initial inhomogeneity. One more relation results from the potential continuity condition at $\rho = a$. Hence the perturbed electric field in the inner zone $E_1' = E_1 - E_0$ is related to the electric field in the outer zone $E_2' = E_2 - E_0$ as

$$E_2' = -\left(E_1' \frac{a^2}{W^2}\right)^* \; ; \; |W| = a \,.$$

Employing the last two relations, we find

$$E_1 = E_0 \frac{\Sigma_2 + \Sigma_2^*}{\Sigma_1 + \Sigma_2 + \Sigma_2^*},$$

or for the real components

$$E_{1x} = -\frac{2\Sigma_{2\perp}\Sigma_{1\wedge}E_0}{(\Sigma_{1\perp}+\Sigma_{2\perp})^2+\Sigma_{1\wedge}^2};$$
$$E_{1y} = \frac{2\Sigma_{2\perp}(\Sigma_{1\perp}+2\Sigma_{2\perp})E_0}{(\Sigma_{1\perp}+\Sigma_{2\perp})^2+\Sigma_{1\wedge}^2}. \tag{11.72}$$

The field Eq. (11.72) depends only on the Hall conductivity of injected ions; the Hall conductivity of the ambient ions is absent since the corresponding ambient Hall fluxes are divergence free. For the case of a pure plasma, $\hat{b}_1 = \hat{b}_2 = \hat{b}$, which corresponds to the one-layer model of the ionosphere, the Hall component of electric field E_x reaches its maximum value at A~1 and δ~1, where $\delta=(b_{e\wedge}-b_{\wedge})/b_{\perp}$. In this case the angle θ between the perturbed electric field $E_1'=E_1-E_0$ and the y axis is also maximum, Fig. 11.8.

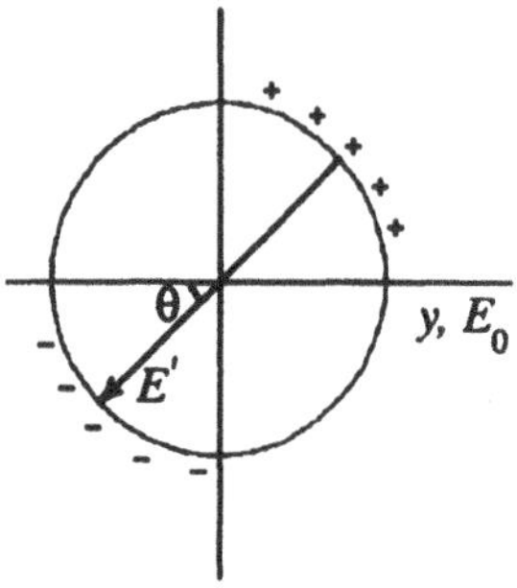

Figure 11.8. The perturbed electric field in the presence of Hall current in pure plasma for A~1, δ~1.

The general expression for this angle has the form

$$tg\theta = \frac{E_{1x}}{E_{1y}'} = \frac{2\Sigma_{2\perp}\Sigma_{1\wedge}}{\Sigma_{1\perp}(\Sigma_{1\perp}+2\Sigma_{2\perp})+\Sigma_{1\wedge}^2}.$$

For A>>1 and A<<1 the angle $\theta\to 0$. If $\Sigma_{2\perp}>>\Sigma_{1\perp}$, then $tg\theta=-\Sigma_{1\wedge}/\Sigma_{1\perp}=\delta$. For the unmagnetized ions, x_i<<1, we have $\delta=(1+x_i^2)/x_i$. The physical mechanism will be discussed in Section 11.8 for the 3D case. Similar effect for the unmagnetized plasma ($x_i x_e << 1$) is studied in the next Section.

In the two-layer model of the ionosphere for inhomogeneities above 120km it is possible to neglect the Hall conductivity in the cloud of injected ions. On the contrary, the Hall conductivity is important in the ionosphere below the cloud. The modelling of the cloud evolution in the two-layer model with account of Hall conductivity was performed in [18].

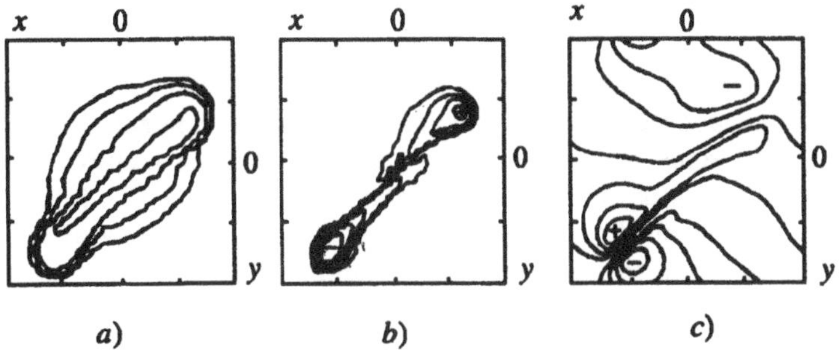

Figure 11.9. Equidensities for injected (*a*), ambient (*b*) ions and equidensities (*c*) obtained in the numerical modelling of the barium cloud evolution in the two-layer model of the ionosphere [18]. The coordinate system moves with the velocity E_0/B along the x axis. The ratio $b_{1\perp}/b_{2\perp}=1.25\times10^{-2}$ corresponds to the cloud height~200km; A=8; the distance is measured in units $2a$=8km; time after the injection t=0.8t_1.The depletion of the ambient plasma is situated in the left low corner in Fig. b.

For $1<<A<<b_{2\perp}/b_{1\perp}$ initial inhomogeneity extends in the direction which forms an angle with the $\vec{E}_0\times\vec{B}$ direction, Fig. 11.9.The reason, as in the absence of the Hall current, is connected with the perturbation of the ambient plasma density. Indeed, at $t<t_1=a/(b_{1\perp}E_0A)$ the injected cloud moves in almost unperturbed ambient electric field. At $t>t_1$ the ambient density becomes strongly perturbed, and, hence, the divergence of the Hall currents becomes important. Since at $t\sim t_1$ the perturbation of the ambient density $\Delta N_2\sim N_2^{(0)}$, for estimate of the perturbed electric field we replace n_1 in Eq.(11.72) by $\Delta N_2\sim N_2^{(0)}$. We obtain then $\text{tg}\theta\sim2\delta/(3+\delta^2)$, and for $\delta\sim1$ the perturbed electric field E' is turned by an angle of the order of 45° with respect to the y direction. As a result, the regions with positive ambient density perturbation, $y<0$, move slower.

In the reference frame moving with the drift velocity $\vec{V}_0=[\vec{E}_0\times\vec{B}]/B^2$, the corresponding part of the cloud drifts in the direction $[\vec{E}'\times\vec{B}]$, Figs. 11.8 and 11.9, which forms an angle θ with the y axis. At $y>0$ the depletion regions are formed, where for an estimate one can put $\Delta N_1 \sim -N_2^{(0)} < 0$. According to Eq. (11.72), the component E_x here changes its sign, while the component of the perturbed electric field E_y' is positive inside the depletion region and thus amplifies the net electric field (see also Section 11.2).

In other words, the electric field in the depletion region is counter-directed to the field shown in Fig. 11.8. Therefore, these parts of the cloud move in the direction, which forms the angle θ with the y axis, towards larger values of x and y. Finally, the cloud becomes stretched, as shown in Fig. 11.9. For $A \gg b_{2\perp}/b_{1\perp}$ the impact of the Hall currents on the evolution pattern is insignificant.

Drift and deformation of barium clouds. Drift and deformation of strong plasma inhomogeneities were studied in many experiments where the barium (and some other) clouds were injected into the ionosphere [17]. Here we shall discuss briefly the experiments "Spolokh" [19] - [20] which illustrate the obtained results. In these experiments the density of injected ions of Ba^+ was so high that the condition $\Sigma_{1\perp} \gg \Sigma_{2\perp}$ was satisfied. Therefore, the main fraction of the injected ions for a long time remained at rest in the reference frame of the neutral wind (the main cloud). Simultaneously part of the injected ions was outflowing in the $\vec{E}_0 \times \vec{B}$ direction creating a tail.

In Fig. 11.10 the photos of the barium cloud from different observation points are presented.

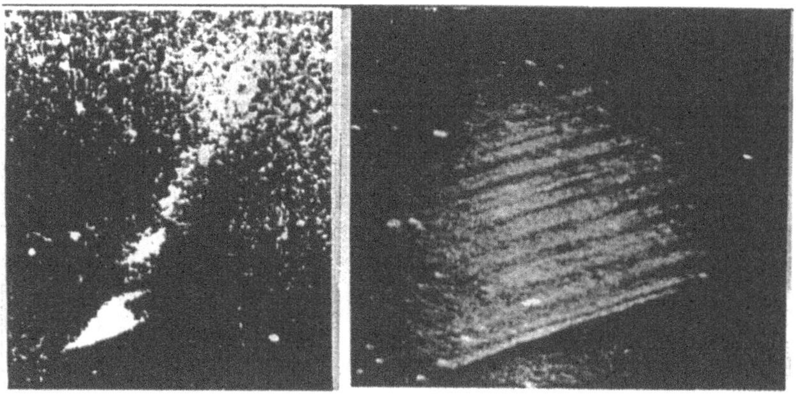

a) *b*)

Figure 11.10. Photo of the barium cloud evolution in the "Spolokh-1" experiment made from the different points of observation. Plasma is separated into striations stretched along the Earth magnetic field. In Fig. *b* the line of sight is perpendicular to the outflow direction.

The results of photometric processing of the resonance radiation of Ba^+ ions are shown in Fig. 11.11.

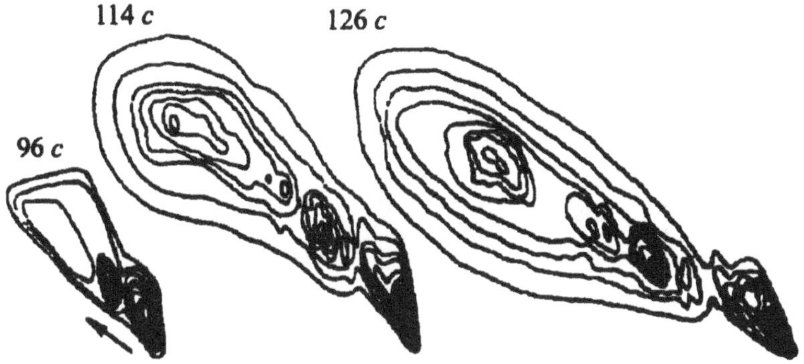

Figure 11.11. Barium cloud in the experiment "Spolokh-1" in the different moments: lines correspond to the constant densities integrated along the line of sight; by arrow is shown the Hall direction (in the reference frame of the neutral wind); the Hall shock is formed at the rear side of the cloud.

The Hall shock is distinctively seen on the rear side of the main cloud. The outflowing tail of injected ions turns out to be split into striations stretched along the Earth's magnetic field. On the contrary, the gradient drift instability apparently did not develop on the rear side of the main cloud, since shock there is rather sharp. The main cloud in this experiment was optically thick, so the number of injected particles in it was rather large, much larger than in the tail.

The total number of particles outflowing to the tail grew linearly with time, while the integrated density N_1 of injected ions Ba^+ in the tail remained of the order of the ambient density integrated over the region under the cloud.

The motion of different parts of the Ba^+ cloud, as well as the motion of the neutral Ba cloud, is shown in Fig. 11.12. It can be seen that the velocity of the main plasma cloud coincided with the velocity of the neutral cloud (the wind velocity) both in horizontal and vertical planes. In contrast, the tail moved perpendicular to the magnetic field direction and thus deviated downwards (magnetic field at the latitude of the experiment forms angle 23° with the vertical direction). At later stages the initial cloud of injected ions elongated in the Hall direction. This fact is connected with its relatively modest transverse scale of the order of 300m.

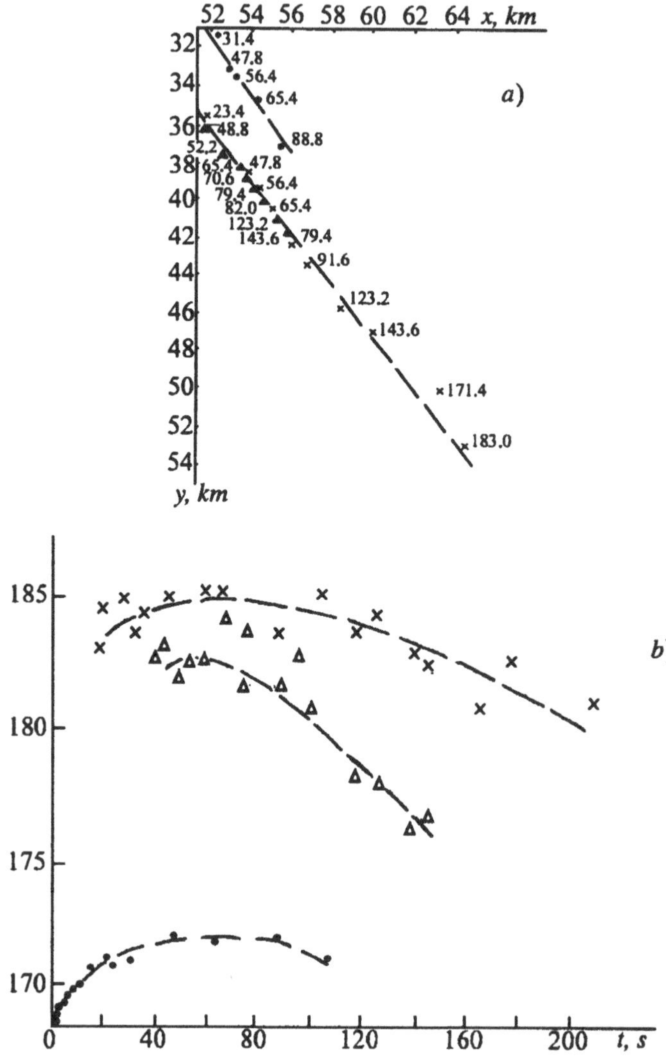

Figure 11.12. Position of different parts of the clouds in the experiment "Spolokh-1": *a*) the horizontal projection, *b*) the height variation; •-center of the spherical neutral cloud, moving with the wind velocity; ×-center of the main ion cloud; Δ-center of the last visible striation, which determines the position of the tail end.

As follows from the analysis of the 3D evolution of Section 11.9, the magnetic field lines in this case cannot be considered as equipotentials. Moreover, the ionosphere below approximately 120km, where the main

Hall current flows, is switched off from the current system. Therefore, the effects connected with the Hall currents were absent in this experiment. As a result, in the course of the evolution the cloud was transforming into the band stretched in the $\vec{E}_0 \times \vec{B}$ (in the neutral wind reference frame) direction.

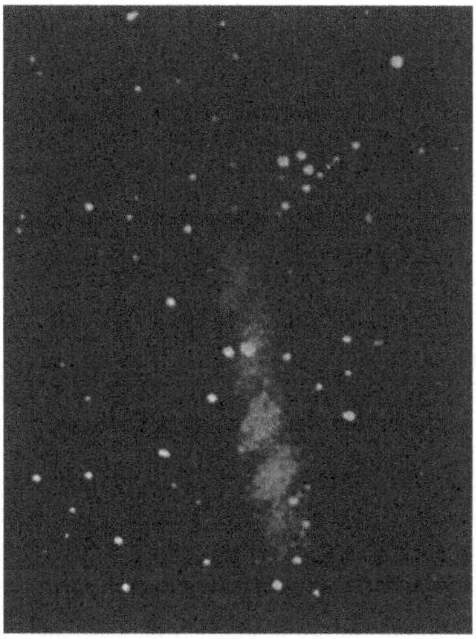

Figure 11.13. Photo of the barium cloud at late stage of its evolution; the injection position is situated in the left upper corner of the picture.

The longitudinal spreading of the injected strong cloud is determined by the parallel ambipolar diffusion with the coefficient $D_{a||}=(1+T_e/T)D_{I||}$. The observed longitudinal profile in the barium cloud experiment [21] had Gaussian shape. In Fig. 11.14 the temporal dependence of the longitudinal cloud scale is shown.

The density of injected ions n_I can be thus easily connected with the integrated density N_1 which was introduced in the equations of the two-layer model. For the initial Gaussian profile we have

$$n_I = \frac{N_1 \exp\left[-\dfrac{z^2}{a_\parallel^2 + 4(1+T_e/T)D_{I\parallel}t}\right]}{\sqrt{\pi}\sqrt{a_\parallel^2 + 4(1+T_e/T)D_{I\parallel}t}} \qquad (11.73)$$

In the experiments [21] $A=N_1^{(\max)}/N_2^{(0)}$ was of the order of unity (approximately 2), so the Hall shock at the rear side of the cloud was not observed. The characteristic time t_1 of the beginning of the cloud deceleration here was of the order of 300÷400s.

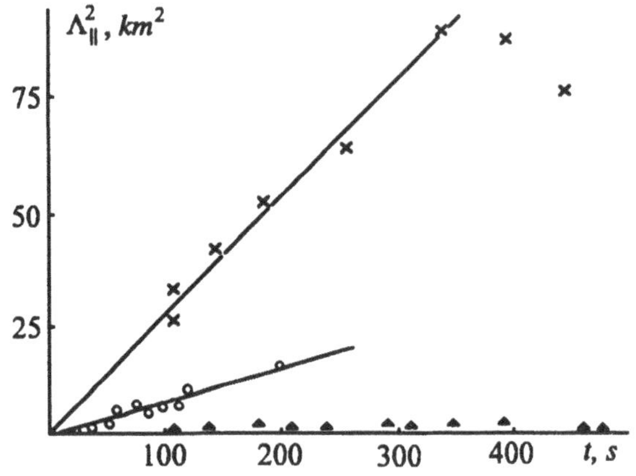

Figure 11.14. Temporal dependence of the square of the cloud half-width: ×-longitudinal scale of the ionized barium cloud; o -longitudinal scale of the neutral barium cloud; Δ-transverse scale of the ion barium cloud. The faster longitudinal expansion of the ionized cloud is connected with the factor $(1+T_e/T)$ in the diffusion coefficient $D_{a\parallel}$. The decrease of $\Lambda_{\parallel}^2$ at later stage is apparently connected with the influence of ambient brightness.

It is very difficult to observe the depletion regions in the ambient ionosphere, since the corresponding perturbations are usually rather small and are often masked by the ionization processes. However, these regions were found in [22] and [23].

11.5 EFFECTIVE CONDUCTIVITY ACROSS WEAK MAGNETIC FIELD. EFFECT OF CONDUCTIVITY RECOVER

In this book we consider mainly the case of the magnetized plasma where $x_i x_e >> 1$. This Section presents the only exception, we shall analyze here the conductivity across a weak magnetic field $x_i x_e << 1$ when plasma is unmagnetized. In other words, the perpendicular mobility of ions $b_{i\perp}$ is assumed to be negligible with respect to the perpendicular electron mobility $b_{e\perp}$. On the other hand, magnetic field is supposed to be strong enough to magnetize electrons, $x_e >> 1$, and thus $b_{e\perp} << b_{e\parallel}$. Such a situation is typical, for example, for a plasma of MHD generators and for some other laboratory applications.

The main physical effect of conductivity recover can be understood from an example in a simplest geometry shown in Fig. 11.15.

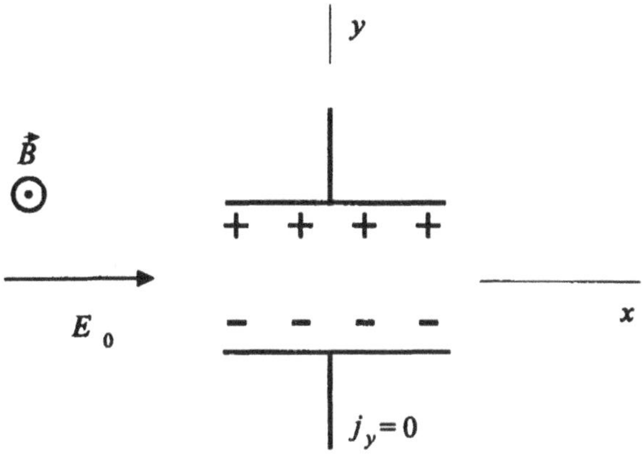

Figure 11.15. Polarization in the channel with dielectric surfaces in the unmagnetized plasma.

Here the current, initially driven by the electric field $\vec{E}_0 \| \vec{x}$, flows in the x direction across a magnetic field. The current channel is restricted by two infinite dielectric surfaces parallel to the x axis. Therefore, the boundary condition at the surfaces is $\Gamma_{ey}=\Gamma_{iy}$ or $j_y=0$. The initial electric field results in the Hall current of electrons in the y direction and in their current along the x axis in the Pedersen direction. The ion fluxes both in the Hall and the Pedersen directions are negligible with respect to the corresponding electron fluxes due to the condition $x_i x_e << 1$, and do not contribute to the current. Since the current in the y direction cannot flow through the boundaries and thus should equals zero everywhere, the polarization arises which is shown in Fig. 11.5. For simplicity we shall neglect the diffusive electric field (of the order of T/eL) with respect to E_0. Then the perturbed electric field E_y can be easily obtained from the condition

$$j_y = -e(-b_{e\wedge} n E_0 - b_{e\perp} n E_y) = 0 . \qquad (11.74)$$

The perturbed electric field is thus negative and is given by

$$E_y = -E_0 b_{e\wedge} / b_{e\perp} . \qquad (11.75)$$

Since the electrons are magnetized, the ratio $b_{e\wedge}/b_{e\perp} = x_e >> 1$, and the perturbed electric field is much larger than the initial electric field E_0. The field E_y causes the drift of electrons in the x direction which, together with the small current driven by the E_x component, determines the current in the system:

$$j_x=-enE_y b_{e\wedge}+enE_0 b_{e\perp}=eE_0 n(b_{e\wedge}^2/b_{e\perp}+b_{e\perp})=enE_0 b_{e\parallel}. \qquad (11.76)$$

We use here the elementary theory model for the electron mobility tensor, Eqs. (2.43) and (2.44). The effective conductivity (coefficient of proportionality between j_x and E_0) thus coincides with the conductivity $\sigma_{e\parallel}=enb_{e\parallel}$ parallel to the magnetic field. In other words the effective conductivity is recovered as if magnetic field is completely absent. The main contribution to the effective conductivity gives the polarization field E_y which x_e times exceeds the field E_0 and which causes large Hall drift of electrons.

The obtained result directly follows from the boundary condition $j_y=0$ at the surfaces. In the case of the short-circuited metal boundaries the current can flow between the electrodes, and the effective conductivity does not recover. Indeed, this situation corresponds to the absence of the polarization field $E_y=0$, while the current in the y direction can flow freely: $j_y=enE_0 b_{e\wedge}$. In the absence of the Hall electric field E_y, the x component of the current is simply $j_x=enE_0 b_{e\perp}$, and the effect of conductivity recover is absent. Between the two extreme cases $E_y=0$ and $j_y=0$ the intermediate regimes are possible when the resistivity is switched between the electrodes. This load determines the value of j_y, and hence the value of the field E_y, and, finally, the effective conductivity. The discussion of applications to the MHD generators one can find in [24] - [25].

The strong polarization electric field can be generated by plasma inhomogeneities even in the absence of the boundaries, similarly to the situation analyzed in the preceding Section. Such strong electric field can produce different effects, and, in particular, cause the ionization instability [26] - [27]. This phenomenon could be understood from the following example. Consider a homogeneous plasma with a current j_x perpendicular to the magnetic field and a positive density perturbation δn, infinite in y direction, Fig. 11.16. In this 1D geometry the current j_x should remain constant:

$$j_x=-enE_y b_{e\wedge}+enE_x b_{e\perp}=const. \qquad (11.77)$$

The y component of the electric field is also constant owing to the potential character of the field ($\partial E_x / \partial y = \partial E_y / \partial x = 0$). Therefore, the inhomogeneity is polarized as shown in Fig. 11.16, so that the perturbed electric field is perpendicular to the equidensities.

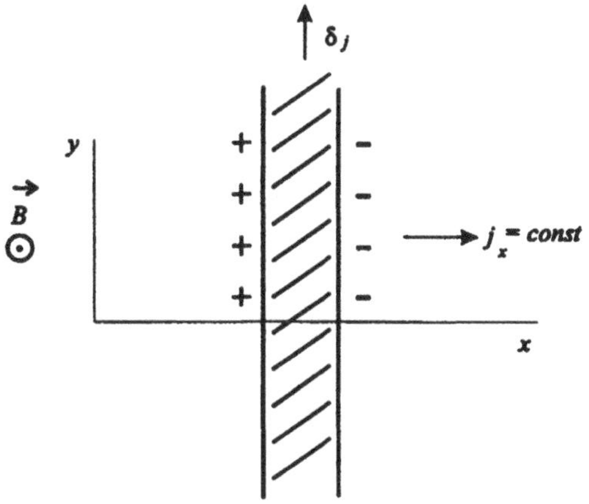

Figure 11.16. Additional current caused by the positive density perturbation.

From Eq.(11.77) and the condition E_y=*const* it follows

$$\delta E_x = -\frac{\delta n}{n_0} x_e E_y . \qquad (11.78)$$

In the absence of the density perturbation the main electric field was *y*-directed, and the main current was caused by the Hall mobility. The density perturbation results in the field amplification and rotation. The perturbed electric field drives rather large current along the equidensities

$$\delta j_y = e\delta n b_{e\wedge} E_x + e n_0 b_{e\wedge} \delta E_x = e\delta n b_{e\wedge} E_x - e\delta n b_{e\wedge} x_e E_y . \qquad (11.79)$$

It is determined mainly by the last term in the r.h.s. of Eq. (11.79). In other words, the density perturbation results in the additional current along the equidensities which, by approximately the factor $x_e \delta n / n_0$, deviates from the unperturbed current j_x. This additional current may result in heating and ionization of the plasma thus causing the ionization instability. The analysis can be easily generalized to the case of arbitrary angle between the unperturbed current and density gradient of the perturbation.

In [28] the effective conductivity across the magnetic field in a turbulent plasma has been discussed. It has been demonstrated that the polarization of the stochastic density perturbations results in the formation of current loops and, finally, in the reduction of average current with respect to Eq.(11.77). This effect was observed experimentally [28].

11.6 THREE-DIMENSIONAL EVOLUTION OF SMALL POINT PERTURBATION IN UNBOUNDED WEAKLY IONIZED PLASMA

Consider the evolution of a linear point perturbation of the weakly ionized plasma ($x_e x_i$>>1) under the same assumptions as in Section 6.1. The basic equations have the form

$$\frac{\partial n}{\partial t} - \vec{\nabla}\cdot(\hat{D}_e\vec{\nabla}n - \hat{b}_e n\vec{\nabla}\tilde{\varphi} + \hat{b}_e n\vec{E}_0) = 0;$$
$$\frac{\partial n}{\partial t} - \vec{\nabla}\cdot(\hat{D}_i\vec{\nabla}n + \hat{b}_i n\vec{\nabla}\tilde{\varphi} - \hat{b}_i n\vec{E}_0) = 0 \quad , \tag{11.80}$$

where $\tilde{\varphi}$ is the perturbed potential; the boundary condition at infinity is $\tilde{\varphi}(\vec{r}\to\infty)=0$. The impact of the ambient electric field $\vec{E}_0$ (in the reference frame of the neutral gas) results in the following expression for the Green's function $G(\vec{r},t)$ and the perturbed potential (compare with the Section 6.1):

$$G(\vec{r},t) = \frac{1}{(2\pi)^3}\int \exp[i\vec{k}(\vec{r} - \vec{V}(\mu^2)t - D(\mu^2)k^2 t]d\vec{k}\ ;$$
$$\delta n(\vec{r},t) = NG(\vec{r},t)\ ; \tag{11.81}$$

$$\tilde{\varphi}(\vec{r},t) = \tilde{\varphi}_1 + \tilde{\varphi}_2$$
$$= \frac{N}{(2\pi)^3 n_0}\int(\tilde{\varphi}_{1\vec{k}} + \tilde{\varphi}_{2\vec{k}})\exp[i\vec{k}(\vec{r} - \vec{V}(\mu^2)t - D(\mu^2)k^2 t]d\vec{k}\ ;$$
$$\tilde{\varphi}_{1\vec{k}} = \frac{D_{e\parallel}\mu^2 - D_{i\perp}(1-\mu^2)}{b_{e\parallel}\mu^2 + b_{i\perp}(1-\mu^2)}; \qquad \tilde{\varphi}_{2\vec{k}} = -i\frac{\vec{k}(\hat{b}_e + \hat{b}_i)\vec{E}_0}{\vec{k}(\hat{b}_e + \hat{b}_i)\vec{k}}\ , \tag{11.82}$$

where N is the total number of injected particles, $D(\mu^2)$ is defined by Eq. (5.32) (or by Eq.(6.5) in weakly ionized plasma for $T_e=T_i=T$), $\vec{V}(\mu^2)$ is given by Eq. (11.7) (by Eq. (11.8) in weakly ionized plasma). The Green's function and the expression for the perturbed potential were analyzed in [1], [29] - [31].

Asymptotic form of the Green's function. Transforming in Eqs. (11.81) and (11.82) to spherical coordinates in $\vec{k}$ space, and integrating over the modulus of $\vec{k}$, in the coordinate system moving with the neutral gas we find the Green's function and the potential perturbation to be

$$G(\vec{r},t)=\frac{1}{32\pi^{5/2}t^{3/2}}\int_{-1}^{1}\frac{d\mu}{D^{3/2}(\mu^2)}\int_{0}^{\pi}d\vartheta \times\left[1-\frac{\rho^2F^2}{2D(\mu^2)t}\right]\exp\left[1-\frac{\rho^2F^2}{4D(\mu^2)t}\right]; \tag{11.83}$$

$$\tilde{\varphi}(\vec{r},t)=\frac{N}{32\pi^{5/2}t^{3/2}n_0}\int_{-1}^{1}\frac{d\mu}{D^{3/2}(\mu^2)}\int_{0}^{\pi}d\vartheta \times\left[1-\frac{\rho^2F^2}{2D(\mu^2)t}\right]\exp\left[1-\frac{\rho^2F^2}{4D(\mu^2)t}\right] \times\left[\frac{D_{e\parallel}\mu^2-D_{i\perp}(1-\mu^2)}{b_{e\parallel}\mu^2+b_{i\perp}(1-\mu^2)}+\frac{\rho F(\vec{k}/|\vec{k}|)(\hat{b}_e+\hat{b}_i)\vec{E}_0}{b_{e\parallel}(\mu^2+\mu_0^2)}\right]; \tag{11.84}$$

where $\vec{\rho}=\vec{r}-\vec{V}(\mu^2)t$; $F=\cos(\vec{k},\rho)$; ϑ is the azimuth angle; polar axis is parallel to the magnetic field. To find the asymptotic form of the Green's function for $r>>V(\mu^2)t$; $(D(\mu^2)t)^{1/2}$, we must convert in Eq. (11.83) to an integration with a new polar axis along the radius vector $\vec{r}$ of the observation point:

$$G(\vec{r},t)=\frac{1}{32\pi^{5/2}t^{3/2}}\int_{0}^{2\pi}d\Psi\int_{0}^{1}\frac{dA}{D^{3/2}(\mu^2)} \times\left[1-\frac{\rho^2F^2}{2D(\mu^2)t}\right]\exp\left[1-\frac{\rho^2F^2}{4D(\mu^2)t}\right]; \tag{11.85}$$

where

$$\mu=A\cos\alpha+(1-A^2)^{1/2}\sin\alpha\cos\Psi. \tag{11.86}$$

Here α is the angle between $\vec{r}$ and $\vec{B}$, Ψ is the azimyth angle, $A=\cos(\vec{k},\vec{r})$. The new coordinate system is shown in Fig. 11.17.

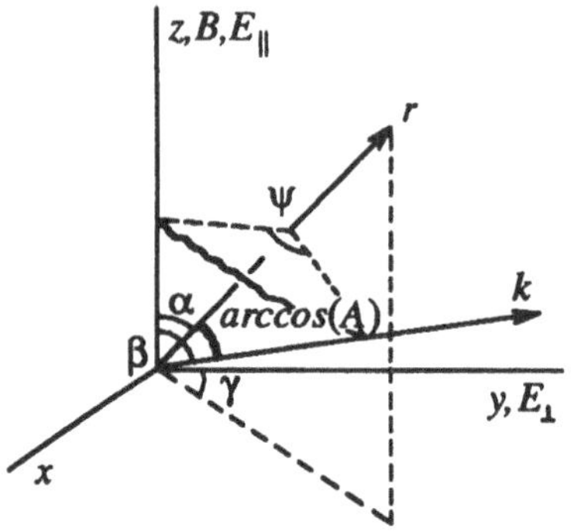

Figure 11.17. Coordinate system for calculation of the Green's function.

Let us express ρF through A:

$$\rho F = \frac{\vec{k}[\vec{r} - \vec{V}(\mu^2)t]}{|\vec{k}|} = |\vec{r}|A - \frac{\vec{k}}{|\vec{k}|}\vec{V}(\mu^2)t\,. \tag{11.87}$$

Projecting vectors on the axis of the new coordinate system, we obtain

$$\begin{aligned}\rho F &= r[A - \frac{\mu V_z t}{r} - \frac{t}{r}(V_x \cos\gamma - V_y \sin\gamma)\sqrt{1-A^2}\sin\Psi \\ &+ \frac{t}{r}(V_y \cos\gamma + V_x \sin\gamma)(A\sin\alpha - \sqrt{1-A^2}\cos\alpha\cos\Psi] \\ &\equiv r\Phi(A)\ ,\end{aligned}$$

where γ is the angle between $\vec{E}_{0\perp}$ and the projection of radius vector $\vec{r}$ on the plane (x,y), Fig. 11.17. After changing of variables in Eq. (11.87), $Z = r\Phi(A)/[2(D(A)t)^{1/2}$, we have

$$\begin{aligned}G(r,\alpha,\gamma,t) &= \frac{1}{32\pi^{5/2}t^{3/2}}\int_0^{2\pi} d\Psi \int_0^1 dZ \exp(-Z^2) \\ &\times \frac{1-2Z^2}{D^{3/2}(A)Z'(A)}\,.\end{aligned} \tag{11.88}$$

Since the main contribution to the integral gives the region $|Z|\sim 1$, for large values of r it is possible to extend the integration to infinity in the inner integral. The function $\Phi(A)\sim A$ and for $Z\sim 1$ the main contribution corresponds to $A \ll 1$. Therefore, expanding the denominator in Eq.(11.88) in series in A, and then reexpanding it in series in Z, neglecting the terms of the order of $(Vt/r)^2$, we find

$$G(r,\alpha,\gamma,t)=\frac{3t}{4\pi^2 r^4}\int_0^{2\pi}\sigma d\Psi = g(\alpha,\gamma)\frac{t}{r^4}, \tag{11.89}$$

where

$$\sigma_1(E_{0\parallel}) = -\frac{1}{2}(V_z)''_A\cos\alpha - \frac{1}{6}(V_z)'''_A\sin\alpha\cos\Psi + \frac{1}{2}(V)'_z\sin\alpha\cos\Psi\ ;$$

$$\begin{aligned}\sigma_2(E_{0\perp}) &= \frac{1}{6}[-(V_x)'''_A\cos\gamma + (V_y)'''_A\sin\gamma]\sin\Psi \\ &+\frac{1}{2}[(V_x)'_A\cos\gamma - (V_x)'_A\sin\gamma]\sin\Psi - (V_y)''_A\cos\gamma\sin\alpha \\ &-(V_x)''_A\sin\gamma\sin\alpha + (V_y)'''_A\cos\gamma\cos\alpha\cos\Psi \\ &-(V_x)'''_A\sin\gamma\cos\alpha\cos\Psi - \frac{1}{2}(V_y)'_A\cos\gamma\cos\alpha\cos\Psi\ ;\end{aligned}$$

$$\sigma=\sigma_1(E_{0\parallel})+\sigma_2(E_{0\perp}).$$

All the derivatives are calculated at $A=0$.

Similar expression for the potential is derived analogously. At large distances the potential has the dipole character

$$\tilde{\varphi}(\vec{r},t)=\tilde{\varphi}_2(\vec{r},t)=\tilde{\varphi}(\alpha,\gamma)/r^2. \tag{11.90}$$

The perturbed potential is determined mainly by the second term in Eq. (11.84) which is proportional to E_0. Similarly to the case $E_0=0$ considered in Section 6.1, at large distances the main role play the field-driven fluxes $\hat{b}_{e,i}n_0\vec{\nabla}\tilde{\varphi}$. The asymptotic form of the Green's function is thus determined by the potential perturbation: $\partial G/\partial t \sim \Delta\tilde{\varphi} \sim r^{-4}$. Hence, G should be equal to t/r^4 multiplied by expression which depends on angles only. Since the parallel and perpendicular components of the electric field make additive contribution to the asymptotic behavior, we can treat the cases $\vec{E}_0 \| \vec{B}$ and $\vec{E}_0 \perp \vec{B}$ separately.

For $\vec{E}_0 \| \vec{B}$, substituting μ as an independent variable in Eq. (11.89) by means of Eq. (11.86), after changing the variables $\mu = \sin\alpha\cos\Psi$, for $r \gg b_{e\|} E_{0\|} t$ we obtain

$$G(r,\alpha,t) = -\frac{3t\cos^3\alpha}{2\pi^2 r^4} \int_{-\sin\alpha}^{\sin\alpha} [\frac{8}{3}\mu^4 (V_z)'''_{\mu^2} + 8\mu^2 (V_z)''_{\mu^2}(1-\frac{\mu^2}{\cos^2\alpha}) + 2(V_z)'_{\mu^2}(1-\frac{3\mu^2}{\cos^2\alpha})]\frac{d\mu}{\sqrt{\sin^2\alpha-\mu^2}} . \quad (11.91)$$

The asymptotic form of the density (and potential) perturbation in this case is thus asymmetric with respect to the replacement of α by π-α. For angles $\alpha \ll 1$ and $T_e = T_i = T$ Eq. (11.91) yields

$$G_1(r,\alpha,t) = \frac{3b_{i\perp} E_{0\|} t\cos^3\alpha}{4\pi\mu_0^4 r^4} \frac{(3\sin^2\alpha/\mu_0^2 - 2)}{(\sin^2\alpha/\mu_0^2+1)^{7/2}} , \quad (11.92)$$

where $\mu_0 = (b_{i\perp}/b_{e\|})^{1/2}$. The sectors $\alpha < \mu_0(2/3)^{1/2}$ and $\pi/2 < \alpha < \pi - \mu_0(2/3)^{1/2}$ are the depletion regions , the density perturbation in these regions is negative.

The Green's function G vanishes in the plane perpendicular to the magnetic field. Consequently, for angles close to $\pi/2$ the asymptotic behavior is determined by $\sigma_2(E_{0\perp})$ and is (for $T_e = T_i = T$ and $x_i \gg 1$)

$$G_2(r,\gamma,t) = \frac{3b_{i\perp} E_{0\perp} t\mu_0 \cos\gamma}{4\pi r^4} , \quad (11.93)$$

This expression is antisymmetric with respect to $\cos\gamma$ and symmetric with respect to $\cos\alpha$. The asymptotic behavior of the perturbation according to Eqs. (11.92) and (11.93) is controlled only by electric fields in plasma.

In general, if the angle between the ambient electric and magnetic fields is not close to zero or to $\pi/2$, the asymptotic behavior is given by the sum of Eqs. (11.92) and (11.93). The term G_2, Eq.(11.93), determines the asymptotic only for angles α approximately equal to $\pi/2$ (since $|V_z|/|V_\perp| \gg 1$ for $E_{0\|} \sim E_{0\perp}$).

Green's function and solution in the neighbouring zone. For short times the solution for a small perturbation is the same as that found in Section 6.1. For larger times

$$V(\mu^2)t >> (D(\mu^2)t)^{1/2}$$

the integral Eq. (11.85) is small everywhere except near the extremum line $\rho=0$, i.e.,

$$\vec{r} = \vec{V}(\mu_m^2)t, \tag{11.94}$$

where $\mu_m{}^2$ takes all values from zero to unity (in partially ionized plasma the condition $\rho=0$, instead of the straight line, Eq. (11.94), determines the curve). The projection of the extremum line onto the $(\vec{E}_0, \vec{B})$ plane is segment AB in the schematic diagram in Fig. 11.18.

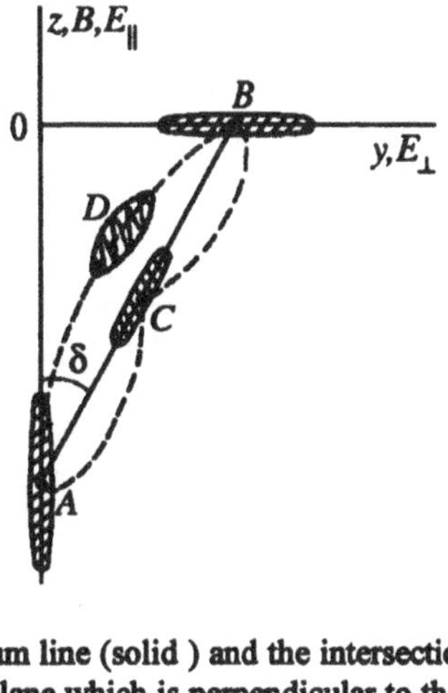

Figure. 11.18. The extremum line (solid) and the intersection of the group-velocity surface (dashed) with the plane which is perpendicular to the x axis and moves with the velocity $[\vec{E}_0 \times \vec{B}]/B^2$.

The angle δ is related to the unperturbed electric field by

$$\text{tg}\delta = \mu_0{}^2 E_{0\perp}/E_{0\|}. \tag{11.95}$$

Point A corresponds to $\mu_m=0$, B to $|\mu_m|=1$, while point C is determined in accordance with $|\mu_{01}|=\sin\delta$.

In order to calculate the density on the extremum line, we notice that for $V(\mu^2)t >> (D(\mu^2)t)^{1/2}$ the main contribution to the integral Eq. (11.85) make the regions in $\vec{k}$ space close to the cone $\rho=0$ and the plane $F=0$. Everywhere below in this Section we assume $T_e=T_i=T$ for simplicity. If the corresponding surfaces intersect, then the main contribution is made by the intersection region. By saddle-point method it can be shown that the density in the vicinity of the part of extremum line, which corresponds to the intersection of the surfaces $\rho=0$, $F=0$, is ($x_i>>1$)

$$G[\vec{V}(\mu_m^2)t,t]=\frac{(\mu_0^2+\mu_m^2)\sin\delta}{4\pi^2t^2b_{i\perp}E_{0\perp}D(\mu_m^2)\mu_0^2\mu_m(1-\mu_m^2-\cos^2\delta)}. \tag{11.96}$$

The perturbed density falls of with distance from the extremum line with the scale of the order of $\sqrt{8D(\mu_m^2)t}$. Such dependence is valid in the vicinity of the segment AC (Fig. 11.18).

For points on line segment BC (here the surfaces $\rho=0$, $F=0$ do not intersect) the density perturbation can be calculated as the sum of two contributions: from the cone $\rho=0$ and from the plane $F=0$. Let us calculate the contribution from the cone I_1. The integrand in Eq. (11.85) is small far from the points $\mu=\pm\mu_m$. Therefore we shall divide the integral in Eq. (11.83) into sum of two integrals. Near $\mu=\mu_m$ we introduce the new variable

$$Z=\frac{t^{1/2}C(\mu)F}{2D^{1/2}(\mu^2)},$$

where

$$C(\mu)=\frac{\left|\vec{V}(\mu^2)-\vec{V}(\mu_m^2)\right|}{\mu-\mu_m};$$

$$I_1=\frac{1}{32\pi^{5/2}t^{3/2}}\int_0^{\pi}d\theta\int_0^{\infty}\frac{dZ}{D^{3/2}(\mu^2)}\frac{\exp(-Z^2)(1-2Z^2)}{Z'_\mu}. \tag{11.97}$$

Expanding the denominator in Eq. (11.97) in series in (μ-μ_m) up to the second order, reexpanding this series in Z, after integrating in the inner integral we obtain

$$I_1=-\frac{1}{4\pi t^3}\int_0^{2\pi}\frac{d\theta}{(CF)^3}\left\{6\frac{[(CF)'_\mu]^2}{(CF)^2}-\frac{3}{2}\frac{(CF)''_\mu}{CF}\right\}. \tag{11.98}$$

where

$$CF = \frac{b_{i\perp}(\mu+\mu_m)}{(\mu_m^2+\mu_0^2)(\mu^2+\mu_0^2)}(\mu E_{0\parallel}+\sqrt{1-\mu^2}\,E_{0\perp}\mu_0^2\cos\theta).$$

The derivatives are taken at $\mu = \mu_m$. The integral over the region in the vicinity of $\mu = -\mu_m$ coincides with I_1. The contribution from the plane $F=0$ is calculated in the same way and is equal to $2I_1$. While calculating it is convenient to use the expression for the Green's function with the polar axis along the extremum line. Finally,

$$G(\vec{V}(\mu_m^2)t,t] = 4I_1\,. \tag{11.99}$$

Eqs. (11.98) and (11.99) can be simplified substantially (x_i>>1) for a purely longitudinal electric field ($E_{0\perp}$=0)

$$G(\vec{V}(\mu_m^2)t,t] = -\frac{3(\mu_m^2+\mu_0^2)^4\mu_0^4}{\pi t^3 E_{0\parallel}^3 b_{i\perp}^3\mu_m^8}\,. \tag{11.100}$$

Evidently, the segment *BC* in Fig. 11.18 corresponds to a depletion region; the perturbation falls off in proportion to t^3 in this region. The spatial scale of the perturbation decay increases as *t* with distance from the extremum line.

The points *A*, *B* and *C*, at which $|\mu_m|$=0; 1; μ_{01} accordingly, and plasma density reaches maxima, should be examined separately. The following expressions are valid at late stage when the characteristic scales of the inhomogeneities exceed the corresponding diffusive scales. At the point *A* the density is (x_i>>1)

$$G(\vec{V}(0)t,t] = \frac{8.7\cdot 10^{-3}}{D_e^{5/4}t^{7/4}(b_{e\parallel}E_{0\perp})^{1/2}}\,. \tag{11.101}$$

This plasmoid is stretched along the magnetic field, the corresponding scale is $(b_{e\parallel}E_{0\perp}t)^{1/2}(8D_{e\perp}t)^{1/4}$. The transverse scale of the plasmoid $(8D_{e\perp}t)^{1/2}$ is governed by the electron diffusion across the magnetic field. The potential perturbation in the plasmoid is negative and is determined by the second term in Eq. (11.82). With the accuracy of two numerical coefficients B_1, B_2, which are of the order of unity, the potential perturbation is

$$\tilde{\varphi}[\vec{V}(0)t,t] = -\frac{(B_1E_{0\parallel}+B_2\mu_0^2E_{0\perp})N}{32\pi^{5/2}\mu_0^2 D_{e\perp}^{3/4}t^{5/4}n_0H^2(0)}, \tag{11.102}$$

where

$$H(0)=\frac{d}{d\mu^2}\left|\vec{V}(\mu^2)-\vec{V}(0)\right|_{\mu=0}.$$

At the point B the density is calculated analogously to Eq. (11.98). The difference consists in the fact that in expansion in (μ-μ_m) it is sufficient to leave the linear term, since the saddle point coincides with the end of the interval. In the maximum we obtain

$$G[\vec{V}(1)t,t]=\frac{1}{16\sqrt{2}\pi^{3/2}t^{5/2}b_{i\perp}D_{i\parallel}^{1/2}b_{i\parallel}E_{0\parallel}^2}. \tag{11.103}$$

This plasmoid is compressed along the magnetic field; the corresponding scale is governed by parallel ion parallel diffusion and is $(8D_{i\parallel}t)^{1/2}$. In contrast, the scale across the magnetic field is determined by ambient electric field and ion mobilities and grows linearly with time: $(b_{i\perp}b_{i\parallel})^{1/2}E_{0\parallel}t$. The potential perturbation in the plasmoid is positive and falls off as $t^{-3/2}$. It depends only on ion parallel and perpendicular diffusion coefficients:

$$\tilde{\varphi}[\vec{V}(1)t,t]=\frac{N}{8\sqrt{2}\pi^{3/2}t^{3/2}D_{i\perp}D_{i\parallel}^{1/2}n_0}, \tag{11.104}$$

and is also determined by the second term in Eq. (11.82).

At the point C where $\vec{r}=\vec{V}(\mu_{01}^2)t$, the cone ρ=0 and the plane F=0 are tangent. Hence, $|\mu_{01}|=\sin\delta$. Choosing the direction of the polar axis along the extremum line, we obtain

$$G[\vec{V}(\mu_{01}^2)t,t]=\frac{1}{32\pi^{3/2}t^{3/2}}\int_0^{2\pi}d\tilde{\theta}\int_{-1}^{1}\frac{dF}{D^{3/2}(\mu^2)}$$
$$\times\left[1-\frac{t\left|\vec{V}(\mu_{01}^2)-\vec{V}(\mu^2)\right|^2F^2}{2D(\mu^2)}\right], \tag{11.105}$$
$$\times\exp\left[-\frac{t\left|\vec{V}(\mu_{01}^2)-\vec{V}(\mu^2)\right|^2F^2}{4D(\mu^2)}\right];$$
$$\mu=F\cos\delta+\sqrt{1-F^2}\sin\delta\cos\tilde{\theta}.$$

The main contribution makes the vicinity of the point of tangency of the cone ρ=0 and the plane F=0 where μ is close to μ_{01}. Expanding Eq. (105) in series in $(\mu-\mu_{01})$, expanding again $(\mu-\mu_{01})$ in series in F, we have

$$G[\vec{V}(\mu_{01}^2)t,t]=\frac{1}{16\pi^{3/2}t^{3/2}}\int_0^{2\pi}d\tilde{\theta}\int_{-1}^{1}\frac{dF}{D^{3/2}(\mu^2)}$$
$$\times\left[1-\frac{tC^2(\mu_{01})F^2}{2D(\mu_{01}^2)}(F\cos\delta-\frac{1}{2}\tilde{\theta}^2\sin\delta)^2\right] \quad , \qquad (11.106)$$
$$\times\exp\left[-\frac{tC^2(\mu_{01})F^2}{4D(\mu_{01}^2)}(F\cos\delta-\frac{1}{2}\tilde{\theta}^2\sin\delta)^2-\right];$$
$$\mu=F\cos\delta+\sqrt{1-F^2}\sin\delta\cos\tilde{\theta}\ ,$$

where $C(\mu_{01})$ is calculated at $\mu_m\to\mu_0$. Dividing the integral into two integrals over [-1,0] and [0,1], introducing the new polar coordinates

$$R=|F|\cos\delta+\frac{\tilde{\theta}^2}{2}\sin\delta; \qquad \Theta=arctg\frac{\tilde{\theta}\sqrt{tg\delta}}{2\sqrt{F}}\ ,$$

we find

$$G[\vec{V}(\mu_{01}^2)t,t]$$
$$=\frac{(1+\sqrt{2})\Gamma(3/8)K(\pi/4)}{16\sqrt{2}\pi^{5/2}D^{9/8}(\mu_{01}^2)C^{3/4}(\mu_{01})t^{15/8}\sin^{1/2}\delta\cos^{1/4}\delta}\ , \qquad (11.107)$$

where K is the complete elliptic integral of the first kind, Γ is the gamma-function. This plasmoid is elongated along the extremum line (C in Fig. 11.18). The longitudinal dimension is

$$C^{3/4}(\mu_{01})D^{1/8}(\mu_{01}^2)\cos^{1/4}\delta\sin^{1/2}\delta t^{7/8}\ ,$$

and the transverse dimension is governed by diffusion, having the value $(8D(\mu_{01}^2)t)^{1/2}$. The perturbed potential at the maximum is negative, if $|\mu_{01}|<\mu_0$, or positive, if $|\mu_{01}|<\mu_0$. It falls off with the time scale proportional to $t^{-15/8}$.

The cases $E_{0\perp}$=0 and $E_{0\|}$=0, with $|\mu_{01}|$=0; 1, in which the central maximum merges with one of the outer maxima, deserve special study. In this situation the density in the maxima increases. For the first case (parallel electric field), using the method of steepest descent we find

$$G[\vec{V}(0)t,t]=\frac{\Gamma(1/6)\mu_0^{2/3}}{72\pi^{3/2}t^{5/3}b_{e\parallel}^{1/3}D_{e\perp}^{4/3}E_{0\parallel}^{1/3}}\,. \tag{11.108}$$

The longitudinal scale of the plasmoid is $(b_{e\parallel}D_{e\parallel}E_{0\parallel}t^2)^{1/3}\mu_0^{2/3}$, and its transverse dimension is $(8D_{e\perp}t)^{1/2}$. The potential in it remains of the order of given by Eq. (11.102).

If, on the other hand, the electric field is perpendicular to the magnetic field, then (x_i>>1)

$$G[\vec{V}(1)t,t]=\frac{\Gamma^2(7/6)}{8\cdot 2^{5/6}\pi^{5/2}t^{11/6}D_{i\parallel}^{7/6}C^{2/3}(1)}, \tag{11.109}$$

where $C(1)=2b_{i\perp}E_{0\perp}\mu_0^2$. The longitudinal scale of the plasmoid is $(8D_{i\parallel}t)^{1/2}$, and its transverse dimension for the magnetized ions is $(b_{i\perp}\mu_0^2E_{0\perp}D_{i\parallel}t^2)^{1/3}$. The potential in it coincides with Eq. (11.104).

Comparing the expressions for the density at the maxima and comparing the scales of the plasmoids, we see that the number of particles in each of the three plasmoids is comparable to the total number of particles injected into the plasma.

Qualitative pattern of plasmoids motion. The motion of the plasmoids in the case of parallel electric field, $\vec{E}_0 \| \vec{B}$, can be described qualitatively as follows. The injected electrons move with a velocity $\vec{V}(0)=b_{e\parallel}\vec{E}_{0\parallel}$ in the ambient electric field, and the quasineutrality of the corresponding plasmoid is maintained by ion fluxes across the magnetic field. Depletion regions are formed in the ambient plasma and move along with the plasmoid in the magnetic field direction with the same velocity. Since both the velocity and spatial dimensions of this plasmoid are governed by electrons, and the potential of this plasmoid is negative, we call such a plasmoid, similar to the case of diffusion in Section 6.1, an 'electron plasmoid'.

The spatial dimensions of the second plasmoid are governed by ions, and the potential in it is positive, so we call it an 'ion plasmoid'. It should be noted, however, that injected ions move with a velocity $b_{i\parallel}\vec{E}_{0\parallel}$, while the ion plasmoid remains at rest. The essence of the matter is that the transverse electron mobility is negligible with respect to the ion mobility, and the transverse electron fluxes are insignificant. The plasma motion along the magnetic field is thus governed by the longitudinal fluxes, i.e. the motion is the same as without the magnetic field. In other words, as in Section 4.2, the maximum of the plasma density remains at rest, and its scale along $\vec{B}$ is determined by the ambipolar diffusion.

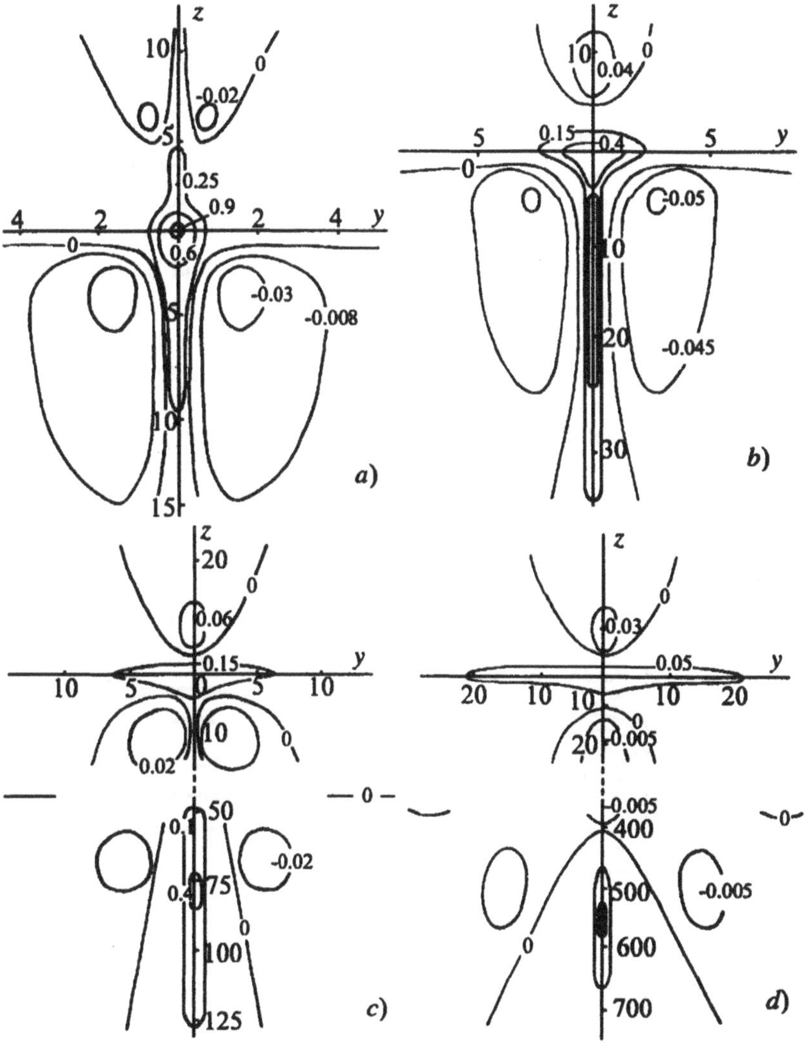

Figure 11.19. Equidensities for the case $\vec{E}_0 \| \vec{B}$; x_e=30; x_i=0.3; the time $\tilde{t}$ is expressed in units of $8D_{i\|}/(b_{e\|}E_{0\|})^2$, the distance in units of $D_{i\|}/(24b_{i\|}E_{0\|})$; the density in units of $2^{-9}\pi^{-3/2}b_{e\|}^3E_{0\|}^3\tilde{t}^{-3/2}D_{i\|}^{-3/2}(D_{i\|}^{-1/2}D_{i\perp}^{-1}+D_{e\|}^{-1/2}D_{e\perp}^{-1})$ is divided by the density at the origin for the case of pure diffusive spreading: *a*) $\tilde{t}$ =9; *b*) $\tilde{t}$ =100; *c*) $\tilde{t}$ =900; *d*) $\tilde{t}$ =10^4.

In the transverse direction the ion plasmoid spreads out linearly as a function of time because of the arrival of electrons along $\vec{B}$ from the depletion regions, which move along the magnetic field with the velocity of electron plasmoid.

In the case $\vec{E}_0 \perp \vec{B}$ the ion plasmoid moves along ambient electric field $\vec{E}_0$ (in the Pedersen direction) with the ion velocity $b_{i\perp}\vec{E}_0$, while the velocity of the electron plasmoid in this direction is zero. The transverse scale of the electron plasmoid $(8D_{e\perp}t)^{1/2}$ is governed by the transverse ambipolar diffusion, which as in the 1D case is controlled by electrons, and its longitudinal scale increases fast with time. The velocities of the plasmoids in the Hall direction coincide with the Hall velocities in the ambient electric field of electrons and ions, i.e. $b_{e\wedge}\vec{E}_{0\perp}$ and $b_{i\wedge}\vec{E}_{0\perp}$ for the electron and ion plasmoids correspondingly.

In the general case the components $V_z(0)$ and $V_x(0)$ are the same as the corresponding components of the electron velocity in the unperturbed electric field, while $V_y(1)$ and $V_x(1)$ are the same as the components of the ion velocity, so that even in the general case it is meaningful to label the first maxima (*A* in Fig. 11.18) the electron maximum and the second (*B* in Fig. 11.18) the ion maximum.

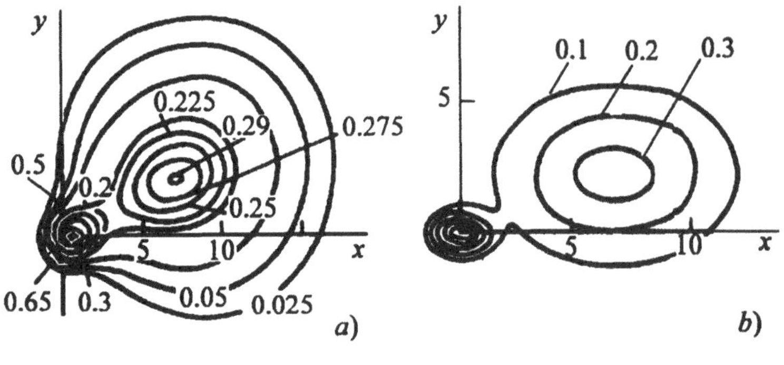

Figure 11.20. Equidensities in *xoy* plane for the case $\vec{E}_0 \perp \vec{B}$ in the vicinity of the extremum line in the reference frame moving with the velocity $[\vec{E}_0 \times \vec{B}]/B^2$; x_e=50; x_i=0.5; $t=20D_{i\|}B^2/E_0^2$; the distance is expressed in units of $2D_{i\|}B/E_0$; the density is divided by the density at the origin for the case of pure diffusive spreading: *a*) simulation [1]; *b*) analytical calculation according to Eq. (11.133); the equidensities corresponds to the values which differ by 0.1; the density in the center of the electron plasmoid is 0.69; in the center of the ion plasmoid the density is 0.32.

In Fig. 11.19 the density evolution in the parallel electric field $\vec{E}_0 \| \vec{B}$ calculated numerically in [31] from Eq. (11.83) is shown. In the initial stage when $\sqrt{D_{el}t} \geq V_z t$ (Fig. 11.19*a*) the density perturbation is governed by the diffusion, and the spreading is qualitatively the same as that treated in the Section 6.1. Later (Figs. 11.19*a*, *b*, *c*) the plasmoids separate; simultaneously the configuration of the depletion regions is changed.
The density evolution in *xy* plane near the extremum line for the case $\vec{E}_0 \perp \vec{B}$ was simulated in [1], Fig. 11.20*a*. In Fig. 11.20*b* the same equidensities are calculated from Eq. (11.133) which will be obtained in Section 7.9. The corresponding 3D pattern is shown in Fig. 11.21*a*.

Finally, we can conclude that in the presence of external electric field the density perturbation, as in the case of diffusion, is separated into the electron and ion plasmoids. However, in contrast to the case of the diffusion, the plasmoids are spatially separated and move with the velocities which are close to the unipolar velocities of electrons and ions in the ambient electric field correspondingly. In other words, in contrast to the traditional pattern of the ambipolar diffusion, the quasineutrality does not impose any substantial restriction on the plasmoids motion. The longitudinal scale of the electron plasmoid and the transverse scales of the ion plasmoid grow with time faster than proportional to $t^{1/2}$ due to extension by the perturbed electric field.

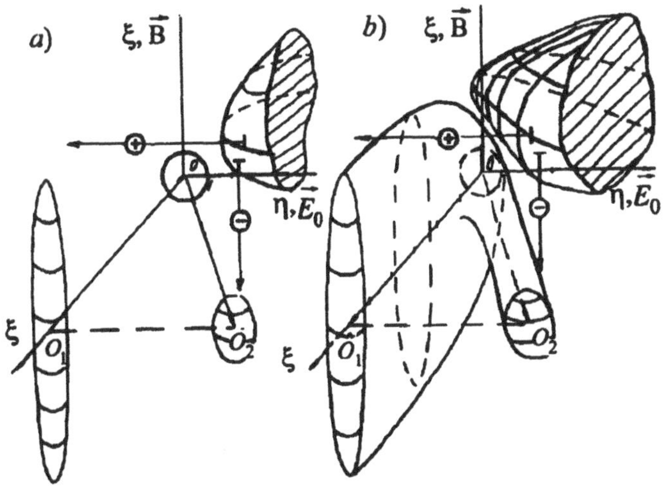

Figure 11.21 The evolution of the density perturbation for the case $\vec{E}_0 \perp \vec{B}$ in the regimes of small, intermediate (*a*) and strong (*b*) nonlinearities: OO_1 and OO_2 are directions of the motion of the electron and ion plasmoids; dashed are the depletion regions.

As a result, the density perturbation in each plasmoid decreases faster than as $t^{-3/2}$. Since such a fast spreading of the plasmoids is connected with the dispersion law of small signals $\vec{V}(\vec{k})$, Eq. (11.7), the corresponding mechanism was called in [1] the dispersion mechanism. The physical reason for almost unipolar motion of the electrons and ions of the initial perturbation is connected with the flowing of the short-circuiting currents in the ambient plasma and the formation of the depletion regions in the ambient plasma.

Evolution of wave packets. Many of the results discussed above (the non-Gaussian asymptotic behavior, existence of the extremum line, formation of the depletion regions and the plasmoids) are consequences of the behavior of the Fourier components of the Green's function G at small values of wave numbers k, where the functions $D(\mu^2)$ and $\vec{V}(\mu^2)$ are singular. The reason is that the perturbation of electric field and the short-circuiting currents at large distances are determined solely by the total number of particles injected into the plasma, i.e. by the component $G_{\vec{k}}$ corresponding to k=0. The perturbation for which $G_{\vec{k}=0} = 0$, i.e. the perturbation of the density profile for a fixed number of particles, should accordingly behave in a different manner.

Let us examine the more general (linear) problem of the spreading of a Gaussian wave packet [31],

$$\delta n(\vec{r},t) = \frac{1}{(2\pi)^3}\int A(\vec{k})$$
$$\times \exp\{i\vec{k}[\vec{r} - V(\mu^2)t] - k^2 D(\vec{k})t\}d\vec{k} \;; \tag{11.110}$$
$$A(\vec{k}) = \exp\left[-\frac{(\vec{k}-\vec{k}_0)^2}{(\Delta\vec{k})^2}\right].$$

When $\Delta k \gg k_0$ the results are the same as those obtained above with the substitution $D(\mu^2)t \to D(\mu^2)t + (\Delta k)^{-2}$. If, on the other hand, $\Delta k \ll k_0$, the integration range in Eq. (11.110) can be split into two parts: $k \gg \varepsilon$ and $0 < k \ll \varepsilon$. Using a series expansion of the exponential function in the first integral, and extending the integration to infinity, for $D(\vec{k}_0)k_0\Delta k t \ll 1$, $V\Delta k \ll 1$ (under this condition the damping of different harmonics in the packet is similar, so that the packet shape does not change), we find the usual result

$$\delta n_1(\vec{r},t)=\left(\frac{\Delta k}{2\sqrt{\pi}}\right)^3 \exp\{i\vec{k}_0[\vec{r}-V(\vec{k}_0)t]$$
$$-k_0^2 D(\vec{k}_0)t-\frac{(\vec{r}-\vec{V}_g t)^2(\Delta k)^2}{4}\}, \tag{11.111}$$

which represents the propagation of the packet at a group velocity (D in Fig. 11.18)

$$V_g(\vec{k})=\frac{\partial}{\partial\vec{k}}[\vec{k}\vec{V}(\vec{k})].$$

In the second integral we introduce an exponential cut off factor $\varepsilon<(\Delta k/k_0)^2$ and extend the integral to infinity:

$$\delta n_2(\vec{r},t)=\frac{1}{(2\pi)^3}\exp\left[-\frac{k_0^2}{(\Delta k)^2}\right]$$
$$\times\int d\vec{k}\exp\{i\vec{k}[\vec{r}-\vec{V}(k)t]-k^2/\varepsilon^2-D(\vec{k})k^2 t\}. \tag{11.112}$$

When $r\gg\max(Vt,\ (Dt)^{1/2})$, $\varepsilon\gg r^{-1}$, the integral in Eq. (11.112) is independent of ε and equal to product of $\exp[-(k_0/\Delta k)^2]$ and Eq. (11.83). Since Eq. (11.111) falls off exponentially with distance from the group-velocity surface,

$$\vec{r}=\vec{V}_g(\vec{k})t,$$

the solution at large distances is governed solely by the perturbation of the total number of particles, i.e. by the value of $A(0)$ in Eq. (11.110). The contribution from the vicinity of the point $k=0$ is also relatively large near the extremum line and near the points $\vec{r}=\vec{V}_g(\vec{k})t=\vec{V}(\vec{k})t$ (A,B and C in Fig. 11.18). If $Vt\Delta k>k_0/\Delta k$, the wave packet in Eq. (11.111) and the density perturbation Eq. (11.112), due to a change of the total number of particles, are spatially separated.

In the absence of an external electric field similar result can be obtained: the wave packet remains at rest and is damped by the diffusion. The power-law asymptotic behavior, which is exponentially small in terms of $k_0/\Delta k$ and which is proportional to t/r^5, is due to the perturbation of the total number of particles.

Density perturbation caused by ions of a different species. In the experiments on ionospheric probing by means of artificial inhomogeneities the perturbation of the plasma density is produced by injecting of ions of different species. Accordingly, we treat the case in which a small plasma inhomogeneity is caused by the injection of ions which mobility is different from the ions of homogeneous background plasma. If $\delta n_I << n_i$, the equations for injected particles are linear, and the perturbation of electric field does not affect the profile δn_I. Hence, in a homogeneous plasma we have

$$\delta n_I(\vec{r},t) = \frac{N}{(4\pi t)^{3/2} D_{I\perp} D_{I\parallel}^{1/2}} \times \exp\left[-\frac{(z - b_{I\parallel}E_{0\parallel}t)^2}{4D_{I\parallel}t} - \frac{(\vec{\rho} - \hat{b}_I \vec{E}_0 t)^2}{4D_{I\perp}t} \right], \tag{11.113}$$

where N is the total number of injected particles, $\vec{\rho}$ is the radius vector in the plane perpendicular to the magnetic field. The equations for the density perturbation of the ambient plasma $\delta n_i(\vec{r},t)$ have their usual form with the initial condition $\delta n_i(\vec{r},0) = 0$. Using Eq. (11.113) and the quasineutrality condition, we find that the Fourier component of the perturbation is [31]

$$\delta n_{i\vec{k}} = NA(\mu^2)\{\exp[-D(\mu^2)k^2 t - i\vec{k}\vec{V}(\mu^2)t] - \exp[-(\vec{k}\hat{D}_I\vec{k})t - i\vec{k}\hat{b}_I\vec{E}_0 t] , \tag{11.114}$$

where

$$A(\mu^2) = \frac{\vec{k}(\hat{D}_e - \hat{D}_I)\vec{k} - i\vec{k}(\hat{b}_e + \hat{b}_i)\vec{E}_0}{k^2 D_i(\mu^2) - \vec{k}\hat{D}_I\vec{k} + i\vec{k}[\vec{V}(\mu^2) - \hat{b}_I\vec{E}_0]} \times \frac{\vec{k}\hat{D}_i\vec{k}}{\vec{k}(\hat{D}_e + \hat{D}_i)\vec{k}},$$

and the functions $\vec{V}(\mu^2), D(\mu^2)$ are defined for the ambient ions in accordance with Eqs. (11.7) and (5.32).

The asymptotic form of the perturbation can again be presented in the form $g(\alpha,\gamma)/r^4$. The positions of the extremum line and the density maxima A, B and C on this line are governed by the argument of the exponential function in Eq. (11.114) and are independent of the mobility ratio. A fundamentally new feature consists in the appearance of a fourth extremum in the electron density at a position coinciding with the maximum of injected ion density: $\vec{r} = \hat{b}_I\vec{E}_0 t$. At this point the value of $\delta n_i(\vec{r},t)$ is negative, and the sign of the resultant electron density perturbation depends on the ratio of the mobilities of the ambient and injected ions. In the particular case of equal mobilities, the perturbation of the ambient ion density compensates completely Eq. (11.113), so that this extremum vanishes. The situation resembles the case in the absence of the magnetic field analyzed in Section 9.1.

11.7 INTERMEDIATE AND STRONG NONLINEARITIES

For a nonlinear perturbation of small dimensions the situations are extremely diverse, depending on the effectiveness of the mechanism of current short-circuiting through the ambient plasma [31]. The criterion, as in the case of a pure diffusion (Section 6.2), is the relative depth of the depletion regions h in the linear problem. From Eqs. (11.92) and (11.93) it follows that h is given by (x_i>>1)

$$h = \frac{N}{n_0 l_\parallel l_\perp^2}, \tag{11.115}$$

where

$$l_\parallel = b_{e\parallel} E_{0\parallel} t + b_{i\perp} E_{0\perp} t / \mu_0;$$
$$l_\perp = b_{i\perp} E_{0\perp} t + b_{e\parallel} E_{0\parallel} t \mu_0$$

are the distances along and across the magnetic field, starting from which the asymptotic expression for the density perturbation is valid. If h<<1 the solution can be sought in the form of the series expansion with h as a parameter. The second order correction to the density perturbation for $V(\mu^2)t >> (D(\mu^2)t)^{1/2}$ is

$$\delta n^{(2)}(\vec{r},t) = \frac{N^2}{(2\pi)^6}$$
$$\times V.P. \int d\vec{k}_1 d\vec{k}_2 \exp\{i(\vec{k}_1 + \vec{k}_2)\vec{r} - it[\omega(\vec{k}_1) + \omega(\vec{k}_2)$$
$$\times \frac{(\vec{k}\hat{D}_i\vec{k})(\vec{k}\hat{D}_e\vec{k}_2) - (\vec{k}\hat{D}_e\vec{k})(\vec{k}\hat{D}_i\vec{k}_2)}{[\vec{V}(\vec{k}_1)\vec{k}_1 + \vec{V}(\vec{k}_2)\vec{k}_2][(\vec{k}\hat{D}_e\vec{k}) + (\vec{k}\hat{D}_i\vec{k})]}$$
$$\to \frac{}{-[(\vec{k}\hat{b}_i\vec{E}_0)(\vec{k}\hat{D}_e\vec{k}) - (\vec{k}\hat{b}_e\vec{E}_0)(\vec{k}\hat{D}_i\vec{k})]}$$
$$\times \frac{\vec{k}_2(\hat{b}_e + \hat{b}_i)\vec{E}_0}{(\vec{k}_2\hat{D}_e\vec{k}_2) + (\vec{k}_2\hat{D}_i\vec{k}_2)},$$

where $\vec{k} = \vec{k}_1 + \vec{k}_2$, $\omega(\vec{k}) = \vec{k}\vec{V}(\mu^2) - iD(\mu^2)k^2$.

Examining the regions in $\vec{k}_1$ and $\vec{k}_2$ spaces, which give the main contributions to the integrals, we find that when $\vec{k}_1$ is almost parallel to $\vec{k}_2$ the integral is small since the factor $(\vec{k}\hat{D}_i\vec{k})(\vec{k}\hat{D}_e\vec{k}_2) - (\vec{k}\hat{D}_e\vec{k})(\vec{k}\hat{D}_i\vec{k}_2)$ tends to zero. In contrast to the case of diffusion, the crossed contributions (when μ_1 corresponds to one of the density maximums and μ_2 to the other maximum) are exponentially small because the maxima A, B and C are spatially separated. Therefore, the parameter of the series expansion turns out to be h. Accordingly, if $h \ll 1$ the density profile in this case of the intermediate nonlinearity is the same as in the linear problem (with the distinction that there is no ambipolar peak, as in the diffusion case, Section 6.2), although the value of the perturbed density can be much higher than n_0. The summation of the series for the potential within terms of the order of h can be performed explicitly, so that near the density maxima A, B and C we have

$$\tilde{\varphi}(t) = \Phi(t)\ln(n / n_0)\,; \qquad \Phi(t) = \tilde{\varphi}^{(lin)} / \delta n^{(lin)}\,, \tag{11.116}$$

where $\tilde{\varphi}^{(lin)}, \delta n^{(lin)}$ are the solutions of the linear problem.

If $h \gg 1$, the influence of the fluxes in the ambient plasma on the evolution of the injected plasma cloud becomes unimportant.

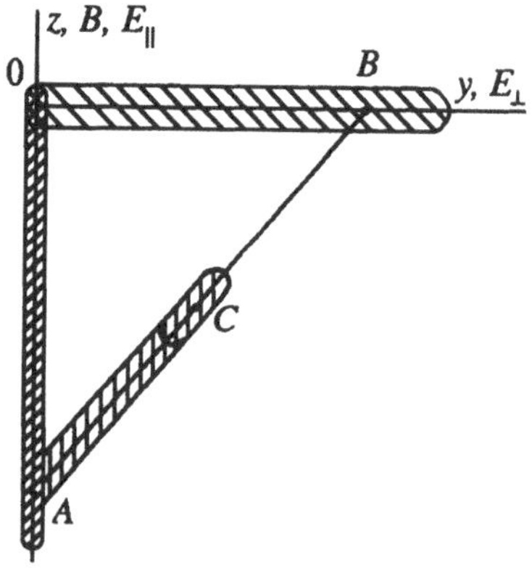

Figure 11.22. Spreading of a very nonlinear perturbation (h>1). The regions in which the plasma density is far above the background density are hatched. The ambipolar peak is at the origin. Branches OA and OB (projection on ZOY plane) correspond to the inipolar motion of the injected electrons and ions respectively.

The ambient electric field inside the initial inhomogeneity is almost completely balanced by the perturbed electric field, and the spreading is governed primarily by anisotropic ambipolar diffusion. The small fraction (of the order of h^{-1}) of the particles injected into the plasma, on the other hand, can move in a unipolar manner as before, forming plasma 'branches', as shown schematically in Fig. 11.22. (*OA*, *OB*).

The short-circuiting currents in the ambient plasma lead to the formation of branch *AC*. Since the perturbed volume of the ambient plasma extremely exceeds the volume of the injected plasma cloud, this regime of the strong nonlinearity exists only for very intense plasma perturbations.

11.8 INHOMOGENEITY OF FINITE DIMENSIONS IN $\vec{E}\times\vec{B}$ FIELDS

In the two preceding Sections the self-similar solutions were addressed, which are valid at late stages when the initial shape and dimensions of inhomogeneities are unimportant. However, for real inhomogeneities the main density decrease takes place at the initial stage when the self-similar solutions are not applicable yet. In this Section we shall analyze the general 3D evolution of inhomogeneities focusing on this non-self-similar stage. We restrict ourselves to the case of $\vec{E}_0\times\vec{B}$ which is very important for the ionospheric application. The mobilities of the injected and ambient ions are considered to be different.

The starting equations are

$$\begin{aligned}
&\frac{\partial n}{\partial t}-\vec{\nabla}\cdot(\hat{D}_e\vec{\nabla}n-\hat{b}_e n\vec{\nabla}\tilde{\varphi}+\hat{b}_e n\vec{E}_0)=0\,;\\
&\frac{\partial n_i}{\partial t}-\vec{\nabla}\cdot(\hat{D}_i\vec{\nabla}n_i+\hat{b}_i n_i\vec{\nabla}\tilde{\varphi}-\hat{b}_i n_i\vec{E}_0)=0\,;\\
&\frac{\partial n_I}{\partial t}-\vec{\nabla}\cdot(\hat{D}_I\vec{\nabla}n_I+\hat{b}_I n_I\vec{\nabla}\tilde{\varphi}-\hat{b}_I n_I\vec{E}_0)=0\,,
\end{aligned}\qquad(11.117)$$

where $n_e=n_i+n_I$, subscripts i, I correspond to ambient and injected ions accordingly. The boundary conditions are

$$n_I(\vec{r}\to\infty)=0\,;\; n_i(\vec{r}\to\infty)=0\,;\; \tilde{\varphi}(\vec{r}\to\infty)=0\,,\qquad(11.118)$$

where $\tilde{\varphi}$ is the perturbed potential. Consider the Gaussian initial inhomogeneity

$$n_I(\vec{r},0) = An_0 \exp[-z^2 / a_\|^2 - (x^2 + y^2) / a_\perp]. \qquad (11.119)$$

The z axis is parallel to the magnetic field, $\vec{y} \| \vec{E}_0$, the x axis coincides with $\vec{E}_0 \times \vec{B}$ direction. Eqs. (11.117) with the boundary conditions Eq. (11.118) and the initial condition Eq. (11.119) have been solved numerically in [32] for $a_\|=a_\perp$ and for the model values of transport coefficients. Comparing the numerical solutions obtained for different sets of transport coefficients with the analytical expressions which are put forward below, and choosing the proper numerical coefficients, we shall find thus the expressions which are valid for the real plasma parameters.

Perturbed electric field. Let us obtain the expression for the perturbed electric field at the initial moment t=0 [32]. In the absence of ambient electric field the potential perturbation is determined by the diffusion and has a quadrupole character, see Section 6.1. The current which flows through inhomogeneity in the presence of ambient electric field causes dipole polarization, as it was shown in Section 11.6. If the ambient electric field $\vec{E}_0$ is sufficiently large, so that $eE_0a_\perp \gg T$, it is possible to neglect the diffusive terms in Eq. (11.117) as it was done in Chapter 9 for the case of the absence of the magnetic field. The exception is the region of diffusive shock which is formed progressively at negative values of x. Therefore, at t=0 for $eE_0a_\perp \gg T$ the perturbed electric field should have the dipole character.

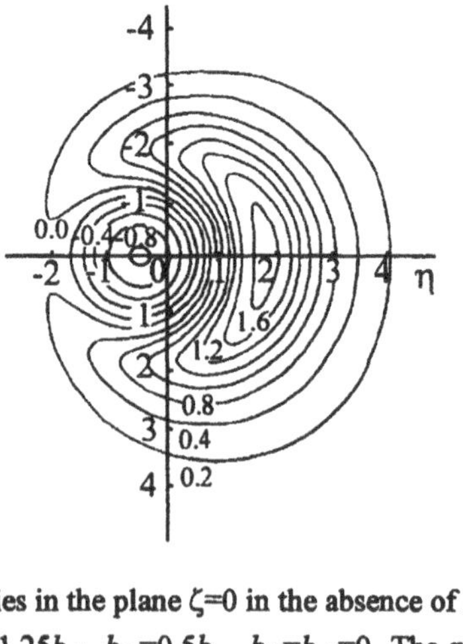

Figure 11.23. Equidensities in the plane ζ=0 in the absence of Hall current (δ=0) for t=0, A=25, $\tilde{E}_0=1$, $b_{e\|}=1.25b_{i\|}$, $b_{i\perp}=0.5b_{i\wedge}$, $b_{i\|}=b_{e\perp}=0$. The potential values at the extrema are: $\Phi_+=1.74$; $\Phi_-=-0.84$.

For the case when the Hall mobilities of ions and electrons coincide (x_i>>1) the calculated perturbed potential is shown in Fig. 11.23.
Here $\zeta=z/a_\perp$; $\eta=y/a_\perp$; $\xi=x/a_\perp$; $\tilde{E}_0 = a_\perp eE_0 / T$; $\Phi = e\tilde{\varphi} / T$. The dipole polarization is caused by the current of injected ions driven by the ambient electric field and leads to partial screening of the ambient electric field in the plasma cloud. From the other side, the perturbed potential causes the longitudinal fluxes of electrons to maintain quasineutrality. Moreover, it is transferred to a large distance $l_\|$ along the magnetic field. This polarization causes also the counter current of the ambient ions at distances of the order of $l_\|$ which closes the electron longitudinal circuit. The characteristic scale $l_\|$ can thus be estimated by equating the divergence of the parallel electron flux and the divergence of the perpendicular flux of ambient ions (in the ambient plasma along $\vec{B}$ away from the initial cloud). In accordance with Eq. (11.117),

$$n_0 b_{e\|}\tilde{\varphi} / l_\|^2 \sim n_0 b_{i\perp}\tilde{\varphi} / a_\perp^2, \tag{11.120}$$

and hence

$$l_\| = a_\perp \sqrt{b_{e\|} / b_{i\perp}} >> a_\perp . \tag{11.121}$$

We see that injection of a plasma cloud causes perturbation of electric field in far larger volume than that of the cloud.

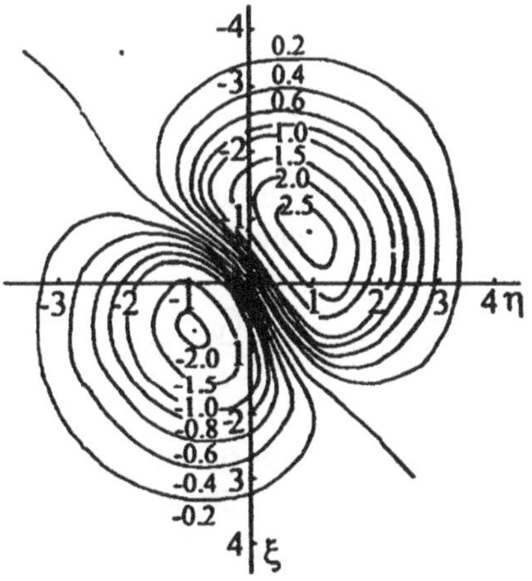

Figure 11.24. Equidensities in the plane ζ=0 in the presence of Hall current (δ=1.6) for t=0, A=5, $\tilde{E}_0 = 5$, $b_{e\|}$=25$b_{i\|}$, $b_{i\perp}$=2.5$b_{i\wedge}$, $b_{i\|}$= $b_{e\perp}$=0. The potential values at the extrema are: Φ_+=2.7; Φ_-=-2.3.

At the first moment in this 'region of influence' only the electric field is perturbed, and later the density perturbation is formed. Inside the initial cloud the ambient electric field also makes contribution to the ion flux divergence, due to inhomogeneity of n_I. While estimating the divergence of the parallel electron flux in this region, it should be taken into account that the electrons are gathered from the distance $l_{\|}$, but are accumulating at a scale $a_{\|}$. Therefore, for $A>>1$ we have

$$\frac{n_I b_{I\perp}(E_0 - E'_y)}{a_\perp} \sim \frac{n_0 \tilde{\varphi} b_{e\|}}{a_{\|} l_{\|}}, \tag{11.122}$$

where $E'_y \sim \tilde{\varphi} / 2a_\perp$ is the perturbed electric field. Combining Eqs. (11.121) and (11.122) we find

$$\frac{E'_y}{E_0} = \frac{1}{1 + K_1 \dfrac{a_\perp \sqrt{b_{e\|} / b_{i\perp}}}{A a_\|} \dfrac{b_{i\perp}}{b_{I\perp}} + \tilde{K}_1} \equiv \frac{1}{1+C}; \qquad E'_x = 0. \tag{11.123}$$

Coefficients $K_1 \approx 1.5$ and $\tilde{K}_1 \approx 3$ are chosen from the comparison of the results of different simulations and also from the comparison with the analytical solution Eq. (11.133) which is obtained in the next Section. Eq. (11.123) becomes more transparent if we introduce the integral Pedersen conductivities of ions

$$\Sigma_{1\perp} = \sqrt{\pi} e a_{\|} A b_{I\perp} n_0 \; ; \qquad \Sigma_{2\perp} = \sqrt{\pi} e l_{\|} b_{i\perp} n_0 \, .$$

Then Eq. (11.123) becomes similar to Eq. (11.58) obtained in Section 11.3 in the two-layer model of the ionosphere,

$$E'_y = \frac{E_0}{1 + K_1 \Sigma_{2\perp} / \Sigma_{1\perp}}. \tag{11.124}$$

If the Hall current is not equal zero (in the weakly magnetized plasma), the perturbed electric field forms a finite angle with the direction of the ambient electric field, Fig. 11.23. The potential minimum shifts in the x (Hall) direction, because electrons move faster than ions along x. The perturbed potential profile looses its symmetry with respect to the origin, the magnitudes of minimum ($\tilde{\varphi}_-$) and maximum ($\tilde{\varphi}_+$) values are no longer equal to each other, not only owing to the diffusive quadrupole polarization, but also because of the Hall current. Inside the initial cloud, instead of (11.122) we have

$$\frac{b_{I\perp} n_I (E_0 - E'_y)}{a_\perp} + \frac{(b_{e\wedge} - b_{I\wedge}) n_I E'_x}{a_\perp} \sim \frac{b_{e\parallel} n_0 \tilde{\varphi}_+}{a_\parallel l_\parallel}; \qquad \frac{b_{I\perp} n_I E'_x}{a_\perp} - \frac{(b_{e\wedge} - b_{I\wedge}) n_I (E_0 - E'_y)}{a_\perp} \sim \frac{b_{e\parallel} n_0 \tilde{\varphi}_-}{a_\parallel l_\parallel}. \tag{11.125}$$

The l.h.s. of Eqs. (11.125) describe the arrival of ions in the regions of the potential maximum and minimum correspondingly, due to the transverse motion. Here $E'_y \sim \tilde{\varphi}_+ / a_\perp$; $E'_x \sim \tilde{\varphi}_- / a_\perp$. In the ambient plasma in the region of influence, which is projected along magnetic field onto the extremuma, Eq. (11.120) is still valid. Hence, the longitudinal scale $l_\parallel$ is determined by Eq. (11.121). Combining Eqs. (11.125), (11.121) and accounting for the fact that the Hall currents are divergence free in the homogeneous plasma, we obtain

$$\frac{E'_x}{E_0} = \frac{\delta C K_2}{(1+C)^2 + \delta^2 K_2^2}; \qquad \frac{E'_y}{E_0} = \frac{1 + \delta^2 K_2^2 / (1+C)}{1 + C + \delta^2 K_2^2 / (1+C)}; \tag{11.126}$$

where $\delta = (b_{e\wedge} - b_{I\wedge})/b_{I\perp}$. Comparison with the results of the simulation (Fig. 11.24) gives $K_2 \approx 1$. The numerical coefficients are introduced is such a way that Eq. (11.126) is transformed into Eq. (11.123) when $\delta \to 0$. It also follows from Eq. (11.126) that $E_x' \to 0$, and $E_y' \to E_0$ when $\delta \to \infty$ or $A \to \infty$.

Character of density evolution. Now we discuss a temporal evolution of the density profile. Consider for simplicity the case $b_{e\wedge} = b_{i,I\wedge}$; $\hat{b}_i = \hat{b}_I$. The situation is essentially different depending on the parameter A, which characterises the excess of the inhomogeneity density over the ambient density.

If

$$a_\parallel A << \sqrt{b_{e\parallel} / b_{i\perp}}\, a_\perp, \tag{11.127}$$

so that the integral of the ambient plasma density over the region of influence $l_\parallel$ exceeds the integral density in the cloud, the perturbed electric field remains small with respect to E_0. In the same time, the local density perturbation can be large: $A >> 1$. Such regime of the intermediate nonlinearity for the case of point like perturbation has been considered in the previous Section. The most of injected ions starts to move in the Hall direction with the velocity $V_x=E/B=(E_0-E_y')/B$ and in the Pedersen direction with the velocity $V_y=b_{i\perp}E$, where the perturbed electric field is given by Eq.(11.123). Injected ions form the ion plasmoid. Electrons of the initial inhomogeneity spread along magnetic field over the distance $l_\parallel=a_\perp(b_{e\parallel}/b_{i\perp})^{1/2}$ and form the electron plasmoid with the dimensions ($a_\perp$, $a_\perp$, $l_\parallel$), which moves in the Hall direction with the velocity $V_x=E/B$. The ambient ions come to the electron plasmoid across the magnetic field owing to the divergence of their transverse flux. The depletion regions in the ambient plasma are thus created. Electrons from these regions escape along magnetic field to maintain quasineutrality in the moving ion plasmoid. The corresponding processes are shown schematically in Fig. 11.21*a*. Therefore, after characteristic time t_1 electrons and ions of the initial cloud are separated in space forming two plasmoids. The electron plasmoid is biased negatively, and the ion plasmoid is positively biased. The moment t_1 corresponds to the shift of injected ions to distance $a_\perp$ in the Pedersen direction:

$$t_1 \sim a_\perp/(b_{i\perp}E).$$

The relative depth of the depletion regions at $t \sim t_1$ is

$$\delta n / n_0 \sim h = A\sqrt{b_{i\perp} / b_{e\parallel}} \ll 1 .$$

At $t>t_1$ parameters of the electron plasmoid can be estimated as follows. Ambient ions are gathered by the potential $\tilde{\varphi}_-$ of the electron plasmoid from the distance $b_{i\perp}E_0t$ across the magnetic field. They are accumulated in much shorter region with the scale $l_\perp^{(e)}=\sqrt{a_\perp^2+8D_{e\perp}t}$ ($T_e=T_i$). The same potential causes the spreading of the electron cloud along $\vec{B}$. Comparing the divergences of the electron and ion fluxes we find

$$\frac{b_{i\perp}\tilde{\varphi}_- n_0}{b_{i\perp}E_0 t l_\perp^{(e)}} \sim \frac{b_{e\parallel}\tilde{\varphi}\delta n_-}{(l_\parallel^{(e)})^2} \sim \frac{\delta n_-}{t}, \qquad (11.128)$$

where δn_- is the density perturbation in the electron plasmoid. Since the total number of particles in each plasmoid is of the order of the total number of particles N injected into the plasma, we have the simple relation

$$\delta n_- l_\parallel^{(e)}(l_\perp^{(e)})^2 \sim N \sim A a_\perp^2 a_\parallel n_0 \quad . \qquad (11.129)$$

From Eqs.(11.128) and (11.129) we obtain

$$\tilde{\varphi}_- \sim \frac{E_0^{1/2} N}{n_0 (b_{e\parallel} t)^{1/2} (l_\perp^{(e)})^{3/2}} ;$$
$$l_\parallel^{(e)} \sim b_{e\parallel}^{1/2} (l_\perp^{(e)})^{1/2} E_0^{1/2} t^{1/2} ; \qquad (11.130)$$
$$\delta n_- \sim \frac{N}{(b_{e\parallel} E_0 t)^{1/2} (l_\perp^{(e)})^{5/2}} .$$

These expressions at $t \sim t_1$ tend to Eqs. (11.121) and (11.123). At larger times

$$t >> t_2 = a_\perp^2/(8D_{e\perp}) >> t_1$$

the evolution is described by the self-similar solution analyzed in the preceding Sections, when the profiles are independent of the initial cloud size $a_\perp$. At $t \sim t_2$ these estimates transform into Eqs. (11.101) and (11.102).

The longitudinal scale of the ion plasmoid is determined by the parallel ambipolar diffusion:

$$l_\parallel^{(I)} = (a_\parallel^2 + 8D_{I\parallel} t)^{1/2}.$$

Its perpendicular scale at late stage $t >> t_3$ can be obtained from Eq. (11.109):

$$l_\perp^{(I)} \sim (b_{i\perp}^2 E_{0\perp} D_{i\parallel} t^2 / b_{e\parallel})^{1/3}.$$

The moment $t \sim t_3$ corresponds to $l_\perp^{(I)} \sim a_\perp$, and hence is

$$t_3 \sim a_\perp^{3/2} E_0^{-1/2} b_{e\parallel}^{1/2} D_{i\parallel}^{-1/2} b_{i\perp}^{-1}.$$

As an example we estimate here the characteristic moments for the ionosphere at the height 200km, the initial cloud size $a_\perp$=100m, and the ambient electric field E_0=1mV/m: t_1=100s; $t_2=10^4$s; $t_3=10^4$s. Consequently, for the ionospheric irregularities of relatively large transverse scales the perpendicular sizes of the plasmoids for a long time ($t \lesssim t_2$, t_3) remain of the order of $a_\perp$.

The estimates Eq. (11.130), as well as the expressions Eqs. (11.101), (11.102) and (11.109), are valid if the characteristic sizes of the plasmoids which are determined by the dispersion mechanism are larger than the corresponding diffusive scales. For the electron plasmoid the criterion has the form

$$b_{e\parallel}^{1/2} (l_\perp^{(e)})^{1/2} E_0^{1/2} t^{1/2} >> (8D_{e\parallel} t)^{1/2},$$

and, finally, neglecting the numerical coefficient,

$$E_0 l_\perp^{(e)} >> T/e. \tag{11.131}$$

In the opposite case the density in the electron plasmoid is determined by the electron diffusion. Similar condition for the ion plasmoid leads to a rather severe inequality

$$E_0 (D_{i\perp} t)^{1/2} >> (T/e) b_{e\parallel} / b_{i\parallel}. \tag{11.132}$$

In our numerical example, for $a_\perp < 100$m inequalities (11.131) and (11.132) are not satisfied. The motion and spreading of such small-scale inhomogeneities for $h << 1$ can be described similar to the case of pure diffusion (Section 6.3):

$$\delta n(\vec{r},t) = A a_\parallel a_\perp^2 n_0$$

$$\times \left\{ \frac{\exp\left[-\frac{z^2}{a_\parallel^2 + 4(1+T_i/T_e)D_{e\parallel}t} - \frac{(x - b_{e\wedge}E_0 t)^2 + y^2}{a_\perp^2 + 4(1+T_i/T_e)D_{e\perp}t} \right]}{[a_\parallel^2 + 4(1+T_i/T_e)D_{e\parallel}t]^{1/2}[a_\perp^2 + 4(1+T_i/T_e)D_{e\perp}t]} + \frac{\exp\left[\begin{array}{l} -\frac{z^2}{a_\parallel^2 + 4(1+T_e/T_i)D_{i\parallel}t} \\ -\frac{(x - b_{i\wedge}E_0 t)^2 + (y - b_{i\perp}E_0 t)^2}{a_\perp^2 + 4(1+T_i/T_e)D_{i\perp}t} \end{array} \right]}{[a_\parallel^2 + 4(1+T_e/T_i)D_{i\parallel}t]^{1/2}[a_\perp^2 + 4(1+T_e/T_i)D_{i\perp}t]} \right\}. \tag{11.133}$$

In Fig. (11.20) the equidensities calculated according to this expression are compared with the numerical simulation [1] for the conditions when inequalities inverse to Eq.(11.131) and (11.132) are satisfied. One can see that Eq.(11.133) gives rather satisfactory description of the evolution process.

In the case of very strong nonlinearity

$$a_\parallel A > \sqrt{b_{e\parallel} / b_{i\perp}}\, a_\perp \ ; \ h > 1 \tag{11.134}$$

electric field inside the initial cloud is almost completely screened. Furthermore, the main cloud which contains most of injected particles remains practically at rest, while its spreading along magnetic field is governed by the ambipolar diffusion. The fraction of electrons and injected ions which are able to outflow from the main cloud, as in the case of pure diffusion (Section 6.2), is determined by the ambient plasma density. The density in the depletion regions becomes much smaller than n_0 (so that $|\delta n|/n_0 \approx 1$), and the plasma branches shown in Fig. 11.21 are formed. At the edge of the main cloud, where the branches are generated, the longitudinal scale of the potential perturbation is $l_{||}=a_{\perp}(b_{e||}/b_{i\perp})^{1/2}$, so that the maximum density in the ion branch should be of the order of $n_0 l_{||}/a_{\perp}=n_0(b_{e||}/b_{i\perp})^{1/2}$. Since the ambient ions are gathered to the electron branch in the Pedersen direction from the distance ~ $a_{\perp}$, the density in the ion branch should be of the order of n_0, and its longitudinal size should be of the order of $l_{||}$. The outflux rate (number of particles ejected from the main cloud per second) to the branches is thus can be estimated as $n_0(b_{e||}/b_{i\perp})^{1/2} a_{\perp} E_0/B$. Hence, the maximum density in the main cloud decreases as

$$\frac{n_{\max}}{n_0} = \frac{A - K_3(E_0 / B)t(\sqrt{b_{e|} / b_{i\perp}} + \tilde{K}_1) / a_{|}}{(1 + 4(1 + T_e / T_i)D_{i||}t / a_{||}^2)^{1/2}} \; ; \qquad (11.135)$$

$$K_3 \approx 1.8 \; ; \quad \tilde{K}_1 \approx 3 \, .$$

The density in the region of influence remains of the order of n_0 because electric field inside the main cloud is still approximately given by Eq.(11.123). Outside the main cloud electric field is only slightly perturbed, and, therefore, the rear side of the cloud (in the Hall direction) becomes steeper while the front side becomes more gradual. The sharp density drop is formed at the cloud rear side which can be considered as a shock in the Hall direction, compare with Section 11.4.

The situation is more complicated when the mobilities of injected and ambient ions are essentially different. If $\Sigma_2^{(0)}>\Sigma_1^{(0)}$, and the integral numbers of particles satisfies $\Sigma_2^{(0)}/b_{i\perp}<\Sigma_1^{(0)}/b_{I\perp}$, the depletion regions become rather deep, so that Eq.(11.123) is valid only at the first moment. The transverse conductivity quickly decreases here, and the perturbed electric field E_y' increases. The Pedersen outflux of injected ions to the ion branch becomes density dependent, and shock in the Pedersen direction is formed provided the condition $eE_0a_{\perp}>>T$ is satisfied.

Results of numerical simulations. Evolution pattern discussed above has been checked in the simulations [32]. The potential profile has been calculated at t=0 for various sets of transport coefficients. It was found out that in accordance with Eqs. (11.123) and (11.126) the polarization field depends only on parameters h and δ. As it was predicted above, $E_x'\to 0$ and $E_y'\to E_0$ for $\delta\to\infty$ or $h\to 0$. The values of the numerical coefficients were obtained: $K_1\approx 1.5$; $\tilde{K}_1 \approx 3$; $K_2\approx 1$. The value of electric field inside the cloud has been calculated as $(\tilde{\varphi}_+ - \tilde{\varphi}_-)/|\tilde{r}_+ - \tilde{r}_-|$. The difference never exceed 20% and was mainly connected with the Hall current. For δ=0 Eq. (11.123) coincides with the numerical results with the accuracy 5%. The length of the region of influence is also consistent with Eq. (11.121).

In Fig. 11.25 examples of the potential profiles along the lines parallel to the magnetic field and passing through the perturbed potential extrema are shown for the strongly nonlinear case A=25 and δ=0. One can distinguish between two different scales: $l_{\parallel}=a_{\parallel}$ and $l_{\parallel}=a_{\perp}(b_{e\parallel}/b_{i\perp})^{1/2}$. This is the result of interference of two different mechanisms: polarization generated by the diffusion and that caused by current flowing through the inhomogeneity. The diffusive potential in the absence of current in ambient plasma has been dicussed in Section 6.1. This type of the potential is proportional to T/e and has a quadrupole character.

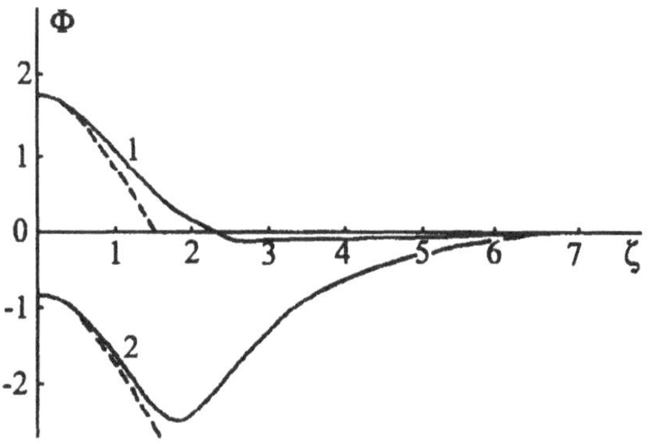

Figure 11.25. Potential profiles along $\vec{B}$ at t=0 in the absence of Hall current (δ=0), A=25, $\tilde{E}_0=1$, $b_{e\parallel}=1.25b_{i\wedge}$, $b_{i\perp}=0.5b_{i\wedge}$, $b_{i\parallel}=b_{e\perp}=0$: 1-along the line ξ=0, η=1.75 (passing through the potential maximum); 2-along the line ξ=0, η=-0.35 (passing through the potential minimum); dashed line is Boltzmann diffusion potential $\ln(n/n_0)$.

Near the center of the cloud its longitudinal profile is characterized by the scale $a_{\parallel}$ and corresponds to the Boltzmann distribution for electrons. The elongated part of the profiles with the scale $a_{\perp}(b_{e\parallel}/b_{i\perp})^{1/2}$ can be associated with the polarization owing to the ambient electric field.

Across the magnetic field the potential profiles also can be treated as the superposition of two contributions. Roughly speaking, the perturbed potential in xoy plane is the sum of the quadrupole diffusive potential and of the dipole part caused by the ambient electric field, Fig. 11.26.

Since the Hall current in this variant was assumed to be zero, the potential profile is symmetrical with respect to the x-axis and is analogous to the diffusive profile. Near the cloud centre it is close to the Boltzmann distribution for ions. Asymmetry of the profile with respect to the y axis is caused by the ambient electric field.

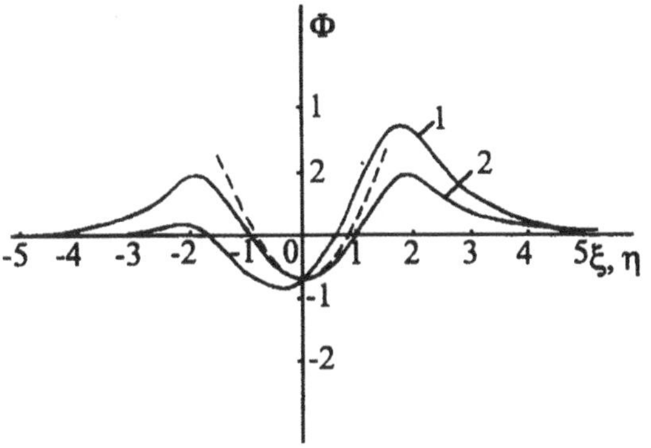

Figure 11.26. Potential profiles along the axis η (1) and the axis ξ(2) at ζ=0 and t=0, the parameters are the same as in Fig. 11.25. Dashed line is the Boltzmann distribution for ions: $-\ln(n/n_0)$.

In Figs. 11.27 - 11.29 the density profiles are shown at the moment $t=1.6a_\perp/(b_{i\wedge}E_0)$ or $\tau=tb_{i\wedge}E_0/a_\perp=1.6$.

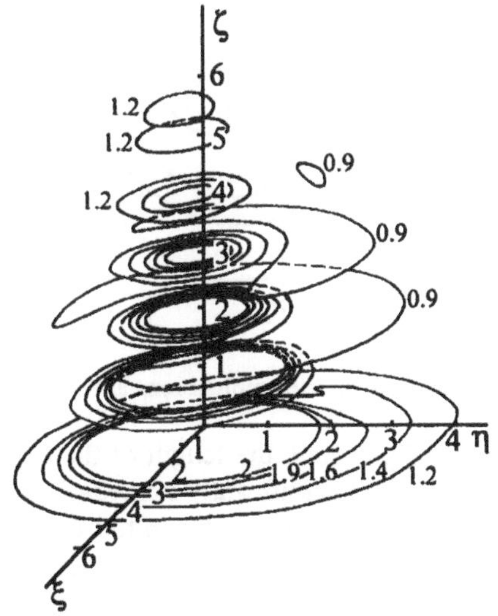

Figure 11.27. The equidensities for the electrons at $\tau=tb_{i\wedge}E_0/a_\perp=1.6$ at ζ=1.05m (m=0; 1; 2; 3; 4 ;5).

The perturbed potential profile, as at the initial moment, is a combination of the dipole and of quadrupole electric fields. The dipole perturbed electric field corresponds to the approximately full screening of the ambient electric field. The mean electric field is slightly larger than that at $t=0$, due to deformation of the profile. The moment chosen is close to t_1. In the linear case the moment t_1 corresponds to the separation of the electron and ion plasmoids. On the contrary, in the calculations this effect is not observed because of the strong nonlinearity of the problem. The maximum of injected ions practically does not move (in the linear case the shift would be 1.6 along ξ and 0.9 along η). The outflow of injected ions both in the Hall and Pedersen directions should start at density $n_0 a_\perp (b_{e\|}/b_{i\perp})^{1/2}/a_\|$ which for the parameters chosen is 1.6 n_0. The region in the right low corner in $\xi 0 \eta$ plane in Fig. 11.27 corresponds to the ion branch.
At the opposite side of the n_I profile shock is formed. Ions of the ambient plasma are redistributed, Fig. 11.28, forming the electron branch and the depletion region.

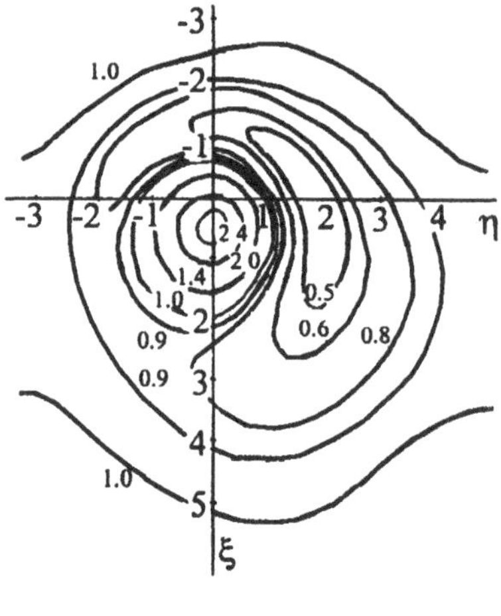

Figure 11.28. The equidensities for the ambient ions at $\tau=1.6$ in $\xi 0 \eta$ plane.

As one can see from Fig. 11.29, the length of the electron branch in the direction of the magnetic field significantly exceeds the logitudinal length of initial cloud $a_\|$. The density decay in the initial cloud is consistent with Eq.(11.135) up to times less than $\tau = t b_{i\wedge} E_0 / a_\perp = 1.6$. Later the influence of the mesh boundaries becomes significant. Depletion regions are also clearly seen in Fig. 11.29.

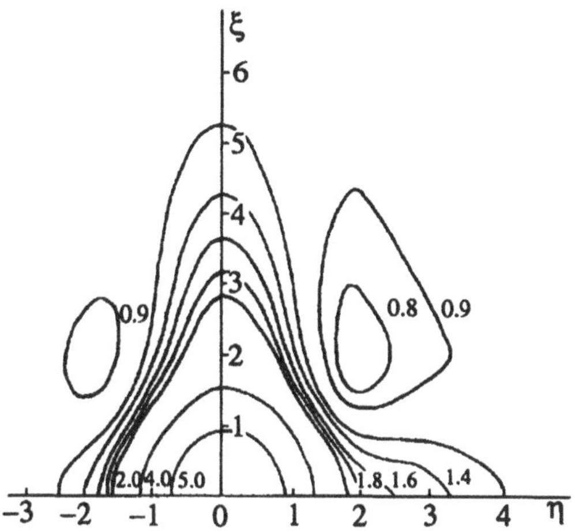

Figure 11.29. The equidensities for the electrons at τ=1.6 in $\zeta 0 \eta$ plane.

11.9 TRANSFORMATION OF TRANSPORT EQUATIONS AND 3D MODELLING OF INHOMOGENEITY EVOLUTION

The main problem in the 3D modelling of the plasma cloud evolution in a strong magnetic field (e.g., the evolution of the ionospheric irregularities) is connected with the fact, that the diffusion and mobility coefficients along and across the magnetic field often differ by orders of magnitude. The corresponding scales for different plasmoids, as we have seen in the preceding Sections, also strongly diverse. Therefore, the direct modelling of the initial transport equations often becomes rather complicated and expensive procedure; it is difficult to trace the full time history of the plasma perturbation, etc. Therefore, we consider here the analytical method of the transformation of the transport equations, suggested in [33], which allow to perform an effective modelling of the plasma evolution in the strong magnetic fields. Examples of the 3D modelling based on the reduced set of transport equation will be presented in the next Section. On the other hand, some analytical solutions based on the reduced equations are obtained below.

Consider the ambient plasma which is nonuniform in $\vec{B}$ direction: $n_0=n_0(z)$; $\hat{b}_\alpha = \hat{b}_\alpha(z)$. Analytical and numerical solutions obtained both for the case $\vec{E}_0 \perp \vec{B}$ in this Chapter, and for the case of the absence of ambient electric field in Chapter 6, demonstrate that the perturbed potential $\tilde{\varphi}$ can be sought in the form

$$\tilde{\varphi} = (T_e / e)\ln n_e / n_0 + \Psi(x,y,z). \tag{11.136}$$

Function Ψ has the longitudinal scale $l_{||} >> a_{||}$, if

$$a_{||} << a_{\perp}\sqrt{b_{e||} / b_{e\perp}}. \tag{11.137}$$

Indeed, as has been shown in the preceding Section, in the unbounded plasma the longitudinal scale $l_{||}$ of the potential perturbation is simply $a_{\perp}\sqrt{b_{e||} / b_{i\perp}}$ at t=0, and further increases with time. In the nonuniform ambient plasma the scale $l_{||}$ can be also determined by the variation of $n_0(z)$ and/or of $\hat{b}_\alpha = \hat{b}_\alpha(z)$. The subdivision of the perturbed potential Eq. (136) can be also confirmed *a posteriori* since the reduced set of equations for the ambient plasma has only large longitudinal scale $l_{||}$.

We consider the nonlinear perturbation A>>1, so that in the vicinity of the injected cloud n_I=n_e. We start with the current continuity equation

$$\vec{\nabla}\cdot\vec{\Gamma}_e = \vec{\nabla}\cdot\vec{\Gamma}_i + \vec{\nabla}\cdot\vec{\Gamma}_I. \tag{11.138}$$

In the ambient plasma, far from the injected cloud $|z|>>l_{||}$, by equating n_i=n_e and accounting for Eq.(11.136), we obtain

$$\begin{aligned} &\vec{\nabla}\cdot[(\hat{D}_i + T_e\hat{b}_i / e)\nabla n_i + (\hat{b}_i + \hat{b}_e)n_i(\vec{\nabla}\Psi - \vec{E}_0)] \\ &= \vec{\nabla}\cdot[(\hat{D}_e - T_e\hat{b}_e / e)\vec{\nabla}n_i]. \end{aligned} \tag{11.139}$$

Further transformation depends on the angle between the electric and magnetic field.

Ambient electric field perpendicular to magnetic field. Let us introduce the dimensionless variables compressed in z direction,

$$\xi = \frac{x}{a_\perp}; \quad \eta = \frac{y}{a_\perp}; \quad \zeta = \frac{z}{a_\perp}\frac{b_{i\perp}^{1/2}(0)}{b_{e||}^{1/2}(0)}.$$

Here $b_{i\perp}(0)$ and $b_{e||}(0)$ are the values of the mobilities at z=0 where the center of the injected cloud is situated. Neglecting the longitudinal mobility of injected ions with respect to the electron mobility, from Eq. (11.139) we obtain the equation for the perturbed potential at a given density profile,

$$\frac{\partial}{\partial\zeta}(b'_{e\parallel}\tilde{n}_i\frac{\partial\Psi}{\partial\zeta})+\vec{\nabla}_\perp\cdot[(\hat{b}'_i+\hat{b}'_e)\tilde{n}_i(\vec{\nabla}_\perp\Psi-\vec{E}_0a_\perp)]$$
$$=-\vec{\nabla}_\perp\cdot\left\{[(\hat{D}'_i-\hat{D}'_e)+(T_e/e)(\hat{b}'_i+\hat{b}'_e)]\vec{\nabla}_\perp\tilde{n}_i\right\}, \qquad (11.140)$$

where

$$b'_{e\parallel}=\frac{b_{e\parallel}(z)}{b_{e\parallel}(0)};\quad b'_{\alpha\perp}=\frac{b_{\alpha\perp,\wedge}(z)}{b_{i\perp}(0)};\quad D'_{\alpha\perp,\wedge}=\frac{D'_{\alpha\perp,\wedge}(z)}{D_{i\perp}(0)};\quad \tilde{n}_\alpha=\frac{n_\alpha}{n_\alpha(0)}.$$

We assume $m_e\nu_{ei}<<\mu_{iN}\nu_{iN}$, so that according to Eq. (2.89) $D_{e\parallel}=(T_e/e)b_{e\parallel}$. Now we substitute Eq. (11.136) into Eq. (11.138) and integrate Eq. (11.138) along the magnetic field from $-z_0$ to z_0, where the value of z_0 is chosen so that $a_\parallel<|z_0|<<l_\parallel$. Inside this region the ambient ion density can be neglected with respect to the injected ion density, and Ψ can be considered as an independent on z function. On the other side, the main part of the integral of injected ion density is located in this region, and the value of n_I can be assumed zero at $z=\pm z_0$. After integration the value of z_0 can be considered as zero for the large scale problem of the ambient plasma evolution. Therefore, we find ($\hat{b}_1\equiv\hat{b}_I$)

$$\tilde{\Gamma}_{e\parallel}|_{\zeta=+0}-\tilde{\Gamma}_{e\parallel}|_{\zeta=-0}=-\vec{\nabla}_\perp\cdot[(\hat{b}'_i+\hat{b}'_e)\tilde{N}_1(\vec{\nabla}_\perp\Psi-\vec{E}_0a_\perp)],$$
$$-\vec{\nabla}_\perp\cdot\left\{[(\hat{D}'_i-\hat{D}'_e)+(T_e/e)(\hat{b}'_I+\hat{b}'_e)]\vec{\nabla}_\perp\tilde{N}_I\right\}|_{\zeta=0}, \qquad (11.141)$$

where

$$\tilde{\Gamma}_{e\parallel}=\tilde{n}_i\frac{\partial\Psi}{\partial\zeta};\quad \tilde{N}_1=\int_{-\infty}^{\infty}\frac{n_Ib_{i\perp}^{1/2}(0)}{a_\perp b_{e\parallel}^{1/2}(0)n_0(0)}dz=\frac{N_1b_{i\perp}^{1/2}(0)}{a_\perp b_{e\parallel}^{1/2}(0)n_0(0)}.$$

Eq. (11.141) should be treated as a mixed boundary condition for Eq. (11.140) at $\zeta=0$. Potential at $\zeta=0$ remains a continuous function, while the parallel electric field is discontinuous.

Equation for the evolution of the integrated density of injected ions is obtained by integration of the third of Eqs. (11.117). Assuming again Ψ=const(z), we find (in the dimensionless ξ, η, ζ variables)

$$a_\perp^2 \frac{\partial \widetilde{N}_1}{\partial t} \qquad (11.142)$$
$$= \vec{\nabla}_\perp \cdot [(\hat{D}_1 + T_e \hat{b}_1 / e)\vec{\nabla}_\perp \widetilde{N}_1 + \hat{b}_1 \widetilde{N}_1 (\vec{\nabla}_\perp \Psi - \vec{E}_0 a_\perp)].$$

This is the required 2D equation. The local density can be reconstructed from N_1 in the practically important case when

$$\partial \Psi / \partial z|_{z=0} << (T_e / e)\partial \ln n_I / \partial z \sim (T_e / e) / a_{||}.$$

In this situation the evolution along the magnetic field is governed by the first term in the r.h.s. of Eq. (11.136), i.e. corresponds to the longitudinal ambipolar diffusion, and the density n_I satisfies Eq. (11.73).

The ambient density is perturbed in the region with the longitudinal scale $l_{||}>>a_{||}$. Neglecting the parallel shift of the ambient ions, i.e. assuming $4D_{i||}t<<l_{||}^2$; $b_{i||}|\partial \Psi / \partial z|t << l_{||}$, we obtain

$$a_\perp^2 \frac{\partial \widetilde{n}_i}{\partial t} = \vec{\nabla}_\perp \cdot [(\hat{D}_i + T_e \hat{b}_i / e)\vec{\nabla}_\perp \widetilde{n}_i + \hat{b}_i \widetilde{n}_i (\vec{\nabla}_\perp \Psi - \vec{E}_0 a_\perp)]. \qquad (11.143)$$

This is also a 2D equation with the parametric dependence on ζ.

The reduced set of Eqs. (11.140), (11.142) and (11.143) with the boundary condition Eq. (11.141) and the boundary conditions at the infinity (absence of the densities and the potential perturbations) can be considered as a starting system for the 3D simulations. It is much more convenient than the initial system of Eqs. (11.117). Moreover, equations for the density evolution are the 2D. On the contrary, the ratio of the mobilities $b_{e||}$ and $b_{i\perp}$ in the initial current continuity Eq. (11.138) may be extremely large (in the ionosphere, for example, this ratio may reach 10^5), which makes difficult the direct simulation of this equation in the combination with the 3D equations for the densities evolution.

For the large scale inhomogeneities in the strong electric field $eE_0a_\perp>>T$ the diffusion terms may be neglected (as in the absence of the magnetic field, Chapter 9), so that the reduced system is simplified:

$$\frac{\partial}{\partial \zeta}(b_{e|}' \widetilde{n}_i \frac{\partial \Phi}{\partial \zeta}) + \vec{\nabla}_\perp \cdot [(\hat{b}_i' + \hat{b}_e')\widetilde{n}_i(\vec{\nabla}_\perp \Phi - \vec{E}_0 / E_0)] = 0;$$

$$\frac{\partial \widetilde{n}_i}{\partial \tau} = \vec{\nabla}_\perp \cdot [\hat{b}_i' \widetilde{n}_i (\vec{\nabla}_\perp \Phi - \vec{E}_0 / E_0)];$$

$$\frac{\partial \tilde{N}_1}{\partial \tau} = \vec{\nabla}_\perp \cdot [\hat{b}_i' \tilde{N}_1 (\vec{\nabla}_\perp \Phi - \vec{E}_0 / E_0)], \qquad (11.144)$$

where

$$\tau = t b_{i\perp} E_0 / a_\perp \; ; \qquad \Phi = \Psi / (E_0 a_\perp) \; ;$$

$$\tilde{n}_i(0)\left(\frac{\partial \Phi}{\partial \varsigma}\Big|_{\varsigma=+0} - \frac{\partial \Phi}{\partial \varsigma}\Big|_{\varsigma=-0}\right)$$

$$= \left[\frac{\partial}{\partial \varsigma}\left(b_{1\perp}' \tilde{N}_1 \frac{\partial \Phi}{\partial \varsigma}\right) + \frac{\partial}{\partial \eta}\left(b_{1\perp}' \tilde{N}_1 (\frac{\partial \Phi}{\partial \eta} - 1)\right)\right]\Big|_{\varsigma=0} \; .$$

We also took into account that at z=0 the Hall current is absent, and neglected the electron transverse mobility. The most transparent is the problem of determining the perturbed potential at t=0, when the ambient plasma is unperturbed. Equation for the potential in this case is of the Laplace type,

$$\frac{\partial}{\partial \varsigma}(b_{e\parallel}' \tilde{n}_0 \frac{\partial \Phi}{\partial \varsigma}) + \vec{\nabla}_\perp \cdot (b_{i\perp}' \tilde{n}_0 \vec{\nabla}_\perp \Phi) = 0 \, . \qquad (11.145)$$

Examples of analytical solutions. The approach addressed above turns out to be rather effective for obtaining analytical results. Consider as an example a homogeneous unbounded magnetized plasma with $b_{e\wedge}$=$b_{i\wedge}$=$b_{I\wedge}$ in a strong electric field which is perpendicular to the magnetic field. Eq. (11.145) now coincides with the Laplace equation. Furthermore, the boundary condition at z=0 corresponds to the disk with the dielectric permittivity $\varepsilon = \tilde{N}_1 b_{1\perp} / b_{i\perp}$ which is inserted into the electric field of unit magnitude parallel to the η axis. Hence, for the initial condition

$$\begin{aligned} \tilde{N}_1\Big|_{t=0} &= A b_{i\perp}^{1/2}(0) / b_{e\parallel}^{1/2}(0), && \xi^2 + \eta^2 \le 1 \; ; \\ \tilde{N}_1\Big|_{t=0} &= 0, && \xi^2 + \eta^2 > 1 \, , \end{aligned} \qquad (11.146)$$

we can use the well-known expression for the electrostatic field of the biased dielectric ellipsoid [15]. In the initial variables the perturbed electric field inside the initial cloud is constant ($\vec{E}_y' = -\vec{\nabla}_\perp \Psi$):

$$\frac{E'_y}{E_0} = \left[1 + \frac{4a_\perp b_{i\perp} / b_{1\perp}}{\pi a_\parallel A (b_{i\perp} / b_{e\parallel})^{1/2}} \right]^{-1} . \tag{11.147}$$

This result only by numerical factor of the order of unity differs from Eq. (11.123) obtained for the Gaussian initial profile by the comparison with the direct simulation. At large distances the potential has the dipole character and in the initial variables is

$$\tilde{\varphi} = \Psi = \frac{N(b_{1\perp} / b_{i\perp})(b_{i\perp} / b_{e\parallel})^{1/2} E_0 y}{4\pi n_0 \left(1 + \dfrac{\pi A a_\parallel b_{1\perp}}{4a_\perp b_{i\perp}} \sqrt{\dfrac{b_{i\perp}}{b_{e\parallel}}} \right)} \cdot \frac{1}{(x^2 + y^2 + z^2 b_{e\parallel} / b_{i\perp})^{3/2}} , \tag{11.148}$$

where $N=4\pi(3a_\perp^2 a_\parallel A n_0)$ is the total number of injected particles.

It follows from Eq. (11.148) that the asymptotic behavior of the perturbed potential is independent of its shape and magnitude and is proportional to the total number of injected particles provided

$$a_\parallel A \ll a_\perp (b_{1\perp}/b_{i\perp})(b_{e\parallel}/b_{i\perp})^{1/2}. \tag{11.149}$$

The condition Eq. (11.149) corresponds to the case of intermediate nonlinearity when the ambient plasma is only slightly perturbed. In this case Eq. (11.148) is time-independent and thus remains valid in the process of the evolution. Since the density perturbation δn_i decreases with distance faster than the potential perturbation $\tilde{\varphi}$, it is possible to use Eq. (11.148) for the calculation of the asymptotic behavior of the density perturbation, and put $n_i=n_0$ in Eq. (11.144). Substituting Eq. (11.148) into Eq. (11.144) yields

$$n_i - n_0 = \frac{3b_{1\perp} E_0 t N y (x^2 + y^2 - 4z^2 b_{i\perp} / b_{e\parallel})(b_{i\perp} / b_{e\parallel})^{1/2}}{4\pi (x^2 + y^2 + z^2 b_{i\perp} / b_{e\parallel})^{7/2}} . \tag{11.150}$$

Eqs. (11.148) and (11.150) coincide with the exact solutions, Eqs. (11.90) and (11.93).

For the case of the strong nolinearity, when the inequality inverse to Eq. (11.149) is satisfied, the asymptotic behavior of the potential is independent of A and is determined by the perpendicular scale of initial cloud $a_\perp$. For a long time, during which $a_\perp$ does not change significantly, the asymptotic behavior of the density perturbation is given by

$$n_i - n_0 = \frac{4n_0}{\pi} \frac{b_{i\perp} a_\perp^3 E_0 t y (x^2 + y^2 - 4z^2 b_{i\perp} / b_{e\parallel})}{(x^2 + y^2 + z^2 b_{i\perp} / b_{e\parallel})^{7/2}}. \tag{11.151}$$

Ambient electric field parallel to magnetic field. Similar expressions could be obtained for the case $\vec{E}_0 \| \vec{B}$. The potential, as in the previous case, can be sought in the form of Eq. (11.136). In the 'stretched' variables

$$\xi = \frac{x}{a_\perp} \frac{b_{e\parallel}^{1/2}(0)}{b_{i\perp}^{1/2}(0)}; \quad \eta = \frac{y}{a_\perp} \frac{b_{e\parallel}^{1/2}(0)}{b_{i\perp}^{1/2}(0)}; \quad \zeta = \frac{z}{a_\perp}$$

for the strong electric field the equation for the potential is again reduced to the Laplace type equation in the medium with nonuniform permittivity, Eq. (11.145). For the initial condition $\delta n = An_0 = const$ inside the ellipsoid with the semiaxes $a_\parallel$, $a_\perp$, $a_\perp$, the parallel perturbed electric field is homogeneous inside the cloud, and

$$\vec{E}_\parallel = \vec{E}_{0\parallel} + \vec{E}'_\parallel = \frac{\vec{E}_{0\parallel}}{1 + A}. \tag{11.152}$$

This expression is valid if $a_\parallel << a_\perp (b_{e\parallel}/b_{i\perp})^{1/2}$. It has transparent physical interpretation: the longitudinal current at $t=0$, when the ambient plasma is unperturbed, should conserve in the vicinity of the injected cloud. At large distances

$$\tilde{\varphi} = \Psi = \frac{N b_{e\parallel} E_{0\parallel} z}{4\pi n_0 b_{i\perp} (1 + A)[(x^2 + y^2) b_{e\parallel} / b_{i\perp} + z^2]^{3/2}}. \tag{11.153}$$

Until the density in the initial cloud does not change significantly (for $A<<1$ it holds at all stages), the asymptotic behavior of the perturbed density is determined by the equation

$$\frac{\partial n_i}{\partial t} = -\frac{\partial}{\partial z}\left(n_0 b_{e\parallel} \frac{\partial \Psi}{\partial z} \right). \tag{11.154}$$

Substituting Eq. (11.153) we find

$$n_i - n_0 = -\frac{3}{4\pi}\frac{Nb_{i\perp}E_{0}t(b_{e\parallel}/b_{i\perp})^4[3(x^2+y^2)(b_{e\parallel}/b_{i\perp})^2-2z^2]}{(1+A)[(x^2+y^2)(b_{e\parallel}/b_{i\perp})^2+z^2)]^{7/2}}. \tag{11.155}$$

This expression is more general than Eq. (11.92).

11.10 EXAMPLES OF 3D MODELLING OF PLASMA CLOUD EVOLUTION IN IONOSPHERE

The simulation of the reduced set of transport equations (11.144), examined in the preceding Section, has been performed in [33] and [34] for the case $ea_\perp E_0 >> T$, $\vec{E}_0 \perp \vec{B}$. The magnetic field is assumed to be vertical, and the typical parameters of the polar ionosphere in the day time were chosen. The following boundary conditions were imposed

$$n_i(\xi,\eta\to\infty)\to n_0(\zeta)\ ;\ n_I(\xi,\eta\to\infty)\to 0\ ;$$
$$\Phi(\xi,\eta\to\infty)\to 0\ ;\ \frac{\partial\Phi}{\partial\zeta}\Big|_{\zeta=\zeta_1,\zeta_2} = 0. \quad , \tag{11.156}$$

Here the coordinates ζ_1 and ζ_2 correspond to the low and high boundaries of the ionosphere, where the Pedersen conductivity is negligible. The last condition in Eq.(11.156) is thus equivalent to the absence of vertical current at the chosen boundaries of the ionosphere. The corresponding heights were taken: h_1=90km; h_2=250-300km, while the maximum of initial cloud (ζ=0) was located at 160km$\le h_0 \le$250km. Initial Gaussian profile of injected cloud, Eq. (11.119), has been used in the simulation with $a_\parallel = a_\perp = a$ and $1 \le A \le 100$ (for $a_\parallel \ne a_\perp$ the results are identical for an amplitude $\tilde{A} = Aa_\perp / a_\parallel$). The perpendicular scale of the injected cloud was chosen in the interval 100m$\le a_\perp \le$5km.

The typical results of the calculations for h_0=160km are displayed in Fig. 11.30 for the initial moment t=0. The perturbed perpendicular electric field is plotted here at $\xi=\eta=0$ as a function of the height for different sizes of initial clouds. It is clearly seen that for the inhomogeneities with large perpendicular dimension $a_\perp \ge$5km the field E_y' remains practically the same in the whole ionosphere.

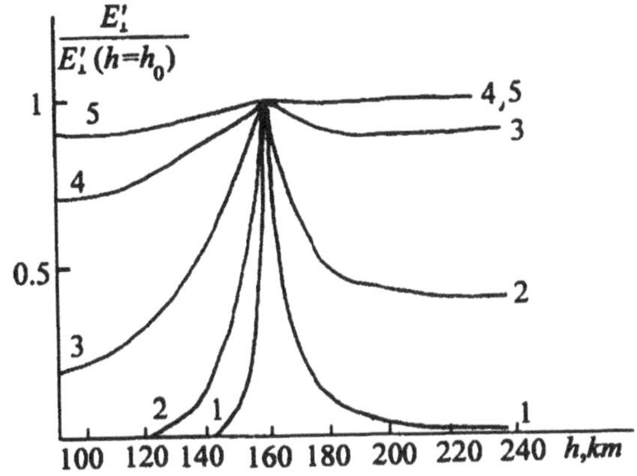

Figure 11.30. The height dependence of the perturbed electric field inside the cloud ($\xi=\eta=0$). The perturbed field is normalized to the value at the injection height h=160km; A=10. 1 -$a_\perp$=0.1km; 2 - $a_\perp$=0.3km; 3 - $a_\perp$=1km; 4 - $a_\perp$=2.5km; 5 - $a_\perp$=5km.

This result means that the approximation of the two-layer model of the ionosphere, examined in the Section 11.4, is rather satisfactory for the case of such large perpendicular scales. The absolute value of E_y' can be described by

$$\frac{E_y'}{E_0}=\frac{\Sigma_{1\perp}}{\Sigma_{1\perp}+2\Sigma_{2\perp}}\ ;$$
$$\Sigma_{1\perp}=e\int_{-\infty}^{\infty}b_{I\perp}(z)n_I(z)dz\ ;\ \Sigma_{2\perp}=e\int_{-\infty}^{\infty}b_{i\perp}(z)n_i(z)dz\ , \qquad (11.157)$$

which coincides with Eq.(11.72) (the Hall conductivity at h=160km is negligible). For $a_\perp$=5km the results of the calculations coincide with Eq.(11.157) with the accuracy 1%. For $a_\perp$=1km the discrepancy reaches 40% due to the decrease of the perturbed electric field in the low ionosphere.

As a result, the low ionosphere with the highest Pedersen conductivity is switched off from the zone where the short-circuiting current is generated. For $a_\perp$<100m the perturbed electric field becomes almost symmetrical with respect to up and down direction. Its value is consistent with Eq. (11.147) (or with Eq. (11.123)). For example, for $a_\perp$=100m, A=10 and h_0=160km the value of the electric field obtained by simulation is $E_y'/E_0=5.1\times10^{-2}$, while from Eq. (11.147) we obtain the value 6.8×10^{-2}.

Note that calculation in the framework of the two-layer model Eq. (11.157) in this case gives erroneous value 5.1×10^{-3}.

Further evolution of the plasma cloud also depends on its initial size and amplitude of the plasma density perturbation. The calculations demonstrated that for $a_\perp$<1km the evolution pattern is similar to the case of unbounded plasma. In particular, if

$$\tilde{N}_1 = \sqrt{\pi} a_{\|} A b_{i\perp}(0) / [a_\perp b_{e\|}(0) n_0(0)] \ll 1,$$

i.e. if the depletion regions in the ambient plasma are shallow, the motion of the injected cloud is almost unipolar. Then, in accordance with Eq. (11.133),

$$N_1(x,y,t) = \frac{N_1(0,0,0)}{1+8D_{i\perp}t/a_\perp^2} \times \exp\left[\frac{-(x-b_{I\wedge}E_0t)^2-(y-b_{I\perp}E_0t)^2}{a_\perp^2+8D_{I\perp}t}\right]. \tag{11.158}$$

At $\tilde{N}_1 = \sqrt{\pi} a_{\|} A b_{i\perp}(0) / [a_\perp b_{e\|}(0) n_0(0)] > 1$ the injected cloud motion is decelerated significantly and the ion branch is created, see the Sections 11.8, 11.9. Several examples of the evolution of such strong clouds are shown in Figs. 11.31, 11.32. for E_0=1mV/m.

In several papers [35] - [38] the direct simulation of the initial transport Eqs. (11.117) were performed for the ionospheric parameters. The results are consistent with the modelling of the reduced equations addressed above.

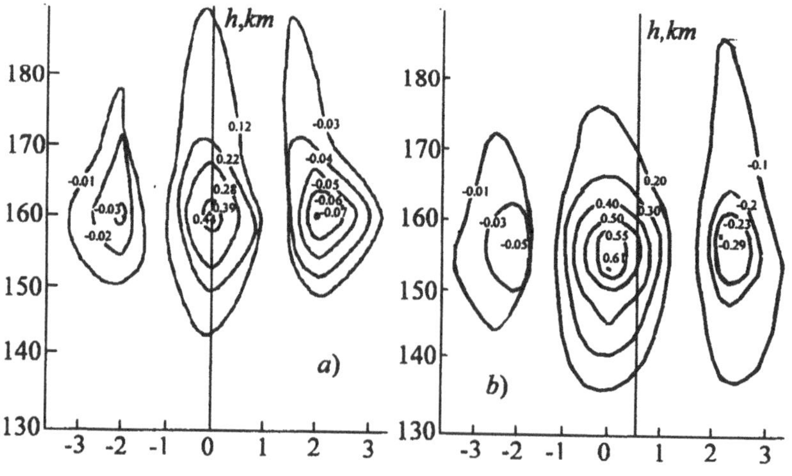

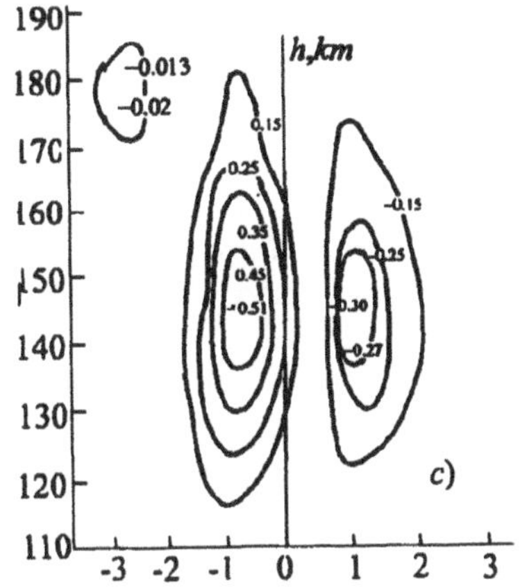

Figure 11.31. Ambient ions equidensities $[n_i - n_0(z)]/n_0(z)$ in the $\zeta\eta$ plane in which the maximum of $\widetilde{N}_1$ is situated: *a)* $a_\perp$=0.1km, A=100, τ=2.0, ξ=1.5; *b)* $a_\perp$=0.3km, A=500, τ=1.0, ξ=0.25; *c)* $a_\perp$=1km, A=100, τ=2.0, ξ=1.

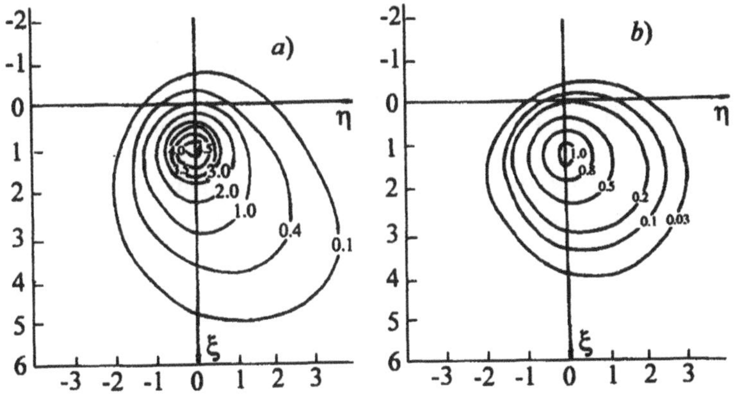

Figure 11.32. Equdensities of injected ions ($\widetilde{N}_1$) in the $\xi 0 \eta$ plane. *a)* $a_\perp$=0.3km, A=500, τ=3.0; *b)* $a_\perp$=1km, A=100, τ=2.0.

REFERENCES

1. A. V. Gurevich, E. E. Tsedilina, *Sov.Phys. Uspekhi* **10**: 214-236 (1967).

2. A. V. Gurevich, *JETP Lett.* **8**: 115-119 (1968).
3. V. A. Rozhansky, L. D. Tsendin, *Geomagnetism and Aeronomia* **24**: 345-349 (1984).
4. V. A. Rozhansky, L. D. Tsendin, *Geomagnetism and Aeronomia* **24**: 491-493 (1984).
5. M. I. Pudovkin, A. M. Lyatskaja, I. V. Golovchańskaja, *Proc. General Assembly IAGA, Prague,* 245-260 (1985).
6. H. Rishbath, O. K. Garriot, *Introduction to Ionospheric Physics* (Academic Press: New York, London, 1969) 235.
7. V. B. Lyatsky, Yu. P. Maltsev, *Results of Researches on the International Geoph. Projects, Magnetosphe-IonosphereInteraction* (Nauka: Moscow, 1983) 192.
8. V. A. Rozhansky *Geomagnetism I Aeronomia* **26**: 780-783(1986)(in Russian).
9. M. I. Pudovkin, A. M. Lyatskaja, I. V. Golovchanskaja, *Geomagnetism I Aeronomia* **26**: 767-773 (1986) (in Russian).
10. I. P. Zolotarev, *Geomagnetism I Aeronomia* **26**: 569-575 (1986) (in Russian).
11. I. P. Zolotarev, *Geomagnetism I Aeronomia* **28**: 228-233 (1988)(in Russian).
12. K. H. Lloyd, G. Haerendel, *J. Geoph. Res.* **78**: 7389-7416 (1978).
13. V. A. Rozhansky, *Geomagnetism and Aeronomia* **25**: 762-765 (1985).
14. G. Haerendel, R. Lust, R. Rieger, *Planet. Space. Sci.* **15**: 1-18 (1967).
15. L. D. Landau, E. M. Lifshits, Electrodynamics of Continuous Media (Pergamon Press: Oxford, 1963) 418.
16. N. J. Zabusky, J. H. Doles III, F. W. Perkins, *J. Geoph. Res.* **78**: 711-724 (1973).
17. T. N. Davis, *Rept. Progr. Phys.* **42**: 1565-1604 (1979).
18. A. J. Scannapieco, S. L. Ossakow, Book D. L. et al, *J. Geoph. Res.* **19**: 2913-2916 (1974).
19. N. I. Dzubenko, A. P. Zhilinsky, I. A. Zhulin et al, *Planet. Space Sci.* **31**: 849-858 (1983).
20. L. A. Andreeva, I. S. Ivchenko, G. P. Milinivsky et al, *Planet. Space Sci.* **32**: 1045-1052 (1984).
21. V. A. Anoshkin, A. P. Zhilinsky, G. G. Petrov et al, *Geomagnetism and Aeronomia* **19**: 712-715 (1979).
22. W. Stoffregen, *J. Atm. Terr. Phys.* **32**: 171-177 (1970).
23. W. Narcisi, E. P. Szuszczewicz, in *Active Experiments in Space, Proc. Intern. Symp. Alpbach*, 299-304 (1983).
24. G. W. Sutton, A. Sherman, *Engineering Magnetodynamics* (New York: McGraw-Hill, 1965): 548.
25. M. Mitchner, C. H. Kruger, *Partially Ionized Gases* (New York:Wiley-Interscience Publ., 1973): 518.

26. E. P. Velikhov, A. M. Dykhne, *Proc. 6th Conf. Phenomena in Ionized Gases*, Paris, 511 (1963).
27. J. L. Kerrebrok, *AIAA Journal*, **2**: 1072-1076 (1972).
28. P. Velikhov, V. S. Golubev, A. M. Dykhne, *Atomic Energy Rev. Vienna: IAEA*: **14** 325-335 (1976).
29. A. V. Gurevich, E. E. Tsedilina, *Geomagnetism I Aeronomia*, **6**: 255-265 (1966) (in Russian).
30. A. V. Gurevich, E. E. Tsedilina, *Geomagnetism I Aeronomia*, **7**: 648-654 (1967) (in Russian).
31. V. A. Rozhanskii, L. D. Tsendin, *Sov. Phys. Techn. Phys.*, **22**: 1173-1179 (1977).
32. S. P. Voskoboynikov, V. A. Rozhansky, L. D. Tsendin, *Planet.Space Sci.*, **35**: 835-844 (1987).
33. S. P. Voskoboynikov, V. A. Rozhansky *Geomagnetism I Aeronomia*, **27**: 359-362 (1987).
34. V. A. Rozhansky, I. Yu. Veselova, S. P. Voskoboynikov, *Planet. Space Sci.*, **38**: 1375-1386 (1990)
35. J. F. Drake, M. Mulbradon, J. D. Huba, *Phys. Fluids* **31** 3412-3424 (1988).
36. N. Sh. Blaunstein, E. E. Tsedilina, L. N. Mirsoeva, E. V. Mishin *Geomagnetism I Aeronomia*, **30**: 799-805 (1990) (in Russian).
37. T. -Z. Ma, R. W. Shunck, *J. Geoph.Res.*, **96**: 5793-5810 (1991).
38. N. A. Gastonis, D.E. Hastings, *J. Geoph.Res.*, **96**: 7623-7638 (1991).

Chapter 12

Electron Energy Transport

Since Eqs. (2.12) and (2.16) which describe the energy transport are rather cumbersome, these processes in plasmas are, generally speaking, far more diverse and complicated than the transport of charged particles. From the preceding Chapters it follows that a great number of problems exists, in which the particle transport can be analyzed at the uniform or prescribed partial temperatures. In contrast, the particle fluxes which accompany the energy transport, are, as a rule, considerable. Also significant is the redistribution of the partial densities in the course of relaxation of the temperature inhomogeneity. Accordingly, the problems of the influence of the self-consistent electromagnetic fields on the energy transport are intricate, and even rough attempts to classify the mechanisms of this influence, especially in magnetic field, are practically absent. The situation is simplified by the fact that in the weakly ionized plasmas the main part of the internal energy is connected with the neutral component, and the balance of the neutral temperature T_N can be analyzed in the traditional manner. The energy exchange between the heavy particles (neutrals and ions) occurs practically during the collision time ν_{iN}^{-1}. It makes it senseless to distinguish between T_N and the partial ion temperatures. On the other hand, since in most of the applications we meet non-equilibrium plasmas, the situations when the electron temperature (mean energy) considerably deviates from T_N, are rather typical. Consequently, the most interesting problem concerns the partial electron energy transport. These processes play an important role if the energy sources and sinks are spatially non-uniform, with the characteristic scale L which is small with respect to the electron energy relaxation length $\lambda_e^{(\varepsilon)}$, Eq. (2.20). In this case the electron thermal conductivity dominates. During the energy relaxation time $(\delta_{eN}\nu_{eN}+\delta_{ei}\nu_{ei})^{-1}$

the electron thermal conductivity levels off the electron temperature T_e over the scale $\lambda_e^{(\varepsilon)}$. In the opposite case $L >> \lambda_e^{(\varepsilon)}$ the local electron energy balance prevails, and the energy transport processes are unimportant.

In the weakly ionized plasma with $\nu_{ee} << \nu_{eN}$ (see Section 2.1) the situation depends whether the collisions between electrons are frequent enough to Maxwellize the distribution function of electrons, $\nu_{eN}^{(\varepsilon)} << \nu_{ee} << \nu_{eN}$. If not, the relaxation length of the electron distribution function Eq. (2.59) coincides with the electron energy relaxation length $\lambda_e^{(\varepsilon)}$. This means that at distances at which the electron energy transport processes play a really important role, the energy has no time to redistribute itself between all the electrons in a given place, and energy transport by every part of the electron distribution is practically independent. This means that a consistent kinetic analysis is necessary. The fluid approach which we use in this book is misleading in this case and can result not only in numerical inaccuracy, but even in qualitative errors. Especially important are the kinetic effects in the case when the electron inelastic collisions with large energy losses are important. As it was shown in Section 2.9, in this case the relaxation length of the electron distribution (starting from which it is only possible to use the fluid description) can greatly exceed the energy relaxation length $\lambda_e^{(\varepsilon)}$.

If the electron-electron collisions are frequent enough, $\nu_{eN}^{(\varepsilon)} << \nu_{ee} << \nu_{eN}$, the situation is less dramatic. If the characteristic scale exceeds the relaxation length of the electron distribution Eq. (2.59), which in this case is small compared to $\lambda_e^{(\varepsilon)}$, the scale interval remains, in which the electron energy transport is important, and, nevertheless, the fluid description is valid.

In this Chapter we restrict ourselves to several simple examples of energy transport processes. In the first Section the phenomena of the electron thermal conductivity and the electron cooling in the course of ambipolar plasma diffusion are chosen to illustrate specifics of the energy transport in currentless plasmas. In the second Section the propagation and amplification of ionization waves (striations) in a *dc* positive column at large currents, when the fluid approximation is valid, is presented as an example of the effects in which the influence of the self-consistent electric field on the electron energy transport in current-carrying plasmas results in non-trivial physical phenomena.

12.1 ELECTRON ENERGY TRANSPORT IN THE CURRENTLESS PLASMA

Electron thermal conductivity. If a small disturbance of the electron temperature is created in uniform isothermal plasma, the redistribution of the partial densities is also small, and in the energy balance equation the density can be treated as unperturbed. In spite of the fact that the particle fluxes are also small, their contribution to the energy balance equation can be, in principle, appreciable. However, their inverse influence on the energy transport can often be neglected. Thus, the problems of energy and particle transport in this case can be treated separately. This fact was used in [1] and [2] to measure the electron thermal conductivity.

From Eqs.(2.12), (2.16) it follows that in uniform plasma the electron temperature profile in the case of small disturbance of the electron temperature satisfies

$$\frac{\partial T_e}{\partial t} = \vec{\nabla}\hat{\chi}_e\vec{\nabla}T_e - (\delta_{eN}\nu_{eN} + \delta_{ei}\nu_{ei})(T_e - T_N), \tag{12.1}$$

where the tensor

$$\hat{\chi}_e = \hat{\kappa}_e / C_{Ve}$$

is determined according to Eqs.(2.30) and (2.70). The temperature profile in the case of a 1D slab stationary source localized at x=0 in the absence of a magnetic field is

$$\begin{aligned} &T_e - T_N = A\exp(-x/\lambda_{e\parallel}^{(\varepsilon)}); \\ &\lambda_{e\parallel}^{(\varepsilon)} = [\chi_{e\parallel}/(\delta_{ei}\nu_{ei} + \delta_{eN}\nu_{eN})]^{1/2}. \end{aligned} \tag{12.2a}$$

Analogously, across a strong magnetic field

$$\begin{aligned} &T_e - T_N = A\exp(-x/\lambda_{e\perp}^{(\varepsilon)}); \\ &\lambda_{e\perp}^{(\varepsilon)} = [\chi_{e\perp}/(\delta_{ei}\nu_{ei} + \delta_{eN}\nu_{eN})]^{1/2}. \end{aligned} \tag{12.2b}$$

The average collision frequencies are determined according to Eqs.(2.23), (2.24).

On the other hand, the solution of a non-stationary 1D problem for heat propagation from a pulsed localized source is given by a well-known expression:

$$T_e - T_N = \frac{B}{(4\pi\chi_e t)^{1/2}} \exp[-(\delta_{ei}\nu_{ei} + \delta_{eN}\nu_{eN})t - x^2/(4\chi_e t)]. \quad (12.3)$$

In the cylindrical and spherical cases the power in the denominator instead of 1/2 is to be taken equal to 1 and to 3/2.

It follows from Eqs. (12.2) and (12.3) that the electron thermal conductivity can be found independently both from stationary and pulsed experiments. These measurements were reported in [1] and [2]. Experiments were performed in decaying isothermal plasma. Since the plasma scale considerably exceeded energy relaxation length $\lambda_e^{(\varepsilon)}$, and the energy input was small, it was possible to neglect the density disturbance and to consider the plasma density as uniform. In Figs. 12.1 and 12.2 the experimental results of measurements [2] of the electron heat conductivity in *Ar* both along and across a strong magnetic field are presented.

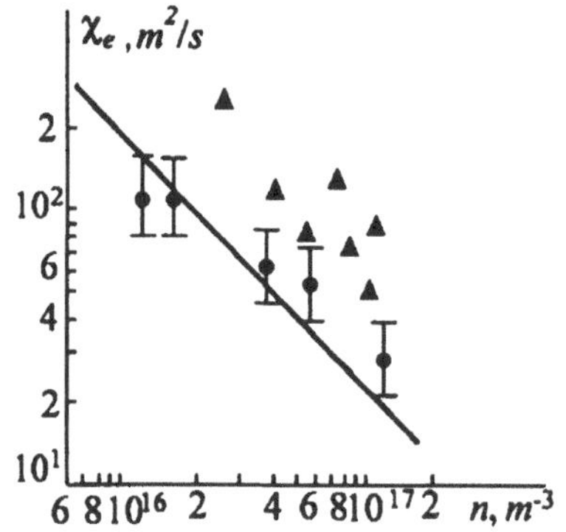

Figure 12.1. Dependence of the electron temperature conductivity χ_e along the magnetic field on the plasma density n in *Ar* at p=27 Pa, T_N=300 K [1], [2]. Circles corresponds to pulsed regime, Eq. (12.3); $\vec{B} = 0$; triangles to stationary regime, Eq. (12.2) for $\vec{B} \neq 0$; the full line corresponds to calculations according to Eq. (2.86).

Since T_e was rather low - of the order of the room temperature, and at such energies the Coulomb cross-sections by orders of magnitude exceed the gas kinetic cross-sections - the collisions between the charged particles dominated in the experimental conditions even at a rather low degree of ionization. Accordingly, the calculated values of heat conductivity obey the dependence Eq. (2.86) for fully ionized plasma. The agreement between the measured and calculated values seems to be satisfactory.

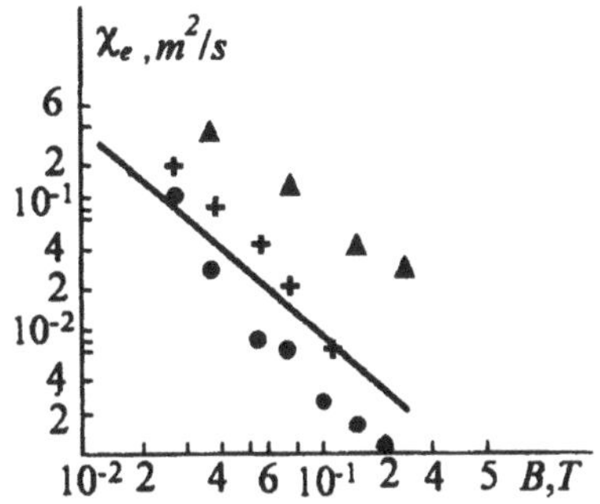

Figure 12.2. Dependence of the transversal electron temperature conductivity in *Ar* on magnetic field [2]. Pressures are: circles - 4 Pa; triangles - 6.6 Pa; crosses - 13.3 Pa. The full line is calculated according to Eq. (2.86).

Diffusive cooling. In the course of the diffusion process the particle flux is always accompanied by energy flux. In mixtures of neutral gases this energy flux is of the order of $\vec{\Gamma}T$. This diffusive thermal effect is discussed in numerous textbooks (see, e.g., [3], [4] and [5]). If the potential difference in plasma exceeds T_e/e, the self-consistent electric field results in far more intense thermal effects. Such a situation arises, for example, in course of the ambipolar diffusion in the absence of magnetic field. According to Eqs.(3.25), (4.21), the potential drop in plasma and sheath considerably exceeds T_e/e. This ambipolar potential hinders the electron motion, and during the diffusive expansion towards the vessel walls the electron gas overcomes this field, works against it, and loses its thermal energy. The energy which is lost by electrons is transferred by the electric field to ions. They dissipate this energy in the plasma by means of friction against neutral gas. The energy which is lost by electrons when they overcome the potential drop in the sheath, is transferred to ions and is dissipated in the ion bombardment of the vessel walls. The energy of the electrons which reach the wall is determined by the electron distribution tail and is of the order of (or less than) T_e.

This effect of the diffusive cooling of electrons is most pronounced in decaying plasma at the last decay stage, when all additional energy sources are practically absent. As a result, the value of T_e can be considerably lower than T_N [6], [7], [8].

Substituting into the equation of the electron energy balance (2.16) expressions for the friction forces, Eqs. (2.22), (2.26), for the heat flux, Eqs. (2.29), (2.30), (2.34) and for the collisional heat production, Eqs. (2.38), (2.39), we obtain, using the momentum balance equation (2.10):

$$
\begin{aligned}
&3/2n[\frac{\partial T_e}{\partial t}+\vec{u}\vec{\nabla}T_e]-\vec{\nabla}\cdot\kappa_e\vec{\nabla}T_e+\vec{\nabla}\cdot[(1+C_{eN}^{(T)})nT_e\vec{u}_e]= \\
&=3/2(\delta_{ei}\nu_{ei}+\delta_{eN}\nu_{eN})(T_N-T_e)-en\vec{u}_e\vec{E}-C_{ei}^{(T)}n\vec{u}_i\vec{\nabla}T_e \\
&-\vec{\nabla}\cdot[C_{ei}^{(T)}(\vec{u}_e-\vec{u}_i)nT_e]-c_{ei}^{(\vec{u})}m_e\nu_{ei}n\vec{u}_i\cdot(\vec{u}_e-\vec{u}_i).
\end{aligned}
\quad (12.4)
$$

The ambipolar diffusion in a dielectric vessel implies

$$\vec{u}_e=\vec{u}_i=-D_a\vec{\nabla}\ln n;\quad \vec{E}=-(T_e/e)\vec{\nabla}\ln n. \quad (12.5)$$

Since the diffusion process is controlled by ions, it occurs usually considerably slower than the electron energy relaxation. In such a situation the electron energy balance Eq. (12.4) can be considered as quasistationary. If the electron energy relaxation length $\lambda_e^{(\varepsilon)}$, Eq. (2.20), is small with respect to the characteristic scale of the density profile a/ξ_1 (a is tube radius, ξ_1=2.405), the electron cooling is significant only in a small region in the vicinity of the wall of the order of $\lambda_e^{(\varepsilon)}$ In the opposite case $\lambda_e^{(\varepsilon)}>>a/\xi_1$ the electron thermal conductivity levels up the T_e profile over the tube cross-section. In such a situation it is possible to average Eq. (12.4) over the tube cross-section, assuming uniform T_e profile. This case closely corresponds to the experimental conditions in [7]. In these experiments the recombination in the plasma volume was negligible, and the sheath was collisionless. The terms which are responsible for the traditional diffusive thermal effect which exists in the absence of the electric field are of the order of $\vec{\Gamma}_e T_e$. Neglecting them with respect to the Joule heating in Eq. (12.4), we obtain

$$T_N-T_e=e\left\langle n\vec{u}_e\vec{E}\right\rangle/\left\langle\frac{3}{2}n(\delta_{ei}\nu_{ei}+\delta_{eN}\nu_{eN})\right\rangle. \quad (12.6)$$

The integration in the nominator of Eq. (12.6) is performed over the whole tube cross-section, including the potential drop in the sheath. The energy lost by electrons consists of two parts. The first of them corresponds to the work of an electron against the electric field in the plasma volume, and the second, in the space charge sheath. Since the expression for the ambipolar field Eq. (12.5) is valid up to distances of the order of the ion mean free path λ_{iN} from the wall, every electron which escapes to the wall loses energy $e\Delta\varphi_{fl}\sim T_e\ln(a/\lambda_{iN})$. According to Eq. (3.25), the energy loss in the sheath adjacent to the insulating surface (which is automatically at the floating potential) is equal to $\sim(T_e/2)\ln(m_i/m_e)$. This last expression is valid, as emphasized in Section 3.6, only when the collisions between electrons are frequent enough, $\nu_{ee} \geq \nu_{eN}$, to restore the Maxwellian distribution at the distribution tail, which is responsible for the escape of the fast electrons to the vessel wall. In weakly ionized plasma, especially at the wall, the electron density is low, and the depletion of the electron distribution at the escape energies is, as a rule, considerable. If $\nu_{ee} \leq \delta_{eN}\nu_{eN}$, these kinetic effects strongly suppress the sheath potential with respect to Eq. (3.25) up to the value $e\Delta\varphi_{fl} \sim T_e$, i.e. of the order of the terms which were omitted in Eq. (12.6). Since both logarithms in the expressions for the electron energy losses are more or less comparable, the kinetic effect of the electron distribution depletion considerably affects the diffusive electron cooling. In Fig. 12.3 the experimental results of [7] are shown.

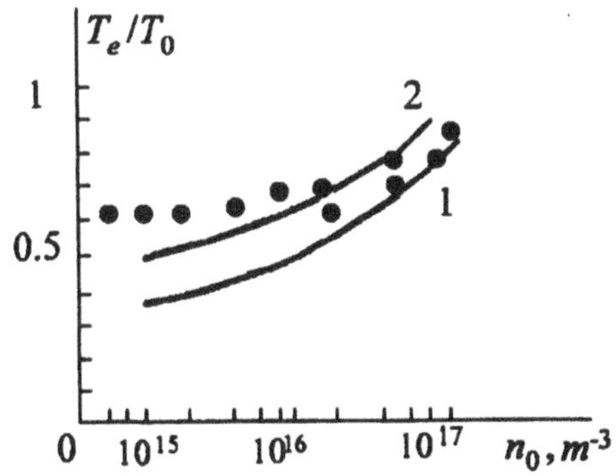

Figure 12.3. The diffusive electron cooling in the decaying Ar plasma [7] versus the plasma density at the tube axis. The conditions are: p=26.6 Pa; a=1.5 cm. The full line 1 is calculated according to Eqs.(12.4), (12.5) with the Langmuire value of $e(\Delta\varphi)_{sh}$, Eq. (3.25); the line 2 corresponds to $e(\Delta\varphi)_{sh}$=0. Different points correspond to different experimental methodics.

It is easy to see that the calculation with $\Delta\varphi_{fl}$=0 more satisfactorily fits the experimental data, especially at low plasma density.

The diffusive cooling becomes even more significant, if the electron current to the conducting surface exceeds the ion saturation current $j_i^{(sat)} = eD_a|\nabla n|$. Up to the wall potential when the electron current is small with respect to the electron saturation current $j_e^{(sat)} = eD_e|\nabla n|$, the potential profile is close to Boltzmann's for electrons. On the other hand, the electron current and the diffusive cooling, are $j_e/j_i^{(sat)}$ times larger than in the case of ambipolar diffusion [9], [10].

12.2 ELECTRON ENERGY TRANSPORT IN CURRENT CARRYING PLASMA. PROPAGATION OF THE IONIZATION WAVES.

The processes of electron energy transport are important only if a problem is characterized by a small spatial scale L with respect to the energy relaxation length $\lambda_e^{(\varepsilon)}$, Eq. (2.20). From Eqs. (2.20), (9.15) for the case when $T_e >> T_N$ due to the Joule heating of electrons in dc electric field E, we have $\lambda_e^{(\varepsilon)} \sim T_e/(eE) = l_T$. In the opposite case $L >> \lambda_e^{(\varepsilon)}$ the energy balance is local. It means that the energy is deposited in the electron subsystem by an external agent (say, by the Joule heating) and is redistributed between other species (predominantly transferred to the neutrals) in the same place; the energy transport processes are of little importance in this situation.

The propagation of self- or externally-excited ionization waves (striations) presents one of the most striking manifestations of the phenomena in which the energy transport processes dominate, and the rather subtle properties and complex interaction of these processes with the processes of particle transport result in macroscopically observable consequences.

It became clear relatively long ago [11], [12], [13] that the mechanism of striations is intimately connected with the transport and ionization-recombination processes. However, first of all it remained unclear, how the combination of these dissipative processes can result in the propagation of the wave-like signals. The second difficulty was connected with the fact that the striations represent spatially non-uniform and/or non-stationary combination of the plasma density, electron temperature, and electric field disturbances. In the local description it is practically impossible to explain the existence of this phenomenon. The gas discharge plasma is maintained due to the ionization by electron impact. Since the ionization rate exponentially depends on the electron distribution function form (on the electron temperature in case of the Maxwellian distribution), all such disturbances are damped extremely fast. Any local increase of the plasma density results, due to the dc current conservation, in a decrease of the field strength and thus of T_e, and in an exponential drop in the ionization rate. This means that such a widely met phenomenon is connected only with short-wave disturbances of the plasma parameters. As it was formulated in Sections 2.4 and 2.9, this problem can be described in the fluid approximation which we use throughout this book only at high plasma density, when the electron distribution function relaxation length λ_{ef} Eq. (2.59) is small with respect to the energy relaxation length $\lambda_e^{(\varepsilon)}$. In such a situation the collisions between electrons is the main factor which controls the electron distribution function. On the one hand, disturbances with a characteristic spatial scale L which satisfies $\lambda_{ef} << L < \lambda_e^{(\varepsilon)}$, can be described in the framework of the fluid approximation. On the other hand, in such small-scale inhomogeneities the electron energy balance is controlled by the transport processes (by the diffusion and electron thermal conductivity). These mechanisms suppress the damping of plasma density perturbations caused by the above-mentioned reduction of the ionization rate in places with high plasma density, and make possible the propagation of weakly damping signals, or even their generation.

In the opposite case, when the Coulomb collisions between electrons are infrequent, $\lambda_{ef} \sim \lambda_e^{(\varepsilon)}$, the electron spatial transport occurs practically independently over different parts of the electron distribution function. Hence, a consistent kinetic description of striations is necessary at low plasma density (i.e., at low discharge current), and the fluid approach is in principle inadequate to analyze such a situation. Examples of the strikingly nonhydrodynamic transport behavior of the electron gas in the simplest systems of such a kind are presented in [14] and [15].

1D Approximation. The characteristic frequency of all the ionization processes (and also of the striations) equals $\nu^{(ion)}$, Eq. (4.18). This value corresponds to the ionization frequency by the electron impact averaged over the electron distribution function. For the Maxwellian distribution function it depends only on T_e. Since in a stationary state the ionization rate is balanced by the relatively slow processes of volume recombination and/or of the ambipolar diffusion, the ionizing electron collisions, which are vital for the maintenance of discharge itself, also represent relatively infrequent events. This means that the electrons which are capable of ionizing collisions constitute only a small fraction of the total electron population; the electron temperature is considerably lower than the threshold energy of the ionization eU_i (U_i being the first ionization potential). In other words, only the high-energy tail of the electron distribution determines the ionization rate; the typical value of the ratio eU_i/T_e in standard gas discharge conditions is of the order of ten.

The energy relaxation time Eqs. (2.19), (2.110) considerably exceeds the equilibrium value of $\nu^{(ion)}(T_e)$, which is equal to the mean recombination lifetime (Eq. (4.19), if recombination at the tube wall dominates). It follows from the fact that the electron energy loss for excitation of electronic states of neutral atoms and molecules is of the order of the ionization energy. On the other hand, at energies close to the thresholds of the corresponding processes (which are only of importance in the standard conditions) the excitation cross-sections far exceed the ionization ones, and the energy loss in one ionization event considerably exceeds the mean electron energy. It means that the electron energy relaxation occurs quickly with respect to the variation of plasma density, and in the analysis of the latter process the electron temperature can be considered as quasistationary. Hence, in the standard situation $T_e >> T_N$ for slow plasma motions (with respect to the electron energy relaxation time $\tau_e^{(\varepsilon)} \sim Ej/(nT_e)$) in the electron energy balance Eqs. (2.12), (2.16) we can neglect the derivatives in time, retaining the Joule energy input, energy loss in quasielastic and inelastic collisions, and the energy fluxes.

We shall start with the simple 1D approach for the case of a cylindrical, longitudinally uniform and diffusion dominated positive column (Section 4.1), with the z axis along the cylinder. We restrict ourselves to pure, weakly ionized, unmagnetized plasma. The momentum balance Eq. (2.40) can be transformed to

$$eE_z = \frac{j}{b_e n} - \frac{T_e}{n}\frac{\partial n}{\partial z} - (1 + c_{eN}^{(T)})\frac{\partial T_e}{\partial z}. \tag{12.7}$$

This relation can be interpreted as follows. The electric field in the plasma is established to maintain the quasineutrality, or to compensate the action of mechanisms which tend to charge separation in an inhomogeneous plasma. According to Eq. (12.7), there are three such mechanisms. The first of them (the first term in the r.h.s of Eq. (12.7)) is connected with the inhomogeneity of the plasma conductivity (enb_e); the corresponding field arises in order to maintain uniform Ohmic current and to avoid space charge accumulation. The second mechanism arises due to the diffusive electron flux, and the third mechanism of the charge separation is due to thermodiffusion. The first (Ohmic) field dominates in large-scale disturbances $L > \lambda_e^{(\varepsilon)}$, and the second (diffusive) term in the Eq. (12.7) gives the main contribution to the field in the short-wave case $L << \lambda_e^{(\varepsilon)}$.

From the energy balance Eq. (12.4) using Eq. (12.7) for $T_e >> T_N$ we have

$$\frac{\partial}{\partial z}\left(\frac{e\kappa_e}{j}\frac{\partial T_e}{\partial z}\right) + \frac{3}{2}\frac{\partial T_e}{\partial z} = -\frac{T_e}{n}\frac{\partial n}{\partial z} - \frac{j}{nb_e} + 3enT_e(\delta_{eN}\nu_{eN} + \delta_{ei}\nu_{ei})/(2j). \tag{12.8}$$

The terms in the left-hand side of this equation correspond to the electron thermal conductivity and to the convective energy transport with the electron flux; the terms in the r.h.s. describe the Joule heating of electrons in the diffusive and Ohmic electric fields, Eq. (12.7), and the collisional energy losses. In long-scale inhomogeneities the electron temperature is determined by balancing the last two terms in the r.h.s. of Eq. (12.8). It is maximum in the minima of the plasma density, and vice versa. As a result, the recombination dominates in the regions where the plasma density exceeds the equilibrium value, the ionization dominates in the rarefactions, and the density profile is levelled fast. In small scale inhomogeneities the temperature is determined by the thermal conductivity and by the heating in the diffusive field, which equals zero in the plasma density extrema.

The approximate 1D equation of plasma balance, according to Section 4.1, is:

$$\frac{\partial n}{\partial t}+b_i\frac{\partial (nE_z)}{\partial z}=n\nu^{(ion)}(T_e,n)-n/\tau, \tag{12.9}$$

where $1/\tau=(\xi_1/a)^2+K^{(rec)}n$ - is the inverse mean lifetime of the plasma particles.

The diverse mechanisms can lead to the dependence of $\nu^{(ion)}$ on the plasma density n. First of all, such an effective dependence can result from the more complicated character of the plasma generation processes. For example, very often the charged particles in extremely non-equilibrium gas discharge plasmas are mainly generated by means of the step-wise ionization of excited neutral molecules [16]. Both the excitation and the step-wise ionization rates are, as a rule, controlled by the collisions of the neutral particles with electrons.

If, from the other side, the electrons do not play a significant role in the removal processes of the excited neutrals, the effective ionization rate ($n\nu^{(ion)}$) is proportional to n^2. This fact makes the ionization instability possible [17], [18].

Nevertheless, taking into account the additional degrees of freedom which are connected with the excited species of the neutral particles results in new modes of plasma density evolution, and makes analysis considerably more complicated. Such an analysis of striations can yield, in principle, abundant information about the ionization kinetics in nonequilibrium gas discharge plasmas. At high discharge currents, when the fluid approach is only valid, the excited particles are fast destroyed, as a rule, in the collisions with electrons [16], [19], [20]. This mechanism restores the linear dependence of the ionization rate on the plasma density, and suppresses the ionization instability.

The second simple kinetic mechanism which results in the superlinear dependence of ($n\nu^{(ion)}$) on the plasma density is connected with the influence of electron-electron collisions on the tail of the electron distribution function [21], [22]. At a very high ionization degree (i.e., at very high discharge current) the whole electron distribution is Maxwellian, and such a dependence is absent. Nevertheless, the formation of different parts of the electron distribution function, especially in atomic gases, is controlled by quite different mechanisms (see Section 2.9). At low electron energies $\varepsilon<\varepsilon_1$ (ε_1 corresponds to the excitation energy of the first excited atomic level which is of the order of the ionization energy eU_i) the energy losses are strongly suppressed. They are characterized by a very large energy relaxation time $\tau_e^{(el)}(\varepsilon)$. In contrast, the energy relaxation at the distribution tail $\varepsilon > \varepsilon_1$ occurs considerably faster: the corresponding times are

$$\tau_e^{(inel)}(\varepsilon)=[\sum_k \nu_k(\varepsilon)]^{-1};\ \tau_e^{(el)}(\varepsilon)=\min\{(\delta_{eN}\nu_{eN})^{-1};\ \varepsilon_1^2/D_\varepsilon\}.$$

If the time of the Maxwellizing collisions between electrons in the vicinity of the ionization threshold $\tau_{ee}(\varepsilon=eU_i)$ satisfies

$$\tau_{ee}(eU_i)<<\tau_e^{(el)}(eU_i),$$

the distribution tail is density-independent, and $(n\nu^{(ion)})$ is proportional to n. The same dependence holds, if

$$\tau_{ee}(eU_i)>>\tau_e^{(el)}(eU_i),$$

and the distribution is fully Maxwellian. The situation at the intermediate currents which satisfies

$$\tau_e^{(el)}(eU_i)<<\tau_{ee}(eU_i)<<\tau_e^{(inel)}(eU_i) \tag{12.10}$$

can be understood as follows: the Coulomb Maxwellizing collisions between the electrons are accompanied mainly by extremely small energy exchange. In other words, their impact on the distribution function corresponds to the diffusion both in momentum and energy. It means, that the coefficients D_ε, W_ε in the kinetic equation (2.104) for the isotropic part of the distribution function are complemented by the items which are proportional to the electron density n, and account for the Coulomb collisions between electrons (see, for example, [21], [22]). This condition (12.10) implies that at the distribution tail the energy diffusion coefficient is determined mainly by the electron-electron collisions, and the distribution tail is strongly depleted with respect to the Maxwellian form at energies above ε_1.

In other words, the problem in the energy space is identical to the problem of diffusion in the vicinity of a strongly absorbing media which is located at $\varepsilon > \varepsilon_1$. The diffusive flux to such an absorbing region is equal to the total absorption rate (i.e., to the total rate of the inelastic collisions in the our case), and is proportional to the diffusion coefficient D_ε, which is determined by the Coulomb collisions. It follows that the total excitation rate per electron is proportional to the electron density. In the case of high current, which is only of interest for us, the excited atoms are usually created and destroyed in collisions with electrons, and step-wise ionization dominates. This being the case, the ionization rate ($n\nu^{(ion)}$) is proportional to n^2. The rate of direct ionization is controlled by the decay of the distribution tail in the energy interval

$$\varepsilon_1 < \varepsilon < eU_i,$$

which is proportional to

$$\exp\{\int_{\varepsilon_1}^{eU_i} D_\varepsilon d\varepsilon\}.$$

The ionization rate in this case depends exponentially on the plasma density.

The dependence of the ionization frequency on n, T_e is often described by the logarithmic derivatives

$$Z_T = \frac{\partial \ln \nu^{(ion)}}{\partial \ln T_e} \approx \frac{eU_i}{T_e} >> 1;$$
$$Z_n = \frac{\partial \ln \nu^{(ion)}}{\partial \ln n}. \tag{12.11}$$

Sketch of the dependence $Z_n(n)$ is presented in Fig. 12.4.

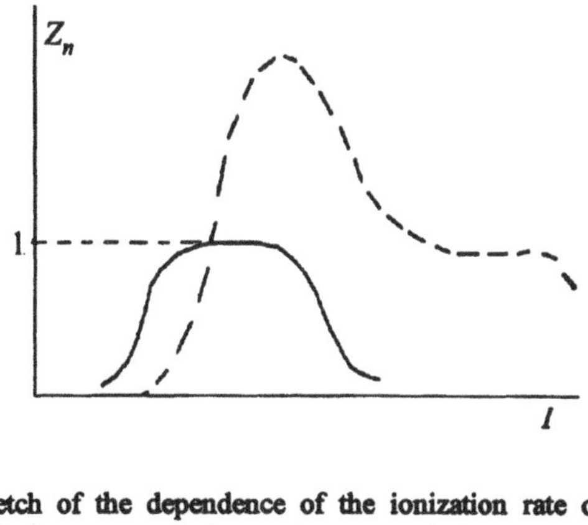

Figure 12.4. Sketch of the dependence of the ionization rate on the discharge current. The full line corresponds to the case of the step-wise instantaneous ionization; the broken one to the direct ionization.

For the small sinusoidal disturbances proportional to $\exp(ikz-i\omega t)$, linearizing Eq. (12.7) for the electric field for short-wave perturbations we obtain [23]:

$$\widetilde{E} = -iK\widetilde{n} - \widetilde{n} - iK(1 + c_{eN}^{(T)})\widetilde{T}_e >> \widetilde{n}. \tag{12.12}$$

The ratios of the small perturbations to the equilibrium values of the corresponding variables are denoted with a tilde; the dimensionless wave number $K=kT_e/(eE_0)$ is of the order of the ratio of the energy relaxation length $\lambda_e^{(\varepsilon)}$ to the wave length.

Linearizing the energy balance equation (12.8), one can obtain analogously,

$$\widetilde{E} = -iK\widetilde{n} - \widetilde{n} - iK(1 + c_{eN}^{(T)})\widetilde{T}_e >> \widetilde{n}. \tag{12.13}$$

The dimensionless electron thermal conductivity is of the order of unity: $\alpha=e\kappa_e/(nT_eb_e) \sim 1$. From Eqs. (12.12) and (12.13) it follows that in short-wave oscillations the diffusive electric field dominates, and the temperature perturbations are strongly suppressed by the electron thermal conductivity. The relative magnitude of oscillations and the phase shifts between them in such sinusoidal oscillations are shown in Fig. 12.5.

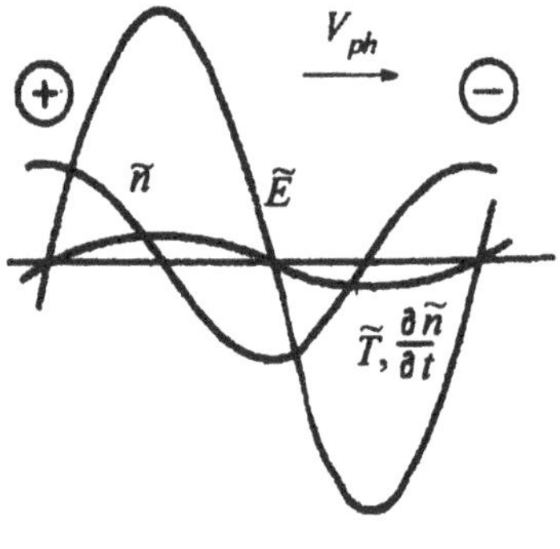

Figure. 12.5. Relative size of the plasma perturbations and the phase shifts between them in the ionization wave at $K>>1$; $Z_n=0$.

Substituting Eqs.(12.12), (12.13) into Eq. (12.9), we find the expression for the dispersion law

$$\mathrm{Re}\omega = eE_0(Z_T - 1)/(\alpha\tau kT_e). \tag{12.14}$$

This dependence is typical for the running striations which are observed at large currents. The phase velocity ω/k is cathode directed, and the group velocity $d\omega/dk$ is directed to the anode and is close to the phase velocity by an absolute value. The propagation mechanism of the running striations can be seen in Fig. 12.5. In order to correspond to the propagating plasma density wave, the sinusoid of $(\partial\tilde{n}/\partial t)$, which is generated by plasma in response to the density perturbation $\tilde{n}$, is $\pi/2$ shifted with respect to $\tilde{n}$. It follows from the fact that the field oscillations, according to Eq. (12.12), are shifted by $\pi/2$ with respect to $\tilde{n}$ towards the cathode. The oscillations of $(\partial\tilde{n}/\partial t)$ are connected mainly with the variation in the electron temperature $\tilde{T}_e$. The latter is determined by the balance between the Joule heating in the diffusive field, and the electron thermal conductivity. It is also shifted by $\pi/2$ towards the cathode with respect to $\tilde{n}$, and decreases with K, as K^{-1}, Eq. (12.14).

The damping of the running striations, and the possibility of their amplification and generation is controlled by imaginary part of the dispersion law

$$\mathrm{Im}\,\omega = -D_a k^2 - [(Z_T - 1)(2 - 3/(2\alpha))/(\alpha K^2) - Z_n]/\tau. \quad (12.15)$$

This dependence of decrement on the wave number is characterized by sharp maximum at

$$\begin{aligned} K = K_0 &= [(\xi_1 T_e/(\alpha a e E_0)]^{1/2}[(2\alpha - 3/2)(Z_T - 1)]^{1/4}; \\ \mathrm{Im}\,\omega(K_0) &= Z_n/\tau - 2D_a K_0^2 (eE_0/T_e)^2. \end{aligned} \quad (12.16)$$

The short waves $K>K_0$ are strongly damped due to longitudinal ambipolar diffusion. The damping of the longer waves occurs due to the fact that if K decreases, the phase shift between the oscillations $\tilde{T}_e$ and $\tilde{n}$ exceeds $\pi/2$ and tends to π at $K\to 0$. It results from the Joule heating by the Ohmic field, Eq. (12.7), and from the convective heat transport with the electron flux. The wave number of the striations which spontaneously exist in a positive column in the case of unstable plasma, when $\mathrm{Im}\,\omega > 0$, are close to K_0. In a stable column the striations with $K \sim K_0$ are also generated in response to an external disturbance.

From Eq. (12.16) it follows that, if the wall recombination dominates, such ionization waves are possible only in rather thin discharge tubes, when the 'resonant' value K_0, Eq. (12.16), satisfies $K_0 \gg 1$. The condition for the tube radius is $\Delta = [eE_0/(\xi_1 T_e)]^2 \ll 1$ - the tube radius is small with respect to the energy relaxation length. The 'resonant' wavelength satisfies $k_0 a \sim \Delta^{1/2} < 1$, i.e. it is small with respect to the tube radius. These waves can be unstable only due to the dependence of the ionization frequency $\nu^{(ion)}$ on electron density n. In noble gases at high currents, when such ionization waves are widely observed, the step-wise ionization of metastable and resonant atomic levels dominates. The excited atoms are both generated and removed in the electron collisions; the ionization process of an excited atom demands relatively small energy with respect to eU_i, and occurs quickly. These factors justify applicability of the developed model.

According to Eq. (12.16) (see also Fig. 12.4), the instability increment Z_n/τ vanishes at large currents when the Maxwellian form of the distribution tail is restored by the collisions between electrons. This means that at currents above a critical value the spontaneously existing striations vanish. In the figures 12.6; 12.7 and 12.8 the results of the calculations [23] and [24] for the dispersion law Eq. 12.16; for the spatial amplification of waves $d = \mathrm{Im}\,\omega(d\,\mathrm{Re}\,\omega\,/\,dk)^{-1}$, and for the upper critical current of existence of the running striations are compared to the experiment [24]. The agreement seems quite satisfactory. The nonlinear theory of ionization waves in the 1D fluid approximation was developed in [25].

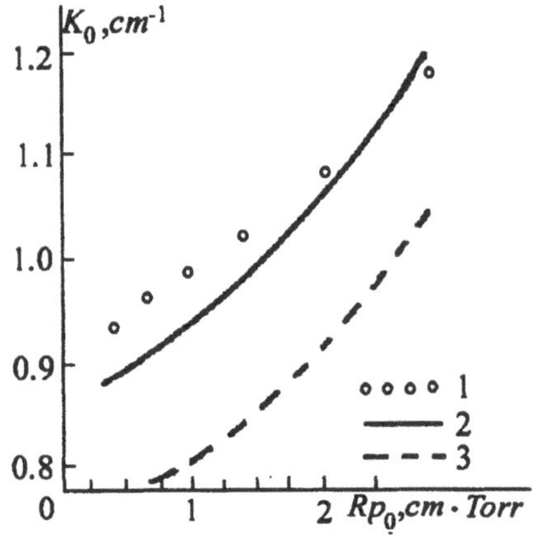

Figure 12.6. Dispersion law of the ionization waves in *Ar* for large currents at higher (with respect to the current density) boundary of the ionization instability. The points are experiment [27]; the broken line corresponds to calculation according to Eq. (12.14) with $\alpha=1$; the full line is calculated with $\alpha'=1.35$, according to Eq. (12.25) accounting for the radial column inhomogeneity. Experimental conditions are: $a=1.65$ cm; $p=60$ *Pa*; the discharge current is 3.6 A. Parameter Δ was of the order of 0.05.

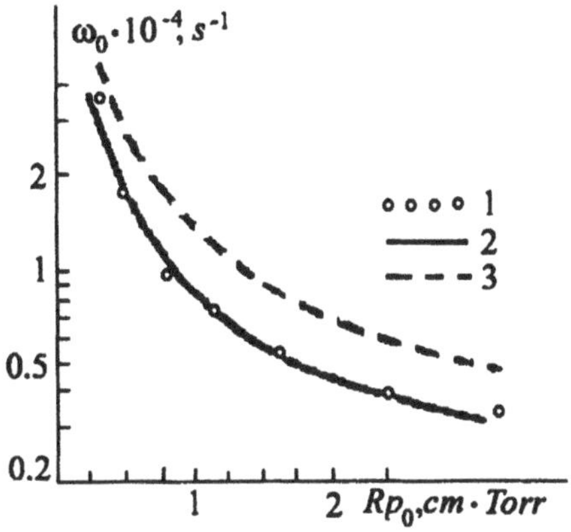

Figure 12.7. Spatial amplification of the ionization waves. Notations are the same, as in Fig. 12.6.

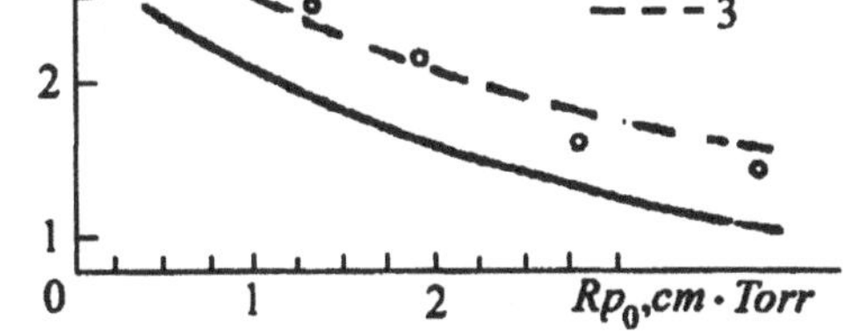

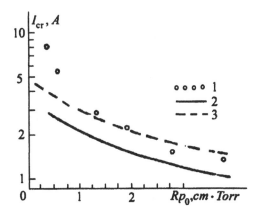

Figure 12.8 The upper (with respect to the current) boundary of existing of the spontaneous striations in *Ar* [27]. The curve is calculated according to Imω=0, Eq. (12.15).

The Running Striations In A Radially Inhomogenous Column. The plasma of the positive column is immanently inhomogeneous. However, in the derivation of Eqs.(12.7) - (12.16) the mean lifetime τ was assumed to be equal to the ambipolar value, Eq. (4.17). This assumption implied that in the real 2D problem of the propagation of longituional ionization waves, the transverse diffusion remained ambipolar. In other words, we have assumed that because the electron and ion fluxes are equal at the tube wall they are equal everywhere in plasma. As it was demonstrated in Chapter 4, it is true in pure isothermal plasma. In the presence of 2D temperature inhomogeneity this assumption is violated, and an eddy current (predominantly of electrons) arises in the plasma volume. Qualitatively the situation reminds us of that analyzed in Chapters 6 - 8, the case of the development of short-circuiting eddy currents in magnetized plasma. We shall discuss below the non-trivial consequences of this fact, following [23].

It is easy to see that in a 2D inhomogenous plasma where both the density and the electron temperature are disturbed, it is impossible to create a potential (electrostatic) electric field which equalizes the electron and ion fluxes everywhere in a plasma volume. This means that a considerable radial electron eddy current arises which does not obey the ambipolarity condition $\Gamma_{er} = \Gamma_{ir}$. As in the 1D analysis above, we shall restrict ourselves to the case of short (K>>1) waves in narrow (Δ<<1) tube in the vicinity of the 'resonant' wavelength K~K_0, Eq. (12.16). The 2D equations for slow motions with respect to the energy relaxation time follow from Eqs.(12.4), (12.9) and (12.10):

$$\begin{gathered}\vec{\nabla}\cdot(en\vec{u}_e) = \vec{\nabla}\cdot[nb_e(e\vec{E} + T_e\vec{\nabla}\ln n + (1 + c_{eN}^{(T)})\vec{\nabla}T_e)] = 0; \\ (5/2 + c_{eN}^{(T)})n(\vec{u}_e\vec{\nabla})T_e - \vec{\nabla}\cdot(\alpha n b_e T_e / e\vec{\nabla}T_e) + \\ en(\vec{E}\vec{u}_e) + n\delta_{eN}\nu_{eN}T_e = 0; \\ \frac{\partial n}{\partial t} + \vec{\nabla}\cdot(n\vec{u}_i) = \frac{\partial n}{\partial t} + \vec{\nabla}\cdot(n\vec{u}_e) = n\nu^{(ion)}(T_e, n).\end{gathered} \tag{12.17}$$

At equilibrium the profiles of T_e, as well as of the field E_x, are radially independent. It follows from the fact that the transverse energy flux, which is connected with the ambipolar diffusion (see the preceding Section) is negligible. Hence, even if the excitation and ionization processes are important in the electron energy balance, the transverse electron thermal conductivity levels up the T_e profile with accuracy up to D_a/D_e.

The boundary conditions at the tube wall, $r=a$, demand regularity of the particle and energy fluxes, and fulfillment of the condition $\Gamma_{er} = \Gamma_{ir}$. In the plasma bulk the ambipolarity condition $\Gamma_{er} = \Gamma_{ir}$ is absent, and a radial electron current arises which far exceeds the ion current. In order to suppress the electron thermal diffusion and to maintain quasineutrality, an additional electric field

$$\delta\varphi(r) = (T_e\tilde{T}_e / e)\ln n \tag{12.18}$$

arises, which confines electrons in the maxima of T_e, and forces them out towards the tube wall in the minima (Fig. 12.9).
The longitudinal electric field caused by this potential perturbation gives rise to an additional longitudinal electron current. The perturbation of the longitudinal electron velocity in it at the tube axis is of the order of $\delta u_{ez} \sim b_e k\delta\varphi(r=0) \sim D_e k\tilde{T}_e$. The total longitudinal current (which is predominantly an electron one) remains constant. This means that this additional current is an eddy current. It is short-circuited by the electron radial currents, as shown in Fig. 12.9. The radial component of the electron velocity in this eddy current is of the order of $\delta u_{er} \sim ka\delta u_{ez} \sim k^2 a D_e \tilde{T}_e$. It is small (of the order of $(ka)^2$) with respect to the velocity in the field given by Eq. (12.18). This fact justifies our assumption that the radial field, Eq. (12.18), corresponds practically to the Boltzmann's distribution for electrons. Nevertheless, the Joule work which performs the equilibrium ambipolar electric field over the radial component of the electron eddy current $\sim(T_e/a)\delta u_{er}$ is of the order of the work which is performed by the diffusive longitudinal electric field Eq. (12.12): $b_e E_z^2 \tilde{E}_z$, and has the opposite sign (see Figs.12.5; 12.9). In other words, accounting for the radial Joule work results in a decrease of the oscillations of $\tilde{T}_e$. This effect can be interpreted as the effective increase of the longitudinal electron thermal conductivity α.

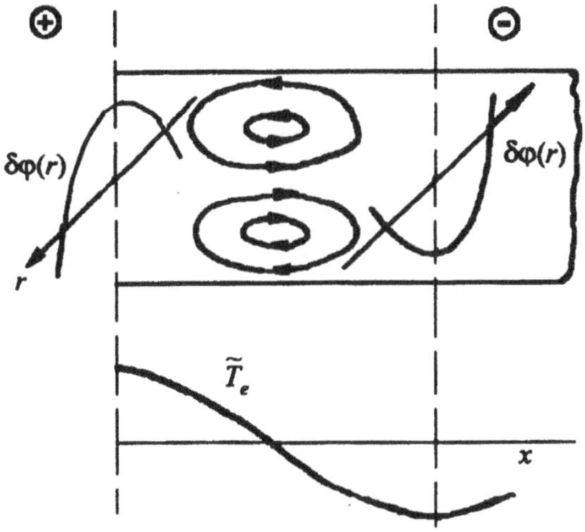

Figure 12.9. The eddy currents in the ionization wave which is propagating in the plasma, which is inhomogeneous in the transverse direction.

The longitudinal electric field caused by this potential perturbation gives rise to an additional longitudinal electron current. The perturbation of the longitudinal electron velocity in it at the tube axis is of the order of $\delta u_{ez} \sim b_e k\delta\varphi(r=0) \sim D_e k\tilde{T}_e$. The total longitudinal current (which is predominantly an electron one) remains constant. This means that this additional current is an eddy current. It is short-circuited by the electron radial currents, as shown in Fig. 12.9. The radial component of the electron velocity in this eddy current is of the order of $\delta u_{er} \sim ka\delta u_{ez} \sim k^2 aD_e\tilde{T}_e$. It is small (of the order of $(ka)^2$) with respect to the velocity in the field given by Eq. (12.18). This fact justifies our assumption that the radial field, Eq. (12.18), corresponds practically to the Boltzmann's distribution for electrons. Nevertheless, the Joule work which performs the equilibrium ambipolar electric field over the radial component of the electron eddy current $\sim(T_e/a)\delta u_{er}$ is of the order of the work which is performed by the diffusive longitudinal electric field Eq. (12.12): $b_e E_z^2 \tilde{E}_z$, and has the opposite sign (see Figs.12.5; 12.9). In other words, accounting for the radial Joule work results in a decrease of the oscillations of $\tilde{T}_e$. This effect can be interpreted as the effective increase of the longitudinal electron thermal conductivity α.

Since the ambipolar potential has a singularity at the tube wall, it is useful to introduce a variable instead of it

$$W = \tilde{E}_z - iK[(1+c_{eN}^{(T)})\tilde{T} + \tilde{n} + \int_0^r \tilde{T}_e \frac{d\ln n}{dr'} dr']. \tag{12.19}$$

The boundary conditions demand regularity of W, $\tilde{n}, \tilde{T}_e$ at r=0, a. Introducing $\rho=r/a$, from Eqs.(12.17) we have

$$\hat{L}W \equiv \frac{1}{\xi_1 \rho n}\frac{d}{d\rho}\rho n\frac{dW}{d\rho} = K^2\Delta(\tilde{n} + W + \\ + iK\int_0^\rho \tilde{T}_e \frac{d\ln n}{d\rho'} d\rho'); \tag{12.20}$$

$$[3/2iK + \alpha K^2]\tilde{T}_e - \alpha\hat{L}\tilde{T}_e - 2W + iK\tilde{n} - \\ -2iK\int_0^\rho \tilde{T}_e \frac{d\ln n}{d\rho'} d\rho' + i\frac{dn}{d\rho}\frac{dW}{d\rho} / [(\xi_1 a)^2 K\Delta n]; \tag{12.21}$$

$$(i\omega\tau - K^2\Delta)\tilde{n} + \hat{L}\tilde{n} - \frac{1}{\xi_1^2 n}\frac{dn}{d\rho}\frac{d\tilde{n}}{d\rho} + \nu^{(ion)}\tau(Z_T - 1)\tilde{T}_e = 0. \quad (12.22)$$

At $K \sim K_0$ the variables W, $\tilde{n}, \tilde{T}_e$ are almost radially independent. The radially dependent corrections are small:

$$\delta W(\rho)/W \sim \delta\tilde{n}(\rho)/\tilde{n} \sim \delta\tilde{T}_e/T \sim \Delta^{1/2} << 1. \quad (12.23)$$

From Eq. (12.20) it follows

$$\frac{d\delta W}{d\rho} = i\frac{\xi_1^2 K^3 \Delta}{\rho n}\tilde{T}_e \int_0^{\rho} \rho' n(\ln n - \langle \ln n \rangle) d\rho', \quad (12.24)$$

where the radial averaging is to be performed with factor (ρn). Substituting Eq. (12.24) into Eq. (12.21) and integrating it over the tube cross-section, we obtain that due to the last term in the r.h.s. of Eq. (12.21), i.e., due to the electron Joule cooling in the unperturbed ambipolar field, the resulting equation for $\tilde{T}_e$ coincides formally with the 1D Eq. (12.13). The only distinction consists in the replacement of the dimensionless thermal conductivity α by

$$\alpha' = \alpha + \frac{1}{\xi_1^2 K^3 \Delta \tilde{T}_e}\left\langle \frac{d\ln n}{d\rho}\frac{d\delta W}{d\rho} \right\rangle. \quad (12.25)$$

For the Bessel profile $n(\rho)$, Eq. (4.19), radial averaging results in $\alpha'=\alpha+0.35$. For *Ar*, in which the experiments [24] were performed, the widely used approximation for the transport collision frequency is $\nu_{eN}(V) \sim V^3$. It corresponds to $\alpha=1$. From Figs. 12.6 and 12.7 it is easy to see that replacing of $\alpha=1$ by $\alpha=1.35$ due to the radial plasma inhomogeneity results in considerably better agreement between the calculations and the experiment.

REFERENCES

1. L. Goldstein, T. Sekiguchi, *Phys. Rev.* **109**: 625-630 (1958).
2. A. P. Zhilinsky, I. F. Liventseva, *Sov. Phys. Techn. Phys.* 17: 1116-1122; 1888-1890; 2034-2036 (1972).

3. S. Chapman, T. G. Cowling, *The Mathemathical Theory of Non-Uniform Gases*, 2nd ed. (Cambr.: University Press, 1952): 350.
4. J. O. Hirschfelder, C. F. Curtiss, R. B. Bird, *Theory of Gases and Liquids*, corrected print (New York: Wiley, 1967): 1249.
5. L. D. Landau, E. M. Lifshitz, *Fluid Mechanics* (Oxford: Pergamon, 1987): 539.
6. M. Biondi, *Phys, Rev.* **93**: 1136-1140 (1954).
7. A. P. Zhilinsky, I. F. Liventseva, L. D. Tsendin, *Sov. Phys. Techn .Phys.* **22**: 177-183 (1977).
8. H. J. Oskam, V. R. Mittelstadt, *Phys. Rev.* **132**: 1435-1445 (1963).
9. F. G. Baksht, G. A. Djugev, S. M. Shkolnik, *Sov. Phys. Techn. Phys.* **47**: 1319-1322 (1977).
10. *Thermoionic Convertors and Low-Temperature Plasma*, Editors: K. Hansen, B. Ya. Moyzhes, G. Ye. Pikus (Washington: Tech. Inf. Center/US Dept. Energy, 1978): 484; F. G. Baksht, V. G. Yurjev, *Sov. Phys. Techn. Phys.*, **21**: 531-548 (1976); **24**: 535-557 (1979).
11. A. V. Nedospasov, *Sov. Phys, Uspekhi* **11**: 174-187 (1968).
12. L. Pekarek, *Sov. Phys, Uspekhi* **11**: 188-208 (1968).
13. N. L. Oleson, A. W. Cooper, *Adv. Electronics Electron Phys.* **24**: 155-160 (1968).
14. T. Ruzicka, K. Rohlena, *Czech. J. Phys.* **B22**: 906-919 (1972).
15. L. D. Tsendin, *Sov. J. Plasma Phys.* **8**: 96-109; 228-233 (1982).
16. L. M. Biberman, V. S. Vorob'ev, I. T. Yakubov, *Kinetics Of Nonequilibrium Low-Temperature Plasma* (New York: Plenum, 1987): 483.
17. V. S. Golubev, M. M. Malikov, A. V. Nedospasov, *Teplofizika Vysokih Temperatur* **8** 1265-1271 (1970) (in Russian).
18. L. D. Tsendin, *Sov. Phys. Techn. Phys.* **16**: 1226-1231 (1971).
19. H. E. Petcek, S. Byron, *Ann Phys.*, **1**: 270-315 (1957).
20. Yu. M. Kagan, R. I. Lyagushchenko, *Sov. Phys. Techn. Phys.* **9**: 627-631 (1964).
21. Yu. M. Kagan, R. I. Lyagushchenko, *Sov. Phys. Techn. Phys.*, **6**: 321-329 (1961).
22. R. I. Lyagushchenko, *Sov. Phys. Techn. Phys.*, **17**: 901-911 (1972).
23. L. D. Tsendin, *Sov. Phys. Techn. Phys.*, **14**: 1013-1019 (1969); **15**: 1245-1252 (1970).
24. K. Wojaczek, *Beitrage Plasmaphysik* **1**: 30-43 (1960); **6**: 319-330 (1966).
25. M. S. Gorelic, L. D. Tsendin, *Sov. Phys. Techn. Phys.* **18**: 479-481; 1007-1011 (1973).

Index